Table of Contents

B. Fission Energy

C. Solar Energy

D. Fusion Energy

INDEX

ENERGY II: Use Conservation and Supply

A special
SCIENCE
Compendium

Edited by
PHILIP H. ABELSON
ALLEN L. HAMMOND

American Association for the Advancement of Science

Printed in the United States of America

The material in this book originally appeared in *Science*, the weekly journal of the American Association for the Advancement of Science. The articles were published over a period ranging from 11 July 1975 to 14 April 1978.

International Standard Book Number (CB) 0-87168-300-8; (PB) 0-87168-237-0
Library of Congress Catalog Number 78-59190
AAAS Publication 78-9

Inquiries should be addressed to:

AMERICAN ASSOCIATION for the ADVANCEMENT OF SCIENCE
1515 Massachusetts Avenue, N.W.
Washington, D.C. 20005

An Interim Look At Energy: Signs of Progress

Despite continuing problems in establishing an effective energy policy for the United States, two significant areas of strength are evident. One is conservation. Insulation sales are at an all-time high. Improvements in the mileage efficiency of the U.S. car fleet, mandated by the Congress, have already been substantial and are accelerating. Industrial waste-cutting and the more gradual introduction of improved process equipment have led to less energy use per unit of output. In fact, the energy efficiency of some 48 major industries improved by an average of 9.2 percent between 1972 and 1977, with even larger gains in some of the most energy-intensive industries—10.8 percent in paper, 11.8 percent in chemicals, 14.4 percent in refining, and 24.9 percent in foundries. With progress in all major sectors of the energy economy, from residential to transportation and industrial, conservation is clearly beginning to have some effect. Evidently the engineering credo that a Btu saved is easier than a Btu earned is true in a sociopolitical sense as well.

A second major area of progress is that of research. Energy problems now engage the attention of a significant fraction of the scientific and engineering talent in this country. The federal government is spending more than $3 billion annually on energy research, by far the largest commitment of resources in the world, and substantial private sums are being committed as well. This vast public and private research effort is not without problems, but it is beginning to produce results in the form of significant new ideas and inventions, in analytical studies, and in engineering development. These activities have not resulted in significant new supplies of energy, but they hold such considerable promise of providing new options for the future that the nation is already far better prepared for a transition to new energy sources than it was at the time of the Arab oil embargo. This volume provides a sampling of the intellectual activity underway within the energy research community, some perspective on transitions in energy use and supply yet to come, and some clues to future energy sources.

The eventual and inevitable shift away from oil and natural gas as the predominant sources of energy to coal, to nuclear power, and to solar energy may take some unexpected forms, if history is any guide. Just a century ago, wood was still the dominant fuel in the United States. It gave way to coal in the last half of the 19th century and coal gave way to oil in the first decades of the 20th. Charles Berg reminds us that the relative cost of two competing fuels was not the sole force governing these transitions, and may not be in the future. Rather, he suggests, the combination of new energy sources and the scientific and technological talent attracted to them led to novel applications of energy and new industrial processes that ultimately proved more important. This point of view might imply, for example, that energy research should focus on creating not just replacement sources of energy but also on new ways to accomplish given tasks.

The energy problem impinges on different nations in different ways. In the future, there will not be any single answer with as universal an impact as that of oil today. Rather each country and region will seek solutions that reflect their own resources and opportunities. In particular, the options of the United States are not the same as those of Europe and Japan, which in turn differ from those available to developing continents such as South America and Africa. Four articles explore these contrasting international perspectives.

The more carefully the energy problem is examined, the clearer the constraints on unlimited expansion of energy supplies become. A key constraint will be water. Carbon dioxide emitted to the atmosphere when fossil fuels are burned may be another. Two articles offer new information on these potential constraints.

Conservation, as already noted, is becoming a significant factor in the energy equation. John Gibbons and his coauthors suggest that the potential is far from exhausted, and five additional articles explore what might be done.

Liquid fuels are perhaps the most critical element of the energy supply picture—it is hard to imagine how our present economy could function without them. But after oil, what? And when? In Canada, oil is now being produced commercially from tar sands, but the extent to which the technology can be applied to U.S. deposits of this material is not yet known. There is new optimism about the activity in oil shale, but uncertainties persist about the scale that the industry could attain. Liquids from coal would tap a far larger source of energy. In the past year and a half, two significant government-industry cooperative projects to test the H–Coal and Donor Solvent processes have gotten under way which could add coal liquefaction to the nation's energy options as early as 1990. L. E. Swabb, Jr., describes the problems—not just technical ones—that must be overcome before additional energy from this source becomes a reality.

Solar energy is not yet a major contributor to U.S. energy supplies, but it has emerged as a growth area of research in the past few years. The abundance and variety of new ideas being put forth is a strong sign of the scientific talent being attracted to work on solar technologies and of the richness of the field itself, both of which suggest that solar energy may contribute substantially in the foreseeable future. Particularly explosive development is occurring in photovoltaic technologies, surveyed here by Henry Kelly. Other articles explore heating and cooling, biomass energy, and the problems associated with using solar energy in a utility network.

Nuclear fission energy is going through a period of reassessment and redirection of research efforts in the United States. J. A. L. Robertson offers a perspective on a different approach to nuclear power, that of Canada. Research on nuclear fusion, however, is making rapid progress toward demonstrating the scientific feasibility of the process and at the same time coming to grips with the more difficult problem of designing a practical reactor—difficulties, as W. E. Parkins points out, that are far from being overcome.

The size and vigor of the research effort illustrated in these articles is good reason to hope that the technological aspects of the nation's energy dilemma will become increasingly tractable in coming years. Still lacking, however, seem to be a willingness to face up to an era of more expensive energy and a political consensus on how to deal with our changed energy circumstances.

—Allen L. Hammond

I

Energy Use in Transition

Process Innovation and Changes in Industrial Energy Use

Charles A. Berg

As industry developed in the United States through the late 18th and 19th centuries, it initially employed wood and some coal as fuel. Beginning in the mid-1850's, the use of coal as an industrial fuel grew rapidly, and between 1850 and 1870 there was a nearly fivefold increase in the production of bituminous coal. By 1885 coal production surpassed firewood production and continued to grow vigorously, while production of firewood declined, gradually (*1*). By 1930, the production of bituminous coal provided somewhat more than half of the fuel used in the United States, and alone represented twice as much fuel as the total of all fuels consumed in this country during 1900. In about 1920, the production of oil and natural gas, and their use as industrial fuels, began a growth that ultimately exceeded even the previous growth in the use of coal. As coal had replaced wood as the principal industrial fuel, so oil and gas came to replace coal.

Summary. American industry in the 19th century switched from wood to coal as its primary energy resource. The history of this switch is reviewed, along with the history of preceding similar trends in Europe and later trends in the switch from coal to oil and gas. Important conceptual advances in the technology of such basic processes as glassmaking, cementmaking, and steelmaking emerged as the switch from wood to coal proceeded in the United States. These advances may have been more important than the relative prices of wood and coal in motivating the conversion of American industry to the use of coal. The historical role of process advances in determining the choice of energy resources suggests that the physics and chemistry of industrial processes may be as important an area of energy research as the various technologies of energy conversion.

Efforts are now being made to predict the possible patterns of industrial energy use in the future. In this context it is of interest to consider the profound changes which have taken place in industrial energy use over the past century. To enter this subject, let us consider the circumstances under which coal largely displaced wood as an industrial fuel.

Early Use of Coal in the United States

The earliest recorded coal mining in the United States occurred in 1701, near Richmond, Virginia, but commercial mining of coal did not begin in this country until 1745. Coal was discovered in Ohio in 1755, and in 1770 George Washington commented on an Ohio coal mine he had seen. Yet, with concentrated deposits of coal available for exploitation, at presumably very low cost, coal still did not make significant inroads into the market for fuels. In fact, even with American deposits of coal having been identified, most of the coal used in America up until the Revolution was imported from England or Newfoundland. The shortage of coal occasioned by the break with England spurred the growth of American coal mining during the Revolution. Government requisitions of coal in Pennsylvania and Maryland, to support the manufacture of munitions, stimulated the beginning of American coal mining as an industry. Thus, the American coal mining industry began, not as a national response to price, but as the result of a shock—the loss of English coal during the Revolutionary War.

A great discovery of coal was made in 1810 when an unusually violent freshet unearthed a huge coal seam, now believed to have been the Pittsburgh seam, near the town of Barton. Coal from this seam was hauled by wagon as far east as Romney and Winchester. Later it was hauled overland to Westernport, where it was placed on barges and rafts, and shipped to Washington. Yet, even with these early discoveries of coal in rich deposits, where fuel could be picked from the surface of the ground, coal production did not make significant inroads in the market for fuels. It was not until 1850 that coal production reached 10 percent of the fuel provided by firewood. Why was it that it took so long for coal to displace firewood? Why did coal displace firewood?

It has been tempting to dismiss these questions by saying that wood must have become scarce and expensive as newly discovered coal offered a cheaper substitute fuel. Therefore, coal displaced wood. The argument based upon price alone is plausible, but misleading. It neglects other important aspects of the competition between wood and coal, aspects which had a strong influence upon the outcome of that competition and which may well have important implications on the development of future patterns of industrial energy use.

Let us consider the notion that resources of fuel wood were near depletion or were in some sense scarce, as compared with the national demand for fuels during the period 1830 to, say, 1850, when the transition from wood to coal became evident. Between 1850 and 1870 the annual production of fuel wood rose from approximately 2×10^{15} British thermal units to nearly 3×10^{15} Btu (*2*). It then declined again to about 2×10^{15} Btu in 1900.

After 1900 production of fuel wood declined steadily, even during the fuel shortage of the two world wars. However, as this decline in fuel wood production occurred, the total annual volume of wood harvesting rose rapidly. From 1869 to 1923 annual lumber production rose from 12.7×10^9 board feet to 41×10^9 board feet (*2*). These data cannot be

Dr. Berg, the former chief engineer of the Federal Power Commission, is now a private consultant who resides in Buckfield, Maine.

translated directly into an equivalent production of fuel wood because they reflect finished lumber taken from the select marketable bole of the tree, rather than the whole of the wood mass (including slash and nonmarketable species, for example) which could be taken for fuel. Nonetheless, the figures do reflect the fact that the wood available, for whatever purpose it might be used, had not been depleted at the time of the interfuel competition of the early 1800's (*3*). To estimate the magnitude that the supply of fuel wood might have attained, one may note that between about 1830 and 1900 approximately 250 million acres of forest land were cleared for farming in the eastern United States (*4*). Most of this clearing occurred after 1850, with the growth of settlements. If one assumes that the cutting of these lands would have yielded ten cords per acre, the total fuel value of firewood which would have been taken as a by-product of clearing would have been approximately 60×10^{15} Btu (*5*). This is somewhat more than twice as much fuel as the entire United States consumed between 1850 and 1900. Indeed, there was no shortage of fuel wood during the competition which was won by coal.

Why, then, was the firewood not used? Why was coal able to capture the industrial fuel market in the face of vast surpluses of firewood? One might suppose that the difficulties of transportation could have made wood a more costly and less attractive industrial fuel. However, difficulties of transportation did not deter those who exploited the forests for logs. They got their merchandise to the very industrial centers in which coal was displacing firewood, so that it could be used for construction. Also, we may note that given large supplies of fuel available at low price, industry might have located so as to exploit it. Few industries moved to the expanding boundaries of the newly cleared farmland.

The price of coal versus the price of fuel wood, as influenced by the costs of gathering and transportation, certainly exerted some influence on the decisions of industry in the period of the interfuel competition. However, the incomplete historical data available to us concerning the price of these two commodities reflect some periods when fuel wood was significantly cheaper than coal.

In 1880 the price of coal, FOB (freight on board) at the mine was $1.25 per ton. It was not until 1899 that this price fell to $0.80, after which year it has risen steadily. Data concerning costs of transportation and distribution of coal in this period are not readily available, nor are data concerning retail prices of coal. However, we may assume that the costs of transporting and distributing fuel wood would not have differed greatly from those of coal, if a wood fuel industry had existed. So, to gain some notion of how the relative price of coal and fuel wood may have influenced the interfuel competition, one might estimate what the value of fuel wood stacked, say, at the roadside might have been, had one been able to sell the heating value of the wood at the same price which was obtained for coal. Taking into account the fact that both the coal and wood must be dried to some extent to yield their maximum heating value, the calculation indicated above will show that fuel wood would have had a value of about $1.20 per cord, stacked at roadside. That is to say that the fuel wood disposed of in clearing farmland would have represented a value of about $12 per acre cleared. Not only was this an extremely high value during the time the clearing took place (1850 to 1900), it even approaches the land prices in the northeastern United States recorded as recently as 15 years ago. One might suppose then, that had there been an industry organized to exploit fuel wood, industry could have sold fuel at a significantly lower price than coal. In fact, one can suppose that *some* fuel wood probably *was* sold at a significantly lower price than coal. Nevertheless, coal prevailed.

How Fuels Are Chosen

Let us return to our original question: Why did coal displace wood as an industrial fuel? There was not an absolute shortage or depletion of fuel wood, which necessitated a switch from wood to coal. Also, price was neither the only nor perhaps even the principal factor in the contest.

Let us reexamine this question. To suppose, as is now fashionable, that industrial fuels will be chosen principally on the basis of price, one must presuppose that the fuels are equivalent in a number of other important respects. For example, one must presuppose that the fuels can be used the same way in the same equipment to carry out the same processes. One must also presuppose that when a transitional competition between two or more fuels arises, such questions as may not be strictly economic in nature (for example, possible modifications of processes) will be addressed with similar energy and intelligence by parties on all sides of the competition. These presumptions are seldom justified.

Consider, for example, the wave of invention which preceded the growth of coal in the United States. Darby, in 1735, invented the first blast furnace in which high-quality iron could be made in large quantity. The basic secret to Darby's furnace was the use of coke produced from coal. Prior to Darby, iron was reduced in furnaces which employed charcoal obtained from wood. But the logistical problems of gathering sufficient charcoal to run such a furnace had always kept the furnaces small (*6*). In the mid-1800's, Bessemer perfected his steelmaking process which further exploited the chemical properties of coal, as well as its heating value. As the technology of coal grew, opportunities to control combustion were developed and exploited. Pulverized coal burners were in use before the turn of the century, affording control of heat release and temperature, which was crucial to development of many industrial processes.

Portland cement, which was invented and patented in 1824 by Aspdin, requires ultimate processing at 1550°C. Coal combustion offered the opportunity to produce Portland cement in large quantity.

The use of coal as a fuel in glassmaking also made possible the execution of controlled processes of mass manufacture, which were not possible in (then) wood-fueled glassmaking (*7*).

Another major aspect of the use of coal as an industrial fuel is that it lends itself to well-controlled gasification. This, in turn, permits even more refined process control than can be attained by way of solid fuel combustion. The importance of the use of gasified coal as an industrial fuel in stimulating the development of new processes may have been immense; in any event, it merits historical study. It is doubtful that the first process for mass manufacture of glass containers, the Owens process, could have been made to work effectively (or at all) had it not been for the use of controlled combustion of gasified coal at one critical stage of the process.

Coal, then, offered not merely the heating value required to run industrial equipment. It offered the opportunity to develop new and more productive processes. This, together with inventive minds which were attracted to examine the possible uses of coal and their implications, had much to do with the growth of coal as an industrial fuel.

This brings one to another aspect of the use of coal. In 1856 Perkins, in England, discovered aniline dyes in coal. This set off not only the chemical industry based on coal products (which soon flowered in Germany) but also created the basis for the growth of the entire field of industrial organic chemistry. The scientific discoveries in organic chemistry attracted talented and motivated people like a magnet (*8*). Although the total fraction of coal production given to chemical products has never been very large, the influence of the chemical industry in attracting human intelligence to the questions of how coal might be used was immense.

So, coal won the contest to replace wood as an industrial fuel because there simply was no contest; the use of coal had on its side the focused intelligence of a large number of the world's most talented scientists, engineers, inventors, and entrepreneurs, all striving to find new and more productive ways to use coal. In addition, the extraction and transportation of coal lent itself to organization as an industry. Fuel wood, on the other hand, was established. It was not organized as an industry, and it attracted no innovative champions to meet the challenge of coal.

That the use of coal attracted talented people does not in itself show that talent was more important than price in determining the outcome of the interfuel competition between coal and wood. One school of thought in economic history holds that it was because of the low price of coal that talent was attracted to its use (*9*). However, this argument is inconsistent with the theory that price is the chief determinant in setting the course of commodity use. If coal had been cheap and if its cheapness had caused its ascendancy over wood fuel, then one would expect that this new cheaper energy source would have been used with profligate inefficiency. But just the opposite was true. The world's genius was focused, in fact, upon finding new, more efficient and more productive means of using coal.

Early Use of Fuel in France and England

England, from the late 16th through the 18th centuries, was beset by deforestation (*10*) and was thus faced with an absolute shortage of domestic fuel wood. The English response to this crisis was to turn to the use of domestic coal. Although coal had been exported from England to the Continent since 1200, it was apparently the shock of the absolute shortage of fuel wood which finally drove the English to exploit their own coal as fuel. There followed a wave of innovation, including the invention of the Darby blast furnace, the construction of inland canals for transport of coal and other goods, the construction of the early steam engine of Newcomen (which was designed to pump water from mines), and other such advances, which collectively gave rise to the British industrial revolution.

During the same time France also experienced difficulties in obtaining sufficient fuel wood. However, these difficulties were associated with practices of wood harvesting and transportation, rather than with absolute deforestation. Although, by the time of Louis XIV, the court believed that a concentrated effort should be made to develop France's coal deposits for fuel, France ultimately did not follow the British example.

French industry, particularly ironmaking and glassmaking, was reorganized to follow retreating forest boundaries (*11*). The forest sites most favorable for such purposes were those near water for power and transportation. These same sites were desired by the navy because they facilitated transportation of the ship timbers which the navy required. Urban growth, much of which followed industry, also drew heavily on these same forest areas for lumber and fuel wood. Thus, although France had abundant forest lands, only limited forest areas were exploited, and three sectors of society, industry, the military, and the citizenry, competed for the wood of these limited areas.

By the early 16th century some French towns had petitioned the king to protect their firewood against the competition of industry. In 1723 Versailles issued an edict which attempted to limit industrial fuel consumption, to protect the (not yet named) consumer. The law provided that no new forge, glass factory, or wood-burning manufacturing plant could be established, nor could an existing one be expanded, near towns or navigable rivers, without a favorable report from the provincial intendant and permission of the court role-general (*12*). Since enforcement of this law was lax, the volume of petitions to the king became immense and, as a result, there developed a complex regulatory system for the allocation of forest resources. At the same time, substantial tracts of forest lands were cleared for farming, near the very towns which had depended on these tracts for fuel wood. There is no evidence that transportation systems to bring fuel wood (or food) from more distant tracts were seriously considered by the government.

Under these conditions the wood supplies diminished, so that by the late 1700's cries of "wood famine" were raised throughout France. Of the three sectors of French society which required wood, the navy was most seriously hurt. The fir species required for mast timbers had been ruined by nearly a century of grazing in the forests; mature oak, required for hulls, had virtually vanished.

The citizenry was the next most severely affected by the dearth of wood. Firewood prices in France climbed 87 percent between 1730 and the eve of the Revolution. Most often the rise in the price of firewood bespoke receding forest boundaries, in the absence of any effective system of transportation, rather than genuine deforestation.

Industry, on the other hand, was little affected by the scarcity of wood. It did not need the high-quality, mature fir and oak, as did the navy. In addition, industry was organized to follow the forest lines, so that costs of transportation, which contributed the greatest component to the price of urban firewood, had less effect on industrial fuel. And, industry used much wood in the form of charcoal, the price of which was determined mainly by costs of processing, rather than by the cost of feedstock.

Finally, insofar as manufacturers' fuel bills were rising somewhat, a number of scientists and entrepreneurs devised improved uses of wood and charcoal in industry; they also enabled industrialists to make use of previously untapped forests. The Marquis de Court found means to raise the yield of iron smelted from rocks by 19 percent, using the same amount of fuel. Buffon devised a forge with natural draft. This eliminated the need for water power to operate bellows, and thus permitted the operation of forges in forest areas far removed from the streams.

In general, the response of French industry to the fuel wood crisis was, in spirit, the same as the approach of "energy conservation" as it is pursued today. This is to eliminate obvious wastes of fuel, to relocate plants, and to apply some innovations to ancillary aspects of industrial processes to improve the use of fuel. But it does not include making major conceptual changes in the processes themselves.

As the French observed the changes occurring in English industry, where coal was introduced to replace wood, they conducted an internal debate on the priorities for the use of fuel. The clergy maintained that the wood required to heat homes should be secured, even if this required denying fuel to industry. Industrialists considered the possibility of switching from fuel wood to coal but concluded that the cost of changing equipment was too high. The French tended toward decision by the (not yet formulated) science of economics. How much this tendency was reponsible for their reluctance to take to great risks reflected by the British action, as opposed to how much this reluctance reflected the absence of an absolute necessity for change, remains an open question (*13*).

The English Experience

It is interesting to examine the history of certain processes through which coal came to be the principal industrial fuel in England in the 19th century. Glassmaking was advanced substantially in France during the last part of the 16th century, and the English and the Dutch both sought French artisans for their glassworks. (Technology transfer it seems, then, as now, is best effected via transfe of persons.) Glassmaking at this time was entirely based on the use of fuel wood. The great glassmaking tradition of England began in 1567, with the construction of two glassworks in Fernfold Wood, Sussex, staffed by imported labor and located so as to exploit the then plentiful fuel wood of the region. Additional glassworks soon were built in other townships in the southeast of England. Each was established to exploit foreign labor and local fuel wood. This pattern of development quickly produced its own problems.

Beginning in Sussex, Surrey, and Kent, where wood for fuel was plentiful, the foreign glassworkers and their descendants migrated from place to place, always driven by the fuel demands of their furnaces. Their progress, and the progress of English deforestation, can be traced by cullet heaps, the ruins of furnaces, and by the distinctive names of foreign artisans recorded in parish registers. The commercial benefits of glassmaking in England, which contributed strongly to the flourishing growth of the British economy under Queen Elizabeth I, soon became an important part of the problem of British deforestation, which the great queen bequeathed to her troubled successors.

British response to this problem was swift. In 1610 a patent had been granted to Sir W. Slingsby for burning coal in glassworks furnaces. In 1615 all patents for glassmaking were revoked, and a new patent was issued for making glass with coal as fuel, in the names of Mansel, Zouch, Thelwall, Kellaway, and Percival (*14*). Simultaneously, the use of wood for melting glass was prohibited. The importation of glass from abroad was also prohibited. In about 1617, Mansel acquired the sole right of making glass in England, and this he retained for more than 30 years.

Ideas Versus Economics as the Force for Change

English decisions were clearly not based on carefully drawn and economically justified plans. The elements in the decision to prohibit the use of wood in glassmaking appear to have been desperation, invention, and arbitrary action, accompanied by a not unpredictable wave of greed (*15*).

The production of glass with coal as fuel was originally a "batch" process, just as was production with wood fuel. The full economic advantages of coal-fired glass production were yet to be developed, and there was still a field of competition between coal and wood as a fuel for glassmaking. The innovations that were to establish the undisputed predominance of coal as the principal source of energy for glassmaking came in the form of broad advances that also affected the use of fuel in other industrial processes. The patents and the royal decree of 1610 compelled the development of a technology for making glass without chemical or mechanical defects by means of coal fuel, and this prepared the way for later and richer innovation.

One major technical innovation in the use of coal was gasification. In the first practical application of gasification, Murdock, between 1792 and 1802, demonstrated the large-scale production of illuminating gas from coal (this art dated back to 1691). In 1799 Lebon obtained a French patent for a process to make illuminating gas from wood; he demonstrated his process in 1802. But wood gas was never to become a major fuel either for illumination or for power.

The first coal gas producers for industrial fuel were built in 1836 in Magdesprung, Germany. These were relatively crude, inexpensive units that produced "low Btu" gas at high temperature and were suitable for use with relatively small furnaces. In 1856, the Siemens company in Germany introduced an improved unit capable of delivering coal gas to several furnaces and suitable for a wide range of industrial uses. In the same year, Perkins in England discovered aniline dye in coal tars, an event that had such an explosive effect on 19th-century science and technology.

The year 1856 marked the emergence of the Siemens brothers as a major source of industrial innovation. From the Siemens company came the regenerative glass-melting tank; this was an adaptation of the Siemens furnace for steelmaking, in which then novel techniques of heat recovery by regeneration permitted, at the same time, (i) maintenance of the high temperatures (~ 2700°F) required for glass melting, (ii) vastly improved efficiency of fuel use and improved control of quality, and (iii), most important of all, reliable continuous operation. Of all of these advantages, continuous production was the key. By 1952, when Professor F. H. Norton (*16*) published *Elements of Ceramics*, the Siemens' coal gas generators, in combination with Siemens' regenerative furnaces, had become the standard to which all discussions of glass melting referred.

Coal in Iron- and Steelmaking

The first iron foundry in Pittsburgh, established in 1793 by Anshutz, made stove castings. It used local water power and wood fuel. The iron ore it used was transported across the Allegheny Mountains to the site. Although this particular foundry failed within 2 years, by 1840 iron and steel founding had become Pittsburgh's principal industry. However, despite the well-known coal deposits of the region and the existing British technology for using coal-based coke, Pittsburgh iron was smelted primarily from charcoal. It required over 100 years for Pittsburgh to adopt the coal-based technology of ironmaking begun by Darby.

In the mid-19th century two developments in the technology of iron- and steelmaking served to reshape not only the iron and steel industry, but also the market for coal in the United States and abroad. One of these was the invention by K. W. Siemens in England, in 1856, of the regenerative furnace for open-hearth steelmaking and continuous glass melting. Siemens was, according to his biographers, fascinated with Stirling's earlier work on the hot-air engine and, in particular, with the thermal regenerator which Stirling introduced. It was this fascination that stimulated Siemens to apply Stirling's concepts to his own work. The first known application of Siemens' regenerative furnace to glassmaking was in Chances glassworks at Birmingham, in 1861. Because of the combined merits of continuous production, high fuel efficiency, and control of quality, worldwide adoption of the Siemens' glass-melting system came fairly rapidly. With it came a new and unprecedented market for coal.

The application of Siemens' furnace to open-hearth steelmaking came equally rapidly. By 1867 Siemens was successfully producing mild steel in regenerative open-hearth furnaces in his own steel works. Worldwide imitation also followed this success, but not quite at the same rate as in the case of glassmaking. Ironically, the second great technical surge came close on the heels of Siemens' invention and, for a time, eclipsed Siemens' open-hearth process.

The second development, also in 1856, was the Bessemer steelmaking process. It was not until 1865, however, that Bessemer developed his to the point that he could reliably produce specimens of steel and malleable iron. The early response to his process had been cautiously optimistic and encouraging, with due expressions of skepticism by metallurgists. The response to his demonstrations was electric. Andrew Carnegie, upon witnessing Bessemer's demonstration, hurried home to Pittsburgh to announce to his partners, "The day of iron is past! Steel is king" (*17*). Carnegie built a Bessemer plant near Pittsburgh, at Braddock. He and his partners attained phenomenal rates of production with their ever-expanding Bessemer plants, and by 1881 the partnership became incorporated, at a value of $5 million. The Bessemer steelmaking boom, set off by Carnegie, gave tremendous impetus to the coal industry near Pittsburgh (*18*).

The basic concept that made it possible to produce Portland cement continuously was the use of heat recuperation (preheating combustion air, while quenching the "clinker") in a rotary kiln. The additional concept of using pulverized coal as fuel to obtain the localized high-temperature flame required in the final stages of calcining was also important. Both these concepts were embodied in the design of the rotary cement kiln by Hurry and Seamen in about 1895. The rotary, coal-fired kiln revolutionized cementmaking and further increased the demand for coal.

Processes and Fuels

The history of the adoption of coal as the principal industrial fuel shows that the advent of new and more effective processes, which happened to be designed about the use of coal, had at least as much to do with the industrial switch from wood to coal as did the relative prices of these two fuels. The switch in fuels can be said to have been economically motivated, to be sure. However, to the degree that it was, the motivation was based upon holistic considerations, rather than upon the price of the fuel itself. The advantages of being able, for example, to produce glass continuously, in a process fueled by coal, was so important that the price of coal itself was probably of little consequence. It was the widespread adoption of these processes that led to the rapid growth in the use of coal as an industrial fuel after 1850.

The processes that resulted in the adoption of coal as the principal industrial fuel may have had their roots in a period of desperation (the English dearth of fuel) when certain elements of basic science and technology were established. Nevertheless, they appear to have arisen chiefly in a burst of discovery and invention during which the attention of a large number of talented persons was drawn toward coal chemistry and other aspects of the use of coal. The motivation of these persons may, in part, be said to have been economic. The potential, revealed by Perkins' discoveries for the use of coal in applications that had not previously been suspected, certainly must have led many to speculate on the possible economic advances and personal gains to be made. But those who actually developed the new processes seem to have been as strongly influenced by fascination with new ideas as by the possible economic consequences of their work. There is little evidence to suggest that the heroic inventors of 1856 were even aware of the price of coal, much less that they were motivated by price, to direct their thoughts toward the use of that fuel.

There are striking parallels between the later growth of petroleum and the growth of coal. In the same way that the discovery of coal tar dyes by Perkins seems to have galvanized the scientific world and focused the world's talented minds on coal science, the refining of petroleum to provide illuminating oil, light motor fuel, and similar substances seems to have caught the imagination of a later generation of talented persons. The demands for motor fuel, especially from 1915 on, were accompanied by the improvement of refining techniques, the discovery of petrochemicals which had not been suspected to exist, and, in fact, the development of modern day chemical engineering. Coal science could not compete with the attraction of the newly developing fields of discovery offered by petroleum for the talented minds of the world. The full effect of the decline of coal science can be seen, for example, in the history of this field in the United States over the past 30 years. In 1973, one was hard pressed to find academic scientists in this country who were familiar with the properties of coal or the potential for production of synthetic chemicals from coal. Yet, coal is this country's major energy resource.

Edison's invention of electric illumination and the consequent growth of the electric utility industry gave coal a buffer from the effects of its declining attraction for talent. The uses of electricity beyond illumination also attracted great numbers of talented people. But insofar as the use of coal itself in producing electricity was concerned, the chief interests seem to have been combustion and economics. These interests were pursued by those whose basic concern was production of electricity, not coal science. As Sporn (*19*) has pointed out, the cushion which supplying fuel to the growing electric utilities and to the railroads gave to the coal industry was probably detrimental to that industry, and to coal science, in the long run. The latter fell behind as progress toward process innovation based on discoveries related to petroleum science and electricity proceeded rapidly.

The Initiation of Change

It is evident that the changes from one form of energy to another are more often motivated by the overall aspects of a change in process than by the price of energy or other individual economic aspects. Industry changed from the use of wood to coal principally because new and superior processes, which happened to be based on coal, became available. Industrial changes from coal to oil were, most often, similarly motivated. Changes from combustion to the use of electricity in industrial heating (for example, industrial direct heating furnaces or steel preheating by induction) are often motivated by the availability of a superior process, or an entirely new process, rather than by the price of fuel versus that of electricity. In 1927, Trinks (*20*), in his classical work on industrial furnaces, noted that even though electricity is more costly than fuel for heat, "the electric type [of furnace] is preferred because it offers advantages that cannot be measured in terms of fuel cost."

Decisions to change either the form or the mode of industrial energy seem to have a set of common characteristics.

1) The basic motives are usually holistic. To secure the widespread interest required to cause a significant change in the way in which industry uses energy, it is usually not sufficient to offer merely a

means of improving the efficiency of energy use. Such means usually entail trade-offs between fuel costs and other costs (capital, labor, administration, for example) of making the improvement. They have, therefore, natural limits; they can be pursued only to the point at which the trade-off "breaks even." However, when a new process emerges to offer the possibility of producing, for example, some substance such as cement, steel, or glass in a radically improved way, the process may be adopted with enthusiasm. Such processes usually entail advances in the efficiency of the use of all factors of production, including energy. They may also entail changes in the form of energy to be used (for example, electricity versus fuel). The price of energy itself may not be an important consideration in the decision to adopt the new process. Industry is always more attracted by processes of this kind than by opportunities merely to substitute one form of additional expense (capital) for a decrease in another (energy).

2) Those basic process innovations which have led industry to switch from one form of energy to another have been followed by further innovations which increase the efficiency of the process—including the efficiency of energy use—even as the price of the energy used has fallen in response to the broader market created for it by the new process. The efficiency of the use of coal, for example, increased radically after 1856, as coal prices fell. There have been important instances in which process innovation resulted in significant improvements in the efficiency of energy use without affecting in the least the form of energy used. Often these have been motivated by concerns that had little to do with energy directly. One of the most important of these was the relatively recent development by Sir Alaistair Pilkington of the float process for manufacturing flat glass. This process is so much more efficient overall that it has entirely displaced its predecessor, the plate process. The float process saves vast quantities of fuel, labor, capital, material, and all other factors used in the manufacture of flat glass. Its developers were motivated by a desire to do just that. Those who adopted it were motivated by a desire not to become obsolete.

3) One may note that from time to time something happens, be it a crisis, a discovery, a war, or some other compelling event, to draw the attention of numbers of talented and imaginative people to the study of some subject. The subject might be the use of coal, the use of heat, petrochemistry, electricity, glass-making, or something else. Whatever the subject, it seems that the focused attention of talented and imaginative people can be relied upon to produce innovative advances, advances such as Siemens' regenerative furnace, or Bessemer's convertor, or Daimler's engine, or Darby's blast furnace. These are advances which, because of their overall superiority to previous practices, can secure rapid and widespread adoption. The opportunities embodied in these advances exert a much more powerful influence than do opportunities merely to substitute one thing for another.

Conclusions

There are a number of things about energy research today which one may think represent misplaced emphasis. It seems to be of highly questionable value to attempt to decide which forms of energy to develop from the resources to be available in the future without giving significant attention to the processes for which these forms of energy might be used. One cannot assume that energy will be used in 2010 in the same way it is today. The history of industrial processes seems to show that each one is susceptible to improvement through basic innovation. One may have to wait for a Siemens or a Perkins or a Watt or an Edison to provide the innovation. But, such persons do appear, and, given a reasonable chance, they cause progress, the nature of which is usually unsuspected in advance. The proper field of endeavor in industrial energy conservation would seem to be the physics of processes rather than the use of fuel in existing equipment.

We are about to have to change the basic energy resources on which we operate many of our basic production processes (for example, coal or solar energy will have to be used to replace natural gas and oil in some applications). We will be faced with a wide range of forms in which energy might be brought to a process (for example, coal might be used for direct combustion, or as synthetic gas, or as electricity, or even in some other form). In considering how various new forms of energy might be used to operate a process, especially one that has been conceptually unaltered since, say, 1856, one often finds opportunities to make fundamental conceptual advances. These entail not only possibilities to improve directly the use of energy in the process, but also opportunities to advance overall productivity. The latter category includes opportunities to reduce losses in manufacturing yields, to advance control of quality, and to devise equipment that can operate flexibly according to the demands for production, rather than on essentially a full-time schedule.

It may be useful to give some examples here of apparently promising directions for technical efforts on process improvements.

In metal fabrication it is not uncommon for 50 to 70 percent of the workstock to be cut away as the final part is formed. Alternative techniques, which include electroforming and ionarc spraying, permit one to form metal parts directly to shape. The perfection of technology for fabrication of this type could both save fuel and lead to immense gains in productivity.

In many industries (for example, electronics, stainless steel), high-temperature furnaces that are used only, say, 40 hours per week must remain energized around the clock (this is because of the peculiarities of traditional furnace design). The time is overdue to develop lightweight furnaces with fast response times so that one can shut the unit off when it is not in use. (The furnace must be capable of quickly returning to the desired operating temperature, and there must be suitable control over temperature.) Heating equipment of this sort would be most valuable.

If, in addition to providing fast response time, the high-temperature heating systems suggested above were also to offer advances in the refinement with which one could control temperature, they might lead to substantial gains in munufacturing yields. In certain critical processes, such as production of electronic components, this could be of immense importance. It would provide large indirect savings in fuel (by reducing wastes of material) and, even more important, would provide direct gains in overall productivity.

In specialty steel production, the very reheating furnace that is energized around the clock, to be used only 40 to 80 hours per week, is also the site at which about 4 percent of the entire product of a plant is lost as oxide scale. Techniques of reheating and forming might be developed that would prevent this loss. For example, the stock could be heated and formed under an inert atmosphere or under a vacuum.

High energy lasers can be used for hardening metals. Laser techniques can reproduce the effects of thermochemical hardening (for example, carburizing) and mechanical treatment (for example, shot peening). The laser techniques can be applied virtually instantaneously and they can be applied to highly localized zones on large parts and even to zones of completed structures.

These examples, which offer only an imperfect representation of what may result from a basic reexamination of industrial processes, all entail the application of known effects to already perceived opportunities. What we should expect, on the other hand, is that the problems of energy should motivate the search for opportunities to make conceptual advances in processes. Such advances could range well beyond direct concerns with energy. Thus the emergence of conceptually advanced processes will probably set the future trends in industrial energy use.

References and Notes

1. These data are taken from the *U.S. Statistical Abstracts* (Government Printing Office, Washington, D.C.).
2. Conversion factors are: from British thermal units to joules, 1055; from acres to hectares, 0.4047. One board foot is 1 inch thick and 12 inches square.
3. The peak year for lumber production in the United States was 1901. These high volumes of production were the result of overcutting. Conservation policies instituted under the Roosevelt administration introduced systems of forest management under which lumber production was to be limited to annual growth. During the very period that the interfuel competition between wood and coal was under way, the forests were being exploited at a rate which exceeded their natural capacity to replenish themselves. One might suppose that clever or greedy exploiters of forest lands could have gathered fuel wood (slash, limbs, nonmarketable species, for example) to sell to industry, while taking other more valuable trees for lumber. The marginal cost of gathering otherwise useless wood for fuel, once one had established a basic system for gathering logs of high value, would probably have been rather low; the fuel wood might have been a by-product of the rapidly growing logging industry.
4. B. E. Fernow, *History of Forestry in Europe, the United States and Other Countries* (Univ. of Toronto Press, Toronto, 1913).
5. According to a number of U.S. Department of Agriculture studies [for example, *U.S. Dep. Agri. Bull. 753* (10 March 1919)] a harvest of 10 cords per acre is reasonable, even in the northeastern United States with its short growing season.
6. Approximately 3 acres of trees were required to produce sufficient charcoal to make a ton of iron.
7. The first coal-fueled glass plant in the United States was, in fact, built in Pittsburgh in 1808, by Blakewell. Details of the design of the furnaces and other equipment are not available; it was a batch process plant and the glass produced there was noted for its exceptional quality.
8. The influence of the rapidly developing field of coal chemistry on intellectual life in the second half of the 19th century was far-reaching. This subject is discussed even in the works of radical political writers of the day, including the works of Lenin.
9. According to this school, talent follows money. An alternative view is that money is created by talent.
10. A visit to Nottingham today will illustrate this. The forest that shielded the legendary Robin Hood from his pursuers has been reduced to a few specimen trees.
11. See, for example, D. B. L. Young, "A wood famine? A question of deforestation in the old regime of France," *Forestry* (1976).
12. The notion of the Environmental Impact Statement evidently goes back a way in history. The attempt by the French crown to curtail industrial fuel also might be compared with U.S. Federal Power Commission Order 467 of 1973.
13. Young (*11*), among others, suggests that France's economic and social reorganization to exploit its ready availability of traditional fuels (wood and charcoal) retarded French economic development.
14. J. G. Percival is credited with the introduction of covered crucibles to protect the glass from the products of combustion of the coal.
15. During the English Protectorate patent rights lapsed and certain glass products were once more imported from the continent. In 1663 the Duke of Buckingham obtained a ban on glass imports but he failed to obtain the monopoly over glassmaking which he had sought. Buckingham did operate a glass works which was noted for production of very high quality glass.
16. F. H. Norton, *Elements of Ceramics* (Addison-Wesley, Cambridge, Mass., 1952).
17. S. Lorant, *Pittsburgh—The Story of an American City* (Doubleday, New York, 1964), p. 147.
18. Carnegie was blessed with good luck also. He happened to have available ores that circumvented certain metallurgical problems with Bessemer steel that were not fully understood until the work of Thomas in 1875.
19. P. Sporn, *Technology, Engineering, and Economics* (MIT Press, Cambridge, Mass., 1969).
20. W. Trinks and M. H. Mawhinney, *Industrial Furnaces* (Wiley, New York, 1927), vol. 1.

Industrial Energy in Transition: A Petrochemical Perspective

Ronald S. Wishart

Although the growth of the petrochemical industry in the United States is attributable to several factors over the years, one of its basic strengths has always been the abundance of relatively cheap raw materials. Even though the industry has never used more than 8 percent of the petroleum and natural gas consumed in the United States, it produces 80 billion pounds of chemicals annually. Out of this has come a cornucopia of products that has affected our daily living and, indeed, has influenced the form of our civilization. A world without plastics, synthetic rubber, manmade fibers, and pesticides would be a different world from the one we have enjoyed at relatively low cost during the past three decades.

Although the domestic demand for such products continues to grow (6 to 8 percent per year according to recent estimates), the petrochemical industry can no longer look forward to satisfying this growth by relying solely on traditional raw materials that are approaching economic depletion. True, "economic depletion" is a relative term; today, we routinely recover hydrocarbons that were considered "uneconomic" just a few years ago. But the concept of finiteness is now significant in the energy dialogue because the costs of recovering oil and gas have escalated and are approaching, for the first time in history, the costs of utilizing substitutes. The impact of this fact on the petrochemical industry will be sizable, but we in Union Carbide are convinced that both the resourcefulness and the required technology exist to meet this challenge.

Over the years, industry has demonstrated its sensitivity to the dynamics of economics and technology in making decisions about energy. At Union Carbide our own hydrogeneration projects powered some of our earliest operations, and we used coal in some other locations. For many years we found oil and gas attractive as fuels as well as feedstocks, but we have been switching oil- and gas-fired boilers to coal for some time now, in recognition of the scarcity of oil and gas, their increasing price, and their greatly enhanced value as feedstocks.

Boiler conversion alone, however, is not enough. Even if we conserve all the scarce hydrocarbons that we can through conversion, improved efficiencies, and other conservation measures, the economics of feedstocks will continue to signal the need for more radical actions. We see the costs for oil and gas converging with and eventually exceeding the costs for some substitutes by the year 2000. These signals have been clear for years, and we have oriented our research and development to prepare ourselves to make the adjustments. Our

The author is director for energy policy, Union Carbide Corporation, 270 Park Avenue, New York 10017.

Summary. The future growth of the petrochemical industry depends in part on the industry's ability to improve efficiency in the use of oil and gas feedstocks and to develop promising alternatives. Technological innovation is proving to be the key to the long-term viability of the industry. The next 6 to 7 years will be characterized by the commercialization of new technologies designed to improve the efficiency of petroleum as a feedstock. Union Carbide's advanced cracking reactor, now nearing the demonstration stage, exemplifies this type of effort. The increasing price of oil and gas will make coal-based synthesis gas more attractive as a feedstock, particularly for oxygenated petrochemical products. A further development involves the conversion of biomass, through fermentation, to useful chemical products and the gasification of municipal wastes to raise steam for electricity generation and as a possible, supplemental feedstock. By the year 2000, it is predicted that feedstocks from all sources other than oil and gas may constitute 10 to 14 percent of the total new material requirement for the petrochemical industry.

goal is to minimize the cost of our feedstocks and thus retain the competitive value of our petrochemical products. We are convinced that we have in hand or under development the technologies necessary to achieve this goal.

To understand the role we see for technological innovation in the petrochemical industry of the future, we must first understand the role of oil and gas in the industry today, the economic and technological forces that will be affecting this role in the years ahead, and the paths along which we expect innovation to develop.

Oil and gas hydrocarbons today account for 98 percent of the feedstocks used by U.S. petrochemical producers. The industry is built on oil and natural gas because of their specific hydrocarbon composition, their relative lack of impurities, and their historically attractive price and abundance. Only 2 percent of U.S. feedstocks are now derived from coal.

Recently, however, economic and political forces have combined to make hydrocarbons less attractive as a long run feedstock for the industry. Specifically, the price of oil and gas has gone up and domestic production has dropped. The U.S. dependence on imports has increased, bringing with it greater vulnerability to supply interruptions. In addition, we now recognize that there are absolute limits to economical recovery of remaining hydrocarbon resources.

Because of this, alarmists have envisioned petrochemical production declining over time, with adverse impacts on employment and growth of national income. But these threats are mitigated by the relatively small proportion of oil and gas devoted to feedstock use and the fact that public policy has to date recognized that feedstocks are a preferred use. And—despite the energy intensity of petrochemical and plastics manufacture—we are less impacted by energy induced cost escalation than by metals or natural fibers. The understanding is also spreading that few, if any, nations can afford to displace acreage for food production with natural fiber production.

The alternatives to conventional feedstocks now being explored range from the familiar—synthesis gas from coal—to the exotic—biological synthesis to turn biomass and solid waste into chemicals. Some technologies are nearing the demonstration stage, others are still in more preliminary stages of development. And, although it is difficult to forecast with too much precision, it is possible that the alternatives we are now studying could provide 10 to 15 percent of the feedstocks that will be needed by the year 2000. Under realistic assumptions, however, such alternatives will not constitute more than 25 percent of the feedstock mix by the turn of the century.

The transition will be evolutionary and is more likely to be measured in decades than in years. This is because the petrochemical industry still does not have the proven processes to significantly reduce its short term use of oil and gas without facing severe output reductions. In addition, oil and natural gas are still relative bargains; they are available in the United States at prices below the cost of the next most expensive substitute. Finally, the transition to alternative feedstocks will be evolutionary because most of our recent additions to capacity have used relatively efficient petroleum-based processes rather than processes based on increasingly scarce and expensive natural gas. Thus the feedstock alternatives of the future are more likely to replace obsolete gas-fed plants than newer, more efficient oil-fed facilities.

On the basis of our experience in developing and commercializing new technologies, we foresee three phases of change in the industry and its resources that we believe will bridge from the present to the year 2000.

The first phase, starting now and lasting at least into the 1990's, is essentially a continuation of today's technology with the introduction of new techniques to more efficiently utilize crude oil, so as to produce higher yields of high-quality chemicals from each barrel.

The second phase, which will mature in the late 1980's, will be characterized by increasing production of synthesis gas (carbon monoxide and hydrogen) made from heavy petroleum fractions and coal for producing bulk chemicals now made from ethylene and other olefins—for example, ethylene glycol.

Paralleling these two phases and effectively extending the duration of phase 1, supplemental crude oil derived from shale, tar sands, and coal will begin to appear in the 1980's and become significant in the 1990's. Wide substitution of coal for fuel uses of oil and gas has already begun, and this is even more important in extending phase 1, well into the 21st century.

In the third phase, the production of chemicals by treatment of various forms of biomass will become significant. In a small way, this trend is beginning now, but it is unlikely to be commercially significant until some time in the 2000's.

These phases are not edged with discrete boundaries, but will overlap and blend into one another as the transition period proceeds. Accordingly, Union Carbide is working in all areas at the same time. For instance, we have a substantial ongoing effort in fermentation and other biochemical processes to understand how industrial chemicals can be produced economically from regenerable raw materials.

Let us now consider each stage of development in detail, and Union Carbide's role in it.

Phase 1: More Efficient Use of Petroleum as Feedstocks

During the next 6 or 7 years the petrochemical industry will have to look more and more closely at devices designed to help reduce costs by increasing the efficiency of using crude oil-based feedstocks. Products of crude oil currently provide 35 percent of the industry's feedstock, and are likely to provide a higher share as new gas liquids (which now furnish 39 percent) soar in cost and become harder and harder to find.

Union Carbide's contribution to efficiency of utilization is our advanced cracking reactor (ACR), which has been under development in cooperation with the Kureha Chemical Industries and Chiyoda Chemical Engineering and Construction Company. This reactor can free a chemical company from dependence on specific naphtha, gas oil, or gaseous paraffin feedstocks and permits savings in net processing costs. A \$15-million demonstration plant which is expected to go on stream at the company's Seadrift, Texas, plant in the middle of 1979, will produce 5 million pounds of ethylene a year, enough to allow testing of scale-up parameters in preparation for the construction of a full-scale plant.

The success of the ACR depends on the production of 2000°C superheated steam in a combustion section. Rapid vaporization and cracking of oil droplets occur as the droplets are injected into the high-temperature stream. The cracking time is a few milliseconds, and the steam and reaction products are quenched in a specially designed cooler that prevents further decomposition of the product. The cooling system is based on Japanese patent 90302/1976 issued to partner Chiyoda, and is known as the Ozaki quench cooler heat recovery system. It involves special circulatory features, notably injection of quench oil upstream of the heat exchanger, which prevent fouling of the walls by coke and tar deposition. These features permit efficient waste heat recovery—for example, steam at 600 to 1500 pounds per square inch—from the hot product stream, which is too "dirty" for conventional transfer line heat exchangers.

The ACR has numerous advantages over conventional cracking systems. From the viewpoint of the chemical manufacturer, it is efficient in producing a high yield of valuable chemical intermediates from a unit of crude oil. In fact, 60 to 70 percent of the crude oil barrel is converted to chemicals, as compared to 40 percent in conventional systems. The company can buy crude (a world commodity), or crude oil fractions that may be in surplus because of imbalance in energy end use, and thereby gain a measure of flexibility from oil company supplies. There is also a predicted price advantage in the ACR product estimated to be 3 to 4 cents per pound of ethylene, and a lower sensitivity compared to conventional cracking to variations of price and cost in the marketplace.

The ACR provides flexibility in choice of raw materials for cracking and in product distribution. With respect to raw material, possible feedstocks for cracking extend from naphtha to vacuum gas oil or deasphalted residue, as well as any blend of these, including petroleum distillates and some undistilled crudes. Since such feedstocks produce much the same product spectrum, it is possible to choose the advantageously priced material for utilization in the ACR and not have to accept what is offered. This freedom gives the ACR a substantial edge over conventional technology. Depending on the demand and supply of crude and crude fractions for fuel, the ACR can be fed oil company discards at less than premium prices.

The present concept of the ACR process also includes the recycle of certain reactor products through the system for further cracking. Ethane, propane, and raffinate C_4's can all be utilized to generate more ethylene under favorable economic conditions. Other feedstocks, such as hydrotreated (desulfurized) whole crudes, also would yield more and richer C_4 gas than that obtainable through conventional processes. Sulfur, moreover, is not a restrictive factor in the ACR process; and, indeed, the process train has been designed to remove substantial quantities of hydrogen sulfide (H_2S) and other sulfur compounds.

With respect to the products, for a given raw material the ACR can produce a wide range of products, depending on the "severity" of cracking. This is conveniently measured by the ratio of ethylene product to acetylene (E/A), ranging from 8 or less (high severity) to 25 or more (low severity). At low severities, propylene coproduct approaches in magnitude the amount from conventional olefin plants. At higher severities, propylene falls, but acetylene and aromatics become more important. At $E/A = 10$, for example, production of acetylene from a billion-pounds-per-year ethylene plant would be 100 million pounds, enough for a sizable vinyl acetate or chloride plant. The upgrading of the liquid products of ACR cracking to higher chemical values is an ongoing challenge.

The ability to shift the ratio of propylene to ethylene over a wide range is of special interest. Conventional technology can make only small shifts in yield patterns and excess by-products must be recycled, fueled, or marketed under distress conditions. The ACR gives us a better chance of balancing product slate with demand in the reactor itself, as well as flexibility in the choice of feedstock.

Phase 2: Coal as Feedstock

In the 1980's or 1990's, as the price of oil continues to rise, we expect to see the beginnings of a shift from oil to coal, which is by far the most generally available substitute for petroleum and natural gas. Known recoverable U.S. reserves of coal run to some 250 billion tons, and ultimate reserves may be twice this. Coal is estimated to constitute about 90 percent of the nation's remaining fossil fuel reserves. Coal as mined is neither chemical nor hydrocarbon but a rather intractable rock, anywhere from 10 to 50 percent of which is mineral matter and water. Conversion to chemical intermediates, whether by gasification, carbonization, or hydrogenation, involves heavy capital expenditure and difficult, potentially polluting processes. Today only South Africa is practicing coal conversion to high-grade fuels and chemicals in any large way, although there are smaller installations around the world.

Although coal is plentiful and United States' experience with its utilization has a long history, coal conversion technology is not as simple as the conversion technology of oil and gas. As one of my colleagues puts it, "Anything you can do with oil you can do with coal—only worse." Handling solids is difficult, especially a solid that may become sticky under reaction conditions. There is the ash which has to be removed, and overriding all the other disadvantages is the problem of air and water pollution. Nonetheless the huge amounts of synthesis gas used today could be made from coal. And many chemical products, especially oxygenated products, could be made from the synthesis gas. Some aliphatic derivatives, notably polyethylene, are more difficult to picture coming from syngas; hence, even as coal's importance as a fuel or a feedstock increases, the ability to bid for a share of the remaining oil and gas supplies will be important to petrochemical manufacturers.

Synthesis gas is already produced in the United States in huge quantities (one estimate puts it at 70 billion pounds per year, calculated as a 1 : 1 mixture of H_2 and CO). And purchase of its precursors, mainly methane and naphtha, is becoming more difficult and definitely more expensive. What, then, is the probability of the raw material base moving to coal? The answer appears to us to be "pretty good," although an interim period is probable in which the base shifts to heavier petroleum liquids, that is, residual fuel oils. A probable scenario making synthesis gas available is the following:

1) The need for medium Btu gas (more or less synonymous with synthesis gas) for upgrading and replacement of methane in existing plant furnaces or for synthetic natural gas to fill the pipelines continues to grow.

2) Coal gasification plants (or heavy

oil gasification plants convertible to coal) are built in or near chemical plants.

3) Chemical plants based on synthesis gas from these large units become competitive with plants based on an ever more costly ethylene.

Thus chemicals from coal-based synthesis gas may begin as early as the mid- to late 1980's and become important in the 1990's and beyond.

Processes to make almost any important aliphatic product from synthesis gas can be imagined, although finding an appropriate catalyst is generally the key. Ethylene, for example, can in theory be made from synthesis gas, thus permitting production of the whole spectrum of ethylene products. But, no practicable catalyst directing the synthesis efficiently toward ethylene is available.

The class of organics which has seemed to us to be a most logical candidate for synthesis is the oxygenated aliphatics, especially the polyhydric alcohols. We began to study the feasibility of such processes in the 1960's, after successful development of our low-pressure Oxo process, which makes alcohols and aldehydes. The direct synthesis of ethylene glycol, based on a unique rhodium cluster compound catalyst, has arisen from this work. At present, Union Carbide holds some 15 patents and is involved in extensive development work on the process.

Some time in the 1980's, as ethylene demand catches up to the now overabundant supply, and as oil and gas raw materials continue to rise in price, glycol from synthesis gas should be more attractive than glycol from ethylene. And glycol can join the ranks—with ammonia, methanol, and acetic acid—of bulk commodity chemicals based on synthesis gas.

Synthetic Crudes:

Oil Shale, Tar Sands, and Coal

In our judgment, it is unlikely that synthetic crudes will ever be the dominant feedstock for petrochemicals. Rather, we see such materials contributing to the overall supply of liquid fuels and feedstocks, thereby extending the duration of phase 1, the era of oil and gas.

The estimates of recoverable oil from oil shale run to very large numbers—50 billion, 100 billion, or more barrels depending on economic assumptions used in the analysis. Optimistic estimates of the maximum likely rate of production, however, are about 2 million barrels per day, about 10 percent of present oil demand. Some very optimistic estimates take the figure as high as 4 million barrels a day. Such production cannot possibly be reached before late in this century, at the earliest.

The limitation on shale oil production is not the size of the resource, but rather environmental and logistical constraints. Questions arise as to what can be done with spent shale, where will the necessary process water come from, and how can we avoid contamination of water supplies.

There is no special merit in shale oil as a chemical feedstock. Because of its lower hydrogen content, severe hydrogenation is necessary to upgrade it to feedstock or fuel-use quality. Similar problems exist in the recovery of feedstocks from tar sands. There are, however, two sizable refineries based on the Canadian Athabaska sands, apparently now capable of operating at a profit.

Tars and heavy oils are reported to exist in huge quantities in various parts of the world, often—like the Venezuela heavy oils—at great depths, or in difficult locations. But the logistics, economics, and energy costs of recovering these supplies are unknown. Nor is there any certainty that any net energy could ever be obtained from them. So in the near future, we cannot count on these supplemental sources to augment the world supply of oil.

Synthetic crude oil from coal, however, is a better possibility. Union Carbide hydrogenated bituminous coal in a small plant as long ago as 1955, marketing liquid products made in this process. Currently Exxon, Gulf, Ashland-HRI, and others, supported in part by the Energy Research and Development Administration, are pursuing coal-liquefaction techniques. Liquid products of these processes are even lower in hydrogen content than crude shale oils—totally unsuitable as feedstocks for aliphatic chemicals. However, they would contribute to the overall liquid fuel supply and could also be a potential source of aromatic chemicals such as benzene and phenol. By-product paraffin gases (ethane, propane, and butane) could—if the industry becomes sufficiently large—augment the feedstock supply for olefin plants based on natural gas liquids.

The operation of the first large commercial coal-liquefaction plant seems unlikely before 1990; so too is the development of a sizable industry much before the turn of the century. Therefore, if this happens at all, the impact will be felt mainly in the early 2000's. The same caveat probably applies to various proposals for processes that would make synthetic natural gas liquids (along with methane) from coal—for example, hydrocarbonization and some forms of catalytic reforming of coal with steam. These too would start late in this century and help extend our phase 1 into the 21st century.

Phase 3: Biomass

The ultimate alternative to oil and natural gas is biomass, the world cover of plant life, from weeds to trees. This is already a more significant source of fuel and chemical feedstocks in the United States than most people realize—on a Btu basis, perhaps 2 or 3 percent of the total supply. Like coal, biomass, in the form of starches and sugars, was once an even more important source of chemicals than it is today. The once-sizable fermentation industry, at least as far as bulk products are concerned, has been reduced largely to the production of ethanol and citric acid. While the ethanol is primarily beverage alcohol, a portion reaches the market as a chemical intermediate or solvent. In specific cases, fermentation alcohol already competes with synthetic alcohol from ethylene. A number of other important chemicals, such as glycerol, furfural, sorbitol, and mannitol, are also, entirely or in part, derived from biomass or animal sources.

The attraction of biomass, of course, is that it is the only feedstock which is regenerable. Planners who calculate how much land it would take to supply all the fuel uses of the United States would gag on the total. Assuming 5 tons per acre per year of material that yields 6000 Btu per pound, I calculate the requirement to be 1.27 billion acres or 1.98 million square miles (1 mile = 1.6 kilometers).

But we don't have to gag at biomass as a chemical feedstock. We already produce about half a trillion pounds a year of corn alone, a mass roughly 20 times that of the total ethylene production of the United States. From about 20 percent of the corn crop, we could conceivably make, by fermentation, chemical products equivalent in bulk to the total petrochemical product from ethylene and higher olefins.

Of course, this amount still requires a huge acreage. And corn is a high energy input crop and probably not the best feedstock. Sorghum, hydrolyzed wood, sugarcane, or something new might be better, but I leave this to the agronomists to work out.

As to how much biomass is available to be converted, the Department of Agriculture's Forest Products Laboratory in Madison, Wisconsin, estimates that "the net photosynthetic productivity of the earth,'' that is, the amount of new green plant life products, is 155×10^9 tons per year, with a total heating value of about seven times that of current world fossil fuel usage. Forests account for 42 percent of this total and croplands account for another 6 percent. Worldwide, the net productivity of forest exceeds, in heating value, the annual consumption of fossil fuels. While this is not true within the boundaries of the United States (after all, we use about a third of the total world's fossil fuel production), our forest's productivity is still very large compared to chemical usage.

Union Carbide has studied the potential of a number of U.S. crops (also of waste materials like newspaper) in terms of their availability for feedstock and the relative costs. Assuming it became necessary to use one of these crops to match the products of a billion pound ethylene plant based on oil or gas, we have noted that, of 18 tonnage crops or agricultural product groups examined, four can yield enough of the chemicals we need without significantly affecting the other uses of the product or its price structure. These include:

1) *Fermentable sugars.* While the costs are still marginal to high, they improve with time and improved technology. In developing countries like Brazil, the route to ethylene from sugarcane via ethanol can be competitive now.

2) *Cellulose-rich products and wastes.* The technology of separation from lignins and of hydrolysis must be improved to make a fermentation route competitive. Possibilities of gasification of wood, forestry wastes, municipal trash, and the like are discussed below.

3) *Corn.* Corn as cornstarch is probably a current prime prospect, although its energy-intensive cultivation may make it less attractive in the future. Corn residuals are also a potential raw material, although many believe these are best returned to the ground as soil conditioners.

4) *Other crops grown on "energy farms."* Of the many crops suggested (including alfalfa, algae, kelp, water hyacinths, and many others), sorghum is already a major crop and is considered a practicable source of fermentable material.

Like coal, biomass, in any of its various forms, can be converted into synthesis gas. The chief disadvantages of the natural materials are their low heating value, high water content, and high harvesting costs. Still, this also is a potential route to all the chemical synthesis gas products.

One version of biomass gasification which Union Carbide has explored in depth is the gasification of municipal waste. The result is our PUROX process, based on a unique, oxygen-fed slagging gasifier that has been extensively shown in a pilot plant (yielding 200 tons per day) in South Charleston, West Virginia. The primary product is a medium Btu gas, much the same as the product of coal gasification. This product can be burned to raise steam and produce electric power or it can be converted into any of the chemical products referred to earlier.

Economic feasibility of PUROX, or of any of the competitive pyrolysis or gasification processes now in various stages of development, depends upon the existence of a sizable trash base—meaning a large population center. It would be perfectly feasible to feed to a PUROX furnace other organic materials, such as waste wood, industrial or agricultural wastes, or even the intentionally grown products of a future "energy farm."

The PUROX interest is primarily to supply a clean, economic method for disposing of municipal America's mounting mountain of solid waste. If the economics could be improved by the use of the gaseous product for chemical production, thus contributing to the overall feedstock supply, this improvement would of course be highly desirable. And extension of gasification to forest products—perhaps in timber-rich, fossil fuel–poor areas of developing countries, or even of the United States—is a definite possibility.

We have no doubt that biomass feedstocks will gradually expand in importance as the cost of petrochemical feedstocks grows. Ethanol fermentation will become more important, perhaps by way of such projects as Nebraska's "Gasohol." We in Union Carbide are sufficiently impressed by the potential importance of biomass to dedicate a sizable research and development group to various fermentations and other biochemical conversions. By the year 2000, the amount of biomass conversion should be significant.

These are the paths along which we see alternative feedstock technologies developing. Since oil and gas will dominate the feedstock picture for many years, the first step is to improve the efficiency of their use. But since these are finite resources, the industry is moving aggressively to develop economically attractive alternatives as the supply of oil and gas dwindles and the price continues to go up. Coal-based synthesis gas leads the list because of the abundance of coal and because of long industrial experience in synthesis gas processes. Crude oil from shale, tar sands, and coal will become more attractive as the price of crude from conventional sources goes up. Finally, increasing oil and gas prices and improving biomass and solid waste utilization technologies will make these sources attractive as feedstocks by the 1990's. By 2000, we predict that feedstocks from all sources other than oil and gas will constitute 10 to 15 percent of the total, but certainly less than 25 percent.

This leads to a larger lesson, namely, the roles and uses of technology in meeting the broader energy crisis. Whatever form the National Energy Act takes, we are certain that scientific and technological innovation will be the true driving force of the plan for both conservation and the development of new energy sources. And this must be done within a framework of high but workable standards for environmental quality and public health and safety. This is an imposing challenge, but one that can be met if we use our resources and resourcefulness wisely.

Those of us who take our signals from the marketplace have known for many years that oil and gas would be in a "price crisis" in the 1970's. Accordingly we have utilized research and development to find alternatives to oil and gas in their use as feedstocks. We are confident that our industry can and will use technology to make the transition to alternative resources so that we can continue to provide the products our civilization has come to need and demand. And given the willingness to use technology responsibly, we are confident that all sectors of the nation can do the same for other uses of oil and gas as well.

The Hobbling of Coal: Policy and Regulatory Uncertainties

The increase of controls on production and coal use is a threat to industry expansion.

Richard L. Gordon

In less than a decade, the U.S. coal industry has been transformed from a lightly regulated industry to one that is affected by many government policies. The industry that long waged a lone battle for more comprehensive government planning in energy now decries such intervention. In 1977, the General Accounting Office when appraising the state of the coal industry claimed an inability to determine whether public policy, on balance, helped or hurt coal (*1*). In this article I review these policies and suggest that the probable net long-term effect on coal production is negative. The long term will, however, be a decade or more; and, should serious barriers to coal utilization rise in the interim, energy problems of the United States could be exacerbated.

Summary. The inevitable by-product of societal concern over the side effects of coal production and use is greater than recognized difficulties to meet coal production goals. Thus, society faces the need to make trade-offs it is not adequately anticipating. The primary influences have come from air pollution regulation dealing with sulfur oxides; regulations in this realm were strengthened in 1977. Questions exist about the ability to meet these requirements and about whether tighter controls will be imposed on other pollutants. Problems also exist on the production side—labor unrest, strip-mine reclamation requirements, and delays in leasing federally owned coal.

The policies restraining coal use were designed to attain legitimate social objectives, and this article cannot be expected to resolve the difficult questions as to whether the socially optimum levels of controls have been imposed. Focus is confined to the narrower question of the impact on coal and, in particular, to the suggestion that such effects may be greater than many policy-planners seem to believe.

Two basic observations may be made before I turn to review specific regulation. First, the coal situation should be considered a prime example of how energy policy-making in the United States is, in practice, an effort to avoid explicit reconciliation of our conflicting policy goals—a problem that is by no means confined to the energy realm.

We have simultaneously proclaimed as objectives the need for lessened imports, abatement of environmental damages, protection of the public from high energy prices, greater competition, and prevention of dislocation of industries and those they employ. Aside from the questions that can be raised about the wisdom of some of these goals, the clear conflicts among them must be recognized. We cannot, for example, avoid some environmental degradation if we move to increase significantly domestic energy production, particularly of coal. Policy-making has involved taking strong stands in all areas and forcing the compromises to emerge in the implementation process. As is shown below, this can be an expensive and unsatisfactory procedure. The litigative abilities of the various sides can become more important than the merits of the case.

The second key point is that the view that the fundamental constraints to coal use are on the demand side may be accepted only in a carefully qualified form. It is quite true that the initiating force in inhibiting coal industry expansion is lack of adequate signals about how much market growth will occur. The coal industry is indeed extremely touchy about any hint that it lacks the capability to produce the 1.2 billion tons of coal advocated for 1985 in President Carter's energy program (*2*). Close reading of industry statements make clear, however, that this does not imply confidence that no influences on the supply side, such as continued labor unrest and increased regulation, will impede production increases. The industry simply claims that it is not the source of any supply problems. Many critics of the industry could argue that management must be charged with at least partial responsibility for labor unrest, and those seeing an energy producer conspiracy to monopolize envision an unwillingness to expand (*3*).

The Issues

We may begin by listing the critical policies that affect coal. The most publicized, and probably the most influential, has been air pollution regulations, particularly those affecting sulfur oxides. To alleviate the impacts of these regulations, various existing and proposed laws have been designed to promote coal use. The principal supply issues are health and safety, impacts on land use, and leasing of federal coal rights. Concern, but not action, has arisen over alleged monopolization of the coal industry by oil companies. All these issues affect all methods of coal mining. However, health and safety are generally considered more of a problem with underground mining while it is the disturbances created by surface mining that are the primary land use issues.

Air Pollution Control

Sulfur dioxide emission control is not only the most important policy affecting coal, but probably the most complex. Two different types of difficulties may be distinguished.

First, many different groups have influenced the form and substance of policy. Second, the policies themselves have become increasingly convoluted.

In terms of participants, distinction should be made at least among the actual policy-makers, the lobbyists for controls, and the affected coal-using industries. The second of these three groups needs little treatment; the success of en-

The author is professor of Mineral Economics, College of Earth and Mineral Sciences, Pennsylvania State University, University Park 16802.

vironmentalists in securing legislative and judicial victories is by now familiar.

In contrast, the intricacies of the regulatory structure and the response of regulated users merit fuller review. The principal actors in policy-making have been the U.S. Congress, the U.S. Environmental Protection Agency (EPA), and state governments. (A fourth group—Indian tribes—also has been given some authority.) EPA has naturally pushed vigorously to ensure compliance with congressionally set air pollution control objectives. At one point in the 1970's, EPA did suggest that states retreat from efforts to comply more rapidly than necessary with federal requirements. A clear picture is not available on state practices. However, evidence does suggest a tendency for some state environmental agencies to attempt to outdo federal standards. Such ambitions have at times been thwarted by decisions elsewhere—mainly in the courts—that the plans were impractical. Congressional action has involved most of the expected political conflicts. Members of Congress from the coal states continually try to blunt the impact of air pollution regulations. Further problems seem to arise from disputes about which congressmen will exercise control over policy-making.

The regulated industries characteristically maintain a posture of perennial consternation. This attitude merits more sympathy than is normally granted. Even when the industry objections are misguided, this arises from nothing more sinister than the parochialism of which everyone is guilty. Businessmen, being normal human beings, envision themselves as providing indispensable services to society and have trouble perceiving that their interests may conflict with the public interest. (This is an attitude that should be maintained; businessmen are ill-equipped to venture outside their area of expertise to judge what is good for society.) Moreover, some aspects of policy-making may need reform. Many, apparently including those who drafted the President's energy message, feel that regulations are altered too frequently. Even if rules were constant, complaints could be levied at rule-making procedures. As I argue below, the stress on standards and timetables appears particularly dysfunctional.

When policy-makers impose goals that seem unreasonable to the affected industries and provide no credible incentives for compliance, nonattainment becomes highly probable. The industries recognize, at least subconsciously, that they are considered so essential that extensions will be granted to permit continued operation even if air pollution regulations are violated (*4*). This seems a powerful incentive for dilatory behavior. It also produces complex policies that set stringent goals but add numerous escape clauses.

To understand the problem, it is necessary to view the actual structure of policy on sulfur oxide control. Understanding federal procedures is sufficiently difficult, and thus review may be limited to that area. The first thing to recall is that two somewhat different bases—air quality and emitter performance—are used for policy-making. Air quality refers to the concentration of pollutants in the air; performance refers to the discharges allowed by a particular facility. Conflicts can easily arise between these alternative rules since they are determined by quite different methods.

Two basic concepts exist for air quality—a primary standard to prevent health hazards and a secondary standard to prevent other damages such as that to property. A maximum allowable average annual concentration is defined as well as how much deviation from the annual norm is allowed on any one day and possibly in any 3- or even 1-hour period. In addition, limits are imposed on the increases in pollution allowed in areas with air already significantly cleaner than what is required by the basic standards. This concept of controlled increases is officially termed prevention of significant degradation. This concept created so many administrative problems that in 1977 Congress enacted elaborate rules governing nondegradation (*5*).

Three levels of strictness were defined—classes I, II, and III, with class I facing the most stringent rules—and specific limits were set on the allowable increase in pollution in each class. All land previously set as class I and certain public lands, such as preexisting national parks exceeding 6000 acres in area, had to be designated class I; others, such as a new national park greater than 10,000 acres in area, must be class I or class II. Otherwise, the area is initially put into class II and the governor of the state may petition, if desired, for up- or downgrading. A further change introduced in 1977 was that areas not in compliance with air quality regulations were subject to more stringent controls. New facilities would not be allowed unless sufficient evidence was provided of vigorous efforts to attain compliance (or a vital need for the facility could be proven).

Until 1977, the emission rules for sulfur oxides required that new facilities (those facilities whose planning began after the promulgation of the rules in 1971) burning more than 250 million British thermal units (Btu) per hour limit their emissions to 1.2 pounds of sulfur oxides per million Btu burned. Compliance could be attained by the use of coal naturally low enough in sulfur content to meet the standard, cleaning the coal to acceptable levels before combustion, or by the use of devices called stack gas scrubbers that capture the sulfur oxides after burning and prevented their discharge. In practice, treatment before combustion has not yet been a relevant strategy. The choice has been between naturally low sulfur coal and scrubbers. In 1977, the option of using naturally low sulfur coal was removed by introduction of the concept of best available control technology (BACT). This technology is to involve the maximum feasible reduction in sulfur emission with an expectation that a 90 percent removal goal will be set for facilities initiated after the amendments come into force (*6*).

State of Scrubber Technology

BACT seems to exacerbate a preexisting defect in sulfur oxide control strategies. The evidence suggests that stack gas scrubbers may still not be as effective a control technology as their advocates contend (*7*). This is best seen by reviewing the latest available evidence. EPA has a contractor—PEDCo Environmental—prepare an extensive bimonthly report on the state of scrubber utilization (*8*). These reports present detailed chronologies of the operating history of those scrubbers actually in place. Scrubber advocates tend to report the summary data and ignore the case histories. The summaries show a growing number of "operational" scrubbers (31 as of September 1977). Examination of the details shows that PEDCo uses a generous definition of operational. Units are added to the operational category when test operations begin. No distinction is made about whether the units are operable (as at least two are not), meeting performance standards (which some are not); and the list even includes two small units that are basically used to test scrubber technology.

Most of the successful units are engaged in mild scrubbing (50 to 60 percent removal) of sulfur oxides from low sulfur coal. Those successes that have occurred in scrubbing high sulfur coal appear to have come at high cost. Redundant units are installed so that frequent cleaning of the unit can occur without putting the plant out of service. The cleaning is necessary because operation tends to cause clogging and corrosion in the units. Often long shakedown periods are required to make the scrubbers operational. Even so, frequent outages seem likely. Finally, present scrubbers capture the sulfur oxides in some absorbent such as limestone, and problems arise in disposal of the resulting sludge. In short, scrubbers create cost, reliability, and waste disposal problems that could exceed those associated with nuclear power.

How then did the devotion to scrubbers arise, why the problems, and why the move to increased reliance despite such difficulties? The initial impetus is easy to explain. The policy-makers eagerly accepted the arguments of the equipment manufacturers and the coal industry that scrubbers were a cheap, available solution. Thus, scrubbers seemed an ideal solution—providing speed, low costs, and maintenance of existing regional production patterns. The question of why problems arose is simpler to explain; scrubber advocates failed to anticipate the difficulties.

Charges are often made that electric utilities did not try hard enough to make scrubbers work. Clearly, accusations of this sort are too vague to permit reasoned evaluation. It can be argued, however, that public policy was poorly designed to stimulate vigorous activity on scrubber development. Credible pressures were not being exerted on the electric utility industry. The industry had good reason to expect that the actual effect of failure to perfect scrubber technology would have been extensions of compliance deadlines. The federal government failed to correct this problem either by adopting effective incentives to compliance such as an emission tax or by financing development work on scrubbers.

The refusal to back off from stress on scrubbers can be attributed to the interaction of three quite different forces: the usual reluctance of politicians to confess their errors, the pressures of coal state legislators to protect local coal, and the desires of many environmentalists to limit the development of low sulfur western coal. These efforts failed to stop a shift to western coal, and the immediate justification for BACT is to slow this move to western coal. How successful this move will be in the long run remains more uncertain than is generally recognized, as I indicate below. (BACT also can be justified because economic growth will necessitate tightening controls on individual polluters.)

In any case, the history to date of sulfur oxide control strategy has been one of setting goals that proved unattainable and then backing away from them. Where we are heading remains unclear. One can cite several proposed new coal-burning power plants that are having difficulties securing approval because of their environmental impacts. However, the Clean Air Amendments do include numerous escape clauses that could alleviate the pressures. It seems safe to conclude that, on balance, construction of coal-fired power plants is beginning to involve increases in planning lead times comparable to those already plaguing nuclear power. The open question is whether these constraints will cause the "inadequacy" of capacity expansion so widely feared by the electric utility industry (*9*). Both determinants of the answer—what expansion rate is economically optimal and what the delays will be—cannot be satisfactorily forecast. Matters are even more confusing for other users of coal. We have very little idea about the economics of coal use under BACT for manufacturing plants (*10*).

Production Problems

With respect to these supply problems, the most serious threat has been the sharp rises (probably in the neighborhood of 10 percent per year in constant dollars) in eastern underground mining costs (*11*). The source of the rise can be epitomized as a radical change in labor market conditions. The pool of workers who felt trapped in the coal industry as job opportunities shrank as a result of falling total output and rising output per worker has vanished.

With rising output, falling output per worker, and retirements, the coal industry has turned to recruiting younger workers who demand greater compensation for or protection from work hazards. These demands have been expressed by successful appeals for a more stringent federal Coal Mine Health and Safety Act (passed in late 1969), pressures for tougher bargaining at the national level, and unilateral action through wildcat strikes at the local level.

Again enormous appraisal problems exist. The primary question is over the prospects for alleviating the cost increases. Presumably, once the industry has fully responded to new labor market conditions, no further direct cost increases should be necessary. However, if the new climate retards or eliminates productivity advance, some persistent cost rises can be expected. Real wages rise in line with productivity advance in the economy as a whole; if there is no productivity advance in coal mining even though wages rise, unit labor costs also rise. All this defies forecasting. Thus, while a slowdown in cost increases should occur, no one knows when and to what degree this deceleration will occur.

An even more difficult problem to resolve is the allocation of blame for the cost increases. The Coal Mine Health and Safety Act of 1969 is charged with causing many of the difficulties. The numerous provisions of the act mandated changes in every aspect of mining deemed to affect health or safety. Clearly, the requirements for extra equipment, for devoting more time to health and safety work, and the more frequent interruption of work by mine inspections could only increase costs. However, the magnitude of these effects remains unclear, even to many of the affected coal companies. Few such companies have tried to isolate the effects of the act from other influences on costs. Whatever the causes, the effects—rising costs—are a serious threat to the competitive position of underground coal mines.

Two further problems affect the underground mining sector. First, the growing reach of environmental regulations in general has caused lengthening of the lead times for mine construction. Numerous permits must be obtained from both agencies regulating mining and various environmental authorities such as those charged with water pollution control. Second, the industry has expressed considerable concern over the provisions of the 1977 Surface Mining Act that require control of the surface effects of underground mines.

However, the main effects of the Surface Mining Act clearly are on surface mines. The 25 strip-mine reclamation requirements of the act in section 515b include (*12*):

1) "Maximize" initial recovery so that a second disturbance to recover other seams is avoided.

2) Restore land to a condition that allows at least as good a use as that prior to mining with extra effort required if the land is considered prime agricultural land.

3) Except where impractical, restore land to the original contour.

4) Stabilize areas to avoid erosion.

5) Segregate and preserve the quality of topsoil or a subsoil of better quality than the actual topsoil.

6) Avoid disturbance of hydrologic balance with special emphasis upon alluvial valleys in the West (roughly valleys

structured so that disruption of water flow would interfere with farming).

7) Revegetate the reclaimed land.

The state governments are authorized to prohibit strip mining on lands on the basis both of specific hazards such as danger of creating floods and of mere incompatibility with land use plans. The Secretary of the Interior is required to prohibit new mines in national parks and other classes of federal lands, has authorization to make prohibitions on the basis of the criteria the states may employ, and must consider private requests for bans. The law requires that when the federal coal rights were retained when the surface was sold, the surface owner must give written permission to mining. Moreover, if substantial opposition exists among surface owners to strip mining, the Secretary is encouraged to prevent the mining.

Appraisal of the law obviously differs among observers. A study for EPA by ICF, Inc., of the 1976 strip-mine bill noted that it would have modest incremental impacts over existing state regulations (*13*). ICF qualified this conclusion by warning that there were sufficient ambiguities in the law so that it could be interpreted to produce more severe effects. The coal industry has expected the worst and severely criticized the initial Interior Department implementation program for adopting far more severe interpretations than the law required (*14*).

On its face, the law does seem to involve potential for further disrupting the development of coal production. At the very least, a substantial shakedown period has been the rule with new regulations. The introduction of such provisions as alluvial valley, prime agricultural land, and surface owner protection seems particularly likely to produce implementation problems. Moreover, Secretary of the Interior Andrus has displayed a clear desire to shift the emphasis of his entire Department from its traditional role as promoter of industrial development to a posture of vigorously protecting the environment. This implies that the interpretations will be more stringent than if the Department had a different orientation.

Land Ownership

Other problems affecting western coal are associated with the complexities of land ownership patterns in the region. The federal government is the principal owner of coal rights. Thus, a substantial number of difficulties arise simply from contending with federal policies affecting the leasing and exploitation of government-owned coal. In addition, much of this federal ownership consists of reservation of mineral rights on lands for which the right to use the surface has been sold. The Surface Mining Act contains, as noted, provisions that require the consent of the surface owners before mining can proceed. The potential for obstructionism appears substantial.

Moreover, the ownership of coal is fragmented by such practices as the "checkerboard" pattern of grants of land to railroads (so-called because the split between railroads and the federal government looks like a checkerboard when mapped). Indian tribes and others also own coal rights. Thus, the problems of working with the federal government are compounded by the difficulties of also dealing with owners of surface rights and others who own coal (*15*).

The most pressing problem seems to be the difficulty in securing permission to exploit existing leases. Approval of exploitation plans has been deemed a major federal action requiring an environmental impact statement. Unfortunately, the Bureau of Land Management, which has primary responsibility in this realm, has proved inadequately staffed to complete the reports rapidly.

A long-term problem arises from the combination of a moratorium on new leases that has prevailed since 1971 and 1976 amendments to the leasing laws. The moratorium does allow leasing of limited tracts adjacent to previously leased land when the extra land is needed to permit optimal development of the property. However, it is not clear whether these leases are being granted as frequently as would be socially optimal. An environmental lawsuit has further restrained leasing.

The overall moratorium began because of concern that too few leases were being exploited. The delays in removal were the result of various forces, including a long gestation period for the environmental impact statement and a rethinking of goals initiated by Secretary Andrus. The leasing amendments mandated greater planning by the Interior Department, increased reporting requirements on firms, and a requirement that leases be developed within a decade (*16*). A U.S. government task force has charted the net effect of all these regulations, and as far as I can tell after winding my way through the four gigantic charts (each of which more than covers a large desk), once leasing resumes, it would take a decade to move from initiating a lease sale to production (*17*). In short, production controls have imposed two severe costs on the coal industry—higher direct costs and the cost of delay.

Several problems are associated with all these restraints. One could follow the General Accounting Office and conclude that the mass of restrictions will prevent the opening of enough new coal mining capacity to meet even the billion-ton level widely forecast for 1985 (*18*), let alone exceed this level as President Carter has proposed. Certainly, considerable doubt may be raised about the short- or long-run efficacy of BACT in reversing the shift to western coal. The numerous available surveys of coal industry expansion plans suggest a heavy concentration of new capacity in the western United States (*19*).

Given the prevailing long lead times, it may be difficult to reorient radically immediate expansion plans. When sufficient time exists to shift emphasis back to eastern coal, it is not clear that this shift will be economically attractive. Continued sharp rises in eastern mining costs may erode the incentives provided by imposition of BACT. Western coal may be sufficiently cheaper that it can overcome the cost disadvantage of BACT. Besides it may prove significantly cheaper to apply BACT to low sulfur coals, and much of the western coal is destined for use in the southwestern states. Moreover, it appears that the lead times for a large new coal-based project may soon no longer differ significantly from those of a nuclear plant. If this is true, managers of electric utilities, on the basis of their belief that, if time permits, it is cheaper to build a nuclear plant, will return to the nuclear option.

Coal Use Incentives

In part, to offset the effects of all these policy measures, President Carter has proposed extensive legislation. Therefore, it is worth discussing his efforts to develop a program to provide incentives to coal use.

The coal use incentives in the 1977 energy program proposed by President Carter are a complex set of basic prohibitions, taxes, and subsidies. In addition, long lists are included giving reasons why temporary or permanent relief can be granted from the requirements to use coal or to be subject to taxes. The basic prohibitions are that new electric power plants and new large industrial boilers must use coal and that, after 1 January 1990, gas use in electric power plants should cease. A shift to oil from gas could be allowed only under special circumstances, and earlier conversions could be required if considered feasible (*20*).

The tax-subsidy elements of the program consist first of taxing oil and gas use and then providing tax rebates equal to the amount of investment expenditures made to use either coal or synthetic fuels from coal. The proposed tax system involves several features. First, the oil system is different from that for gas. Second, electric utilities are treated differently from other industries. Third, the rates are adjusted for inflation. Abstracting from inflation adjustments, the electric utility tax system involves a flat tax of 25 cents per million Btu on oil use after 1982. Other taxable oil users pay a tax that starts at 15 cents per million Btu in 1979 and rises in steps to 50 cents in 1985. All these tax rates are to be raised at the same rate as the implicit deflator for the gross national product. The basic gas tax is ultimately set at the difference (on a per-million-Btu basis) between prices for number 2 fuel oil in the region and the actual cost of gas to the consumer. Again a phased introduction is proposed for the taxes. Electric utilities would not be covered until 1983, and from 1983 to 1985 the tax would be 50 cents per million Btu below the level required in the basic formula. The reduction falls to 25 cents in 1987 and 1988 and disappears in 1989. Other users start being taxed in 1979 with a reduction that is set at $1.05 in 1979, falls to 40 cents in 1980, then more gradually goes to 15 cents by 1984 and in 1985 the full tax is applied. The amount of oil and gas taxed would vary with consumption. Those using less than 500 billion Btu would be exempt from tax. The exempt amounts are reduced for larger users and those using 1.5 trillion Btu or more pay the tax on all oil and gas use.

The energy bill that passed in the House included prohibitions close to those proposed by the President. The tax provisions, however, were substantially altered. Essentially the same tax on oil was set for electric utilities. Large industrial boiler fuel users of oil and gas also were subjected to taxes roughly equal to those proposed by the President. However, the House exempted more users and subjected others to lower taxes. A flat cents-per-million-Btu tax starting, without inflation adjustment, as a maximum of 65 cents in 1983 and rising to 75 cents in 1985 was set for electric utility gas use (*21*). This last tax cannot raise costs of gas above the cost per million Btu of fuel oil.

At least three problems threaten to undermine the President's program. First, the proposals were weakened in the House and have done even worse in the Senate. Second, the Administration may have significantly underestimated how much incentive would be needed to encourage coal use in the face of all the difficulties facing the coal industry. Third, there is good reason to suspect that more restrictions on coal production and use will arise. Existing programs could be made more stringent, and action might be taken in other areas—notably, control of air pollution due to nitrogen oxide, carbon dioxide, and trace elements.

Conclusions

In this article, I argue that the prospects for coal use have been hindered by growing restraints on coal production and use. The constraints increase the costs and lead times of coal-related developments. President Carter has proposed measures to offset this process. The potential for enactment, wisdom, and efficacy of the Carter proposals remains very much in doubt. For present purposes, wisdom and efficacy are the principal issues.

One can dispute the wisdom on two levels. First, the whole idea of promoting coal use could be challenged. Better ways may exist to deal with U.S. energy problems. Second, the concept of first adopting ambitious goals for controlling the side effects of coal production and use and then adding another complex program to cushion the shock seems unnecessarily cumbersome. Given the evidence that the Administration does not understand the full impact of the controls being imposed on coal, the efficacy of its incentive program also remains very much in doubt.

The risks of failure are quite serious at least through the 1980's. The failure of the coal option to develop as expected may be very expensive to remedy. Emergency replacements may be very costly. With sufficient time, a more efficient adjustment process can emerge.

Our current knowledge is inadequate to indicate whether or not the controls on coal produce benefits commensurate with their costs. The main conclusion reached here is that the costs of restricting coal production and use have been underestimated. A particular concern is that the expenses associated with learning how to comply with new regulations are inadequately considered. In sum, we and our policy-makers are insufficiently aware of how great a conflict exists between proposals for expanding coal use and policies that hobble coal. As a result, we are not making adequate preparations to resolve the problems that may arise.

References and Notes

1. U.S. Congress, General Accounting Office, *U.S. Coal Development—Promises, Uncertainties* (General Accounting Office, Washington, D.C., 1977), especially p. 1.7.
2. The overall energy goals were reported [U.S. Executive Office of the President, Energy Policy and Planning, *The National Energy Plan* (Government Printing Office, Washington, D.C., 1977), pp. 95–96] in barrels per day of oil equivalent. However, several sources have reported the tonnage assumptions [U.S. Executive Office of the President, Energy Policy and Planning, *Replacing Oil with Coal and Other Fuels in the Industrial and Utility Sector* (Executive Office of the President, Washington, D.C., 1977), p. III-2; U.S. Congress, Congressional Budget Office, *President Carter's Energy Proposal: A Perspective* (Government Printing Office, Washington, D.C., 1977), p. 47; U.S. Congress, Office of Technology Assessment, *Analysis of the Proposed National Energy Plan* (U.S. Government Printing Office, Washington, D.C., 1977), p. 46]. For coal industry reaction, see G. R. Schleede, "The outlook for coal demand and supply" (National Coal Association, Washington, D.C., 1977).
3. For typical attacks, see J. M. Blair, *The Control of Oil* (Pantheon, New York, 1976); R. Engler, *The Brotherhood of Oil* (Univ. of Chicago Press, Chicago, 1977). For a more restrained view, see T. D. Duchesneau, *Competition in the U.S. Energy Industry* (Ballinger, Cambridge, Mass., 1975).
4. For a similar argument applied to automobile pollution, see H. Margolis, "The politics of auto emissions," *Public Interest* **49** (fall 1977), pp. 3–21.
5. Here, as in much of the rest of the article, I have relied on reading the laws when I could obtain copies [U.S. House of Representatives, *Clean Air Amendments Conference Report* (Report 95-564), 95th Congress, 1st session, 1977].
6. *New York Times* of 27 October 1977 reported that new sources were being defined by EPA as those started after March 1978 when the implementation plan is expected to be complete; members of Congress were quoted as feeling that the date the act passed should be the cutoff point. In any case, we now have two vintages of new sources—those covered by the 1971 rules and a later vintage covered by the implementation of the 1977 Clean Air Amendments.
7. See, for example, Sulfur Oxide Control Assessment Panel, *Final Report on Projected Utilization of Stack Gas Cleaning Systems by Steam-Electric Plants* (U.S. Department of Commerce, Washington, D.C., 1973); U.S. Environmental Protection Agency, *Report of the Hearing Panel, National Public Hearings on Power Plant Compliance with Sulfur Oxide Air Pollution Regulations* (Government Printing Office, Washington, D.C., 1974).
8. PEDCo-Environmental, Inc., *Summary Report—Flue Gas Desulfurization Systems—August–September 1977* (PEDCo, Cincinnati, 1977).
9. Adequacy is, like an energy gap, the layman's imperfect description of an excess demand created by price controls. For a typical example of the concern, see *7th Annual Review of Overall Reliability and Adequacy of the North American Bulk Power Systems* (National Electric Reliability Council, Princeton, N.J., 1977).
10. For a review of the problems in appraising these issues, see R. L. Gordon, *Historical Trends in Coal Utilization and Supply* (National Technical Information Service, Springfield, Va., 1976). U.S. Executive Office of the President, Energy and Policy Planning, *Replacing Oil with Coal and Other Fuels in the Industrial and Utility Sector* (Executive Office of the President, Washington, D.C., 1977), was accompanied by a report by Energy and Environmental Analysis, Inc., *Methodology: Replacing Oil and Gas with Coal in the Industrial Sector*, 1977. Unfortunately the only verifiable data in the report are the costs of a representative boiler, and these estimates are suspect. Considerably higher figures are given by M. H. Farmer, E. M. Magee, and F. M. Spooner [*Application of Fluidized-Bed Technology to Industrial Boilers* (National

Technical Information Service, Springfield, Va., 1977). Farmer *et al.* of Exxon Research and Engineering Company prepared the study for EPA, FEA, and ERDA].

11. The starting point for the calculation is an estimate made by C. J. Johnson and me that 1969 prices were in the range of 18 to 20 cents per million Btu's [C. J. Johnson, "Coal demand in the electric utility industry," thesis, Pennsylvania State University, University Park (1972); R. L. Gordon, *U.S. Coal and the Electric Power Industry* (Johns Hopkins Univ. Press, for Resources for the Future, Inc., Baltimore, Md., 1975)]. Comparisons were made on *Coal Week* listings of 1977 quotations for coal selling on long-term contracts, deflating by the implicit deflator for the gross national product.
12. U.S. House of Representatives, *Surface Mining Control and Reclamation Act of 1977 Conference Report*, 95th Congress, 1st session, 1977.
13. ICF, Inc., *Final Report Energy and Economic Impacts of H.R. 13950 (Surface Mining Control and Reclamation Act of 1976)* (ICF, Inc., Washington, D.C., 1977). For an alternative view on total costs, see Energy and Environmental Analysis, *Benefit/Cost Analyses of Laws and Regulations Affecting Coal* (Government Printing Office, Washington, D.C., 1977).
14. The numerous papers making the case were made available to me by the National Coal Association.
15. See W. E. Tyner, R. J. Kalter, J. P. Wold [*Western Coal: Promise or Problem?* (Department of Agricultural Economics, Cornell University, Ithaca, N.Y., 1977)] for a review of these problems, which suggests that ownership patterns and leasing policies hinder both ensuring maximum returns to the federal government and efficient industry development.
16. U.S. Congress, Public Law 94-377, *An Act to Amend the Mineral Leasing Act of 1920, and for Other Purposes*, 94th Congress, 2nd session, 1976.
17. U.S. Mountain Plains Federal Regional Council, "Preliminary draft of permitting procedure for surface mining of coal on federal lands or coal deposits" (Denver, Colo., 1977).
18. See for example U.S. Federal Energy Administration, *National Energy Outlook, 1976* (Government Printing Office, Washington, D.C., 1976).
19. G. F. Nielsen, "Coal mine development and expansion survey . . . 617.3 million tons of new capacity 1977 through 1985," *Coal Age* (February 1977), pp. 83–100.
20. U.S. Executive Office of the President, *National Energy Act, 1977*.
21. U.S. Congress, House of Representatives, *HR 8444*, Report No. 95-543 (1977), especially pp. 429–467.
22. I thank various government agencies that made reports available to me, ICF, Inc., and Robert Kalter who directly supplied their reports, coal industry sources who provided their material, and E. Welch who checked the manuscript for me.

B An International Perspective

Energy Options and Strategies for Western Europe

Wolf Häfele and Wolfgang Sassin

The present world energy situation is characterized by total consumption at an average rate of about 7.5 terawatts (1 TW = 10^{12} watts) of thermal energy; this is roughly equivalent to 8 billion tons of coal equivalent per year. As much as 5.5 of this 7.5 TW is supplied today by oil and gas, and this high proportion has built up through the last 30 years. Oil systems as we know them today have a low capital cost investment [~ \$50 per kilowatt thermal (kWt)]; they are clean, easy to handle, and can easily adjust to market and end-user requirements. Not exactly but essentially the same is true for natural gas. It is therefore quite natural that the energy situation of Western Europe is mostly reflected by the supply of oil and gas. For the European member nations of the Organisation for Economic Cooperation and Development (OECD), the *World Energy Outlook* (*1*) expects a total energy requirement in 1980 of 1400 million tons of oil equivalent, which is practically 2 TW. Of this, about 0.9 TW is expected to be supplied by oil, which amounts to roughly 12 million barrels per day (MBD). Reasonable estimates of the expected oil imports of the United States and Japan are 10 and 7 MBD, respectively, giving a total for these countries and Western Europe of about 30 MBD for 1980. Under present circumstances oil production by OPEC (the Organization of the Petroleum Exporting Countries) might be between 33 and 45 MBD. A recent study by the Workshop on Alternative Energy Strategies (WAES) (*2*) indicates that a ceiling of 45 MBD for OPEC production might lead to a gap between supply and inherent demand by 1990, and a ceiling of 33 MBD might lead to such a gap as early as 1982.

Competition for Buying Energy

If and when world oil demand outstrips supply, who gets the OPEC oil? The answer is, those who can pay. In the WAES study oil prices up to \$24 a barrel are considered; three questions follow from it.

1) Which countries will maintain a strong enough balance of payments to pay for the high-cost oil, and what will be the effect of such high oil prices on the world trade pattern?

2) How much oil will be added to world reserves at a price of \$24 a barrel, and what are the lead times for producing from these additional resources?

3) What is the domestic resource base of each of the nations competing for oil imports and, more generally, for energy imports?

The first question is essentially one for national and international economists; we will deal with it only briefly here. The order of magnitude of the problem is indicated by the fact that in 1970, when the overall energy price was roughly \$3 a barrel, an average of 3 percent of the gross domestic product (GDP) of all countries was given to the energy sector, while the total exchange between world trade partners was about 9 percent of the GDP. Countries that had to import all of the energy they required therefore had to devote one-third of their trade to pay for it. One might naively conclude that at \$24 a barrel, 24 percent of the GDP would have to go into exports in order to buy the imported energy. The picture is more complex, but such energy prices would certainly open a new era of competition in which only a few countries, if any, could win.

The second question is quite important. Oil has always been explored in terms of ratios of reserves to production such that there are enough proved reserves for, say, 10 or 15 years under current circumstances. By contrast, coal has been explored mostly on geological

W. Häfele is deputy director and W. Sassin is senior scientist at the International Institute for Applied Systems Analysis, A-2361 Laxenburg, Austria.

Summary. Western Europe, now largely dependent on oil imports, has to prepare for strong competition for oil and energy imports in general before the year 2000. The more unlikely it is for Western Europe to secure from outside rich supplies of coal or uranium at readily acceptable economic and political conditions, the more serious this competition becomes. Even exceptionally low projections of economic growth and optimistic assumptions about energy conservation urgently call for vigorous and simultaneous development of indigenous coal and nuclear sources, including the breeder. Long-term contracts for the possession and deployment of foreign oil, gas, and coal deposits are mandatory and should be negotiated in view of the possible aggravation of north-south confrontation.

grounds. Present estimates for oil reserves are 90 billion tons, but geological evidence suggests that the figure should be much higher. Experts at the 1977 World Energy Conference in Istanbul estimated recoverable petroleum resources worldwide to be about 260 billion tons, assuming that the present recovery rate of 25 percent will be raised to 40 percent toward the end of the century (*3*). This does not include deep offshore and polar resources, which are still classified as unconventional petroleum; including these the estimate is 300 billion tons. In fact, some estimates by individual experts are higher, up to 600 billion tons. Thus, there may be significantly more oil in the ground, but its recovery is very uncertain. To actually produce such quantities would require industrial activities that are economically and institutionally infeasible today. But this sheds light on the dimension of the problem. Whether Western Europe would be able to participate in such activities is open to question. In any event, they would take several decades. The real question is whether there is enough time.

The third question raised by WAES has to do mostly with coal and uranium. About 85 percent of all coal reserves and resources are in the United States, the Soviet Union, and China. The possibility of intensified use of coal is receiving more attention (*4–6*), particularly in the United States. However, the environmental and social costs of such intensified coal production would be considerable. It is not surprising that there have been no indications so far that any of the above countries would consider or would be capable of assuming a role in coal production like that of OPEC in oil production. Indeed, most of the OPEC countries have little choice but to trade in their natural resources for the industrial goods they need for consumption and development. This is not the case for the United States and the Soviet Union, and probably not for China, which in any event is a special case. What is left for Western Europe therefore is its own coal reserves and resources. West Germany and Britain are the coal countries. The proved reserves of both countries taken together are 70 billion tons of coal equivalent, or roughly 65 TW-years. In view of the total yearly energy consumption of 2 TW today and perhaps up to 4 TW during the next three or four decades, 65 TW-years is an uncomfortably low figure. However, the established resources in the ground are higher by up to a factor of 6. This coal is very deeply located, as deep as 5000 meters. Coal beds extend from the Ruhr area down to the bottom of the North Sea and to the coal mines of Britain. As in the case of additional amounts of oil, actual production of this coal would necessitate industrial activities that are economically and institutionally infeasible today.

While coal reserves and resources are somewhat known, the same is not true for uranium. The amount of available uranium as a function of price is the subject of an ongoing debate. For a quick orientation, it might be stated that roughly 5 million tons of uranium should be available at prices up to $60 a kilogram and perhaps 10 to 20 million tons at prices up to $200 a kilogram. The salient point for Western Europe is that no significant amounts of *cheap* uranium, say up to $60 a kilogram, seem to be available. The amount of energy available from light water reactors (LWR's) using 5 million tons of uranium is roughly 21 TWe-years (TWe, terawatt electric) full load. It should be kept in mind, however, that about 4 TWe-years is already committed for existing and firmly planned power plants. Getting energy from LWR's would imply, as in the case of coal and oil, a competition for uranium imports.

Prospects for Western Europe

What, then, is an appropriate strategy for Western Europe? To elaborate on this question we take as an example the Federal Republic of Germany. Its present yearly consumption of energy is roughly 380 million tons of coal equivalent (MTCE). In 1975 a demand of 700 MTCE was projected for the year 2000. Because of slower economic growth, a declining population, and more rigorous energy conservation measures, experts now project 600 MTCE for the year 2000 (*6a*). Where will 600 MTCE come from? A recent study by German coal experts in collaboration with the International Institute for Applied Systems Analysis (IIASA) (*5*) suggests that 200 MTCE each will come from hard coal and nuclear power, and that the remaining 200 MTCE must continue to come from oil imports. Natural gas supplies of about 50 MTCE may be there in addition. Let us review each of these amounts.

In terms of both supply and consumption, 200 MTCE from hard coal is a high figure. Use of coal in West Germany has been narrowed down to electricity and steel production. In 1980 there might be 35 MTCE for the former and 41 MTCE for the latter, totaling not more than 76 MTCE. Present efforts are being directed to maintaining this level, and it would seem difficult to increase it in view of the realities of the next 10 years. Environmental standards are becoming more stringent, and it is not clear whether plans to restrict sulfur dioxide emissions to 1.5 kg per megawatt-hour (MWh) will not further suppress the use of coal. Public opposition to coal has already led to construction delays in some cases (for instance, at Voerde). In order to concentrate on the physical aspects of the problem, for the rest of this article we will idealistically assume that issues related to public acceptance of coal and nuclear power will be resolved.

Increased use of coal will in any event require a fresh look at this problem and new technological solutions. Fluidized bed combustion for electricity generation offers great improvements with respect to the environment; capital costs of 700 deutsche marks per kilowatt electric (DM/kWe) could lead to applications in the load range of 4000 hours per year and 0.06 to 0.08 DM/kWh beginning in 1985. Unfortunately, the ratio between the deutsche mark and the dollar is constantly evolving, and it is not clear how these costs translate into dollars and mills. For the purposes of this article we assume an exchange rate of 2.5 DM per dollar, so the equivalent values are $280/kWe and 24 to 32 mills/kWh. Process heat for industry at 7000 hours per year could substitute for heavy fuel oil at $85/kWt beginning in 1985, and synthetic natural gas at 7500 hours per year (for example, by the LURGI gasification process) could be used at $350 per 1000 m³ beginning in perhaps 1990. Optimistically, by the year 2000 all these new technologies could lead to a consumption of roughly 120 MTCE. Together with the conventional uses of coal, this could make up the envisaged 200 MTCE. But it will not be easy and will require persistent pursuance of an energy strategy.

Let us also look at the supply side. It is not likely that domestic production can exceed a level of 120 to 140 MTCE as long as the industrial undertakings considered above remain infeasible. There is now increasing awareness within German coal circles that the country may need to import 60 to 80 MTCE annually by the year 2000. One should bear in mind that the present scale of world coal trade is only about 200 MTCE. Also, West Germany and Britain, where similar trends were found, are traditionally looked on as coal exporters rather than importers. The fact that the rest of the Western European countries have an even greater need for imports shows the severity of the situation.

What about the idea of West Germany having a nuclear capacity of 200 MTCE by the year 2000? This corresponds roughly to 130 gigawatts electric (GWe). The country's present nuclear (LWR) capacity is 7.4 GWe, and plants that will produce an additional 14.8 GWe are licensed and under construction. For 1985 it is not realistic to expect a nuclear capacity of more than 25 GWe because of public opposition and other uncertainties. Requiring 105 GWe in 15 years means a rate of 7 GWe per year. This seems physically possible, but our earlier idealistic assumptions about public acceptance should be kept in mind.

For Western Europe as a whole the World Energy Conference estimated two types of energy demand, unconstrained and constrained (that is, without and with energy conservation) (*7*). The projected unconstrained demand for the year 2000 is close to 3.5 TW, while the constrained demand is 3 TW. The partitioning of supply between coal, nuclear, and oil for all of Western Europe differs from that for West Germany. It is given as 21 percent for coal, 17 percent for nuclear, 50 percent for oil and gas, and 12 percent for renewable resources. In absolute amounts this would mean for the case of constrained energy demand: 630 MTCE for coal, 510 MTCE for nuclear, 1500 MTCE for oil and gas (or 21 MBD), and 360 MTCE for renewable resources. The overall misfit is fairly obvious: domestic coal production cannot be developed fast enough in Western Europe, nuclear is lagging behind, and the projected oil and gas imports are more than can be easily borne and perhaps more than will be available, even with an assumed 15 percent conservation potential and an assumed 12 percent share for renewable resources.

Two Inexhaustible Energy Sources

In the very long run there are two essentially inexhaustible energy sources: nuclear breeders and solar power. The case of nuclear breeders has been discussed over and over again (*8*) with regard to the mechanism of fuel supply and nuclear engineering, and we will not repeat it all in this article. The point that must be made here instead concerns the timing of nuclear breeders. For Western Europe they cannot come early enough. Similar to its dependence on oil imports is its situation with respect to uranium, which is worsening the more LWR's are used. The impending renegotiation of the United States, Canada, and the European community for the supply of enriched and natural uranium speaks for itself. Beyond the question of sufficient uranium reserves, the political security of the supply is at stake—which also implies that the related technology must be indigenously available. It is therefore not surprising that all major Western European industrial nations continue to pursue their fast breeder programs vigorously—France with Italy, West Germany with Belgium, the Netherlands, and Britain. It should also be noted that very close cooperation between the continental fast breeder programs was initiated recently and is now materializing (*9*).

The other inexhaustible source is solar power. It is important to make a distinction between production of solar power for local applications such as water and space heating and bulk solar power production for base load purposes. This problem has been studied in detail—for instance, at IIASA for the case of Austria (*10*). It is certainly not the yearly average solar radiation that limits the applications of solar power in Western Europe. As much as 3 kWh/m²-day can be expected for this yearly mean, while in equatorial regions not more than 5 to 6 kWh/m²-day is available. The big difference is the summer-winter fluctuation, which is roughly 8:1 for Western Europe and 2:1 for equatorial regions. The salient point therefore is energy storage, and local applications of solar power crucially depend on the intended level of storage. With storage covering a period of 1 week or so total solar energy system costs might be on the order of $20/MBTU, compared to ordinary heating costs today of about $7/MBTU—a difference of a factor of 3 (some people say it would be greater). This level of solar heating ($20/MBTU) could provide roughly 50 percent of the total energy requirements (20 MWhe) of a one- or two-family house. In Austria the roof area is about 300 km². Assuming that one-third is suited for solar absorbers, the upper limit would be 20 TWh for Austria as a whole, or 10 percent of its present demand for secondary energy. More realistic figures would be closer to 4 percent, however; to provide more than this one would have to go beyond such local applications.

A fairly detailed IIASA study indicates that in Austria an area of about 2500 km² could be available for solar power (*11*). This could provide the technical potential for producing 100 to 400 TWh/year, depending on local conditions, temperatures, and so on—a high figure indeed. However, the seasonal cycle has to be bridged, and this leads to large-scale storage requirements. The study concludes that not more than 60 to 80 TWh/year could be harvested if hydrostorage were used. Thus, even though Austria has many mountains and therefore enjoys this large storage potential, the total solar energy available is constrained by storage and not by land. This constraint is much harder on most other Western European countries. It could only be overcome by the generation of a storable and transportable secondary energy carrier, which couples the development of large-scale solar power for base load purposes to the development of, for instance, a hydrogen technology. What the system costs will turn out to be eventually is not clear. In any event, the determining parameter ultimately will be the feasible rate of introduction should the capital costs be acceptable. Solar power on a significant scale is therefore a long-range option, say beyond the year 2020. It would therefore be prudent to allot to it not more than about 5 percent of the 3-TW energy demand assumed for the year 2000—to many observers this still appears a very big number.

A Strategy for Western Europe

In view of the discussion above, the following strategy seems appropriate for Western Europe.

1) After 1990 sharply increasing difficulties with energy imports should be expected; therefore *all* domestic energy sources should be developed.

2) Until the year 2000, perhaps 300 MTCE of hard coal per year could be made available indigenously. As much coal as possible should be imported. Despite the sometimes overwhelming market difficulties for coal today, it appears prudent to take long-range contracts or options for buying or possessing coal abroad. A reasonable target figure for such coal imports to Europe would be 300 to 400 MTCE/year.

3) Develop nuclear power to the greatest possible extent. A target figure of 500 to 600 MTCE, or 300 to 400 GWe, appears appropriate. At the same time, it is mandatory for Western Europe to develop the nuclear breeder as rapidly as possible and without delays. Appropriate steps must be taken for nuclear power to produce more than just electricity (*11a*).

4) All other renewable energy sources such as solar and wind power must be developed. The key parameter, besides capital cost, is energy storage. It does not appear prudent to expect more than 200 to 300 MTCE from such renewable energy sources by the year 2000.

5) Energy conservation must be given the same priority as energy supply. A conservation potential of perhaps 500 MTCE appears feasible by the year 2000.

6) Steps 1 to 5 will still leave Western Europe needing to import oil and possibly gas at a rate of perhaps 1500 MTCE/year by the year 2000. Economically and politically, it will probably be exceedingly difficult to do this. Prices of up to $25 and possibly $30 per barrel should be expected. Any long-range contract for supply or the possession of oil sources abroad must be made use of. Besides cost, the entrepreneurial, institutional, and political implications of going to unconventional oil resources must be kept in mind.

A Global Constraint

In the future there will clearly be a strong worldwide competition for energy sources. It is therefore appropriate to ask for the reasonable upper potential of global energy production in the next decades. In an IIASA paper for the World Energy Conference (*12*) we estimated this potential to be about 35 TW by the year 2030. Compared with today's figure of 7.5 TW, this means roughly a factor of 4.7, or slightly more than 3 percent growth per year averaged between now and 2030. On a global average this would permit not more than a 3.7 percent economic growth rate. Realistically, one should assume not more than a 3.2 percent upper limit for the latter. For comparison, the scenarios in the recent study by Leontief *et al.* (*13*) for the United Nations and the projections of the New Economic Order give such average growth rates as about 5 percent. The dilemma is that the developing countries will demand growth rates higher than the global average. Corresponding growth rates lower than 3.2 percent for the developed countries would further aggravate their balance of payments problems, which result from increasingly expensive energy imports. Instead, they will try to stimulate economic growth in order to partly compensate for the rising share of energy costs in their balance of payments. In other words, it is possible that the north-south confrontation will become worse. It is against this background that we finally conclude that *every* available energy source, including nuclear and solar power, must be developed as quickly as possible without any further delays. And against this serious background the problem of public acceptance must be assessed.

References and Notes

1. Organisation for Economic Co-operation and Development, *World Energy Outlook* (ISBN 92-64-11595-1, Paris, 1977).
2. C. L. Wilson, *Energy: Global Prospects 1985–2000. Report of the Workshop on Alternative Energy Strategies* (McGraw-Hill, New York, 1977).
3. P. Desprairies, *Report on Oil Resources, 1985–2020, Executive Summary* (Conservation Commission, World Energy Conference, London, 1977).
4. W. Peters *et al.*, *Report on Coal Resources, 1985–2020, Executive Summary* (Conservation Commission, World Energy Conference, London, 1977).
5. W. Sassin, F. Hoffmann, M. Sadnicki, Eds., *Medium-Term Aspects of a Coal Revival: Two Case Studies* (CP-77-5, International Institute for Applied Systems Analysis, Laxenburg, Austria, 1977).
6. W. Sassin and W. Häfele, "The role of coal in the evolution of the global energy system: A reference case," paper presented at the Third IIASA Conference on Energy Resources, Moscow, 28 November to 2 December 1977.

6a. *Energie Programm der Fundesregierung, zweite Fortschreibung vom 14.12.1977* (Federal Ministry for Economics, Bonn, West Germany, 1977).

7. I. J. Bloodworth *et al.*, *Report on World Energy Demand, 1985–2020, Executive Summary* (Conservation Commission, World Energy Conference, London, 1977).
8. See, for instance, *Bericht des Bundesministers für Forschung und Technologie über die Entwicklung des Natriumgekühlten Schnellbrutreaktors* (Federal Ministry for Research and Technology, Bonn, West Germany, 1977).
9. See, for instance, H. H. Hennies and A. Brandstetter, *Atomwirtsch. Atomtech.* **22**, 4 (1977).
10. W. Häfele, "The contribution of solar energy to meet present and future energy demands," paper presented at the BMWF/ASSA Symposium on Solar Energy Research (Sonnenenergieforschung), Vienna, 1977.
11. W. Korzen, in *Third Energy Systems Program Status Report* (International Institute for Applied Systems Analysis, Laxenburg, Austria, 1978).

11a. W. Häfele and W. Sassin, Proceedings of the European Nuclear Conference, Paris, 21–25 April, 1975, page 107ff.

12. W. Häfele and W. Sassin, "A future energy scenario," paper presented at the Tenth World Energy Conference, Istanbul, 19 to 23 September 1977.
13. W. Leontief *et al.*, *The Future of the World Economy* (Sales No. E76.II A6, United Nations, New York, 1976).

Energy and Resources

A plan is outlined according to which solar and wind energy would supply Denmark's needs by the year 2050.

Bent Sørensen

By classifying resources according to their properties, for example, as metals, semiconductors, and insulators, one can obtain broad groups of substitutable materials that are present in the earth's crust, including the water of the oceans, in quantities that probably exceed any foreseeable requirement by man. If one can assume that the use of a resource will rarely result in its being removed from the earth into space, then the term usage will at worst imply dilution to the average concentration of the resource in the earth's crust. Thus, the absolute degree of resource depletion is of minor relevance; rather one must expect to spend an increased effort, related to cost, in order to recover the resources from still lower grade deposits. If used raw materials were returned at maximum dilution, the cost of recovering new raw materials would increase as a function of the integrated usage over the past.

According to Dunham (*1*), for example, the easily accessible deposits of several raw materials may last only 20 to 100 years at the present rate of usage. Thus, it seems reasonable to assume that some groups of resources will undergo sizable cost increases within the time span considered in current planning. The cost of a resource that is extracted by recycling processes will depend on its concentration in the product that is recycled. Today the concentration is often low because of mixing of wastes of different composition. However, if usage of the resource is expanded, the amounts required beyond the recycled quantities must in any case be obtained from the earth's crust at increasing cost.

Energy resources have a particular role in man's activities. The earth receives a continuous flow of energy from the sun, and from solar radiation are derived other energy forms such as wind, currents, and sea temperature gradients. Only minute fractions of these free energy flows have been utilized by man up till now. Instead, mankind exploits finite energy resources such as coal, oil, and fissile material which have been accumulated over periods much longer than it is taking man to deplete them at present rates of consumption. Because the chemical and nuclear energy stored in these resources is released by usage, there is no possibility of recycling. The use of breeder reactors for producing nuclear energy will constitute a type of recycling, but such breeders will still involve resource depletion, for example, of uranium-238.

Energy Policy in Industrialized Countries

Associated with the life-styles of industrialized countries is a strong dependence on manufactured goods with comparatively short lifetimes. Because of the increasing costs of extracting new raw materials and energy from nonrenewable sources, the production of such goods will follow a growth curve of the A type shown in Fig. 1a (*2*). The contribution from the recycling of used goods, shown in Fig. 1b, is included in Fig. 1a. The amounts of change in the level of production and recycling of raw materials are determined from the average profit over 25 years (rather time units) of continued production at the same level.

If the cost of raw materials were to increase while energy prices remained constant, as they might do if energy production by nuclear fusion became safe and acceptable to society, then the production curve would become S-shaped and approach a constant value instead of declining after the peak. However, today's energy policy cannot be based entirely on the unproved potentials of a future resource, and it is necessary to consider the potential of the continuous energy sources. As also shown in Fig. 1, the existence of a continuous flow of energy, of which a fixed amount can be utilized at constant cost (in practice, the ceiling may be far lower than, for example, the total solar energy input), leads to a stable production level. This is preceded by an overshoot phase, when the accumulated use of raw materials is still low, and an adjustment phase, when the recycling is augmented and approaches its maximum value.

In addition to considering the possibility of physical constraints on continued growth of energy and resource consumption, one should consider the constraints imposed by the interconnections between structural changes in society, the systems in use for production and energy conversion, and the economic growth rate with its fluctuations. For instance, the amount of energy spent on raising the gross national product (GNP) by one unit is widely different in different parts of the world. During the last 20 years, Europe and Japan have used only half as much energy as the United States to increase their GNP's per capita by one unit. This indicates a lack of universal relationship between energy consumption and economic growth. Furthermore, it is recognized that the GNP provides a poor measure of the standard of living or quality of life because it includes any activity in society, regardless of whether it benefits that society or not. Statistical data show that in most countries there has been no decline in the amount of energy spent on each new unit of increased GNP per capita. There is evidence that the improvement of living standards (basic needs and luxury) constitutes a diminishing fraction of each new unit of increased GNP per capita; the rest is spent on the structural changes required by the growth itself, on its side effects, and on managing its wastes.

Following this line of thinking, I suggest that the traditional manner of growth, characterized by diminishing returns related to the quality of life, should be replaced by our implementing a policy aimed directly at stimulating growth that will improve the quality of life. This again requires that society increase its influence on the types and durability of goods that are manufactured (for example, by changing tax policy). It also places the choice of production and energy conversion systems in a new perspective that may be summarized by such questions as: "Which system would be most compatible with the formation of a society in which the largest possible fraction of the GNP would be placed at the direct disposal of the population, instead of its being absorbed by the structure of the system itself?"

Many nations might find that, compared with nuclear power, the renewable energy sources such as wind and sun, which favor decentralized utilization, would facilitate development in the direction of placing more emphasis on the quality of life. This

The author is associate professor of physics at the Niels Bohr Institute, University of Copenhagen, Blegdamsvej 17, 2100 Copenhagen, Denmark.

is partly because the reactor industry gives priority to large units and partly because of the uranium supply situation. A nation without uranium mines or without uranium enrichment facilities might fear that the monopolization of uranium supplies outside its control could result in steeply rising fuel costs after it had made a commitment to a number of nuclear reactors. This dependence on an outside uranium supply and nuclear technology is more serious than a dependence on an outside oil supply, because the capital costs of nuclear reactor plants are larger than those of equivalent oil-fired plants. The large amount of capital that would be invested in nuclear energy production would prohibit a swift change to an alternative energy source, should the uranium supply or the safety issues demand such a transition.

Compared with nuclear energy, the harnessing of solar or wind energy requires equipment designed on a different technological scale. Such equipment is less exposed to monopolization than that for the nonrenewable energy sources, which is one additional reason for suggesting that, in making its energy policy, any nation should keep the continuous energy sources well in focus. The first step toward establishing a policy favoring renewable energy sources will be to evaluate the potential of various regions for solar and wind energy. In general, the renewable energy sources are considerably more equitably distributed over the world than the nonrenewable resources. I will attempt to make such an evaluation specifically for Denmark, a country which, excluding Greenland, has practically no nonrenewable energy sources and which, since World War II, has developed a 90 percent dependence on oil from the Middle East.

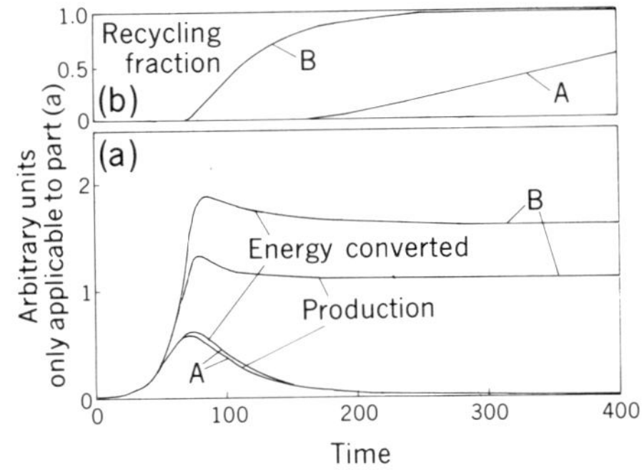

Fig. 1. A simple model of industrialized production. (a) In alternative A the price of new raw materials and energy from nonrenewable sources increases in proportion to the accumulated use. In alternative B the option of using a finite amount of continuous energy sources at constant cost is added. The contribution from recycled used goods, shown in (b), is included in (a).

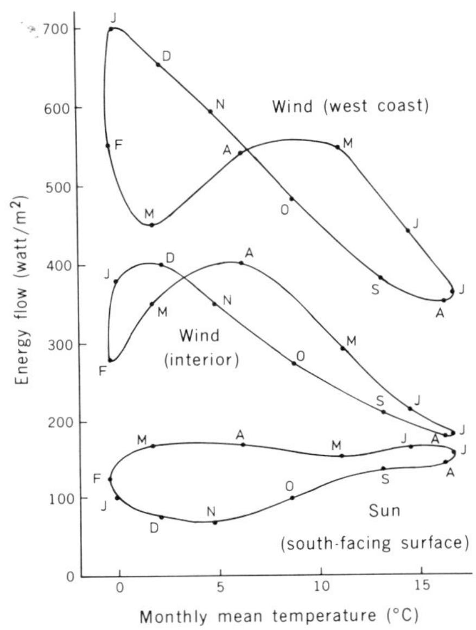

Fig. 2. Monthly average energy flow from continuous sources through a vertical square meter in Denmark, as function of the monthly mean temperature. The sun's height over the horizon at noon is 11° at winter's solstice. The wind data are taken 25 meters above smooth ground.

Potential Energy Resources of Denmark

The average input of solar and wind energy in Denmark is shown in Fig. 2 (*3*). Fluctuations in the wind energy through the year are favorably correlated with the need for space heating, as well as with the variation in electricity use, particularly on the west coast. Solar input is at a minimum in November, but is high in the cold months of February and March when a south-facing, vertical area is considered. The less favorable correlation between solar input and heating requirements during December and January would necessitate the use of additional collecting surfaces in these months, but these surfaces would remain largely unused during the rest of the year.

To evaluate Denmark's total potential of solar and wind energy, one would have to choose a reference height interval for the vertical collectors and establish how close together they could be placed without shading problems. If one used as a reference a 50-meter-high collecting area for either sun or wind energy, there would have to be a distance of about 1 kilometer between solar collectors to allow for shadow lengths in December; the restoration of the wind profile behind a series of 50-meter-high windmills would require about 2 kilometers between rows of mills (for example, if the mills were arranged in rows from north to south to allow for the prevailing westerly winds) (*4*). On the other hand, the spacings perpendicular to the wind direction may be quite small. For the total area of Denmark, 44,000 square kilometers (Greenland excepted), this leads to reference collection areas totaling 2200 square kilometers for solar panels and 1100 square kilometers for windmill swept areas, corresponding to a yearly average of 285 gigawatts of solar energy and 310 gigawatts of wind energy. For the critical months, the solar energy reaching these panels would on average amount to 167 gigawatts (December) and the wind energy to 300 gigawatts (February). Fluctuations between years might alter these values by as much as a factor of 2. For comparison, the average total energy consumption in Denmark is now about 20 gigawatts.

The fraction of the energy estimates made above that could be utilized with known technology would be determined by the efficiencies of the collection and conversion systems. Optimal efficiencies on a yearly average basis are, at present, about 50 percent for solar heat collectors, 5 to 20 percent for electricity-producing solar cells, 40 percent for windmills producing d-c electricity or heat, and 30 to 35 percent for a-c generating windmills that are forced to run at constant angular velocity. Thus, on the basis of the collection areas required, the use of wind energy should be favored for electricity production and possibly also for heat production. However, windmills may require building restrictions in all directions, whereas solar panels can be placed directly on south-facing walls or roofs of buildings.

Solar Energy Systems

The solar heat system I have in mind would consist of a collector (a black absorber with a multilayered glass cover which might be coated to reduce reflection and retain long-wavelength radiation) and a storage tank containing water, salt, or gravel. The storage tank would have a capacity of about a month's usage, and could be common for an entire building block (placed underground), in order to save insulation area. To allow for winters colder than usual or with less sunlight than usual, a cheap auxiliary energy-producing system would have to be added, for example, a system for heating the material in the storage tank by electrical resistance. Solar heat panels of the type described here are estimated to cost up to \$100 for each square meter, depending on the scale of production (1974 prices are used throughout this article). The cost of a storage system equivalent to 1 cubic meter of water per square meter of solar collector could probably be covered within the figure mentioned.

The economic assessment may be based

on a comparison of the cost of the solar collector with the expenditure that would be necessary to generate, by fossil fuel, the same amount of energy as the solar energy system generates over its lifetime, say 25 years. In order to make these two costs comparable, they must be referred to the same time. This is achieved by transforming the future fuel expenditures into a so-called "present value," which is the amount of money that, together with the accumulated interest, would allow payments to be made annually, each payment being adjusted for the actual fuel cost for that year.

Because I am using 1974 prices, the interest and possible fuel cost increase are corrected for inflation. For simplicity the maintenance costs of the solar and fossil fuel systems are assumed to cancel each other. I will consider two alternatives: (i) the interest (corrected for inflation) is 5 percent per year, and the fuel price stays fixed; (ii) the interest is 5 percent per year and the fuel price increases by 5 percent per year. There is a third important alternative, that both interest and fuel price increase are zero, but it gives results identical to (ii).

The two alternatives of fuel price development could be realized by there being either an increase due to the supply and demand situation described earlier, or a fixed cost achieved by political pressure. The alternative of nonzero interest is based on there having been, in the past, marked fluctuations in the difference between the market interest and inflation, with the two most recent decade-averages being close to 5 percent and 3 percent per year, for Denmark. However, the magnitude of the yearly interest should not be allowed to play a decisive role in the long-range planning of a nation. Thus, an energy program with high investment and low operating cost might be subsidized, for example, through low-interest governmental loans, in such a way that the effective interest becomes zero. This is the other alternative concerning interest.

Today, most heating in Denmark is provided by oil-fired boilers, operating at an efficiency of about 65 percent. If one takes the price of fuel oil in Denmark for 1974 to be 54¢ per gallon (12¢ per liter), then to produce 1 kilowatt-hour of heat per year for 25 years, the cost of the fuel, taken as the present value defined above would be 26.5¢ according to alternative (i) (fixed oil price) or 44.7¢ according to alternative (ii) (either increasing oil price or both zero interest and zero fuel price increase).

The alternative cost of a solar panel which on average will produce 1 kilowatt-hour of heat per year is 17.6¢. However, this should be divided by the fractional usage time of the solar energy system, which is the fraction of the energy collected that can be utilized if the collector size and energy storage capacity are dimensioned so that the heating requirements can be satisfied any time during the year (or nearly any time; that is, there should be negligible use of auxiliary electric heating).

For hot water production alone, the requirements are fairly constant throughout the year, and one typically finds a fractional usage time of 0.76, corresponding to a cost of 23¢ for 1 kilowatt-hour per year over 25 years.

For space heating the fractional usage time lies between 0.2 and 0.5, depending on such factors as insulation and the compass direction of windows. For a combined hot water and space heating system, the fractional usage time can be kept above 0.5, leading to an average cost of 35¢ for 1 kilowatt-hour per year for 25 years.

Wind Energy Systems

Wind power may be based on decentralized systems of windmills or on more concentrated windmill rows, typically arranged in north-south directions in order to avoid the shadowing effects caused by the prevailing winds that blow from the west on the west coast of Denmark, and from the west or east in the interior of Denmark. The 1974 cost of a 200-kilowatt experimental a-c producing windmill with fixed blades, which was tested around 1960 at Gedser, Denmark, is estimated to be $260 per square meter of swept area (*5*). Its efficiency was about 25 percent, but this was because its gear was not optimal, and if this was improved, the efficiency would be about 33 percent. Windmills with adjustable blades have reached higher efficiencies, but the improvement has not outweighed the higher cost (*6*).

A 1973 report indicates that a-c generating windmills with vertical axes can be built with efficiencies only slightly below those of the conventional mills with horizontal axes (*7*). Because of their better stability during storms, the vertical axis mills may be preferred for large-scale projects. If only heat production is needed, the angular velocity of the mill can be allowed to adjust itself to the wind, and the electric generator can be replaced by a heat generator (for example, of the friction type, based on the formation of turbulence in a fluid). The cost of such mills may, according to some estimates, be as low as $65 per square meter (*8*). Some authors have even quoted prices as low as $30 to $40 per square meter for either vertical axis rotors (*7*) or one-bladed horizontal axis rotors (*9*).

I will assume, in agreement with the conclusions reached at the recent Stockholm meeting (*9*), that a price no higher than $200 per square meter is realistic for large-scale application, and that the mills have an efficiency of 35 percent for heat production and 30 percent for a-c production. The cost of 1 kilowatt-hour per year over 25 years then amounts to 23¢ for heat and 27¢ for a-c electricity in eastern Denmark, or 13¢ for heat and 16¢ for a-c electricity on the west coast. As in the case of solar energy, these figures must be divided by the fractional usage time.

For heating purposes, windmills are likely to be placed in the interior of the country, and to be equipped with an energy storage system of the same type as for solar energy. With a fractional usage time of 0.7 for hot water production the cost becomes 32¢ for 1 kilowatt-hour per year during the lifetime of the mill, say 25 years. For both hot water and space heating, the fractional usage time becomes high, 0.8, and the cost 29¢ for 1 kilowatt-hour per year over 25 years.

In summary, the production of hot water by solar energy is competitive with production based on oil, even with high interest and fixed oil prices, whereas the feasibility of space heating by sun or wind on a strictly economic basis requires low interest or increasing oil price. For the combined production of hot water and space heating, the windmills are cheaper than the solar collectors, under Danish conditions, but both are economically sound and other criteria for choosing the system may be added.

For electricity production I assume for the moment that the electricity generated by the wind would constitute only a small fraction of the energy flowing through the a-c net into which it would be fed. In this case no storage facility would be required, and the price would be given directly by the average power, amounting to 27¢ (16¢ on the west coast), for 1 kilowatt-hour per year over 25 years. Here the west coast figure of 16¢ might be the relevant one. Typical fuel costs for coal- or oil-fired power plants in Denmark lead to corresponding present values of 27¢ and 45¢ for alternatives (i) and (ii) of fuel price development (based on a 1974 fuel price of $1.60 per gigajoule). If the total electricity requirements did not increase, most of the average usage could be covered by windmills without storage facilities, since the existing production apparatus (in particular the gas turbines) could be used to provide auxiliary power.

For comparison one may look at electricity produced by means of fission reactors. According to a recent estimate (*10*), capital costs of nuclear power amount to

$702 per kilowatt and the variable costs to 0.435¢ per kilowatt-hour. Based on the fractional usage time of 0.8 (see *10*) (for a power plant running at base load) the production of 1 kilowatt-hour of electricity over 25 years amounts to 16.4¢, if one calculates the present value using a 5 percent yearly interest in fixed prices. The actual cost may be higher for a number of reasons: (i) Nuclear fuel prices may increase; for a 5 percent yearly increase, the electricity cost would be 19.5¢. (ii) Fractional usage times may be less than 0.8; Comey quotes the figure 0.54 (*11*), which leads to a cost of 21.2¢ per kilowatt-hour per year over 25 years, with fixed fuel prices. It should be noted that the low capacity factors quoted by Comey are partly due to restrictions on power output imposed by authorities. (iii) The depreciation time may be shorter for nuclear reactors than for windmills, because of their greater complexity and their being subject to radiation damage, for example. If the depreciation time were only half as long, the capital requirement would increase by 50 percent. We see that wind power (produced on the west coast of Denmark) in any case is competitive with nuclear power, and is probably even substantially cheaper.

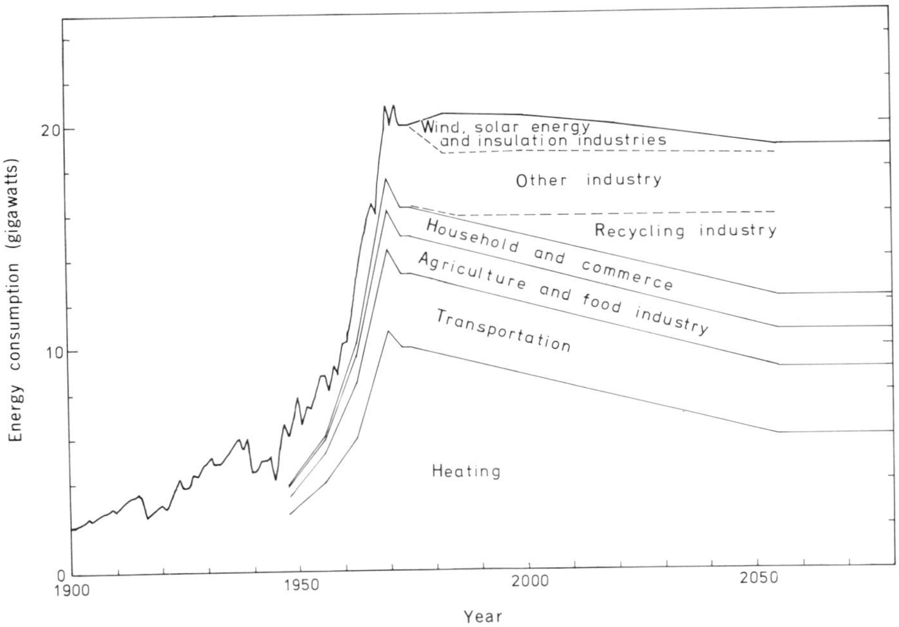

Fig. 3. Net energy consumption in Denmark, shown according to end uses. Up to 1974, actual data are presented (*14*) (yearly averages are given for every year; however, for subdivisions the data are available only for selected years and the corresponding points have simply been connected by straight lines). After 1974, the data show the proposed plan for a transition to a constant level of energy consumption (see text for more details). Energy for heating is shown as a separate entry and is thus excluded from the individual sectors, such as household and industry. Similarly, transportation is taken out of the individual sectors and listed separately. By net energy consumption it is understood that losses during gas and electricity production and exports are omitted. However, waste heat and fuel discarded after incomplete burning at the end-use stage are still included.

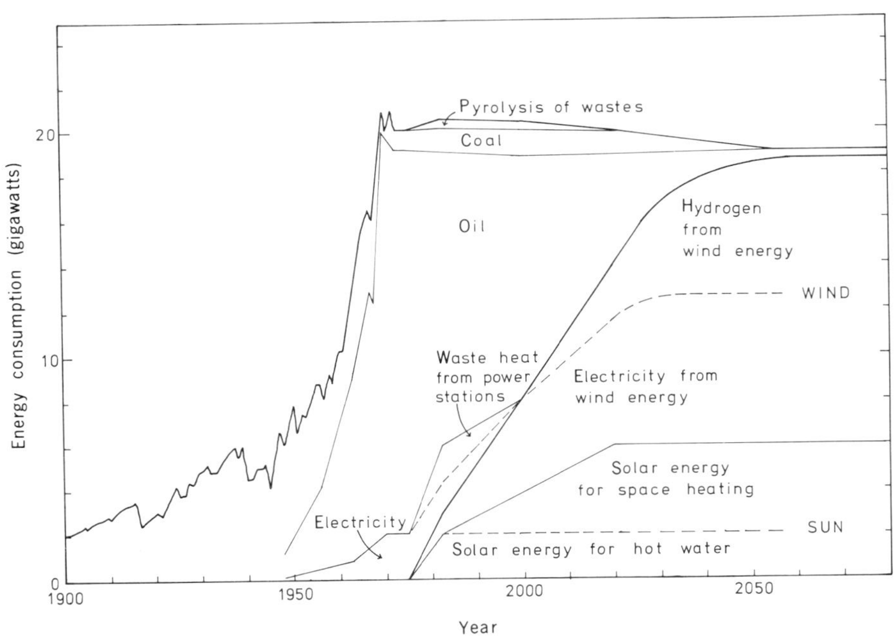

Fig. 4. Net energy consumption in Denmark, shown according to sources. Up to 1974, actual data are shown (*14*); data after 1974 indicate the proposed plan. The relative weighting between solar and wind energy shares might be altered, for example, if a major breakthrough occurred in the development of solar cells, making them competitive to wind-produced electricity under Danish conditions. The heavy, solid line indicates the proposed total share of solar and wind energy.

So far I have assumed that, during an initial phase, wind-based electricity production would never exceed the demand, so that the wind power systems would never be idle, and no storage facility would be required. The same assumptions were made for nuclear power in the economic comparison. In connection with large-scale electricity production by wind, suitable storage systems would have to be added. The topography of Denmark is not suited for pumped water storage, but underground compressed air storage and possibly flywheel storage might be applicable (*12*). Such storage systems already appear to be economically feasible in connection with fuel-based electricity production. A compressed air storage system is being constructed in Schleswig, just south of the Danish-German border, and another is being seriously considered by the Electricity Company of Western Denmark, utilizing the salt domes abundant in Jutland's underground (*12*). For night-to-day storage systems a cost of $117 per kilowatt, which would correspond to 8 percent of the cost of the windmill, has been estimated for a compressed air power unit that could replace the average power of the windmill. Per kilowatt-hour per year this amounts to 1.3¢, to be added to the 16¢ for the mill itself. The actual cost may be somewhat larger, because of the need for larger air storage volumes in connection with windmills, but as long as natural salt domes can be used, the excavation cost is not decisive. A number of other systems (e.g. hydrogen-based) may later become feasible.

Energy Requirements

Estimates of future energy consumption will, to some extent, require a model of future life-styles. Changes in life-styles that cannot be anticipated at present will thus be excluded. I shall base my estimates on a few expected changes in life-styles plus the assumption that increased effectivity of energy conversion will be pursued as a worthwhile effort.

From World War II to 1970 the energy used in Denmark for heating has increased exponentially because of a strong effort to catch up with a housing deficiency and because of increased room temperatures. Future increases in energy consumption for heating would be directed toward improving housing standards that are already quite high, and would therefore be less marked. At present, heating requires an average of 10 gigawatts, and the Danish population is 5 million. For simplicity I shall consider a total of 2.5 million building units, each using 4 kilowatts (yearly average) for space heating and hot water. These units are defined as average family dwellings (three-room apartments or houses plus bathroom and kitchen) or building sections, all with equivalent heating requirements; 1.6 million of these units are used for living, and 0.9 million are used for commerce and industry. Had all of these units been insulated to the best of present standards, the average power requirements per unit would be 2 kilowatts (800 watts for air exchange; 600 watts for losses through walls; and 600 watts for hot water)—that is, only half of the present value. According to my plan, there would be a gradual replacement of most of the present units with optimally insulated ones during the next 75 years. On the assumption that a slight increase in the number of units will be needed (for example, to allow for population increase), say to 3.0 million units each requiring 2 kilowatts, then the energy requirement for heating in the year 2050 will be 6 gigawatts. The size of the unit may also be assumed to increase (the average size of a dwelling becoming four rooms, exclusive of kitchen and bathroom, or 120 square meters of floor area), with some changes in the energy balance: 800 watts for air exchange; 800 watts for heat losses through walls; 1000 watts for hot water; but a gain of at least 600 watts from unshielded, south-facing windows, indoor activities, and possibly heat exchangers recovering heat from return air and water.

The rapid development of advanced telecommunication systems is expected to have considerable impact on all aspects of society and, in relation to energy consumption, particularly on the structure of business, commerce, and transportation. Advanced telecommunication systems are characterized by low energy consumption, and substantial energy savings can be expected when business meetings (all gatherings in which direct human contact is not essential) are based on videotelephone communication rather than on face-to-face communication requiring air and ground travel. Similar savings can be expected when trade on all levels, including shopping by individuals, is based on selecting merchandise from computerized video displays, so that actual transport of goods from place to place can be minimized (by computer), in contrast to the confusion of present-day private car traffic to and from supermarkets and inconveniently located shopping centers. In addition, many jobs could be handled from the homes of employees so that unnecessary commuter traffic could be avoided. On this basis a slight decrease in energy consumption in the transportation sector from 3.3 to 3.0 gigawatts in the year 2050 would seem to constitute an upper limit which would only be reached if the efficiency of energy conversion in the transportation sector were to drop from the present value of about 25 percent. This should not be unexpected if there has to be a conversion (for example, from wind energy) through hydrogen or other synthetic fuels.

The energy requirement in the agriculture and food industry sector is assumed to remain fairly constant at an average of 1.7 gigawatts.

In addition to the energy used for heating, households and commerce today use 1.3 gigawatts for cooking, light, and other electric equipment. This is assumed to rise slightly to 1.5 gigawatts, with a decreasing share going to commerce, because of the more efficient trade pattern mentioned above and the emphasis on producing goods with longer lifetimes and of higher quality, which generally will diminish the need for commerce. A large increase in the number of commodities installed in homes is expected to be partly compensated by the development of more energy-efficient equipment.

Due to increasing costs of raw materials, a substantial increase in recycling is expected to occur. In the year 2050 it is assumed that the recycling industry will use as much energy as is used today by all industries put together (3.7 gigawatts). I also assume that the production industry will, at the same time, have reduced its energy consumption slightly, to 3.1 gigawatts (partly by increased effectiveness and partly by reduced production of superfluous products). The improvements in building insulation that I propose, as well as the establishment of a new energy production system based on solar and wind energy, would require a rapid expansion of the corresponding industries. Not until the end of the 75-year transition period would these new industries have accomplished their task and would then gradually have decreased in size toward the proportions necessary for adequate maintenance of the energy systems; that is, the industries responsible for the repairs and replacements necessary for the energy systems and the buildings associated with them will reach a stage of equilibrium after 75 years of operation. The plan I propose for energy consumption after the year 1974 is summarized in Fig. 3.

Choice of Energy Systems

In deciding the amounts of energy that should be obtained from the different sources, I took into consideration the need to reduce Denmark's dependence on nonrenewable sources (oil in particular) as quickly as possible, and the desirability of basing as large a fraction of the energy consumption as possible on continuous (renewable) sources. My plan is outlined in Fig. 4.

First, solar energy systems are utilized for hot water production. Such systems can be installed in existing buildings, the collectors being placed on the roofs or south-facing walls of the buildings, for example. The use of solar energy for space heating is expected to be introduced more slowly, because new buildings or substantial improvements in insulation will be required. Windmill-based heating could be an alternative during this period. Electricity-producing windmills are first to be placed at the west coast and in other wind-favored locations. As their capacity increases, and the present coal- and oil-based production systems become obsolete, storage systems will have to be added. Electrification is assumed to increase steadily, from the present 10 percent share of the total net energy consumption to about 33 percent in the year 2030. From about the year 2000, the storage systems such as those based on compressed air are expected to become insufficient and a hydrogen-based economy will be introduced, so that by the year 2050 close to the entire energy requirement (including that of transportation) may be obtained by conversion of sun and wind energy. Various types of batteries might take the place of hydrogen as a storage medium.

During the transition phase the desired decline in oil importation could be assisted by short-term solutions, such as the utilization of waste heat from conventional power plants (for local heating) and the pyrolysis of wastes, which may contribute about 0.8 gigawatt [10 gigajoule per ton of waste (*13*)]. As the recycling of used goods increases, the energy content of the wastes will decrease, and this energy source will become negligible.

Summary

Two possible futures for the industrial world may be distinguished: (i) Large amounts of low-cost energy become available and the more energy-intensive methods for extracting resources from lower-grade deposits continue to sustain industrial expansion until either the environmental impact becomes unacceptable or ultimate limits, such as climate disruptions, put an end to such growth. (ii) The cost of nonrenewable energy resources continue to rise, but a fixed amount of energy from continuous sources may be utilized at constant cost. In this case a lower production level may be set by the amount of energy that is available from renewable sources, and society may thus have to be reshaped with energy economization in focus. If it is possible to choose between these two alternatives, the choice should be based on a discussion of the pros and cons of each one, and in particular on the desirability of having to process an increasing fraction of the earth's crust in search of raw materials in order to maintain growth as long as possible. However, the availability of the first option is far from certain and it thus seems reasonable to plan for the second alternative.

I have tried to propose such a plan for a small, homogeneous geographical region, namely Denmark. The ceiling on the consumption of energy from continuous sources is chosen in accordance with the criterion of not having to convert a major part of the land area to energy-collecting systems. The proposed annual average energy consumption of 19 gigawatts by the year 2050 corresponds to solar energy collecting panels (in use only 50 percent of the time) with an area of roughly 180 square kilometers and a windmill swept area of about 150 square kilometers. These (vertical) areas constitute less than 1 percent of the total land area.

The selection of solar or wind energy for different applications has been based on known technology and may be subject to adjustments. The project has been shown to be economically feasible according to estimates of the cost of various alternatives during the 25-year depreciation period adopted. However, the initial cost per energy unit produced is higher than that for most of the alternatives, so that action is not expected to be taken immediately as a result of purely private initiative. In a public economic evaluation, other factors must be considered in addition to the cost of energy per kilowatt-hour. At present, Denmark has over 10 percent of its labor force out of employment and a substantial deficit on its balance of payments, so that an early start on the solar and wind energy project, based on national industry, would have additional payoffs compared with energy systems based on imported technology or imported fuels. Several factories that are now being closed down as a result of the economic crisis could be adapted to the production of parts for solar or wind power systems, and the building industry, badly hit by unemployment, would receive legitimate work.

References and Notes

1. K. Dunham, *Ambio* **3**, 126 (1974).
2. B. Sørensen, "A simple model of economic growth or decline under the influence of resource depletion" (Niels Bohr Institute, Copenhagen, preprint, December, 1974).
3. Solar energy is calculated from Danish Building Research Institute, Report No. 89 (Teknisk Forlag, Copenhagen, 1974), according to the method of N. Robinson [*Solar Radiation* (Elsevier, Amsterdam, 1966)]. Wind energy is calculated from M. Jensen [*Proceedings of the United Nations Conference on New Sources of Energy*, Rome (1961), vol. 7].
4. M. Jensen, thesis. Danish Technical University (1954).
5. Data from the report of the Wind Power Committee, Association of Danish Power Stations, Copenhagen, 1962, combined with industrial price indices, published yearly by Denmark's Statistical Department.
6. Vattenfall Press-Information 1974-06-19, Swedish State Power Board, 1974.
7. P. South and R. Rangi, report from the 1973 Annual Meeting of the Pacific Northwest Region of the American Society of Agricultural Engineers, PNW 73-303.
8. Communication from the Wind Energy Committee, Danish Academy of Technical Sciences.
9. Report of Workshop on Advanced Wind Energy Systems (Swedish Board for Technical Development, Stockholm, 1974).
10. D. Rose, *Science* **184**, 351 (1974).
11. D. Comey, *Bull. At. Sci.* (November 1974), p. 23.
12. R. Post and S. Post, *Sci. Am.* **229** (No. 6), 17 (1973); G. Styrbro, *Elektroteknikeren* **21**, 450 (1972).
13. N. Davis, *New Sci.* **9**, 319 (May 1974).
14. Publications No. 2 (1959) and No. 20 (1967) by Denmark's Statistical Department, Copenhagen; Statistical Yearbooks, Denmark, 1968 to 1973. The final 1974 data are not available as of this writing, but preliminary statements indicate a drop in total energy consumption of at least 10 percent relative to 1973 (this drop is not included in the figures).

Brazil: Energy Options and Current Outlook

J. Goldemberg

Brazil has an area of 8,511,965 square kilometers and a population of approximately 110 million people (*1, 2*). The country was kept dormant as a Portuguese colony for more than 300 years and, after gaining political independence in the last century, remained as a producer and exporter of agricultural products, mainly coffee and sugar, until World War II.

After World War II, Brazil entered a phase of accelerated industrialization which resulted in the growth of very large cities in the southern (and more temperate) part of the country and, consequently, in an exodus of rural populations to urban centers. The fraction of the population living in cities increased from 36 percent in 1950 to 45 percent in 1960 and 56 percent in 1970.

The energy consumption "per capita" of the average Brazilian therefore increased enormously in a brief span of time; in addition, the profile of consumption changed very significantly Figure 1 indicates that by 1976 Brazil had reached a level of consumption comparable to some of the less developed European countries (approximately 10 megawatt-hours per year per capita) (*3*).

Energy consumption as a function of per capita income has increased in recent years in a manner similar to that in developed countries. Thus the country has entered a phase of modernization in which the energy-intensive consumption patterns of the great industrial countries have been adopted, without any critical assessment, through the transplantation of modern, foreign industries. As shown in Fig. 2, the relation between per capita energy consumption (E) and per capita income (I) is practically linear, a characteristic of highly developed countries (*4, 5*).

Dr. Goldemberg is director of the Institute of Physics at the University of São Paulo

The ratio of total energy used to per capita income has also remained almost constant in the last 10 years at the level of 60×10^3 British thermal units (Btu) per dollar. This is approximately two-thirds of the value in the United States (*6*) and indicates that the efficiency of energy use has been slowly improving (Fig. 3); however, less energy is needed in Brazil to produce one dollar of income than in the United States, indicating a smaller use of energy-dissipating devices such as air conditioners and freezers.

The profile of consumption (*3*) changed in a somewhat predictable fashion in the era of cheap and abundant petroleum (Fig. 4). The relative importance of biomass in the balance of the energy consumed decreased dramatically from 1940 to 1975, with a corresponding growth of petroleum consumption.

Coal has had an insignificant role in energy consumption in Brazil, but hydroelectric power has increased its share to 20 percent of the total energy consumption in recent years. The balance is taken by petroleum; its contribution has grown from 9.2 percent in 1941 to 28.0 percent in 1952 and to 44.8 percent in 1972 (Fig. 5). Natural gas consumption has been negligible.

The energy profiles of Brazil and the United States are compared in Table 1. Petroleum represents approximately 44.8 percent of the total energy consumed in Brazil and only 20 percent of it is produced internally. The remaining 80 percent (700,000 barrels per day) is imported at a cost of more than \$3 billion per year. To compensate for this deficit in the balance of trade, the country has to export large quantities of raw minerals, agricultural products, and semi-industrialized goods.

The future prospects for energy consumption are presented in Fig. 6, which shows the official projections for consumption of different sources up to 1985: The contribution of biomass remains approximately constant, whereas hydroelectric production increases significantly but not enough to avoid a strong increase in the use of liquid combustibles (predominantly petroleum, although other sources such as alcohol and shale oil are also considered). Energy consumption is expected to grow at a rate of approximately 10 percent a year; the population growth rate per year is 2.9 percent, which, by itself, represents a great stress on the Brazilian economy.

In this article I review the known resources in Brazil and their possible utilization. As I will show, the prospects for a greatly increased petroleum production in Brazil are limited. Alternatives must therefore be found if an extremely serious situation is to be avoided in the next 10 to 15 years.

Energy Resources in Brazil

The data given below are taken from the compilation made by the Brazilian National Committee of the World Energy Conference and published in (*2*). The data refer in all instances to 1972.

Summary. Brazil's energy options and current outlook are examined, and a summary of known reserves of fossil and renewable energy resources is given. Brazil has abundant renewable energy resources but very modest reserves of fossil fuels. Consequently, the emphasis in the future will have to be on the utilization of solar energy, hydroelectric power, and biomass in a program designed to preserve local traditions and culture.

Crude oil. Sedimentary basins onshore extend over an area of 3,168,000 km^2, plus about 800,000 km^2 offshore on the continental shelf. The first basins to be explored were those in the northeastern region (state of Bahia) near the Todos-os-Santos Bay, and all locally produced oil has come from this region. Since 1939, more than 3900 wells have been drilled all over the country; 1985 of the wells produced oil, and 108 natural gas. The overall results have been disappointing, however; the recoverable oil reserves are only about 106×10^6 metric tons, and the annual production of 8.2×10^6 tons covers only one-fifth of the country's needs, the balance being imported (Table 2).

The remaining possibility for increased oil production in Brazil seems to be the offshore exploration that led to the discovery in 1976 of new fields off the shore of Rio de Janeiro. These reserves increased the figures shown in Table 2 by about 15 percent, and there are hopes for doubling the Brazilian production of oil by 1982.

Natural gas reserves. Natural gas reserves (Table 3) are usually associated with crude oil deposits. Although the production of natural gas was 1.2 km^3 in 1972, only 18 percent was used for energy and industrial uses; the balance was either reinjected into the wells—to increase pressure or for storage—or flared locally. The principal reason for flaring is that the main producing centers are located in Bahia, which is not an industrial state. The recent establishment of a petrochemical industry near the crude oil and gas fields will encourage further use of gas.

Shale oil. Depending mainly on economic factors, the oil recovered from shales could complement conventional crude oil. The known shale oil reserves in the southern region of Brazil are given in Table 4. Most shale deposits contain 4 to 8 percent oil. The recoverable oil, considering only the formations in the southern region, is estimated at about 500×10^9 barrels, almost five times more than the known crude oil deposits.

Studies are under way for the commercial exploitation of shale deposits in Paraná. Operation of the oil-producing facilities is expected to begin in 1978 with a daily production of about 10,000 tons of oil, 900 tons of sulfur, 400 tons of liquefied petroleum gas, and 1.7 million cubic meters of light heating gas. Environmental problems associated with these plans have yet to be solved. For the oil to be recovered, large quantities of rock have to be removed (200,000 to 400,000 tons) daily, and material remaining after extraction is highly acidic and presents a disposal problem.

Coal. Coal deposits are scarce in Brazil and all those that have been discovered are in the southern region (Table 5). The total known reserves are 3256×10^6 tons of which 2014×10^6 tons are of the subbituminous type, 1240×10^6 tons bituminous, and only about 2×10^6 tons anthracite. Coal of a metallurgical grade suitable for coking can only be obtained after complex processing. Prewashed coal has an ash content of up to about 30 percent. The subbituminous coals of the states of Rio Grande do Sul and Paraná are also of poor quality and produce only steam coal for local use. Fluidized bed combustion of poor-quality coal might be used to advantage in Brazil.

Uranium and thorium. Only very recently did prospecting for the so-called fissile fuels begin in Brazil. Coastal monazite-sand deposits containing both thorium and uranium are known to exist, but the cost of exploiting them commercially has not been evaluated.

The only known uranium deposits (Table 6) are in the southeastern region, in the State of Minas Gerais (Pocos de Caldas). They occur in an intrusive pipe of alkaline rocks covering an area of about 900 km^2. The economic thorium reserves (Table 7) come from monazitic

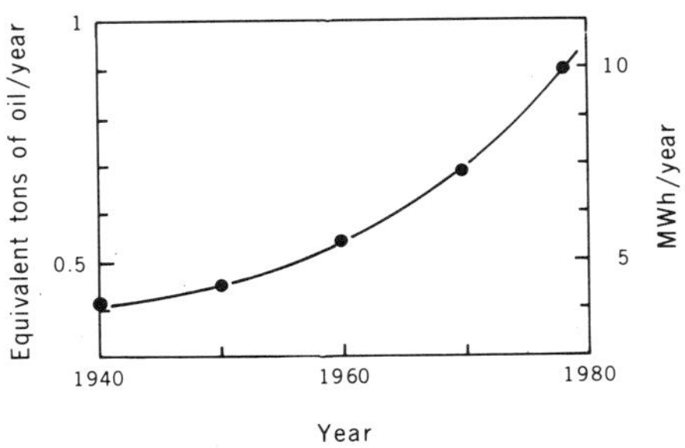

Fig. 1. Brazil's energy consumption per capita in the period 1940 to 1977.

beach sand deposits. The monazite occurs in association with zircon and rutile, and the deposits are located along the eastern and northeastern Atlantic Coast. A total of only 408 tons of thorium was produced in 1972.

Other possible sources of fissile fuels, not yet sufficiently known in terms of economic exploitation, occur in the State of Minas Gerais. These include thorite and thorogummite, in connection with highly decomposed alkalic rocks. In the Araxá niobium deposits, thorium appears associated with monazite. Still other sources could be fluvial monazite-sand deposits and monazite-bearing pegmatites in the states of Minas Gerais, Goiás, and Rio Grande do Norte.

Hydroelectric energy. The geomorphology of Brazil is such that water resources are spread over the whole country, with a certain predominance in the northern region where the world's biggest basin—the Amazon—is located.

Table 8 shows the main characteristics of the local hydroelectric potentials. Figures are presented by geographic region. Of the overall installable hydropower capacity of 118,980 megawatts, approximately 17,000 MW are being used and this number should increase to 35,000 MW by 1985 with the completion of the huge Itaipu central dam, with a capacity in excess of 10,000 MW, on the Paraná River, and a few other dams. At present, only 14 percent of the total hydroelectric power of the country is being used, although in the most populous southeastern region this fraction is 27 percent (Table 8). There remains a potential of almost 50,000 MW mainly in the tributaries of the Amazon River in the northern region (Tocantins, Araguaia, Xingu, Tapajós, Cotinga, and Trombetas).

A serious transportation problem may develop because these sources are located at least 2000 km from the great population centers (Rio de Janeiro and Saõ Paulo) in the southeastern part of the country that has a total population of 40 million people.

Biomass. Wood accounts for approximately 30 percent of Brazil's present energy needs; this represents approximately 170 million cubic meters, most of it used as fuel. These numbers probably represent a lower limit.

The utilization of wood for charcoal is important in the steel industry and it is estimated that 40 percent of the needs of this industrial sector are supplied from this source (approximately 2 million tons yearly).

As indicated in Table 9, forests originally covered 55 percent of the Brazilian territory. The Atlantic and Brazilian pine (*Araucária braziliensis*) forests have now been almost completely decimated, but the Amazon forest remains almost intact (*7*).

Total reserves. The proved recoverable energy reserves of the country for 1972 are summarized in Table 10.

The Nature of the Problems

In the southern part of Brazil the patterns of energy consumption are not very different from those of the great industrial nations. This part of Brazil suffers from crowded cities, roads clogged with automobiles, and air pollution, and there is a general deterioration of the quality of life in the urban centers.

The number of automobiles in Brazil has grown from a few thousand 30 years ago to about 5 million in 1975. For the transportation of goods and services, road transportation has been emphasized, often because of the availability of subsidies for road construction with the result that the energy spent in road transportation is approximately 83 percent of the total transport energy (*8*).

In addition to the gasoline consumed in cars there is an industrial energy-intensive sector that uses large quantities of petroleum derivatives and electricity. The transportation of electricity to these urban centers from distant locations in the Amazon basin poses a serious technological problem, and for this reason the construction of nuclear reactors was considered as a solution to the energy problem. The Brazilian government has now begun an ambitious program of installing several nuclear reactors, mainly in the southeastern part of the country. This program could alleviate the energy shortage in the urban and industrial southeast, but would have little effect on the automobile and the automotive industry, which is the leading industrial sector in Brazil.

Since most of the petroleum consumed in Brazil is imported, the price of gasoline is now among the highest in the world: U.S. $1.80 per gallon (approximately three times the current price in the United States).

Because of strong pressure to reduce consumption of gasoline (and other oil products), the government has made a determined effort to promote the use of ethyl alcohol (ethanol) as a partial (or total) substitute for gasoline in automobiles.

Table 1. Energy resources in Brazil and the United States in 1972.

Source	Brazil (%)	United States (%)
Coal	3.6	17
Petroleum	44.8	46
Gas	0.3	32
Hydroelectric power	20.8	4
Biomass*	30.5	1
Total	100.0	100

*Including wood and sugarcane bagasse.

Table 2. Petroleum reserves (1972) in 10^6 tons.

Locality	Reserves		Past cumulative production
	Total	Recoverable	
Bahia	515.7	83.2	81.0
Alagoas	6.2	0.4	0.3
Sergipe	212.7	20.4	7.4
Offshore	14.9	2.6	
Total	749.5	106.5	88.7

Table 3. Gas reserves (1972) in 10^6 cubic meters.

Locality	Reserves		Past cumulative production
	Total	Recoverable	
Bahia	49.3	28.9	12.5
Sergipe	4.1	1.6	0.1
Alagoas	0.9	0.6	
Offshore	5.9	4.7	
Total	60.2	35.8	12.6

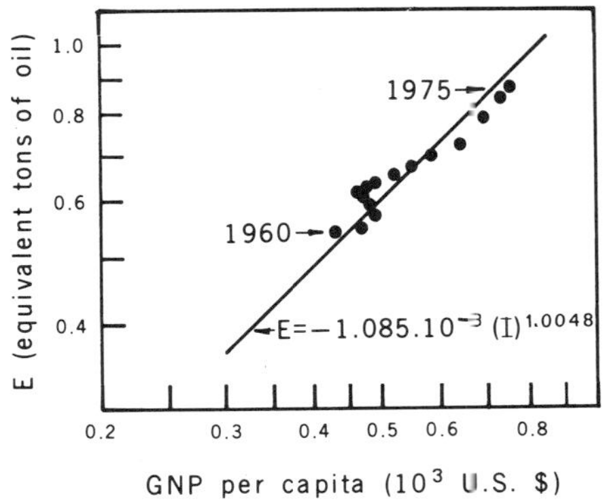

Fig. 2. Energy consumption (*E*) versus income (*I*). The data indicate elasticity *p* close to unity (*4*).

Even if the programs to produce nuclear energy and to substitute alcohol for gasoline succeed, the problem remains that the development patterns which "modernized" the southern part of Brazil—in the sense of introducing the latest gadgets and products of industrialized countries—have benefited only approximately 20 percent of the population concentrated in large cities. Thus a situation has arisen in which there are, in effect, two countries in one: a reasonably well-developed section that includes 20 percent of the population, and an underdeveloped section that includes the remaining 80 percent of the population living either in the northeast under conditions of poverty or in slums in the big cities.

Although the total population growth rate in Brazil is 2.9 percent per year, the growth rate for the rural population is 2.3 percent whereas that for the urban population is 6.13 percent per year. Thus, in addition to meeting the requirements of the high general rate of growth, energy production must meet the requirements of the even higher rate of urban population growth. Up to now, the high rates of growth of energy production (approximately 10 percent per year for all sources; 12.5 percent per year for hydroelectric power) (Fig. 6) have been able to keep pace with population growth, but how long this rapid growth rate can be sustained is a fundamental question because, in developed countries, such growth rates in general are not higher than 5 percent per year.

Table 4. Shale oil reserves (1973).

Locality	Deposit area (km^2)	Oil content (kilograms of oil per 1000 kg of shale)	Total recoverable oil (10^6 tons)
São Paulo	200	40 to 130	288
Paraná			
(Iratí)	82	70	95
Rio Grande			
do Sul	350	20 to 80	114
Total	632		497

Table 5. Coal reserves (1972). The total recoverable quantity is 688 × 10^6 tons.

Locality	Total reserves (10^6 tons)	Annual production (10^3 tons)
São Paulo	1	
Paraná	35	346
Santa Catarina	1200	4536
Rio Grande		
do Sul	2020	978
Total	3256	5860

The possible utilization of wind energy, thermal gradient of oceans, tidal power, and other new sources of energy has been discussed in Brazil, but the work being conducted on these techniques can be generally regarded as following the "state of the art."

Possible Solutions

There are several possible solutions to Brazil's energy problems. Here I will discuss biomass and the alcohol program, hydroelectric power, solar energy, nuclear energy, and conservation measures.

Biomass and the alcohol program. The average intensity of solar radiation over most of Brazil is approximately twice the intensity in the United States, and this, together with other conditions, such as humidity, permits rapid growth of most types of vegetation.

Wood is not being extensively utilized in Brazil for the production of ethanol and methanol. Recently, however, with the alcohol program becoming an important government priority, much research is being conducted on the possibility of growing various plant crops for alcohol production.

Technically, the idea of using alcohol as a gasoline substitute is feasible: present internal combustion engines can use or be converted to run on ethanol, and the huge amounts needed could be produced, mainly from sugarcane, in Brazil (*9*). Alcohol can be mixed with gasoline without any changes in present-day internal combustion engines in any amount up to 20 percent. Although ethyl alcohol has a calorific content that is 39 percent lower per liter than gasoline, there are other compensating factors: ethanol has a higher density than gasoline, and the power of a motor running on alcohol is 18 percent higher than a motor running on gasoline. The overall result is that only 1.4 percent more alcohol is consumed than gasoline. Motors running on 100 percent alcohol can be produced with minor modifications of the present ones (*10*).

Typical goals of the Brazilian alcohol program are given in Table 11 (*11*). The amounts of land that would be needed for some of these scenarios are very large (*12*). Of Brazil's total area of 850 × 10^6 hectares, approximately 70 × 10^6 ha are fertile lands. If one assumes that by the year 2000 the total amount of fertile land will increase to 14 percent (120 × 10^6 ha), approximately 20 percent of that land would be necessary to produce all the alcohol needed by then (almost twice the amount for scenario 4 in Table 11). This would correspond to 3 percent of the Brazilian territory being covered by sugarcane—a difficult goal to achieve and probably an undesirable one.

The overall efficiency of alcohol production from different crops is indicated in Table 12 (*13*). In the case of sugarcane, the energy gain is 2.40 with a total production of 3564 tons per hectare. This corresponds to an efficiency of solar energy utilization of a few tenths of a percent (0.3 to 0.4 percent). It is assumed that the bagasse is used as a combustible in the alcohol distillation plants. There is, however, a surplus of bagasse that could be used to generate electricity as a byproduct in a "cogeneration" process.

Hydroelectric power. The average annual growth of Brazilian hydroelectric consumption in the last 15 years has been 12.5 percent. By 1985 all the reserves of the southeastern part of the country will have been used up and it will be necessary for the southern region

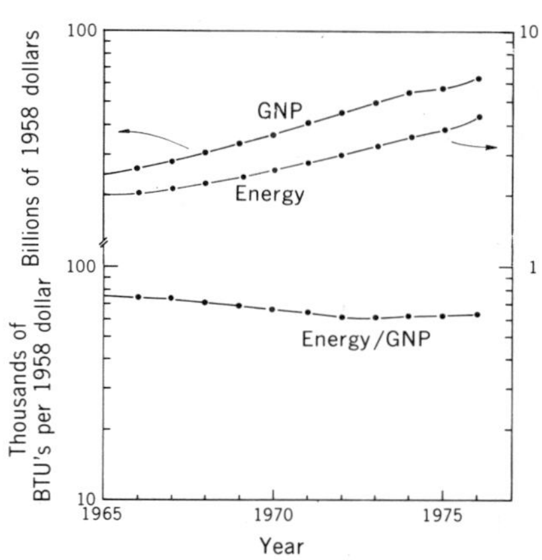

Fig. 3. Energy and gross national product (*GNP*) in the period 1965 to 1975. The ratio of energy to gross national product is approximately constant over the period and somewhat smaller than the equivalent value for the United States.

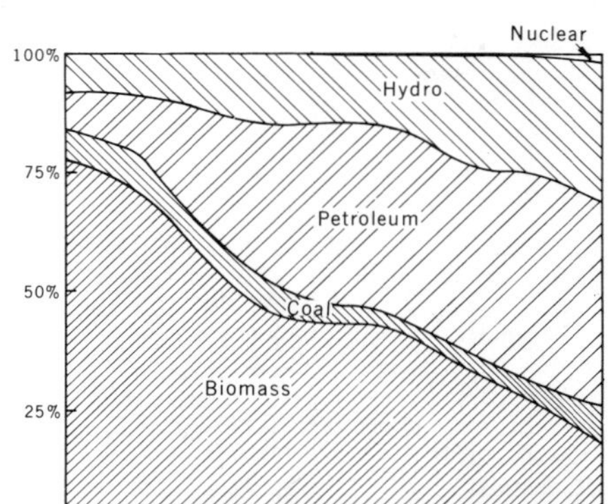

Fig. 4. The energy consumption profile of Brazil indicating the rapid decline of biomass consumption.

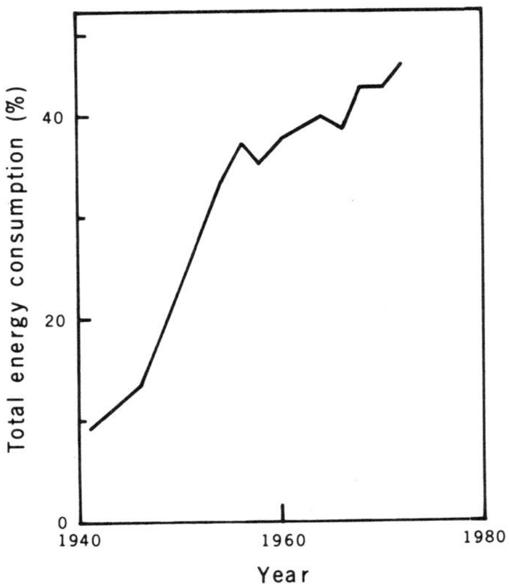

Fig. 5. The rapid rise of petroleum consumption in the period 1940 to 1977.

to consume electricity generated in the Amazon tributaries.

Long distance lines will therefore be needed, and possibly the use of direct-current high voltage transmission to minimize losses that are inevitable with the usual methods of transmission. The construction of a 4.5 million-kilowatt hydroelectric power plant located on the Tucurui River in the Amazon Basin is being considered; the electricity produced would be consumed locally in the production of aluminum from bauxite instead of being transported long distances. Ideas of this type have been studied in some detail by Strout (*14*) and probably deserve further consideration.

Some smaller hydroelectric power plants remain to be built in the State of São Paulo such as Porto Primavera and Ilha Grande on the Paraná River. Water resources might also be used more economically; for example, reversible stations could be constructed on the Serra do Mar where water could be pumped uphill during off-peak hours (*15*). Another 10,000 MW could be made available through this method.

A new idea receiving much discussion involves the use of "bulb type" generators that can work efficiently in small waterfalls (2 to 20 meters) or in fast-moving streams (*16*). Costs of these turbines compete favorably with the conventional ones, and they produce from 10 kW to 10 MW of electricity. Such "mini" hydroelectric generating stations are adequate for decentralized electricity-dispensing with expensive transmission lines. They are economically competitive and have been used for at least 15 years in France, Germany, Japan, England, and a few other countries, but have been used in only a few locations in the United States.

Solar energy. The most significant projects for the use of solar energy in Brazil are those for drying agricultural products and for heating water for industrial or hospital use. In one such project, 1000 m^2 of flat plate collectors are being installed in the roof of the University Hospital at São Paulo (400 beds). The water heated by the collectors will be used to provide steam and for other purposes. The total energy consumption of this hospital is approximately 1800 tons of fuel oil per year, and it is expected that 15 to 20 percent of this fuel oil will be saved by solar energy utilization. The same method will be used in some textile and food industries; it has been estimated that the total energy consumption in the industrial sector in Brazil could be reduced by 5 percent by this method (*17*).

The project for grain drying is being developed at the University of Campinas (*18*) where it has been shown that 10 tons of sorghum per day can be successfully dried under economically competitive conditions.

Nuclear energy. The strong pressure to guarantee the supply and prosperity of the large cities in the São Paulo–Rio de Janeiro area has encouraged the inclusion of nuclear reactors in plans to provide alternative energy sources in Brazil, although paradoxically there are enormous hydroelectric reserves in other parts of the country. After a long period of hesitation, a 624-MW nuclear reactor was purchased in 1969 from Westinghouse for construction in Angra dos Reis, Rio de Janeiro. This pressurized water reactor (PWR), which is due to start operation in 1979, is fueled with enriched uranium, and it was almost totally constructed by U.S. technicians. No transfer of technology was contemplated and participation of Brazilian industry was restricted to the civil works and some low technology equipment. The decision to purchase this reactor was strongly criticized by many scientists and by industry spokesmen because it increased Brazil's dependence on foreign imports. Sensitive to that criticism, the government in 1974 embarked on an ambitious nuclear program in cooperation with West Germany whereby Brazil would achieve complete autonomy in the nuclear field over a period of 15 years.

Eight PWR's (1300 MW each) were scheduled to be installed in Brazil with an increasing index of nationalization that should come close to 100 percent in

Table 6. Uranium reserves (1972). Costs are in U.S. dollars. New findings, raising these numbers to perhaps 10,000 to 20,000 tons, have recently been reported.

Locality or deposit	Range of maximum recovery cost ($/kg)	Quantity recoverable (tons)	Average grade of ore (% of U_3O_8)	Estimated additional resources (tons)
Minas Gerais	Up to 33.0	3195	0.15 to 0.19	4250
Monazite sands	Up to 22.0	49	0.18	37
Total		3244		4287

Table 7. Thorium reserves (1972). Costs are in U.S. dollars.

Locality or deposit	Range of maximum recovery cost ($/kg)	Quantity recoverable (tons)	Average grade of ore (%)	Estimated additional resources (tons)
Minas Gerais	Not available	65,000	0.09 to 1.0	1,200,000
Monazite sands	Up to 20	1,350	5.0	1,000
Total		66,350		1,201,000

Table 8. Hydroelectric potential in Brazil.

Region	Installable capacity (MW)	Potential annual generation* (GWh)	Power generated in 1975 (GWh)	Fraction utilized (%)
North	46,200	202,400	1,020	0.5
Northeast	13,349	58,500	8,800	15
Southeast	39,577	183,700	51,570	27
Central-West	1,164	5,100	1,730	34
South	18,690	67,200	8,390	12.5
Total	118,980	516,900	71,510	14

*Available 95 percent of the time.

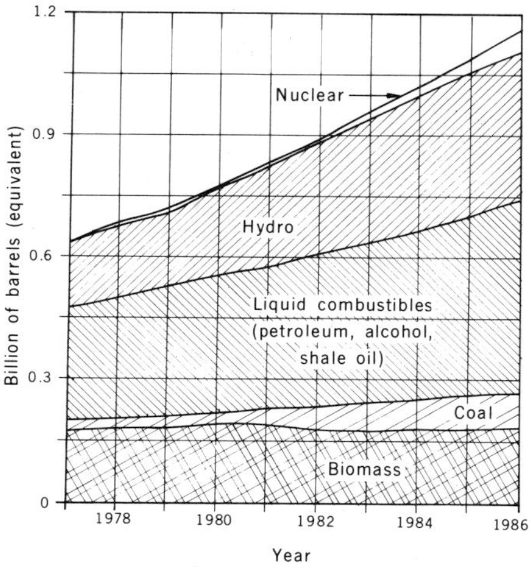

Fig. 6. Projections of energy consumption of different sources in the period 1977 to 1986. All sources converted to oil equivalent.

1990. A semi-industrial uranium enrichment plant (250 tons per year) was to be built in Brazil based on the jet-nozzle method. In addition, a plutonium-reprocessing plant was to be built by German enterprises. In all cases companies were to be formed in which the Brazilian government would be dominant but there would be German associates (and in some cases private Brazilian capital).

Although there was some initial enthusiasm for such a comprehensive nuclear program, the difficulties soon became apparent. First, the feasibility of uranium enrichment (which is essential if any real autonomy is sought) was viewed with strong skepticism by many scientists because the Becker jet-nozzle method to be used was not considered a proved method; for many years to come, therefore, enriched uranium would have had to be purchased from other suppliers. The achievement of "nuclear independence" was therefore in the far distant future.

Second, the transfer of technology was not successful from the very start; disagreements on the role to be played by Brazilian local industry and German industries surfaced and it became apparent that the program would benefit the German nuclear industry in Germany and not in Brazil.

Third, the role of Brazilian scientists and technological institutes was neither properly defined nor given much attention; this strengthened the suspicions that no real transfer of technology was to occur.

Finally, the planned installation of a plutonium-reprocessing plant raised strong objections in the United States because such a plant would constitute an added danger to nuclear proliferation.

These problems caused the program to be delayed, and the urgency of nuclear energy in Brazil is now being reassessed. Most Brazilian scientists and industrialists would probably prefer a more modest program in which they could play a dominant role and acquire the necessary technology. The construction of a prototype nuclear reactor—preferably working on the thorium cycle—would probably be the most appropriate choice from the point of view of Brazilians familiar with nuclear problems.

Conservation measures. Although a number of energy-saving methods have been suggested, the only practical actions taken by the government have been in the direction of reducing gasoline consumption in private automobiles and in trucks. To this effect the following measures were adopted about 1 year ago: (i) limitation of the maximum speed on roads to 50 miles per hour; (ii) gradual banning of cars from the center of large cities to discourage people from driving their own cars to work or to go shopping; (iii) improvement of the bus system with the introduction of "executive" buses and exclusive bus lanes; (iv) closing of gasoline stations in the evenings and weekends; (v) gradual price increases for gasoline; (vi) the addition of up to 20 percent ethyl alcohol to the gasoline used by cars in some cities.

Table 9. Brazilian vegetation. All numbers represent percentages of total territory.

Forest	Primitive	Present
Amazon	40	36
Atlantic	10	1
*Araucaria**	5	0.5
Total	55	37.5

*Brazilian pine.

Table 10. Brazil's energy resources.

Resource	Proved reserves (× 10^6 ton)	Total energy content (10^{15} joules)	Percentage of total
Nonrenewable			
Crude oil	106.53	4,687	11.1
Natural gas*	35.8	1,468	3.5
Shale oil	497.0	21,868	52.0
Mineral coal	688.0	13,760	32.7
Uranium†	0.0032	309	0.7
Total		49,092	100.0
Renewable			
Hydroelectric power (present)		649	
Biomass (present)		954	
Hydroelectric power‡ (potential)		3,023	

*Proved reserves are in cubic kilometers. †Figures are for pressurized water reactor without plutonium recycling. ‡Annual generation based on potential of 119 GW and on historical average hydrological conditions, converted at substitution basis (1 KWh = 12.75 × 10^6 joules).

Table 11. Goals of the alcohol program in Brazil. The scenarios are as follows: 1, 20 percent alcohol added to gasoline plus 10^9 liters for industry; 2, 100 percent alcohol plus 10^9 liters for industry; 3, 100 percent alcohol plus 50 percent of the diesel oil consumption; 4, 100 percent alcohol plus 100 percent diesel oil consumption.

Scenario	Production (liters/year × 10^9)	Cultivated area needed in 1000 ha of sugar cane*
1	4	1000
2	16	4400
3	22	6000
4	33	9000

*The average agricultural productivity was taken as 60 tons per hectare and industrial output as 70 liters per ton. One hectare is equivalent to 10,000 m² or 2.47 acres.

A precise evaluation of these measures has not been made, but a modest decrease (7 percent) in gasoline consumption occurred in 1977 which is encouraging if one considers that the number of cars increased by about 10 percent. Some of the more drastic suggestions made were not adopted, such as allowing the use of cars only on alternate days according to the last digit in their license plates (odd digit, odd days; even digit, even days), or rationing gasoline according to quotas. The suggestion to limit the power of cars produced by manufacturers (which is a better gasoline-saving measure than velocity limitation) was also refused. Many industries and bus and truck companies, however, are shifting away from gasoline to diesel oil, which is cheaper by a factor of 2.

Cars are generally smaller in Brazil than in the United States, but they are not very efficient; this upsets the gain due to the smaller weight. Although detailed data are not available for comparison, the average consumption seems to be 20 to 30 percent higher for cars of corresponding sizes in the United States; this may be related to the lower quality of the gasoline used in Brazil and to less efficient motors. Important improvements could be made on these two counts as in other countries.

It is interesting to point out, however, that price increases alone were not very successful in Brazil in discouraging people to drive cars. This indicates that the class of people owning cars has enough money—and a strong adherence to the established habit of driving—to af-

Table 12. Annual energy balance in alcohol production.

Crop	Agricultural yield (tons/ha)	Alcohol production (liters/ha)	Energy (Mcal/ha)		
			Required	Produced	Gain
Sugarcane	54	3,564	14,952	36,297	2.40
Cassava	14.5	2,523	12,751	18,783	1.45
Sorghum	32.5	3,775	16,554	31,689	1.92

ford price increases; or, perhaps, they give up other goods so that they can continue to drive their cars. People with lower incomes are discriminated against by high prices: the number of prospective car owners is reduced, and this creates a situation that ultimately will affect the economic health of the automobile industry.

Conclusions

It is evident that Brazil has plenty of renewable energy resources but very modest reserves of fossil fuels. The present energy problems of the country are the consequence of industralization having been based largely on imported fossil fuels in an era of cheap and abundant petroleum.

In order to utilize Brazil's renewable energy sources—mainly hydroelectric power and biomass—the people will have to make considerable social adjustments. These adjustments will probably include the relocation of some of the producing centers so that they are closer to the hydroelectric plants, and a general decentralization of the urban centers. This does not mean a return "to the fields" or to a more primitive type of civilization, but a redistribution of the amenities of modern life in a way more compatible with nature. The pollution in the air and waters of the city of São Paulo has already reached such proportions that there is strong political support for decentralization.

To reduce energy consumption, restrictions on the use of automobiles will probably have to be imposed, and there will have to be a stronger emphasis on mass transportation. The use of rail and hydro transportation will have to be encouraged. Solar energy will probably be used directly in water heating, mainly for industrial purposes, and solar dryers may well be used in agriculture. Innovation is very fast in this area and the direct use of solar energy could become important in the future.

Biomass, for the production of alcohol as well as charcoal, will have an enormous role in the future. The historical association of the use of wood with underdevelopment will have to be counteracted by pilot projects and demonstrations of the utility of biomass. The reorientation of the steel industry toward the use of charcoal instead of mineral coal has been done to some extent and should be encouraged, together with a coordinated program of reforestation.

Nuclear energy will have only a secondary role in energy production in Brazil; efforts to promote a widespread use of this form of fossil energy cannot be justified. High-grade uranium resources are small in Brazil, and serious risks would be associated with a new dependence on imports of nuclear fuel.

Although oil shale is abundant, the extraction of the oil poses serious technical and environmental problems, and this resource should probably be used only in the chemical and pharmaceutical industries. Recovery in situ might eventually prove to be an attractive technology.

In the overall picture, Brazil seems to be capable of finding within its own frontiers most of the energy needed to sustain its population.

New patterns of energy consumption will have to be adopted in Brazil. The construction of completely enclosed buildings—requiring constant air conditioning—in Brasilia and other cities in the temperate regions of the country indicated a total disregard for efficient energy use. The Portuguese colonizers some 200 years ago constructed simple but functional buildings and with some effort it should be possible for Brazilians today to design a new type of "tropical civilization" that is appropriate to the environment.

References and Notes

1. Except when specifically noted, data in this article are taken from *Anuário Estatístico do Brasil 1975* (Instituto Brasileiro de Geografia e Estatística, Rio de Janeiro, 1975); *Balanco Energético Nacional* (Ministério de Minas e Energia, Brasilia, 1977); and from (2).
2. *Estatística Brasileira de Energia*, Biannual Bulletin of the Brazilian National Committee of the World Energy Conference No. 19, Rio de Janeiro (1974).
3. J. A. Wilberg, *Energia Elet.* **35**, 25 (1973).
4. C. L. Wilson, *Energy—Global Prospects 1985–2000, Report on the Workshop on Alternative Energy Strategies* (McGraw-Hill, New York, 1977); J. Goldemberg, unpublished data.
5. The exponent p, in the relation $E = k(I)^p$, which is called "income elasticity of energy demand," can easily be shown to be the ratio of the rate of growth of energy consumption ($\Delta E/E$) to the rate of "per capita" income ($\Delta I/I$). For developed countries this ratio is approximately one or even smaller; it is much larger than one for underdeveloped countries.
6. M. H. Ross and R. H. Williams, *Energy and Economic Growth*. Report prepared for the Joint Economic Committee of the U.S. Congress (Government Printing Office, Washington, D.C., 1977).
7. P. A. Beirutti, *Ciênc. Cult. (São Paulo)* **29** (No. 3), 274 (1977).
8. J. Colombi Neto, thesis, Universidade de São Paulo (1975).
9. J. Goldemberg, paper presented at the Fourth Annual College of Biological Sciences Colloquium, Columbus, Ohio, 1977.
10. U. E. Stumpf, paper presented at XI Reunião Anual da Sociedade Brasileira de Física, São Paulo, 1977.
11. A. M. A. Licio, *Etanol: combustível e matéria prima*, Semana de Tecnologia Industrial, Ministério da Indústria e Comercio, Rio de Janeiro (1976).
12. J. Gomes da Silva, paper presented at XI Reunião Anual da Sociedade Brasileira de Física, São Paulo 1977.
13. J. Gomes da Silva, G. E. Serra, J. R. Moreira, J. C. Goncalves, J. Goldemberg, *Science*, in press.
14. A. Strout, "The future of nuclear power in developing countries," unpublished paper. MIT Energy Laboratory, 1976.
15. L. N. Garcez, in *Energia no Brasil*, J. Goldemberg, Ed. (Academia de Ciências do Estado de São Paulo, 1976), p. 73.
16. C. Luca, E. Koelle, J. L. A. Junqueira, in *ibid.*, p. 83.
17. J. V. Martin and G. Goldemberg, unpublished data.
18. G. Roa and L. G. Villa, personal communication.

19. I thank R. H. Williams of the Center for Environmental Studies for a complete revision of the manuscript and for useful discussions.

Solar Energy for Village Development

Norman L. Brown and James W. Howe

Most of the people in the developing countries of Asia, Africa, and Latin America live in rural areas. Progress in these areas depends on finding substitutes for human muscle energy, which is now relied on for many village tasks.

Summary. The National Academy of Sciences held a joint workshop with the Government of Tanzania last August on the potential of solar energy for the villages of that country. Costs of five solar technologies (mini–hydroelectric generators, wind, methane generation from organic wastes, photovoltaic cells, and flat-plate solar collectors) were compared with costs of diesel-generated electricity and with electricity from the national grid. Each of the five technologies is either now competitive with diesel or will be in a few years. Although the figures presented are not conclusive since they are derived from calculations rather than an actual test, the results are encouraging enough to warrant serious testing in Third World villages. (This workshop was sponsored by the Office of Science and Technology, AID).

The prohibitive costs of large central generators and massive transmission and distribution systems, as well as the slow pace of the spread of rural electrification programs, discourage hopes that rural energy needs can be met with a national electric grid. Hence, we have inquired into the potential of small-scale technologies that use renewable energy sources coming from the sun. Current solar energy comes from four major systems: (i) photosynthesis, which is the basis of all life, both plant and animal; (ii) the water cycle, which is driven by the sun; (iii) wind, caused by the atmospheric pressure differences due to changing amounts of solar energy falling on different places; and (iv) direct sunshine. Proven small-scale technologies exist for using each of these solar energy systems.

In photosynthesis, biogas plants use anaerobic bacteria to turn animal, human, and crop wastes into methane gas and, at the same time, leave a residual slurry that is useful for fertilizer. It has been reported that there were 1.2 million biogas plants installed in China in the first 6 months of 1976 alone, and that 4.7 million are now in operation (*1*). The technology appears to be proving itself and is improving all the time. Methane can be used for cooking, crop drying, power generation, and various other purposes.

To make use of the water cycle, a number of very small hydroelectric generators are now being manufactured for as little as $800. Miniature units producing only a few hundred watts or a few kilowatts can operate either with a small dam or simply by the flow of a small stream. In China, it is reported, there are 60,000 such units averaging about 40 kilowatts in capacity that successfully supply most of the electricity used by three-quarters of China's rural communes (*1*, pp. 27 and 28). In the 19th and early 20th centuries, most of New England's commercial power came from small hydrofacilities.

Windmills had long since proved themselves in Holland and the plains of the United States before they were driven out of existence by rural electrification—originally based on coal and later on cheap oil.

At least two small-scale technologies are available to make use of the direct rays of the sun. One is a flat-plate collector for space and water heating but which could also serve for activities such as crop and fish drying, distilling water, and refrigerating. The other technology for collecting the direct rays of the sun to generate electricity is a solar cell or photovoltaic array such as that used to power spacecraft.

Such promising technologies could be useful for a variety of tasks in rural areas. This potential is described in Table 1, where 14 technologies are applied to 15 common village tasks with the result that there are 34 applications that seem useful and 39 applications that seem marginally useful.

But what of costs? Are these not all technologies that require years more of research or the establishment of mass markets to become cost competitive with conventional energy sources? So few good data have been kept on the costs and performance of such technologies in actual village situations that we cannot answer this question completely. However, some preliminary calculations suggest that at least five technologies now "on the shelf" are, or soon will be, cost effective when compared with either diesel electric generation or the existing electric grid in the case of Tanzanian villages. These calculations were made in Tanzania during a 10-day workshop jointly sponsored by the U.S. National Academy of Sciences and the Tanzanian National Scientific Research Council. For a week, this workshop group (*2*) poured over calculations of the costs of performing certain village tasks (i) with diesel motors, (ii) with electricity from the Tanzanian electric grid, and (iii) with five small-scale renewable technologies. The results surprised even the experts. Each of these five technologies appeared to be now or soon competitive with diesel. A number of them compete well with the electric grid for certain village tasks. These results are valid despite the assumption that capital to build the facility costs 10 percent and must be repaid during the life of the equipment. In actual fact, a great deal of capital is often available to developing countries on far more generous terms. This is more advantageous to solar energy over the life of the project than to diesel power because capital costs of the former are high relative to operating costs while the reverse tends to be more true of diesel power. Thus, for example, it may be possible to borrow nearly 100 percent of the total life of project costs for a photovoltaic installation (since such costs consist nearly entirely of initial capital), whereas for the initial capital for diesel one could borrow only about one-eighth of the life of project costs. Hence cheap capital favors solar over diesel energy.

Of course the panelists recognized that

Mr. Howe is a Senior Fellow with the Overseas Development Council, 1717 Massachusetts Avenue, NW, Washington, D.C. 20036. He was chairman of the National Academy of Sciences team that conducted a workshop in Tanzania on solar energy. Dr. Brown is an official in the Office of International Programs of the Department of Energy, Washington, D.C. He was a member of the Tanzanian solar energy workshop.

Table 1. Solar energy applicability matrix. Animals are included as a solar technology (based on photosynthesis). For many villages the use of animals would represent a modernizing step. It includes the use of dung for burning or fertilizing. Source: adapted from a presentation by J. R. Williams at the Tanzanian workshop. Symbols: ++, applicable; +, potentially applicable; −, not applicable.

Solar technology	Energy use													
								Heating						
	Water pumping	Lighting	Cooling	Communications	Water desalting	Spinning	Sawing	Cooking	Space	Domestic water	Grinding	Drying	Transport	Fertilizer
Solar cells (flat plate)	+	+	+	++	−	+	−	−	−	−	+	−	−	−
Flat-plate collectors	+	−	+	−	++	−	−	+	++	++	−	++	−	−
Concentrating collectors	+	+	+	−	+	−	−	+	−	−	−	−	−	−
Solar, Stirling	+	+	+	−	−	+	+	−	−	−	+	−	−	−
Solar, Rankine	+	+	+	−	−	+	+	−	−	−	+	−	−	−
Wind (mechanical)	++	−	−	−	−	+	+	−	−	−	++	−	−	−
Wind generator	++	++	+	++	−	+	+	−	−	−	++	−	−	−
Water (mechanical)	++	−	−	−	−	++	++	−	−	−	++	−	−	−
Hydroelectric	++	++	++	++	−	++	++	−	−	−	++	−	−	−
Bioconversion wood/pyrolysis	−	+	−	−	−	−	−	++	++	++	−	++	−	+
Biogas	+	++	+	−	−	−	−	++	−	+	−	+	+	++
Draft animals	++	−	−	−	−	−	−	+	−	+	++	−	++	++

a final determination of whether a given technology should be applied to a given village situation depends not only on the financial costs, but also on availability of resources, social obstacles, social benefits, institutional barriers, long-range power requirements, and long-range development goals. In the short time available to them, they focused on estimating financial costs of the technologies considered most likely to be applicable with the understanding that this was but a first step in the decision-making process (*3*).

Basis for Cost Comparison

A hypothetical village of 300 families was chosen as a basis for estimating the magnitude of financial investment needed. It was assumed, for these initial installations, that the needs of each family could be met, on the average, by the use of 1 kilowatt-hour per day (300 kWh total) to be applied to lighting, operation of a village television receiver, radio communications, pumping water, or grinding grain. At this rate of energy consumption, Tanzanians receiving central-station generated electricity from TANESCO (Tanzania Electric Supply Company) would pay, on the average, about 0.93 shilling (*s*) per kilowatt-hour (*4*) (the U.S. equivalent is \$0.113/kWh). It was assumed further that the 300 kWh/day could be supplied either on a 5-hour basis, that is, at the rate of 60 kW, or for 20 hours, that is, at a rate of 15 kW.

Technologies of Energy Supply

Five technologies were chosen as illustrative examples, on the basis of their immediate or short-term availability (or both) (*5*).

Photovoltaic power generators. The analysis of the cost of supplying electrical power by a solar-cell array was based on a cost estimate of \$20 (U.S.) per watt-peak (W_p). Although, in a recent large-scale purchase by the U.S. government, the cost was \$15/W_p, a cost of \$20/W_p was used in these estimates on the assumption that the more favorable price would not be available for an initial small-scale purchase.

Cost calculations are summarized in Table 2 (*6*). The size of the array needed to supply 300 kWh daily was based on World Meteorological Organization insolation data for Tanzania (*7*). These figures indicate that an array with a peak power capacity of 1 kW will produce, on the average, 5.3 kWh daily. Thus, for 300 kWh/day, a generous estimate of 60 kW_p was used. An interest rate of 10 percent per annum, the prevailing rate in Tanzania, was used to calculate financing costs. The estimated cost of energy provided by photovoltaics, 11*s*/kWh, is approximately 12 times the average current cost of electricity in Dar es Salaam, as noted above.

In view of the predicted drop in the cost of photovoltaic devices—the Department of Energy looks to a cost of \$0.50/W_p by 1985, or earlier—it is interesting to see how sensitive the electricity cost is to the array cost. For an array cost of 4.1*s*/W_p (\$0.50/W_p) the total system cost for this example would be 528,000*s* with the same battery costs assumed. The annual cost would then be 94,000*s*, and the cost of electricity 0.83*s*/kWh, or 10 percent less than the current cost of electricity from the grid. The "break-even" array cost, that is, the cost of photovoltaics that would enable electricity to be generated at the average current selling price in Dar es Salaam, would be 5.28*s*/W_p or \$0.64/W_p. Thus, the use of photovoltaic devices to generate electricity for the villages of Tanzania is likely to be economically competitive within 10 years (*8*). Moreover, for any village that wants small amounts of electricity now but is not within reach of the grid (discussed later) photovoltaic devices may already be cost effective compared with conventional alternatives.

Small-scale hydropower. In the mountainous regions of Tanzania there are many small streams and rivers with a flow of water sufficient in quantity and reliability to be considered as a source of small-scale hydroelectricity. This technology is a mature one, with devices of a variety of sizes available "off the shelf" not only in the United States, but in many other countries (*9*). With known cost figures as a basis, four cases were considered, as shown in Table 3. These four cases provide for two basic situations—one in which the needs of the village (300 kWh/day) would be supplied directly by the generator during the hours of use (that is, 60 kW), and one in which a smaller generating capacity (that is, 15 kW) would be used with storage batteries to supply the required 300 kWh/day over a period of 20 hours.

The cost calculations for these four ways of supplying hydroelectric power are summarized in Tables 4 to 6. For the purposes of this exercise, it was assumed that dams and penstocks, if needed, could be constructed with local labor and materials (timber and earth) at no significant capital costs (*10*). The water requirement estimates were based on data supplied by manufacturers of turbine or generator (or both) systems of the scale discussed. In all cases, where battery storage was required, it was assumed that the cost of the batteries would be amortized over the 6-year life expected for that subsystem, in order to allow for its replacement as needed, independent of the lifetime or amortization rates (or both) for the other equipment. As with the dams and penstocks, no cost figures were included for protective housing for batteries or power-conditioning equipment, on the assumption that locally available materials would suffice. (For all of these cases, some maintenance would be advisable. At an estimated cost of 1000*s*/month, this would add 0.10*s*/kWh to each of the cost figures.) Any of these assumptions might not be valid in specific circumstances; nevertheless the estimates show that small-scale hydroelectric installations are cost competitive with present large-scale systems—to say nothing of the serious environmental and social consequences of large-scale hydroelectric projects—and are worth serious consideration in rural areas.

As a result, the panelists felt that it would be useful to extrapolate these estimates to the case where a large region of, say, 300 villages would be supplied with perhaps as many as 500 small hydroelectric generating units—assuming, of course, that sufficient rainfall and water flow rates were available. As an example, the total capital requirement for case 1 would be 330 million shillings. If a loan from the International Development Association (IDA) (the "soft-loan window" of the World Bank) were available, the economics of small-scale hydroelectricity applied on a large scale are very attractive.

For such a loan, no payment is required for the first 10 years; the loan is repayable, in equal annual installments over the next 40 years, at an annual interest rate of 0.75 percent. For this example, this would mean a financing cost of 9.6 million shillings per year from

Table 2. Cost of supplying electricity by means of photovoltaic generators. Financing costs were based on amortization of the loan in equal yearly installments.

Base data	
Lifetime of silicon solar cells	20 years
Lifetime of batteries	6 years
Efficiency of batteries	85 percent
Cost of silicon solar cells	160*s*/W_p
Cost of batteries	800*s*/kWh
Equipment costs	
Solar-cell array, 60 kW_p at 160*s*/W_p	9,600,000*s*
Battery storage	
(300 kWh at 85 percent efficiency)	
353 kWh at 800*s*/kWh	280,000*s*
Total equipment costs (approximate)	10,000,000*s*
Financing costs	
Solar-cell array	
9,600,000*s* at 10 percent for 20 years	1,100,000*s*/year
Batteries	
280,000*s* at 10 percent for 6 years	64,000*s*/year
Total cost of financing (approximate)	1,200,000*s*/year
Cost of electricity	
Total energy generated	110,000 kWh/year
Unit cost	11*s*/kWh

Table 3. Types of hydroelectric power equipment considered.

Case	Power: Type generated	Power: Type used	Power: Conditioning required	Storage required	Generating capacity (kW)
1	d-c	a-c	Yes	No	60
2	a-c	a-c	No	No	60
3	d-c	a-c	Yes	Yes	15
4	a-c	d-c	No	Yes	15

Table 4. Water requirements for supplying hydroelectricity by small-scale systems.

Power (kW)	Head (m)	Flow (m^3/min)
60	48	12
	24	25
	12	50
	6	100
15	24	6
	12	12
	6	24

Table 5. Base data for cost of supplying hydroelectricity by means of small-scale systems.

Item	Lifetime (years)	Unit cost (*s*)
Generator/turbine	30	4000/kW
Power conditioning	30	6/W
Batteries (85 percent efficiency)	6	800/kWh

Table 6. Cost of small-scale hydroelectric systems (in shillings). The cost of dams and penstocks was not included.

Item	Case 1	Case 2	Case 3	Case 4
Cost of equipment				
Generator/turbine	240,000	240,000	60,000	60,000
Power conditioning	360,000	0	90,000	0
Installation	60,000	24,000	15,000	6,000
Subtotal*	660,000	265,000	165,000	66,000
Battery storage	0	0	280,000	280,000
Installation	0	0	28,000	28,000
Subtotal	0	0	308,000	308,000
Total (approximate)	660,000	270,000	470,000	370,000
*Financing costs at 10 percent annual interest rate**				
Annual cost in shillings†				
30-year loan	70,000	29,000	89,000	78,000
10-year loan	107,000	44,000	99,000	82,000
Cost of electricity (*s*/kWh)‡				
30-year loan	0.64	0.26	0.81	0.71
10-year loan	0.97	0.40	0.90	0.75

*Based on amortization of the loan in equal yearly installments. †In both cases, it was assumed that the cost of the battery storage would be amortized over 6 years. ‡Based on 110,000 kWh/year.

years 11 through 50. If the energy generated were sold to the villagers at the cost figures calculated in Table 6, the annual income would be 53 million shillings if the loan were a conventional 10-year loan, and 35 million shillings if it were a conventional 30-year loan. This would provide the government with between 25 and 40 million shillings for other projects. On the other hand, if energy were sold at a rate just sufficient to meet the 40-year cost of the IDA loan, the rate would be 0.17*s*/kWh, a figure less than one-fifth the current price of electricity in Dar es Salaam, and a cost well supportable by the villages electrified in this scheme. (This is approximately 1*s* per week.)

The four hydroelectric cases considered were based on a use, by the villagers, of 200 watts of electricity for 5 hours per day. For cases 3 and 4, this is the total capacity of the system. For cases 1 and 2, however, the system would be capable of supplying energy at the same rate (60 kW) day and night for 24 hours. If this capacity were used for more than 5 hours daily, the cost per kilowatt-hour could be reduced by as much as a factor of 4 or 5. That is, with load factors greater than 0.2 (5 hours' use per day), the cost of electricity production by small-scale hydropower systems of the type considered could approach 0.1 to 0.2*s*/kWh.

Biogas generation. The generation of methane from human, animal, and agricultural wastes is a process that is finding more and more use in developing countries, as people grow more aware of the potential of these "waste" materials as a source of both energy and fertilizer (*11*). Indeed, several biogas plants are operating in Tanzania, and a body of experience in their construction and operation is being accumulated (*12*). Thus, it was natural for the workshop participants to consider the use of biogas not only for village lighting and cooking, but for the generation of electricity or mechanical power. For this exercise, the use of cattle dung was assumed, with average production rates and methane concentrations. Furthermore, lighting both by a gas-mantle lantern and with generated electricity was considered. The calculations are outlined in Table 7.

Analysis of the data shows that gas requirements for cooking three times a day are approximately 1.4 m^3/day per family, which could be supplied by dung from three to four cows. In contrast, to supply the electric lighting needs of a family (1 kWh/day) by using the biogas to operate an engine-generator set would require heat energy of 5 kWh/day, which could almost be supplied by one cow. This method of lighting, therefore, is almost four times as efficient, in energy use, as a gas-mantle lantern, albeit it requires a much greater capital investment. Finally, this same amount of gas—the gas supplied by one cow—would provide mechanical power alone at 1 kWh/day.

Another potential use for biogas, as a replacement for wood and charcoal, is in firing clay pottery used for water and cooking vessels. For small-scale production, estimates based on gas-fired firebrick kilns indicate that about 8 m^3 of methane, over a period of about 5 hours, would be required to fire earthenware at 1000°C in a kiln having a volume of about 0.2 m^3 (6 cubic feet). This is the equivalent of about 13 m^3 of biogas, which could be supplied by a 4-day accumulation from the dung of five cows Larger production capacities—for instance, for a village industry—could be available by scaling-up the above estimate accordingly, and would represent another potential use for community digesters.

To summarize then: (i) heat energy, even from a single-family biogas plant, can be supplied at well below (less than one-ninth) the equivalent energy cost of electricity, and (ii) Ujamaa villages lend themselves readily to communal systems. Because of the economies of scale in biogas plant construction such communal systems (for example, schools, community latrines, community cattle-dung digesters in Masai villages) should be seriously considered.

Windmill generators. The use of windmill generators to supply a village's proposed electrical energy needs was considered in this exercise. However, even though the financial costs can be reasonably well estimated, measurements of wind-velocity and duration must be made to determine the suitability of any proposed site before any plans for a windmill project are contemplated.

The estimates that were made (Table 8) assume minimum wind speeds of about 12 miles per hour (5 m/sec) for an average of about 10 hours daily throughout the year. With windmill generators of a size reasonably available (10 kW), this would provide an average of 100 kWh daily, with electricity costs of less than 2*s*/kWh. This means, also, that to provide the proposed 300 kWh daily to a village of 300 families, three such installations would be required. If the task to be performed is pumping water, serviceable wind pumpers can be made with substantial savings by using local materials and skills. Thus, for example, in the Omo River Valley in Ethiopia, simple sail wind pumpers were built with Ethiopian materials following a design long used in Crete. The costs of mills that pumped 800 imperial gallons of water per hour, an elevation of 9 feet in winds of 13 miles per hour, was only $375 for each sail mill (*13*).

Solar refrigeration. Having visited a village where fishing was one of the major activities, the workshop panelists considered the possibility of using solar cooling to enable the villagers to preserve their fishing catch long enough to market the excess over their local needs. The cost of cooling and preserving a catch of 500 kilograms—typical for a day's fishing at the village visited—was

Table 7. Cost of supplying cooking fuel, lighting, mechanical power, and electricity by biogas.

Base data	
Cattle dung	10 kg/day per cow
Biogas production (at 0.06 m^3 per kilogram of dung)	0.6 m^3/day per cow
Energy content of biogas as 60 percent methane*	6.4 kWh/m^3
System lifetime	20 years
Conversion efficiencies	
Internal-combustion engine**	25 percent
Electrical generator	80 percent
Plant cost	
Single-family plant (3 m^3)	6000*s*
Financing costs†	
Single-family plant, 6000*s* at 10 percent for 20 years	705*s*/year
Engine/generator to provide 1 kWh/day, 5400*s*‡ at 10 percent (5 hour/day)	5940*s*/year
Engine to provide 1 kWh/day, mechanical energy, 3200*s*‡ at 10 percent (4 hour/day)	3520*s*/year
Energy costs§	
Cooking and lighting directly by gas (20 kWh/day)	0.10*s*/kWh
Cooking and generation of electricity at 1 kWh/day	18.2*s*/kWh
Cooking and generation of mechanical power at 1 kWh/day	11.6*s*/kWh

*Approximately 4 kWh of biogas energy is available daily from each cow. †Based on amortization of loan in equal yearly installments. ‡Retail costs, Dar es Salaam, August 1977. §Cost of energy production only—does not include cost of appliances or cost of collecting dung. **Assuming peak loading.

calculated, on the basis of the use of local materials and labor for the container and its insulation, and commercially available collector. Estimates were based on the assumption that ammonia-based cooling systems could be used since the units will be located out of doors, and that they could be constructed in Tanzania. The container was assumed to be a 1-m cube, the cost of the chiller unit was estimated from known costs of available machines, and current U.S. costs of collectors were used. The results, based on the assumptions stated, are shown in Table 9. They indicate that solar refrigeration would cost only slightly more than the current cost of electricity alone, assuming a 5-year equipment life.

Table 8. Cost of supplying electricity by means of windmill generators.

Base data	
Useful life of windmill generator	15 years
Lifetime of batteries	6 years
Efficiency of batteries	85 percent
Equipment costs	
Windmill generator (10 kW)	170,000*s*
Wiring, controls	42,000*s*
Installation	42,000*s*
Battery storage—(100 kWh at 85 percent efficiency) 118 kWh at 800*s*/kWh	94,000*s*
Total equipment costs (approximate)	350,000*s*
*Financing costs**	
Windmill generator, wiring, installation, 254,000*s* at 10 percent for 15 years	33,000*s*/year
Batteries, 94,000*s* at 10 percent for 6 years	22,000*s*/year
Total cost of financing (approximate)	55,000*s*/year
Cost of electricity	
Total energy generated	36,000 kWh
Unit cost	1.5*s*/kWh

*Based on amortization of loan in equal yearly installments.

Table 9. Cost of solar refrigeration of fish.

Base data	
Lifetime of equipment	5 years
Weight of fish (M) chilled daily	500 kg
Specific heat of fish (C_p) (estimate)	1 cal/g·°C
Average storage time (t)	3 days
Average ambient temperature	25°C
Cold storage temperature	5°C
Surface area (A) of insulated box	12 m^2
Thermal conductivity (K) of insulation*	10^{-4}cal/sec·cm·°C
Insulation thickness (d)	20 cm
Hours of sunshine (average)	6
Heat requirements	
Cooling (Q_c) $Q_c = M{\cdot}C_p{\cdot}\Delta T$ = 10,000 kcal/day	
Losses (Q_L) $Q_L = K{\cdot}A{\cdot}\Delta T{\cdot}t/d$ = 1,000 kcal/day	
Total cooling requirement	11,000 kcal/day
Total heat requirement @ coeff. of performance = 0.5	22,000 kcal/day
Equipment costs	
Chiller ($^2/_3$ ton ≅ 2,000 kcal/hr)	(estimated 20,000*s*
Collector @ eff. = $^1/_3$ (15 m^2)	(estimated) 17,000*s*
Total equipment cost	(approximated) 37,000*s*
Financing cost†	
37,000*s* at 10 percent for 5 years	9,800*s*/year
Cooling cost	
Energy required for cooling (22,000 kcal/day)	9,200 kWh/year
Unit cost	1.06*s*/kWh

*Based on the use of a material, such as sisal, with a thermal conductivity similar to that of sawdust. †Based on amortization of loan in equal yearly installments.

Conventional Technologies

The preceding analyses, as was noted earlier, compared the cost of generating electricity to the current selling price from the grid. To provide a useful context for this comparison, the costs of supplying electricity by extension of the present grid, and by local diesel generators, were examined.

Distribution grid to a village. This situation is based on either eventual extension of transmission lines to cover the country to the extent of coming within 20 km of every village, or constructing distribution systems to villages within 20 km of present transmission lines. In view of the overwhelming cost of the first alternative, serious thought was given only to the second. On this basis, it was assumed that a village of 300 families would occupy about 16 km^2 at a distance of some 20 km from a 33-kV transmission line. A summary of the analysis is given in Table 10. Transmission line costs (33 to 11 kV with a transformer of 500 kW) and substation (11 kV to 400 V at 5 kW) costs and connection costs came to an estimated total of about 2.18 million shillings. Financing costs, based on a 20-year life for the system, and a 10 percent annual rate of interest, would be about 256,000*s* per year. If this system were fully utilized (that is, load factor of 1.0) the unit cost would be 0.06*s*/kWh. Usually a grid system is poorly matched to village needs, however. In the case considered, for example, a village of 300 families has a 500-kW supply continuously available, much more than it can reasonably use on a continuous basis. A more reasonable load factor to assume would be 0.1, which would make the transmission costs 0.58*s*/kWh. With the average generating cost in the existing grid 0.30*s*/kWh, the cost for delivered electricity by this scheme would be 0.88*s*/kWh—a figure comparable to current consumer prices in Dar es Salaam. This modest cost, it must be remembered, would be available only to those fortunate few villages within 20 km of the existing grid—assuming, of course, that TANESCO would be willing to sell electricity to the distribution point for its cost of generation. A more realistic estimate would take into account the cost of transmission to the distribution point, but this information was not available to the panel.

Small-scale diesel generation. In view of the ubiquity of diesel electric generators, it is instructive to examine the cost of generating electricity by this technology. Table 11 summarizes the cost figures based on current retail prices in Dar es Salaam. At a cost of 2.3*s*/kWh for diesel-generated electricity, it is apparent that, with the exception of photovoltaics at present prices and biogas-generated electricity, the alternative technologies considered in our discussion would be preferable on a unit-cost basis. Even photovoltaic devices will be preferable in a few years.

One of the problems with such solar energy technologies that depend on wind

Table 10. Cost of supplying electricity from existing grid.

Base data	
Grid voltage	33 kV
Distance from grid to village (assumed)	20 km
Distribution voltage to substation	11 kV
Local distribution voltage	400 V
Number of substations assumed*	45
Average hook-up distance for 45 substations and 300 families in 16-km² village	0.2 km
Fixed costs	
High voltage step-down transformer	
33 kV to 11 kV, 500 kW at 1,000,000*s*/500 kW	1,000,000*s*
11-kV transmission line 20 km at 500*s*/km	10,000*s*
Substation transformers	
11 kV to 400 V, 50 kW at 10,000*s*/50 kW	450,000*s*
Local distribution lines (400 V); 300 families at 0.2 km per family and 12*s*/m	720,000*s*
Fixed costs	2,180,000*s*
Financing† costs	
2,180,000*s* at 10 percent for 20 years	256,000*s*/year
Cost of electricity‡	
Total annual capacity, 500 kW × 8760 hour/year	4,380,000 kWh
Unit cost of capacity (load factor = 1)	0.06*s*/kWh
Unit cost at 10 percent load factor	0.58*s*/kWh

*Because local distribution is so expensive, in this example it is cost effective to underutilize the substations by using many of them to reduce local hook-up distance. †Based on amortization of loan in equal yearly installments. ‡This calculation represents the unit cost of transporting electricity from the 33-kV transmission line to the consumer. It does not take into account maintenance costs for the system nor the cost of electricity delivered to the point where the 11-kV line starts.

Table 11. Cost of supplying electricity by small-scale diesel generators.

Base data	
Operation	5 hour/day
Useful life (about 10 years)	20,000 hours
Fuel consumption	0.35 liter/kWh
Overhaul	every 5,000 hours
Fixed costs	
Retail cost, 6-kW diesel generator	29,000*s*
Overhaul costs	15,000*s*
Installation	2,900*s*
Annual costs	
Equipment and installation, 32,000*s* at 10 percent for 10 years*	5,200*s*/year
Overhaul, 15,000*s* at 10 percent for 3 years*	6,000 *s*/year
Maintenance, operator	7,000*s*/year
Total annual costs (less fuel) (approximate)	18,000 *s*/year
Cost of electricity	
Total energy generated annually	11,000 kWh
Unit cost, less fuel	1.6 *s*/kWh
Fuel cost, at 2*s*/liter	0.7*s*/kWh
Total unit cost	2.3 *s*/kWh

*Based on amortization of loan in equal yearly installments.

or sunshine or even flowing streams is that these are intermittent sources of energy, so that in the United States they require a backup system based, for example, on gas, oil, or electricity. The cost of the solar-based system plus the backup system may be prohibitive. However for many village tasks in the developing countries, the backup system may be the traditional energy system. If the wind fails to pump water, the villagers simply revert to carrying it on their heads. If the sun fails to shine to refrigerate fish, they are smoked over a wood fire, and so on. Moreover, a number of tasks can be performed whenever primary energy is available—for example, grain can be ground or water pumped and then stored for later use.

Thus far we have established that there are good technologies, that there is preliminary evidence that some of them may be cost effective and that, in many cases, there is a ready backup system that entails no capital costs. But there is still very little experience with installing such technology in rural villages of the Third World under circumstances where good cost and performance data have been kept. In fact, many of the experiments in placing technology in villages have ended in failure. A number of windmills and a few methane digesters have been tried in African villages, for example, with far from encouraging results. Outside technicians came into the villages with a preselected technology to perform a preselected village task, and without very much interaction with villagers they erected and operated the energy-producing hardware. A few weeks or months later, the visiting technicians left and shortly afterward the technology fell into disuse. But this pattern is not unique to small-scale renewable energy hardware. Whether it be farm machinery, or transport or construction equipment, to provide such capital goods to a village without first training a cadre of people able to operate, maintain, and repair it, and without ensuring that there is local institutional infrastructure to support it, has universally proved to be futile. The latter might be a government institution such as a school, or a cooperative society, or a private entrepreneur. Without trained people and a responsible institution the chances are that energy technology will not be properly or long used. To give such technology a fair chance for success, it is necessary to enlist the enthusiastic cooperation of appropriate villagers. This means villagers must be involved in selecting the village task to be energized, in selecting the technology, and in determining what village institutions should be responsible for its maintenance and use.

The subject of finding renewable energy sources for Third World rural areas is not just a matter of humanitarian concern for those people. It may directly affect the availability of world energy supplies. Third World rural areas are not significant commercial energy users at present. However, influential leaders in the international community have set a target of eliminating the worst aspects of absolute poverty by the end of the century. The Overseas Development Council has examined the energy consumption of the 25 developing countries that have largely achieved this target to gauge what might happen to energy demand if all developing countries were to do so. Three preliminary important conclusions appear warranted. First, those nations (such as Brazil and Mexico) that have emphasized growth alone are relatively much higher per capita energy users than those countries (such as Taiwan) that have pushed growth with equitable distribution of incomes, or those (such as Sri Lanka) that have focused exclusively on equitable income distribution and have not made such headway or growth. Second, even if all Third World countries follow the relatively energy efficient growth with equity strategy, the goals of eliminating the worst aspects of absolute poverty cannot be met without imposing a claim on world oil resources larger than world oil production will permit. Third, therefore these development goals can only be met if nonpetroleum sources of energy are made available to these rural areas. Under these circumstances, exploring the use of small-scale renewable sources of energy appears to be an attractive path.

References and Notes

1. V. Smil, *Bull. At. Sci.* **23** (No. 2), 25 (1977).
2. The U.S. team, in addition to the authors of this article, included Dr. J. R. Williams, associate dean for research, College of Engineering at the Georgia Institute of Technology; Dr. T. Lawand, director of field operations of the Brace Research Institute, McGill University, Montreal, Canada; Dr. B. Williams, director of the Energy Systems Research Laboratory of R.C.A.; and Dr. P. Reining, Office of International Science of the AAAS. The Tanzanian participants included representatives from the sponsoring council, the Prime Minister's office, relevant ministries and interested organizations, and the University of Dar es Salaam.
3. Calculations such as those in the following pages will be included in a forthcoming NAS publication on the proceedings of the workshop. Although these calculations are based on the work done at the NAS workshop, the particular numbers in this article are those of the authors and have not been checked yet with other members of the workshop.
4. The rate is 0.93 shilling per kilowatt hour. Actually the rate is graduated according to the amount used. The exchange rate is approximately U.S. $1 = 8.2 shillings.
5. The estimates made by the panel for the renewable resource technologies aimed at producing electricity as well as for diesel did not include the cost of power distribution in the village. Such costs are treated in the discussion of extending grid to the villages.
6. In view of the uncertainities in the cost estimates, all numbers are given only to two significant figures.
7. World Meteorological Organization, average of data for June 1966.
8. In this analysis, and those to follow, the cost of generating electricity is compared with the current price to the consumer. The principal justification for this approach is the assumption that the communal nature of the Ujamaa villages lends itself to eventual communal ownership of the generating system—hence the important datum for the villagers is the energy cost.
9. National Academy of Sciences, *Energy for Rural Development* (National Academy of Science, Washington, D.C., 1976). See "Hydropower" in part 2, *"Indirect Uses of Solar Energy,"* for a discussion of the growing use of small-scale hydroelectricity in the People's Republic of China.
10. Tanzanian participants thought that this was reasonable since village labor and materials would be volunteered and hence would not need to be included in the funds budgeted for the project. In a similar situation, techniques in which local labor and materials were used to construct dams and penstocks for installations capable of producing power in tens of kilowatts have been developed in Colombia at the Centro de Desarrollo Integrado "Las Gaviotas."
11. See *Methane Generation from Human, Animal, and Agricultural Wastes* (National Academy of Sciences, Washington, D.C., in press).
12. P. S. Mhina, "A brief description of manufacture and utilization of gobar gas in Tanzania," presented as a Seminar on Appropriate Technology in Small-Scale Industries, 15 to 17 December 1976, Dar es Salaam Technical College [Small Industries Development Organization (SIDO), Box 2476, Dar es Salaam]; A. Mzee, *Pilot Project on Utilization of Gobar Gas in Small I.C. Engines* [Ministry of Water, Energy and Minerals, P.O. Box 9153, Dar es Salaam, Tanzania (1977)].
13. P. Fraenkel, *Food From Windmills* (Intermediate Technology, Ltd., London, 1975).

C An Environmental Perspective

Energy and Water

John Harte and Mohamed El-Gasseir

Providing energy for human use consumes water and providing water consumes energy. We discuss here only the consumption of water for energy. Our objective is to assess the constraints that limited and unpredictable supplies of freshwater in the United State may place on energy development.

Summary. The geographic and temporal variability of freshwater supply in the United States constrains the choice and level of use of future energy sources. Ecological criteria for acceptable freshwater consumption, together with hydrological data on stream flow, provide a framework for estimating these constraints. The water consumption requirements for a variety of energy options are presented, and comparative judgments drawn. Attention is focused on problems resulting from synthetic, gaseous, and liquid fuel production. Scenarios describing possible future levels of coal and electricity use are analyzed. They point to the importance of water supply constraints in both the eastern and western United States.

Energy technologies use water resources in numerous ways. For example, the cooling of electric generating plants or coal gasification and liquefaction plants may consume freshwater. Coal and oil shale conversion processes require water as a chemical feedstock. Coal mining and land reclamation subsequent to surface mining require water. Solar bioconversion plantations are likely to require irrigation water. Hydroelectric power consumes water in the sense that artificial lakes enhance evaporation losses. In fact, nearly every imaginable energy system demands water. Because of the limited freshwater supply in many regions of the United States, and because of the unpredictable nature of precipitation, it is important to understand the freshwater requirements for each of the many energy technologies available to society during the next several decades. We will demonstrate here that water consumption requirements place serious constaints on the future level of development of many of this country's energy options.

One energy technology that we will discuss in some detail is the production of synthetic gaseous and liquid fuels. During the coming decades, the United States will have to find energy sources that can replace natural gas and petroleum. Because many end uses, especially transportation and home heating, rely today on these two fuels, it appears that only three paths are available. One option is to adapt such end uses to electricity. The second is to replace dwindling gas and petroleum supplies with synthetic gaseous and liquid fuels. A third path, which could ease the demand on gaseous and liquid fuels for space heating and cooling, is to expand active and passive use of the sun. Part of our concern here will be with the comparative impacts of these three paths on water consumption.

First we set the background for a

Dr. Harte is a physicist in the Energy and Environment Division, Lawrence Berkeley Laboratory, University of California, Berkeley 94720; M. El-Gasseir is a graduate student in the Energy and Resources Program, University of California, Berkeley. This article is based on a study done in connection with the National Academy of Sciences' forthcoming CONAES report.

quantitative discussion of energy and water scarcity and develop a framework for evaluating the impacts of water consumption. In the third section we estimate water requirements, on a per unit energy basis, for those energy options that are candidates for major expansion in the United States and that are likely to require large quantities of water. Based on these estimates, we make some comparative judgments about water impacts of competing technologies. We then analyze a number of energy scenarios, looking on a regional level at the constraints on energy growth likely to be imposed by limited freshwater supplies.

Freshwater Supply and Demand

Withdrawal and consumption. To estimate the consequences of the water requirements for energy production, a distinction must be made between water withdrawal and water consumption. Water withdrawn is water taken from a water supply but not necessarily consumed. Water consumed is water rendered unavailable for specified further uses. The water consumption of a given activity depends on the ways water is used in the activity and the ways it is needed by downstream users, including the spatial distributions and time schedules of all such uses (*1*).

Thus, heavily polluted water that is discharged from a coal gasification plant is consumed water for many competing uses, although not, perhaps, for mine floor wetting. Water evaporated from a wet cooling tower or an artificial lake, or from surface-mined land under reclamation, is consumed water from the viewpoint of other users in the region because the evaporated water cannot be expected to fall as rain on the same region. Also, water used as a source of hydrogen for synthetic fuel production is consumed water—notwithstanding the fact that this water is regenerated when the fuel is eventually burned.

To clarify further our treatment of these issues, consider the following hypothetical and highly simplified case. Assume that a conversion plant takes 10^6 cubic meters of water per year (m^3/year) from a river (*2*). This is the withdrawal rate. The conversion process uses the hydrogen in 10^5 m^3/year during the hydrogenation-methanation steps. Another 3×10^5 m^3/year is lost by evaporative cooling. The remaining and now heavily polluted portion of the withdrawn water is delivered to a treatment facility where 10^5 m^3/year evaporates, 0.5×10^5 m^3/year is disposed of as waste product, and the rest (4.5×10^5 m^3/year) is treated and returned to the river 5 kilometers (*2*) downstream from the intake point. Assuming that the treated water is adequate for all downstream users, and that the outflow from the plant is staged in time so as to be compatible with downstream use, then the consumption rate for downstream users is the sum of what is used for chemical feedstock, plus the evaporated portion, plus what is disposed of as a concentrate, or 5.5×10^5 m^3/year. This remains valid as long as the plant operates in the prescribed fashion.

On the other hand, water could be conserved by eliminating all evaporation and extracting all water from the waste product, properly treating it, and returning it to the river. Then consumption would be minimized to the rate at which water is used as chemical feedstock in the conversion process, or 10^5 m^3/year. However, for those water-dependent activities (including maintenance of ecological habitats) along the river between the point of withdrawal and the point of return, consumption is equal to withdrawal.

The above example illustrates that technical, economic, and policy considerations in the development of new energy sources can change the balance between water withdrawal and consumption. In this work we assume that the rate of consumptive use of water varies from the minimum rate believed to be achievable under strict conservation and purification efforts to the maximum rate where water-conserving practices and adequate water treatment are not even attempted. This maximum consumptive rate, in some instances, would simply be the withdrawal rate.

We are concerned here primarily with freshwater consumption requirements of alternative energy systems. Withdrawal, while less worrisome than consumption, is nonetheless an important environmen-

Table 1. Freshwater use in the United States, 1975, expressed as cubic kilometers per year.

Use category	Withdrawal	Consumption
Municipal use including domestic and commercial	40.0	9.2
Industrial mining and manufacturing	52.0	5.8
Thermal electric power plant cooling	180.0	2.6
Irrigation, livestock, and rural use	200.0	115.0
Evaporation from man-made reservoirs*	18.0	18.0
Total	490.0	151.0

*Data adapted from (*34*); all other data adapted from (*35*).

Table 2. Regional runoff, 1975 consumption, per capita runoff, and consumption per unit runoff. Data adapted from (*35*).

Region	Mean annual runoff (km^3/year)	Data for 1975: Consumption (km^3/year)	Per capita runoff (10^{3}m^3/person/year)	Consumption/mean annual runoff
New England	93.0	0.61	7.9	0.0066
Mid-Atlantic	120.0	2.2	3.0	0.018
South Atlantic Gulf	270.0	5.1	10.2	0.019
Great Lakes	100.0	1.5	4.5	0.015
Ohio	170.0	1.7	8.0	0.01
Tennessee	57.0	0.39	17.0	0.0068
Upper Mississippi	90.0	1.3	4.6	0.014
Lower Mississippi	100.0	7.6	17.0	0.069
Souris-Red-Rainy	8.6	0.17	12.0	0.016
Missouri	75.0	24.0	8.4	0.32
Arkansas	100.0	16.0	16.0	0.16
Texas Gulf	44.0	13.0	4.2	0.30
Rio Grande	6.9	6.0	3.5	0.87
Upper Colorado	18.0	3.4	40.0	0.19
Lower Colorado	4.4	10.0	1.7	2.3
Great Basin	10.0	5.5	7.0	0.55
Pacific Northwest	290.0	18.0	44.0	0.062
California	86.0	34.0	4.1	0.40
Alaska	800.0	0.0077	2000.0	9.6×10^{-6}
Hawaii	18.0	0.77	22.0	0.043
United States	2471.0	151.0	11.0	0.060
United States excluding Alaska and Hawaii	1653.0	150.0	7.8	0.091

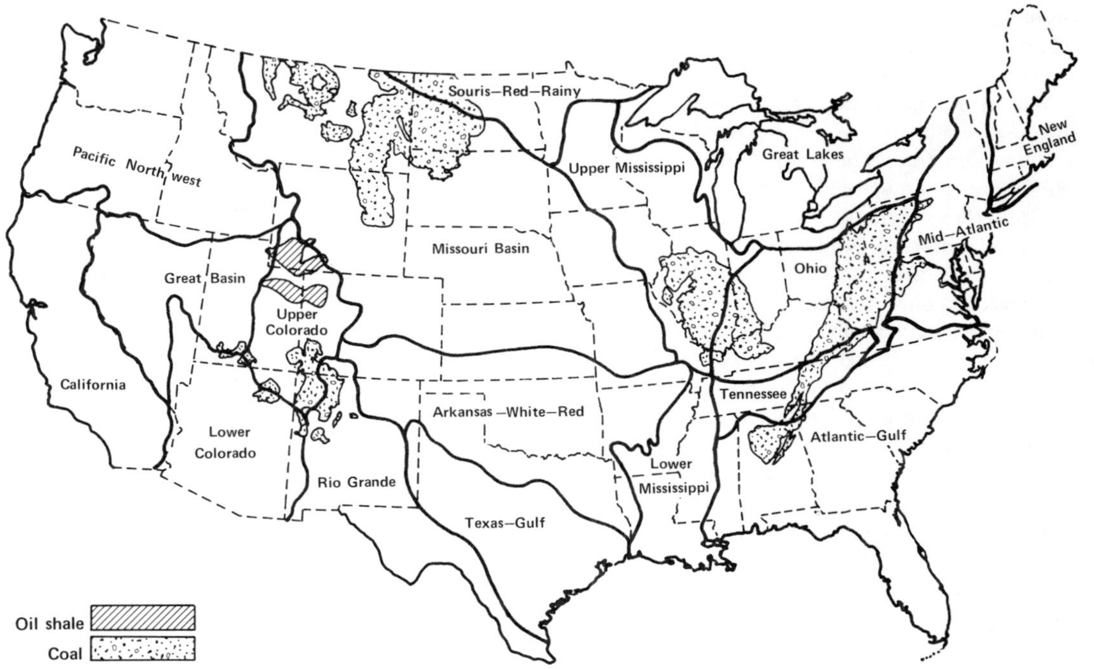

Fig. 1. Map of the conterminous United States showing Water Resources Council regions and major coal and oil shale deposits, adapted from (37, 38, 45). Recently found eastern oil shale deposits of uncertain commercial value are not shown.

tal problem for several reasons. First, the rate at which water is withdrawn provides a rough measure of the rate at which aquatic habitat is temporarily destroyed and aquatic organisms are killed or injured. Organisms, for example, can be killed by entrainment in cooling condensers (3). Second, the larger the withdrawal, the greater is the need for a storage reservoir for operation in times of low flow. Because of the great range and intensity of environmental hazards associated with the damming of rivers to create reservoirs (4, 5), including, though by no means limited to, large consumptive losses caused by excessive evaporation and bottom seepage, the size of withdrawal requirements should not be overlooked in assessing possible future energy sources.

Aggregated supply and demand. A description of the amount of freshwater potentially available to users in the United States should include information about average flows available and also statistical data on regional fluctuations in water supply.

First, let us consider the average, aggregated water supply situation in the 48 conterminous states. Precipitation averages about 5600 cubic kilometers per year (km^3/year), about 70 percent of which either evaporates or is transpired by vegetation before it reaches the oceans (6). The remaining 30 percent, or 1700 km^3/year, is called runoff or stream discharge. Although runoff is the portion of precipitation often considered to be available for human use, it should not be thought of as lost or wasted when not consumed directly by humans; a major part of runoff maintains the health of streams, lakes, and estuaries. Maintenance of this health is likely to be of aesthetic, commercial, and recreational value to man (7) as well as of intrinsic value as ecological habitat. An important issue to which we will return shortly is the question of just what fraction of the runoff can be safely consumed.

The 1975 aggregate water demand in the United States is outlined in Table 1. At first glance, by comparing the averaged annual freshwater runoff of about 1700 km^3/year with the annual consumption of 151 km^3/year, water availability does not appear to be a major problem. Such a conclusion is erroneous, however, because the actual supply and demand of water are highly diverse across time and space. Precipitation and river flow can vary enormously from season to season and from year to year. In much of the West, for example, the precipitation rate for the past 2 years has averaged only about one-half of normal. In addition, time-averaged local runoff is neither distributed uniformly over the United States nor is it distributed in proportion to present-day demand. Finally, the location of many of the country's potential energy resources, such as coal, oil shale, uranium, sunlight, and geothermal energy, in the dry western regions of the country exaggerates the geographic unevenness of future water demand in relation to its supply.

Spatial and temporal variation. The principal water drainage regions of the 48 conterminous states provide a useful starting point for discussing geographic variation in water supply and demand. These regions, 18 in number, are hydrologically distinct entities that are relatively isolated from one another with respect to surface water flow, except for linkages along major rivers (as in the case of the Upper and Lower Mississippi regions) (8). They are shown in Fig. 1. Regional mean annual runoff and 1975 consumption are shown in the first two columns of Table 2; also in Table 2 is a regional breakdown of runoff per person and of consumption per unit runoff. It is interesting that population is distributed more nearly in proportion to runoff than is consumption, a result largely due to heavy irrigation demands in the West. For future reference, note that the major coal deposits in the United States are located in the Missouri, the Upper Colorado, the Upper Mississippi, the Ohio, and the Tennessee regions (see Fig. 1).

The temporal variation of runoff and river flow can also be quite large. A useful statistical quantity, which can be used to describe unusually low flow conditions, is the x-day, y-year low flow. This is defined as the lowest flow rate, averaged over x consecutive days of the year, expected, on the average, every y consecutive years. We denote it by the symbol ${}_xQ_y$ (9). From a table of daily river flow rates over a period of many years, ${}_xQ_y$ is easily computed. One first determines for each year the lowest x-consecutive-day flow rate. For each consecutive y-year period one then takes the lowest of the x-day low flow rates during that period and averages them over all possible consecutive y-year periods for which data are available. Note that $_{365}Q_1$ is the mean annual flow rate.

We have taken five rivers and compiled some illustrative river flow statistics on each (*10*). There are two in the West—the Yellowstone and the Colorado—and three in the East—the Ohio, the Kanawha, and the Wabash. These particular rivers are chosen because they are located in coal-rich regions and, along with nearby and hydrologically similar rivers, are among the likely sources of water for coal-related activities in the United States. Figure 2 shows representative values of $_xQ_y$ at specific U.S. Geological Survey stations on each of the five rivers. These stations were chosen sufficiently upstream to reflect primarily precipitation and watershed conditions, although the presence of man-made storage projects does influence the flows. It can be noticed that the values of $_7Q_{10}$ range from 7 to 16 percent of the mean annual flow. The ratio of 7-day, 10-year low runoff to mean runoff is also roughly in the same range for the 18 hydrological regions (*11*). The ratios of the $_7Q_{10}$'s to the $_7Q_{20}$'s or $_7Q_{40}$'s are fairly uniform from river to river. To the extent that there are variations, it appears that $_7Q_{10}$ is a higher fraction of the mean flow in the West and that $_xQ_y$ is a slightly more rapidly decreasing function of y in the West. Despite the obvious temptation to do so, quantitative extrapolation of curves like these to values of y larger than those for which data are available can be a very ambiguous procedure (*12*).

The importance of the concept of $_xQ_y$ is based on two considerations. First, in siting a water-consuming facility along a river, it is important that not only the mean flow be adequate but also that the actual instantaneous flow be nearly always adequate. The practical meaning of "nearly always" will depend on storage capacity and acceptable shutdown time when drought conditions prevail. For a given acceptable amount of shutdown, knowledge of the $_xQ_y$'s allows the minimum storage capacity to be determined. For example, consider a facility that consumes river water at a rate C. In order for the facility to continue operation through a particular $_xQ_y$ low flow period, where $C \geq {_xQ_y}$, the required storage capacity would have to exceed $(x)(C - {_xQ_y})$. It is clear that when C only slightly exceeds $_xQ_y$, even a small percentage increase in the consumption rate can necessitate the construction of greatly increased storage.

Second, ecological considerations point to the importance of these statistical parameters. Adequate river flow is necessary for maintaining riparian and estuarine habitats. The flushing and transport of minerals and organic materials, the dilution of pollutants, the maintenance of adequate oxygen levels, and the thermal structure of rivers and estuaries are dependent upon the magnitude and timing of river flow (*13*). The taxonomic diversity of river zooplankton (*14*) and the ability of benthic organisms to secure nest sites in river bottoms (*13*) also are flow rate dependent. Lower than normal flows during any stage of the annual flow cycle can cause significant loss of aquatic habitat and increase the level of toxic substances. The river water may become undesirably hot and oxygen levels may drop. Adverse impacts on fish populations are reported (*15*). Hynes (*13*) has thoroughly described the role of river current in maintaining ecological balance as well as the sensitivity of river organisms to a prolonged decrease in river flow. In estuaries, as in rivers, the time dependence and magnitude of freshwater flow is also critical. The life cycles of numerous estuarine organisms are intimately linked to the circulation of freshwater, which in turn is linked to river inflow. Moreover, pollution levels in estuaries are regulated in part by river flow (*16*).

Table 3. Some useful energy quantities for the United States in 1975. Data adapted from (*25*).

Energy category	Energy (10^{18} joule/year)
Total energy consumption	72.7
Liquid fuels consumption	34.5
Natural gas consumption	21.3
Coal consumption	14.1
Steam-generated electricity output	6.1
Energy yield from 1 km² average western surface-mined coal	0.1 to 0.2
Annual average sunlight on 1 km²	0.0056

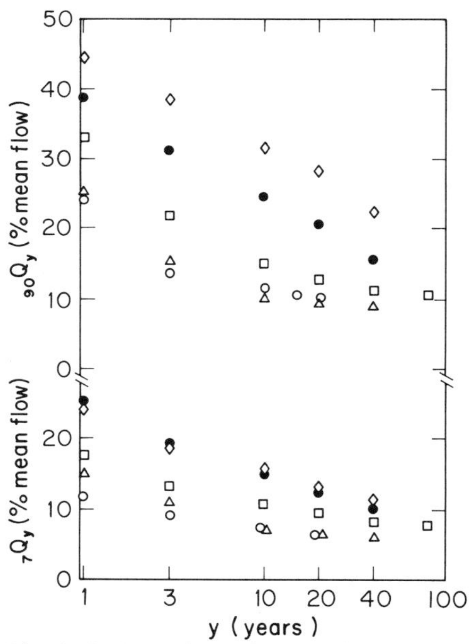

Fig. 2. Values of $_7Q_y$ for five rivers in the United States. The rivers and the U.S. Geological Survey measuring station at which the data were taken from which these flows were calculated are: the Yellowstone River (◇) at Miles City, Montana; the Colorado (●), near Cisco, Utah; the Kanawha (□) at Kanawha Falls, West Virginia ; the Wabash (△) at Mount Carmel, Illinois; and the Ohio (○) at Huntington, West Virginia. The mean annual flows of these five rivers at the designated measuring stations and the period of time over which the daily measurements were taken were as follows: Yellowstone, 10.4 km³/year, 1930 to 1977; Colorado, 6.9 km³/year, 1924 to 1976; Kanawha, 11.2 km³/year, 1878 to 1976; Wabash, 24.1 km³/year, 1929 to 1976; and Ohio, 67.2 km³/year, 1933 to 1968 (*10*).

Water consumption criteria. To protect rivers and estuaries from excess consumption of runoff, criteria must be developed to evaluate how much decrease in natural water flow can be permitted. One type of criterion might allow a fixed percentage of the mean flow to be consumed. The problem with this can be seen by referring to Fig. 2. Suppose that consumption were limited to 15 percent of the mean flow. Then on the Upper Colorado this would allow total depletion of river flow on the average for 90 consecutive days every 40 years, or 7 consecutive days every 12 years. In contrast, the flow of the Wabash could be totally depleted for 90 consecutive days every 3 years, or 7 consecutive days every year. Thus, the criterion would have very different implication for the two regions, both for consumers and ecosystems. On the other hand, if the criterion were formulated in terms of an allowed percentage of some x-day, y-year low flow, where $y > 1$ and $x < 365$, then the criterion would account better, although by no means perfectly, for supply limitations and ecological impacts intrinsic to the hydrological characteristics of the two regions. For example, if consumption were limited to, say, 40 percent of the $_{90}Q_3$ low flow, then flow would be totally consumed on both the Upper Colorado and Wabash rivers for about 7 consecutive days every 30 years.

These considerations, combined with the fact that organism tolerances to stress are often limited to days or weeks rather than to years (*13*), suggests that limits to consumption be based on a percentage of some x-day, y-year low flow.

In a paper on water requirements and water consumption criteria for electric power-plant cooling, Samuels (*11*) reviewed water flow data in the United

States and proposed ecological criteria for permissible water use by nuclear power plants. From these criteria, Samuels then identified those rivers where five or more 1200 megawatt electric (MWe) nuclear plants could be located. Samuels' criteria would permit water use for nuclear power up to a fixed percentage of certain ${}_xQ_y$'s. For rivers without significant water storage facilities, two of his criteria would allow consumption of up to 10 percent of ${}_7Q_{10}$ and withdrawal of up to 15 percent of ${}_7Q_{10}$. Samuels' criterion for rivers with storage assumes that the storage is for seasonal variations only, and states that consumption should not exceed 10 percent of ${}_{365}Q_{20}$. Note that this standard is considerably more lenient than that for the no-storage case.

If the aim is solely to ensure that a shortfall in industrial water supply does not occur too frequently, then this is a reasonable way to determine a standard for rivers with storage. But the ecological effects of storage facilities on a river can vary greatly depending on how the outflow from the reservoir is managed. For a given level of consumption, the presence of a storage facility does not necessarily ensure that downstream flow will approximate the natural flow better than if there had been no storage. For example, if a reservoir is managed for hydroelectric power, the downstream flow will tend to be more uniform throughout the year than the natural flow, leading to increased flows during periods of normally low flow (usually late summer and fall). These increased flows can be destructive to bottom-living organisms that rely on low flow periods to secure their nest sites on bottom materials and can interfere with the incubation habits of certain fish species (*13, 15*). If the reservoir is managed for highly consumptive use at the storage site, then during prolonged dry periods, a self-interested manager might eliminate downstream flow entirely in order to prolong use of the water supply.

On the basis of these considerations, we will not attempt to develop different criteria for regions with differing levels of storage. In our scenario analysis, we simply compare total regional freshwater consumption (from total present use and possible future energy-related activities) in each region with the estimated ${}_7Q_{10}$ low flow in that region. Our decision that x should be 7 days was based on the considerable evidence that aquatic organisms often can tolerate several days of stress but not weeks or months (*13*). Figure 2 indicates that the choice of y is relatively insensitive to regional differences in that the ratios of the ${}_7Q_{10}$'s to the ${}_7Q_{20}$'s are fairly uniform from river to river. Rather than dignify any particular percentage of ${}_7Q_{10}$ as an acceptable level of consumption, we simply compare consumption with this low-flow quantity. Because we express our estimates of water consumption in our scenarios in absolute terms as well as in terms of percentages of ${}_7Q_{10}$ for each region, the interested reader can apply any desired criterion in order to assess the constraints of water supply on various levels and kinds of energy development.

Water Consumption Requirements of Energy Alternatives

In this section we estimate the water requirements for a variety of energy technologies that are candidates for major expansion in the United States. Based on these estimates, judgments are given about the relative impacts on water resources of some technologies which are competitive in the sense that they could provide similar benefits to society. We adopt as an energy reference the quantity, 10^{18} joules. Note that the commonly used unit of energy called the quad (10^{15} British thermal units, Btu) is approximately equal to 1.05×10^{18} joules. Some energy quantities, pertinent to the following discussions, are listed in Table 3.

Coal and oil shale: Mining, reclamation, and conversion to synthetic fuels. Coal and oil shale are the major fossil fuel resources of the United States and potentially form the base for a large and long-lasting energy supply. Most of the explored coal and oil shale deposits are found in five water resource regions (see Fig. 1); about 50 percent of the total recoverable coal reserves and 30 percent of the surface-mineable reserves are in the Ohio and Upper Mississippi regions, which are also close to major demand centers. The remaining coal resources (which happen to be more attractive commercially) and the principal oil shale reserves are found in sparsely populated, arid or semiarid areas of the West, with the oil shales confined to a far smaller region than the coal.

Estimates of water consumption for various shale- and coal-conversion pathways are given in Table 4 (*17*). Clearly the major part of the water consumption occurs at the conversion stage itself. Two other major categories of water consumption are reclamation of surface-mined land and coal transport via slurry pipelines. Unless the water used in a slurry pipeline is adequately treated and returned to its source, it must be considered to be consumed water in its region of origin.

Among the entries in Table 4, one with an especially large range of uncertainty is that for land reclamation in the West. This uncertainty is mostly due to a lack of understanding of environmental factors such as soil-binding properties and the conditions under which detrital and soil microorganism–based nutrient cycles can be reestablished in dry, disrupted terrain (*18*). The uncertainties include not only the unknown requirements for annual irrigation but also the unknown number of years for which irrigation would be necessary for reestablishing a viable ecosystem. Successful revegetation (though not necessarily restoration to original conditions) is likely to be necessary in order to reduce problems of erosion, mine drainage (with subsequent deterioration of downstream water quality), and possibly flooding. The lower value given in Table 4 in our judgment has a low probability of leading to genuine revegetation (*18, 19*). We have not attempted to include in our estimates the additonal water consumption resulting from secondary impacts of erosion, drainage, or flooding, should land reclamation be unsuccessful.

Table 4 shows the water consumption for converting shale to be smaller than the consumption for syncrude production from coal. However, the listed ranges of water consumption mean very different things in the two cases, and the actual situation could turn out to be more complicated than these numbers indicate. The effects of coal mining have long been recognized. Mine drainage, soil erosion, and alteration of runoff characteristics are among the important ones. These effects are also expected from shale mining. However, mining of oil shale results in a volume of processed shale that is about 1.2 times greater than the raw shale. The resulting difficulty of storing the wastes in the excavated areas has led to proposals to use natural canyons as storage space for spent shale. Such action would lead not only to permanent loss of many canyon lands but also to the destruction of natural habitats, many of which are homes for a number of rare and endangered species (*19, 20*), and to an alteration of the hydrologic regime of the region. Furthermore, the stability of the spent shale when subjected to precipitation and snowmelt is questionable (*21*).

These, and also economic, considertions have directed attention toward in situ technology for extracting oil from shale. One water-related problem with in situ processes is particularly worrisome.

The significant shale deposits of the Piceance Basin in Colorado are in themselves an integral part of the mechanism by which groundwater quality and flow are naturally maintained (*22*). A disruption of this system could affect the flow and quality of the White River and ultimately the Green and Colorado rivers by causing the release of artesian, saline groundwater into freshwater systems. Our listed range of uncertainty in Table 4 does not cover the case of aquifer disruption leading to alteration of the White, Green, and Colorado rivers; it includes only the more narrow range of water requirements associated with the range of technological options.

Problems affecting water availability and quality in the Upper and Lower Colorado regions are already serious. Overallocation of water, low-flow conditions, salinity, and erosion are well recognized (*18, 20–22*). An oil shale industry, whether based on surface or deep mining, aboveground or in situ retorting, poses the risk of serious ecological impacts in its competition for water. The geographic confinement of commercially attractive and explored oil shales to this region is in contrast to coal, which is found in significant quantities across a spectrum of meteorological, topographical, hydrological, and ecological conditions. In choosing between coal and oil shale, the greater flexibility of coal mining sites and uncertainties about aquifer disruption from oil shale activities must be considered along with the numbers in Table 4.

Cooling requirements for steam electric plants. The freshwater required for the major ways of cooling steam electric power plants is listed in Table 5. The listed range of requirements reflects variation in regional evaporation rates, differences in the temperature to which the cooling water is heated, and some uncertainties arising from the complex mechanisms by which open water dissipates heat. The thermal efficiency of the electric generating system is assumed to be 38 percent, typical of a modern coal-burning plant.

Table 5 shows that the use of wet cooling towers is not necessarily preferable to once-through cooling. Wet tower cooling reduces the withdrawal requirements, while once-through cooling reduces consumption requirements, provided that additional water storage is not needed to meet withdrawal needs of a once-through system. In areas where water is scarce and river flow is variable, the large withdrawal needs of a once-through system may not be met without providing for additional storage. If storage must be added with a once-through cooling system, then wet cooling is preferable. In this circumstance, wet tower cooling not only reduces water consumption but also avoids problems of thermal pollution in aquatic habitats, as well as the many ecological hazards associated with damming free-flowing rivers (*5*). In circumstances where additional storage is not required but water consumption is a problem (for example, western lakes), the once-through method may be preferable (*23*).

Table 4. Water consumption for the production of synthetic fuels from coal and oil shale in the United States. Data are expressed as cubic kilometers per 10^{18} joules of synthetic fuel product. The data and the references were derived from (*19*). All calculations are based on coal energy content of 28, 22, and 14 million joules per kilogram of bituminous, subbituminous, and lignite coals (*36*) and on conversion efficiencies of 67 to 85, 55 to 67, and 41 to 75 percent for low- and high-Btu gasification and liquefaction, respectively (*37–39*).

	Category of use						
	Mining*	Reclamation†	Transport by slurry pipelines‡	Conversion§	Associated urban‖	Total with slurry pipelines	Total without slurry pipelines
			Low-Btu gas				
Eastern coal:							
Surface-mined	0.0028 to 0.0035	0.0 to 0.030	0.045 to 0.057	0.083 to 0.58	0.018	0.15 to 0.69	0.10 to 0.63
Deep-mined	0.0062 to 0.0078	0.0	0.045 to 0.057	0.083 to 0.58	0.018	0.15 to 0.66	0.11 to 0.61
Western coal:							
Surface-mined	0.0028 to 0.0070	0.0028 to 0.14	0.045 to 0.11	0.083 to 0.58	0.018	0.15 to 0.86	0.11 to 0.74
Deep-mined	0.0062 to 0.010	0.0	0.045 to 0.11	0.083 to 0.58	0.018	0.15 to 0.72	0.11 to 0.61
			High-Btu gas				
Eastern coal:							
Surface-mined	0.0035 to 0.0042	0.0 to 0.036	0.057 to 0.069	0.083 to 0.58	0.049	0.19 to 0.74	0.14 to 0.67
Deep-mined	0.0078 to 0.0095	0.0	0.057 to 0.069	0.083 to 0.58	0.049	0.20 to 0.71	0.14 to 0.64
Western coal:							
Surface-mined	0.0035 to 0.0085	0.0036 to 0.17	0.057 to 0.14	0.083 to 0.58	0.049	0.20 to 0.95	0.14 to 0.81
Deep-mined	0.0078 to 0.012	0.0	0.057 to 0.14	0.083 to 0.58	0.049	0.20 to 0.7	0.14 to 0.64
			Syncrude				
Eastern coal:							
Surface-mined	0.0031 to 0.057	0.0 to 0.048	0.051 to 0.093	0.11 to 0.74	0.029	0.19 to 0.92	0.14 to 0.82
Deep-mined	0.0070 to 0.013	0.0	0.051 to 0.093	0.11 to 0.74	0.029	0.20 to 0.88	0.15 to 0.78
Western coal:							
Surface-mined	0.0031 to 0.011	0.0032 to 0.23	0.051 to 0.19	0.11 to 0.74	0.029	0.20 to 1.2	0.14 to 1.0
Deep-mined	0.0070 to 0.017	0.0	0.051 to 0.19	0.11 to 0.74	0.029	0.20 to 0.98	0.14 to 0.79
			Oil from shale				
Surface technology							
Surface-mined	0.0040 to 0.0056	0.033 to 0.053	NA	0.030 to 0.044	0.0069 to 0.0092	NA	0.074 to 0.11
Deep-mined	0.0041 to 0.0056	0.032 to 0.056	NA	0.030 to 0.044	0.0082 to 0.011	NA	0.074 to 0.12
In situ technology:							
Modified in situ	0.0019 to 0.0026	0.014 to 0.030	NA	0.027 to 0.047	0.0087 to 0.010	NA	0.052 to 0.090
True in situ	NA	0.0 to 0.0077	NA	0.0 to 0.044	0.0088 to 0.010	NA	0.009 to 0.062

*In the East, surface and deep mining consume 2.3 and 5.2 $m^3/10^{12}$ joules of coal mined. In the West, consumption is 2.3 to 4.7 and 5.2 to 6.8 m^3 per 10^{12} joules mined, respectively (*40*). †In the East, land disturbance is 22 to 65 m^2 per 10^2 joules of coal mined (*37*) and annual water consumption is 0 to 0.015 m^3/m^2 over a 1- to 2-year period (*19*). In the West, the corresponding figures are 3.9 to 31 m^2 per 10^{12} joules of coal mined (*37*) and 0.30 to 0.61 m^3/m^2 over 2 to 5 years (*19*). The shale estimates include consumption for revegetation as well as processed shale disposal (*41*). ‡Slurry pipelines consume 38 and 37 to 76 m^3 per 10^{12} joules of coal mined in the East and the West, respectively (*42*). §For coal conversion see (*42, 43*); for shale extraction see (*41*). ‖For coal conversion see (*38*); for shale see (*41*).

Lest it appear that dry cooling is an unqualified blessing because of savings in water, we note that a coal-burning, dry-cooled, electric generating plant is likely to have a thermal efficiency of about 1½ percentage points lower than a plant with a once-through or wet tower cooling (*24*). Thus more fuel will be required for a given electric output, and extra water will be consumed for mining and land reclamation. Consider, for example, two electric power plants producing the same electric output from western surface-mined coal. Assume that one operates at 38 percent efficiency and employs wet tower cooling, while the other operates at 36½ percent efficiency and is dry-cooled. From Tables 4 and 5 it can be calculated that the dry-cooled plant indeed leads to less total water consumption than the water-cooled plant. The dry-cooled plant would consume an additional 0.0005 to 0.0095 $km^3/10^{18}$ joule electric at the mine site whereas the wet tower cooled system would consume an additional 0.4 to 0.6 $km^3/10^{18}$ joule electric at the power plant. However, one must consider that the additional water use at the mine may be environmentally more critical in terms of a possible shortage in local water supply.

Coal and uranium for electricity. Coal and uranium are often viewed as alternative sources of energy for future electric generation. Although these are not the only candidates for meeting future demand for electricity, it is nevertheless interesting to look at the coal-nuclear issue from the perspective of water resources.

The cooling required for a light-water reactor (LWR) is considerably greater than that for a modern coal- or petroleum-fueled plant producing the same electric power. Consider, for example, an LWR with a typical thermal conversion efficiency of 33 percent and a fossil-fuel plant operating at 38 percent efficiency (*24*). For the same power output this difference in efficiency results in the release of about 24 percent more waste heat by the LWR. Because a nuclear plant releases all but from 0 to 5 percent of its waste heat through its cooling condensers, whereas a coal-burning plant typically releases 15 to 20 percent of its waste heat directly into the atmosphere with flue gas (*24*), the LWR actually requires about 39 to 50 percent more cooling water than does the fossil plant. Together, these differences cause an additional consumptive loss of water by the LWR of 0.16 to 0.30 $km^3/10^{18}$ joule electric as compared to the fossil fuel plant, if both employ wet tower cooling. Moreover, with once-through river water cooling, the need for storage reservoirs for the LWR will be greatly increased because of the 39 to 50 percent increased water withdrawal requirement.

Table 5. Water requirements for electric power-plant cooling. Data are expressed as cubic kilometers of water per 10^{18} joules of electric output. It is assumed that the thermal efficiency is 38 percent and that 17 percent of the waste heat is dissipated directly to the atmosphere in the form of hot stack gases (*23*).

Cooling mode	Withdrawal	Consumption
Once-through (no storage)	28.0 to 40.0	0.2 to 0.4
Once-through (storage*)	28.5 to 41.5	0.5 to 1.5
Wet cooling tower†	0.6 to 0.8	0.4 to 0.6

*Reservoir capacity is assumed to meet backup storage requirements of 1000 MWe-sized plants for 90 days; lake surface evaporative loss is assumed to be in the range .75 to 1.5 meters per year. For further assumptions, see King (*23*). †Wet tower consumption is the sum of evaporative loss plus drift; withdrawal is equal to consumption plus blowdown.

The future water needs for uranium mining, even on a per-unit-energy basis, are difficult to predict because of uncertainty over available reserves of high-grade uranium ore. Today, uranium fuel can be obtained from ores containing 50 times the energy content of coal per unit weight (*25*). As long as such rich sources of uranium fuel are available, the water required for uranium mining and reclamation will be considerably less than they are for surface mining of coal. But as these rich supplies dwindle, nuclear reactors will require the use of low-grade ores. One possible ore, the Chattanooga shale, has an energy content per unit weight roughly twice that of coal (*25*). Should such ores be mined, their geographic location and depth would be decisive factors in comparing impacts of water requirements of coal and uranium mining.

One worrisome possibility is that the last remaining rich supplies of uranium ores might happen to lie either in areas of special ecological value or in regions of especially scarce water. Economic pressure to exploit these supplies might be difficult to resist. In contrast, coal, being more widespread geographically, would then offer a wider choice of mining sites. These issues and the actual water impacts of uranium mining will become clearer when the amount and distribution of uranium fuel reserves become better known.

Taking into account the entire fuel cycle, we may question how coal and nuclear electric generation today compare with respect to water consumption. On the one hand, coal stripping and reclamation require an additional 0.004 to 0.09 $km^3/10^{18}$ joule electric of water compared to uranium mining. On the other hand, nuclear plants which are wet tower-cooled required 0.16 to 0.30 $km^3/10^{18}$ joule electric more water than a coal-fired plant. Thus the nuclear plants are more freshwater-intensive. With dry cooling or seawater cooling, the situation is, of course, reversed.

Future efficiencies of power plants are quite uncertain. Pollution control equipment on fossil-fuel plants could reduce their efficiency, but fluidized-bed combustion could eventually provide ways for controlling emissions at efficiencies higher than today's coal-burning plants (*26*). The breeder reactor is likely to have a higher efficiency than the LWR. And finally, cogeneration of process steam and electricity, combined-cycle, fossil-fuel plants, and development of uses for waste heat will make the water bookkeeping more complicated than presented here.

The solar options. The various solar energy technologies differ greatly with respect to water consumption. Because solar radiation is most intense and most predictable in parts of the United States where runoff is lowest and least predictable, water impacts of solar energy technologies must be thoroughly examined.

Several solar options for electricity generation are attractive on this score because the only water they would consume would be that used during the manufacturing of materials and the installation and maintenance of operating facilities (*27*). Wind energy is an example, because wind-generated electricity requires no cooling water. Certain methods of photovoltaic conversion provide other examples. Among the solar thermal conversion systems that have been suggested, either open-cycle Brayton generation (gas turbine) or Rankine-cycle conversion with dry cooling towers would require minimal amounts of water. Although thermal generation of electricity by solar energy is likely to be less than 20 percent efficient, the steam cycle should operate at about the same efficiency as a fossil-fuel plant and therefore water consumption for wet tower or once-through cooling will be approximately the same as for coal-fired plants (*28*).

Bioconversion is a possible means of producing gaseous and liquid fuels. One of the most efficient crops for energy plantations is sugar beets, which could have an annual yield of 10^{18} joules on about 8000 km^2. On this basis, approximately 8½ percent of the land area of the conterminous 48 states would be required to meet all current U.S. energy needs, provided that this land were sufficiently irrigated and fertilized and had high insolation and warm temperatures. Irrigation requirements alone for such a crop are estimated to be 10 $km^3/10^{18}$ joules of biomass (*29*), about half of which would be consumed. Table 1 indicates that the water consumed in meeting the current U.S. annual energy demand of 80×10^{18} joules by bioconversion would exceed *all* current water consumption in the United States by almost a factor of 3. If such bioconversion plantations were located in the Southwest, as would be favored by factors such as climate and land availability, their annual water withdrawal requirements would exceed the mean annual runoff of all rivers in the conterminous United States west of the Mississippi. Evidently, such plantations could be maintained only by a massive system of water imports. Bioconversion schemes using artificial ponds for freshwater algal culture would result in comparable water consumption on a per-unit-energy basis unless evaporation-preventing protective covers were used.

On a smaller scale, however, biconversion systems designed to process agricultural or feedlot wastes or designed in tandem with sewage treatment facilities could actually have a net beneficial effect on water resources, and could make small but useful contributions to U.S. energy needs.

Solar rooftop panels and passive systems for domestic and commercial heating appear quite favorable from the viewpoint of water conservation. Indiscriminate cutting down of trees in the vicinity of houses could lead to greater household water consumption for maintenance of lawns and low shrubs, but coordinated efforts of landscape architects and solar engineers should avoid such problems.

How should coal be used to heat homes? Coal can be used to heat homes directly or by conversion to synthetic fuels or electricity. Direct heating is not environmentally acceptable. Establishment of a major synthetic fuel industry is likely to require massive amounts of natural resources, and it is therefore imperative that a careful assessment of the consequences of such an industry be made. An assessment procedure that avoids some pitfalls of cost-benefit analysis is to estimate and compare the environmental impacts and the consumption of natural resources which will accompany the provision of a given measure of a particular end use via alternative technologies. Here, we compare the amounts of water consumed for two home-heating methods, one using electricity produced from coal and the other using synthetic high-Btu gas produced from coal.

Electricity production is less efficient than synfuel production. However, there is also a considerable gap between the number of joules of electricity and of gas required for space heating at the point of use. This is illustrated in the first two columns of Table 6 which show the energy requirements, at the point of use, for electrically heated and gas-heated model unit houses in two locations, as developed by the Federal Energy Administration (FEA) (*30*). The difference between the number of joules of electricity and gas is attributed to the lower system-efficiency of gas-heated homes. First, at the point of conversion to heat, the gas furnaces of today are less efficient than electric heaters, the difference being about 20 percent. Second, and more important, is the higher heat loss rates in gas-heated homes today, arising from duct and ventilation losses. Electricity also allows for individual zonal or room thermostat settings, in contrast to most gas-heated homes. It is quite difficult to predict improvements in the efficiency of coal-conversion plants or electric generating plants, and also of home-heating systems. In principle, one can build homes so that human warmth and electric lighting suffice for space heating. Concern over indoor air pollution may influence progress toward this ideal by gas-heated homes.

We show our results for three cases in Table 6. In case A, which is our worst case from the viewpoint of water consumption, cooling is carried out by the once-through method with storage; minimal water conservation and treatment is assumed in the production of synfuels (see Table 4); and home insulation and

Table 6. Water and energy consumption for home heating by synthetic gas and electricity derived from coal. Case A denotes little or no conservation of energy or water. Case C represents the other extreme, while case B is intermediate.

Region	End-use energy consumption (10^9 joule/house/year)*		Coal consumption (10^9 joule/house/year)		Water consumption (m^3/house/year)			
					Gas§		Electricity‖	
	Gas	Electricity	Gas†	Electricity‡	Surface mining	Deep mining	Surface mining	Deep mining
				Case A				
East	220	79	390	230	150 to 160	152	88 to 100	89 to 100
West	120	52	210	150	83 to 100	83	58 to 77	58 to 63
				Case B				
East	160	72	280	210	66 to 72	67	25 to 40	26 to 36
West	86	47	150	140	36 to 50	36	16 to 34	16 to 21
				Case C				
East	60	26	97	78	8.8 to 11	9.1	0.77 to 6.2	0.99 to 4.9
West	60	26	97	78	9.0 to 18	9.1 to 9.2	0.64 to 11	0.68 to 3.4

*These are estimates of the energy to be delivered to a single-family, one-story detached house for the purpose of space heating. Cases A and B are based on synthesized (model) demand (*30*). In case A, the demand reflects 1970 conditions. Case B is based on projected reductions of, respectively, 28 and 9 percent in gas and electricity consumption per home (relative to 1970). In case C, the house is designed according to NEMA standards (single thermostat) with net heating requirements amounting to 52×10^9 joules/year and 60×10^9 joules/year for the electric- and gas-heated home, respectively (*30*). The homes are equipped with a gas or an electric heat pump of equal coefficients of performance (COP =2). The gas heat pump has a mechanical efficiency of 33 percent, but half of the heat not converted is recovered. In cases A and B, East denotes a Michigan house dependent on eastern coal, while West refers to a New Mexico location fueled with western coal. East and West in case C denote only the source of coal. †Based on regional distribution and pipeline transport (1600 km average) losses of 0.7 and 7 percent, respectively (*44*). In cases A and B, efficiency of conversion (to high-Btu gas) is assumed to be 61 percent. In C, the efficiency is 67 percent (*19*). ‡A transmission loss of 8.6 percent is assumed (250 km) (*44*). Power-plant thermal efficiency is 38 percent of cases A and B and 36.5 percent for case C (dry tower cooling) (*24*). §Slurry piplines are not included. In case A, conversion water consumption is 0.58 m^3 per 10^9 joules of gas (at the plant). In case C, it is 0.083 $m^3/10^9$ joules. Case B assumes the mean of these two values. Other assumptions are the same as in Table 4 (high-Btu gasification). ‖Cooling by once-through in cases A and B (A uses storage, B does not) and by dry tower in case C. Water consumption estimates include mining and reclamation (see Table 4), cooling (1.0, 0.3, and 0.0097 m^3 per 10^9 joules electric for cases A, B, and C, respectively) [Table 5 and (*24*)], coal cleaning (0.012 to 0.062 m^3 per 10^9 joules electric) in the East and none in the West (*35*), and air pollution control (0 to 0.10 m^3 per 10^9 joules electric) (*24*) (all joules electric refer to the power plant).

heating appliances are typical of those in use today. In case B, cooling is carried out by the once-through method without storage; water consumption in synfuel production is assumed to be midway between the worst and best cases (see Table 4); and home insulation and home-heating appliances are taken from FEA estimates (*30*) of improved 1990 homes. In case C, dry cooling is employed; maximal water treatment and conservation is assumed in the production of synfuels; home insulation is superior to case B (*30*); and home heating is carried out with heat pumps, with one-half of the waste heat from the gas-fired heat pump captured and used in the home.

In case C, which minimizes water consumption, both coal and water needs for home heating are sufficiently low that resource considerations would probably not be an important factor in deciding between the electric and the synfuels path. Where they could be an important factor, in either of the first two cases, the electric path appears to be superior to the synfuels path. From the perspective of water consumption, the use of active or passive solar space heating would be preferable to either of these coal paths.

Table 7. Scenarios for future energy development in the United States based on the use of coal.

Scenario	Total annual U.S. coal consumption (10^{18} joule/year)	Coal mining distribution*	Slurry pipeline† (10^{18} joules/year)	Total annual coal conversion to synfuels‡ (10^{18} joule synfuel/year)	Conversion distribution§
1	16	E	0	0	NA‖
2	16	W	0	0	NA
3	32	E	0	4	B
4	32	W	0	8	A
5	32	W	4	8	B
6	32	E	0	8	B
7	32	W	0	16	A
8	48	W	0	16	A
9	48	W	4	16	B
10	48	E	0	16	B
11	48	W	0	0	NA
12	64	W	8	32	B

*All coal is divided into two classes, eastern and western, where Illinois and other midwestern coals are included under eastern. Coal mining distribution plan W assumes 75 percent of the coal is western and 25 percent is eastern. Plan E assumes 25 percent is western and 75 percent is eastern. All the western coal is surface-mined, while eastern coal is assumed to be half surface-mined and half deep-mined. †Slurry pipeline refers to the use of western water to transport coal away from western mine sites. ‡Total coal conversion produces 50 percent high-Btu gas by energy content and 50 percent liquid syncrude. §In conversion distribution plan A, 50 percent of the conversion is in the East and 50 percent in the West; in plan B, 75 percent is in the East and 25 percent is in the West. ‖Not applicable.

Analysis of Energy Scenarios

Here we describe the demands on freshwater resources which are likely to arise as a consequence of an expansion of certain energy activities in the United States. We specify future energy development in this country by a series of scenarios prepared on a regional basis. These scenarios are intended to portray possibilities, not projections or predictions. Moreover, they do not specify all aspects of future energy development, but only those pertaining to electric generation cooling requirements, in one set of four scenarios, and coal mining, land reclamation, slurry pipeline, and conversion of coal to synthetic fuels, in another set of 12 scenarios.

The numerical specifications of the two sets of scenarios are given in Tables 7 and 8. We assume that the water for mining, land reclamation, slurry pipeline, and coal gasification and liquefaction that take place in the West will be drawn from two hydrologic basins: the Missouri (which includes the Powder River Basin) and the Upper Colorado region. We further assume that water for the coal activities in the East will be drawn from three basins: the Upper Mississippi, the Ohio, and the Tennessee. Table 2 indicates that these three eastern regions yield over 30 percent of the mean runoff in the entire eastern United States. Within the two western regions, and separately within the three eastern regions, we assume optimum geographic matching of water supply and demand in the way that would occur if interregional planning lumped the three eastern regions together and the two western regions together.

In Table 8 we estimate only the water consumption required to meet cooling needs of electric power plants because we do not wish to specify the mix of fuel sources used to produce the electricity, and because the water required for other phases of a fuel cycle generally will be obtained outside the region in which the electricity is produced. While some of the cooling required for future electric power plants is assumed to be met with seawater, not all regions have access to oceans. Therefore, the electric power specified in the scenarios as not produced with seawater cooling (three-fourths of the total) is assumed to be distributed among the regions in proportion to the amount of power now produced in each region by freshwater cooling.

Table 8. Scenarios for future energy development based on electricity production.

Scenario	Total steam-generated electricity (10^{18} joule /year)	Cooling mode*
1	12	A
2	12	B
3	30	A
4	30	B

*Cooling mode A refers to the following mix of cooling methods: once-through cooling (no storage), 25 percent; once-through cooling (storage), 25 percent; wet cooling towers, 25 percent; seawater cooling, 25 percent. In cooling mode B we assume: once-through cooling (no storage), 15 percent; once-through cooling (storage), 10 percent; dry cooling towers, 50 percent; seawater cooling, 25 percent. Included in the range of uncertainty for power-plant cooling requirements will be a range of thermal efficiencies varying from 33 to 38 percent. We further assume that the fraction of waste heat released directly to the atmosphere ranges from 0 to 20 percent.

Table 9 shows the estimated water consumption for the coal scenarios, as listed separately for the eastern and western regions. This consumption is expressed in two ways: (i) as the absolute amount consumed and (ii) as the ratio of the sum of present-day water consumption plus anticipated coal-related water consumption to the low-flow parameter, ${}_7Q_{10}$, for the eastern and the western regions (*31*).

The water problems arising from future coal activities are of a somewhat different nature in the East and West. In the East, water consumption for a major coal-conversion industry becomes a large fraction of total present-day consumption (3.4 km³/year, in 1975). The large relative increase in water consumption for the East would pose problems for water allocation management and could have major economic repercussions. Moreover, in the East the sum of present-day consumption plus the additional water consumption for coal activities would approach the ${}_7Q_{10}$ flow in the scenarios for intensive coal use, and thus the eastern regions will become vulnerable to drought. Before setting forth on a course of massive development of a coal-conversion industry in the United States,

Table 9. Water consumption in coal scenarios. For meaning of East and West, see text. In 1975, total water consumption for all users in the areas of the United States we have denoted East and West was 3.4 km^3 and 27.4 km^3, respectively; $_7Q_{10}$ in East and West is 22 km^3/year and 14 km^3/year, respectively; and mean annual runoff is 317 km^3/year and 93 km^3/year, respectively.

Scenario	Coal-related water consumption (km^3/year)		1975 consumption plus additional coal-related consumption as percentage of $_7Q_{10}$	
	East	West	East	West
1	0.046 to 0.18	0.019 to 0.38	16	196
2	0.015 to 0.056	0.056 to 1.12	16	196 to 203
3	0.48 to 2.4	0.17 to 1.45	18 to 26	197 to 206
4	0.55 to 2.9	0.63 to 5.0	18 to 29	200 to 231
5	0.81 to 4.3	0.52 to 3.9	19 to 35	199 to 223
6	0.87 to 4.5	0.30 to 2.1	19 to 36	198 to 211
7	1.1 to 4.7	1.2 to 7.8	10 to 41	204 to 251
8	1.1 to 5.8	1.2 to 9.0	20 to 42	204 to 260
9	1.6 to 8.5	0.85 to 6.4	23 to 54	202 to 241
10	1.7 to 8.8	0.49 to 3.9	23 to 55	200 to 223
11	0.046 to 0.17	0.17 to 3.4	16	197 to 220
12	3.2 to 17	1.6 to 10	30 to 93	207 to 267

it would be important to explore further the implications of this finding for freshwater and estuarine ecosystems and for present and future human activities that depend on reliable freshwater supplies.

In the West, present-day consumption of water is already a large fraction of $_7Q_{10}$, and even of total runoff. Because the West is already vulnerable to drought, the additional water consumption for scenarios with intensive coal use would greatly exacerbate the existing problem of competition for water rather than create, as in the East, new kinds of problems. It is possible that water for future coal-related activities in the West will be diverted from present consumers of freshwater, in particular from crop and livestock growers (*32*).

We have constructed the scenarios in such a way as to highlight some of the trade-offs that are possible in the production of coal. Table 9 shows that if consumption were at the upper end of the range of uncertainty, major water consumption problems would arise in the East, or the West, or in both, as total synfuel production approached 8×10^{18} joule/year (one-seventh of the present use of oil and gas in the United States). But even if maximum water conservation and water treatment efforts were made in coal conversion (including dry cooling), and if western land reclamation were given minimal effort (leaving little likelihood of successful revegetation), the quantities of water involved in the high coal-conversion scenarios would not be inconsequential compared with $_7Q_{10}$ or with present-day consumption. Table 9 also shows that, while the scarcity of water today is far more critical in the West than in the East, attempts to put more of the water burden on the East by giving it a larger role in mining and converting coal would simply transfer the problem of water supply (compare, for example, scenarios 4 and 6 or 8 and 10). Finally Table 9 indicates that coal mining is far less a water consumption problem than coal conversion, although even without a synthetic fuel industry, water consumption at the upper limit of possible use (reflecting a serious effort at land reclamation) is large enough to be worrisome. Indeed, western production of coal equal to 12×10^{18} joule/year (scenario 2) would require an amount of water that could be considered unacceptable.

The total annual water consumption, by region, for each of the electricity scenarios is given in Table 10, expressed in absolute amounts of water used for cooling. The range in consumption assigned to the various scenarios represents the range of values from Table 5 plus a range of thermal efficiencies ranging from 33 to 38 percent. It is instructive to compare the water consumption projected by these scenarios with total present-day regional consumption for all uses. For those regions and scenarios in which the upper limit for the additional water required for cooling of power plants (upper limit of regional entry in scenario minus present-day regional cooling-water consumption) exceeds 50 percent of present-day total regional consumption for all uses (see Table 2) the entry in Table 10 is marked with an asterisk. Note that all these regions are in the East. An interesting fact is that such regions are generally not the ones with a high ratio of present-day consumption to mean annual runoff (see Table 2). Although the additional water consumed for cooling represents a major increase in water consumption in these regions, the environmental impacts created by such consumption are likely to be of a different nature than those arising in the West. To emphasize this distinction, entries in Table 10 are marked with a dagger when the upper limit for additional water for cooling exceeds 5 percent of the mean annual regional runoff (*33*). Except for the Great Lakes region (which is a special case in the sense that much of its water comes from Canada and is not indicated in Table 2), no overlap is found between regions having a relatively deficient annual flow and regions where the projected demand would substantially exceed present consumption.

Although broad conclusions drawn from the electricity scenarios can be regarded as either vague or indefensible, we venture to conclude (neglecting all facets of the electricity supply problem other than cooling water) that 30×10^{18} joules of electric output per year would be tolerable with cooling mode B (dry cooling dominant) but would pose major unacceptable regional problems with mode A (evaporative cooling dominant). While 30×10^{18} joule/year of electricity may seem to be an absurdly high level to consider (it is a fivefold increase over present levels), we included this level in our scenarios because of our concern with the general problem of finding ultimate replacements for natural gaseous and liquid fuels.

Conclusions

We have examined constraints of freshwater on the expansion rate of particular energy options and have answered specific questions which were posed in terms of rather narrow sets of choices among alternative technological means to common objectives. From technology comparisons and scenario analyses the availability of freshwater is clearly a paramount factor to be considered in setting energy policy. Our conclusions are based solely on the factor of water consumption; numerous other factors, including land use, air and water pollution, economics, and occupational hazards, must be included in any overall planning effort.

Our analysis suggests several conclusions. One is that a coal-conversion industry in the United States supplying as much as 8×10^{18} joule/year of synthetic fuels will be constrained by a scarcity of freshwater. An annual production of 8×10^{18} joules of synthetic fuels is not even enough to replace the present con-

Table 10. Freshwater consumption for electric power-plant cooling in four scenarios (see Table 8 for scenario specification).

Region	1975 freshwater consumption for electric power generation (km³)	Future freshwater consumption for electric power generation (km³/year)			
		1	2	3	4
New England	0.13	0.096 to 0.32	0.029 to 0.11	0.24 to 0.80*	0.072 to 0.28
Mid-Atlantic	0.19	0.34 to 1.2	0.10 to 0.38	0.85 to 2.0*	0.25 to 0.95
South Atlantic Gulf	0.29	0.44 to 1.5	0.13 to 0.49	1.1 to 3.8*	0.33 to 1.2
Great Lakes	0.072	0.58 to 2.0*	0.17 to 0.66	1.5 to 5.1*†	0.43 to 1.6*
Ohio	0.39	0.98 to 3.4*	0.29 to 1.1	2.5 to 8.3*	0.75 to 2.8*
Tennessee	0.081	0.20 to 0.68*	0.062 to 0.23	0.50 to 1.7*	0.16 to 0.58*
Upper Mississippi	0.13	0.42 to 1.4*	0.13 to 0.49	1.1 to 3.6*	0.33 to 1.2*
Lower Mississippi	0.40	0.20 to 0.68	0.062 to 0.23	0.50 to 1.7	0.16 to 0.58
Souris-Red-Rainy	0.0017	0.0058 to 0.020	0.0017 to 0.0065	0.015 to .049	0.0043 to 0.016
Missouri	0.094	0.19 to 0.66	0.056 to 0.22	0.48 to 1.7	0.14 to 0.55
Arkansas	0.13	0.24 to 0.80	0.070 to 0.27	0.60 to 2.0	0.18 to 0.68
Texas Gulf	0.52	0.40 to 1.4	0.12 to 0.46	1.9 to 3.9†	0.30 to 1.2
Rio Grande	0.028	0.042 to 0.14	0.013 to 0.048	0.11 to 0.36†	0.033 to 0.12
Upper Colorado	0.082	0.062 to 0.20	0.018 to 0.070	0.15 to 0.50	0.045 to 0.18
Lower Colorado	0.065	0.054 to 0.18	0.016 to 0.060	0.14 to 0.46†	0.040 to 0.15
Great Basin	0.0079	0.022 to 0.072	0.0064 to 0.024	0.055 to 0.18	0.016 to 0.060
Pacific Northwest	0.012	0.024 to 0.080	0.0070 to 0.027	0.60 to 0.20	0.018 to 0.068
California	0.044	0.0096 to 0.032	0.0029 to 0.011	0.024 to 0.080	0.0073 to 0.028
United States (excluding Hawaii and Alaska)	2.6	4.5 to 15	1.3 to 4.9	11.0 to 37.	3.3 to 12

*See text. †See text.

sumption of natural gaseous and liquid fuels in only those end uses for which direct burning of coal is inappropriate (for example, transportation and home heating). This deficiency, coupled with the low likelihood that bioconversion can meet these present needs in an environmentally acceptable fashion, suggests the importance of directing greater R & D effort toward ultimate end-use modification which would permit the use of electricity in place of natural gaseous and liquid fuels. It also emphasizes the acute need for more stringent energy conservation in transportation and home heating.

A second finding is that production of steam-generated electricity as a substitute for natural gaseous and liquid fuels would cause conflicts in the use of freshwater unless dry cooling were extensively used. Technologies for electricity production that do not depend on water, such as wind and photovoltaics, as well as solar active or passive home heating look especially desirable in this light.

Combining these two observations we conclude that limited availability of freshwater is likely to be a severely constraining factor in future energy development. Even if no overall growth in energy consumption were to take place in this country, the need for substitutes for natural gaseous and liquid fuels could pose staggering problems for water resource management and for natural ecosystems that depend on relatively free flowing freshwater. Overall growth in U.S. energy consumption would, of course, exacerbate these problems.

The degree of dependence of energy development on freshwater hinges on a number of unknown factors: the extent to which water conservation practices, including water pollution treatment, are carried out in coal-conversion plants and mining operations; the economic feasibility of dry cooling or cooling with agricultural wastewater; the economic feasibility of desalination; the results of further research on groundwater and its management as a renewable resource rather than as a commodity to be mined and lost; the results of further experience with land reclamation, especially in areas hard to reclaim such as the northern Great Plains; and the feasibility of piping seawater inland for use in cooling power plants. The consequences to society of use of freshwater for energy will depend also on what the future demand will be in competing sectors of the water economy such as agriculture, municipal use, and industry. Moreover, decisions on acceptable limits of water use for energy will require greater understanding of rivers, lakes, and estuaries and greater knowledge of climatic variability.

Resolving these uncertainties will not be easy. Information on biological and climatic constraints is likely to be especially elusive. Yet planning must proceed, even in the face of uncertainty. Water constraints on energy development are sufficiently great to warrant far more attention. Two broad and urgent needs are identified. First is the need to develop adequate criteria for acceptable water consumption based on considerations of ecosystem balance, human well-being, nonuniform distribution of water, and the vicissitudes of its abundance under a capricious climate. Second is the need to set energy policy and water management on a course compatible with the criteria that are chosen. That course is certain to be characterized by a vital and enormous role for energy and water conservation.

References and Notes

1. For an overview of water use taxonomy, see F. W. Sinden, in *Boundaries of Analysis: An Inquiry into the Tocks Island Dam Controversy*, H. A. Feiveson, F. W. Sinden, R. H. Socolow, Eds. (Ballinger, Cambridge, Mass., 1976), pp. 163–215.
2. One cubic meter is approximately 264 gallons; one acre-foot is approximately 1.23×10^3 m³.
3. J. F. Storr, in *Thermal Ecology* (U.S. Atomic Energy Commission, U.S. Government Printing Office, Washington, D.C., 1973), pp. 291–295.
4. A. Jassby, in (*5*), pp. A1-A26.
5. J. Harte, Ed., *Energy and the Fate of Ecosystems*, report of the Ecosystem Impacts Resource Group of the Risk/Impact Panel of the Committee on Nuclear and Alternative Energy Systems (National Academy of Sciences–National Research Council, Washington, D.C., in press).
6. B. Skinner, *Earth Resources* (Prentice-Hall, Englewood Cliffs, N.J., 1969).
7. Environmental impact assessment that only estimates the stresses on natural systems caused by human activities and the consequences of those stresses to the health of natural systems are incomplete. They stop short of assessing the social costs resulting from environmental degradation. For a survey and evaluation of the ways in which environmental degradation from energy activities can reduce the quality of human life, both materially and intangibly, see J. Harte, in (*5*, pp. 6–103). See also W. E. Westman, *Science* **197**, 960 (1977).
8. An ambiguity in the assignment of runoff to a region arises when one region is downstream from another. This ambiguity is resolved by the U.S. Water Resources Council by assigning the flow entering a downstream region to the region in which the flow originated. Thus, runoff as-

signed to the Lower Colorado region does not include the flow entering that region from the Upper Colorado. Because most of the coal deposits in the West and Midwest are located in "upstream" rather than "downstream" regions, the water resources that appear to be potentially available for coal-related activities may be exaggerated, at least from the downstream user's point of view.
9. See, for example, C. W. Fetter, *Water Resour. Bull.* **13**, 309 (1977).
10. Stream flow data have been obtained from the U.S. Geological Survey computerized data bank with the help of L. Jorgensen and others at the Survey office in Menlo Park, Calif. We are grateful for their cooperation.
11. G. Samuels, *Assessment of Water Resources for Nuclear Energy Centers* (ORNL-5097 UC-80, Oak Ridge National Laboratory, Oak Ridge, Tenn., September 1976).
12. R. K. Linsley, M. A. Kohler, J. L. H. Paulhus, *Hydrology for Engineers* (McGraw-Hill, New York, 1975), pp. 338–373.
13. H. B. N. Hynes, *The Ecology of Running Waters* (Liverpool Univ. Press, Liverpool, 1970), pp. 196–235 and 326–332.
14. J. Cairns, in *The Structure and Function of Fresh-water Microbial Communities*, J. Cairns, Ed. (American Microscopial Society Symposium, Virginia Polytechnic Institute and State University, Blacksburg, 1970), pp. 219–247.
15. C. Hazel, *The Effects of Altered Streamflows on Fish and Wildlife in California* (U.S. Department of Commerce, National Technical Information Service, PB-263 245, Springfield, Va., 1976); M. Brusven and C. MacPhee, *The Effect of River Fluctuation from Hydroelectric Peaking on Selected Aquatic Invertebrates and Fish* (Idaho Water Resources Research Institute, University of Idaho, September 1976).
16. G. Lauff, Ed., *Estuaries* (AAAS, Washington, D.C., 1967). See especially L. E. Cronin, in *ibid.*, pp. 667–689; K. F. Bowden, in *ibid.*, pp. 15–36; J. Hedgpeth, in *ibid.*, pp. 408–419; M. Carriker, in *ibid.*, pp. 442–487. See also B. Ketchum, *Sewage Ind. Wastes* **23** (No. 2), 198 (1951); *J. Mar. Res.* **10**, 18 (1951); G. Gunter, *Limnol. Oceanogr.* **6**, 182 (1961); J. Hedgpeth, in (*5*), pp. G1-G53).
17. A detailed discussion of the environmental impacts associated with the water requirements for coal conversion, including effects on aquatic life which could result from untreated effluent water, are discussed in (*19*).
18. R. Curry, in *Practices and Problems of Land Reclamation in Western North America*, M. K. Wali, Ed. (Univ. of North Dakota Press, Grand Forks, 1976).
19. M. El-Gasseir, in (*5*), pp. EI1-EIX.52.
20. A. McDonald, *Shale Oil, an Environmental Critique* (CSPI Oil Ser. 3, Center for Science in the Public Interest, Washington, D.C., 1974).
21. F. M. Pfeffer, *Pollutional Problems and Research Needs for an Oil Shale Industry* (U.S. Environmental Protection Agency, EPA-660/2-74-067, Washington, D.C., June 1974).
22. G. D. Weaver, *Oil Shale Technology*, Hearings before the Subcommittee on Energy, U.S. House of Representatives, 93rd Congress, 2nd Session, 1974; J. B. Weeks, G. H. Leavesly, F. A. Welder, G. J. Saulnier, Jr., *U.S. Geol. Surv. Prof. Pap. 908* (1974).
23. L. King, in (*5*), pp. K1–K74.
24. T. H. Pigford *et al.*, *Fuel Cycles for Electrical Power Generation*, parts I and II, *Towards Comprehensive Standards: The Electric Power Case* (Teknekron, Inc., Berkeley, Calif., 1974).
25. P. Ehrlich, A. Ehrlich, J. Holdren, *Ecoscience: Population, Resources, Environment* (Freeman, San Francisco, 1977).
26. J. M. Beer, *The Fluidised Combustion of Coal, Sixteenth Symposium (International) on Combustion* (Combustion Institute, Pittsburgh, Pa., 1976), p. 439.
27. A rough estimate of water consumed in the provision of materials for a solar central-receiver system can be worked out from data provided by Davidson and Grether (*28*). Most of the water consumption will result from the cement requirements for the heliostat. Averaged over an assumed 30-year plant lifetime, this water consumption will be about .002 $Km^2/10^{18}$ joules of electric output. We thank B. Wolfson for information on this subject.
28. M. Davidson and D. Grether, *The Central Receiver Power Plant: an Environmental, Ecological, and Socioeconomic Analysis* (LBL-6329, UC-62, Lawrence Berkeley Laboratory, Berkeley, Calif., 1977).
29. C. Calef, *Environment* **18**, 17 (1976).
30. For cases A and B, see Federal Energy Administration, Project Independence Blueprint—Final Task Force Report, *Project Independence—Residential and Commercial Energy Use Patterns 1970–1990* (Government Printing Office, Washington, D.C., November 1974), vol. 1, pp. 78–81. For case C see Energy Utilization Systems, Inc., *Energy Consumption and Life-Cycle Costs of Space Conditioning Systems*, sponsored by Electric Environmental Equipment Section (National Electrical Manufacturers Association, New York, 1976).
31. The value of ${}_7Q_{10}$ in the two western regions and in the three eastern regions is estimated by multiplying mean annual runoff by the ratio of ${}_7Q_{10}$ river flow to mean annual river flow. This we have done for the East and the West, using the Colorado and Yellowstone rivers in the West and the Wabash, the Ohio, and the Kanawha in the East (see Fig. 2). This procedure is reasonable if the river flow is a significant fraction of runoff. In fact, the combined mean annual flow of the three eastern rivers at the chosen U.S. Geological Survey stations accounts for 35 percent of the combined mean annual runoff of the Ohio, the Upper Mississippi, and the Tennessee regions, while in the West, the combined annual flow of the Colorado and Yellowstone at the chosen stations is 20 percent of the combined mean annual runoff of the Missouri and Upper Colorado regions.
32. The fact that an energy producer (or housing developer) can afford to pay more for land and water than a farmer or rancher generates a strong force toward what may be a serious future misallocation of resources. See, for example, J. J. Jacobsen, *A Dynamic Analysis of the Environmental and Social Impacts of Coal Development in the Eastern Powder River Basin of Wyoming, 1960–2010* (Systems Dynamics Group, Thayer School of Engineering, Dartmouth College, May 1975), pp. 31–32. See also W. D. Gertsch, J. Sathay, R. Ritschard, S. Parker, *Water Requirements for Future Energy Development in the West: State Perspectives* (LA-6688-MS, Los Alamos Scientific Laboratory, Los Alamos, N.M., August 1977).
33. Comparison is made with mean annual runoff rather than ${}_7Q_{10}$ because the latter quantity is not well estimated in many of the hydrological regions. When consumption is 5 percent of mean annual runoff, it is very roughly 50 percent of ${}_7Q_{10}$ and thus will generate significant effect on river flow during times of drought.
34. U.S. Water Resources Council, *1975 National Assessment*, preliminary draft.
35. C. R. Murray and E. B. Reeves, *U.S. Geol. Surv. Circ. 765* (1977).
36. V. E. Swanson, J. H. Medlin, J. R. Hatch, S. L. Coleman, G. H. Wood, S. D. Woodruff, R. T. Hilderbrand, *Collection, Chemical Analysis, and Evaluation of Coal Samples in 1975* (U.S. Geological Survey Open-file Report 76-468, 1976).
37. E. M. Dickson *et al.*, *Impacts of Synthetic Liquid Fuel Development—Automotive Market*, Environmental Protection Agency, Interagency Energy-Environment Research and Development Program Report (EPA-600/7-76-0046, Government Printing Office, Washington, D.C., July 1976), vol. 2.
38. U.S. Department of the Interior, Energy Research and Development Administration, *Synthetic Fuel Commercialization Program*, Draft Environmental Statement (ERDA-1547, Government Printing Office, Washington, D.C., December 1975), vol. 4.
39. U.S. Department of the Interior, Bureau of Mines Process Evaluation Group, Morgantown, W. Va. Prepared for the Energy Research and Development Administration, ERDA 76-47, 76-48, 76-49, 76-52, 76-57, 76-58, 76-59, March, 1976 (National Technical Information Service, U.S. Department of Commerce, Springfield, Va., 1976).
40. Committee on Interior and Insular Affairs, U.S. Senate, *Factors Affecting Coal Substitution for other Fossil Fuels in Electric Power Production and Industrial Uses* (The National Fuels and Energy Policy Study, Serial No. 94-17, Government Printing Office, Washington, D.C., 1975).
41. Federal Energy Administration, *Project Independence Blueprint Final Task Force Report*, Project Independence-Interagency Task Force on Oil Shale (Government Printing Office, Washington, D.C., November 1974).
42. Western States Water Resources Council, *Western States Water Requirements for Energy Development to 1990* (Western States Water Resources Council, Salt Lake City, Utah, November 1974).
43. W. H. Smith and J. B. Stall, *Coal and Water Resources for Coal Conversion in Illinois*, State Water Survey State Geological Survey Cooperative Report 4, 1975; Water Resources Council, *Project Independence-Water Requirements, Availabilities, Constraints, and Recommended Federal Actions*, Federal Energy Administration Project Independence Blueprint Final Task Force Report (Government Printing Office, Washington, D.C., November 1974), p. 22.
44. Colorado Energy Research Institute, *Net Energy Analysis: An Energy Balance Study of Fossil Fuel Resources* (Denver, Colorado, April 1976).
45. U.S. Water Resources Council, *The Nation's Water Resources* (Government Printing Office, Washington, D.C., 1968); Federal Energy Administration Project Independence Blueprint Final Task Force Report, *Project Independence—Coal*, November 1974.
46. We thank the Energy Research and Development Administration for their support. We are also grateful for the numerous discussions we have held with many members of the National Academy of Sciences' Nuclear and Alternative Energy Systems Study. L. King, a participant in that study, has especially contributed to our thinking. P. Benenson, P. Fox, J. Holdren, L. King, D. Levy, and especially H. Malde, suggested numerous improvements in both substance and style in an earlier draft of this article.

Predicting Future Atmospheric Carbon Dioxide Levels

The predictions provide a basis for evaluating the possible impact of the continuing use of fossil fuel.

U. Siegenthaler and H. Oeschger

Summary. Results from different models for the natural carbon dioxide cycle are compared. Special emphasis is given to the type of ocean modeling (diffusive deep-sea or two-box ocean), behavior of the biosphere, and value of the oceanic buffer factor against carbon dixoide uptake. According to the most probable models, the fraction of the cumulative production remaining airborne will be between 46 and 80 percent 100 years from now for any realistic assumptions concerning future carbon dioxide production. For a prescribed maximum increase of 50 percent above the preindustrial carbon dioxide level, the production could grow by about 50 percent until the beginning of the next century, but should then decrease rapidly.

Since the beginning of industrialization man has been significantly changing the atmospheric carbon dioxide concentration. Until 1974 the fossil fuel CO_2 input into the atmosphere amounted to roughly 21 percent of the preindustrial atmospheric CO_2 content, and as a result the atmospheric CO_2 level increased by about 13 percent. If we continue to exhaust oil and coal reserves at a faster and faster rate, in a few decades the increase of the atmospheric CO_2 concentration will be of the same order of magnitude as the preindustrial CO_2 concentration itself. There is serious concern about the possible environmental consequences of such an increase. The average surface temperature of the earth is about 35°K higher than its radiation temperature of 253°K as seen from interplanetary space (*1*). This is partly due to the so-called greenhouse effect, caused by atmospheric water vapor and CO_2. With a higher CO_2 concentration the greenhouse effect will be enhanced. Using a three-dimensional general circulation model, Manabe and Wetherald (*2*) found that the mean surface temperature would be 2° to 3°K higher for a doubled CO_2 level, with a strong amplification (8° to 10°K warming) in the polar areas. Although some important phenomena, such as changes in cloudiness, could not yet be considered, the climatic models developed so far clearly indicate that man-made CO_2 could trigger climatic changes with serious consequences for human society. A review of models of the climatic effect has been given by Schneider (*3*).

The CO_2 problem has three aspects (*4*): (i) the prediction of future CO_2 levels, (ii) the climatic consequences of a CO_2 increase, and (iii) the impact a given climatic change will have on man. We intend to discuss here the first aspect only, by means of models for the global CO_2 cycle. Many authors have estimated future CO_2 levels [for example, see (*5–12*)]. Our aim is to compare predictions for various input functions and different model approaches to find the extent to which predictions are possible today and the uncertainties that are involved because of our limited knowledge of the global carbon system. We will not consider basically new models, but instead will restrict ourselves to models for the CO_2 cycle already discussed in the literature.

Model Concepts

On time scales of centuries, atmospheric CO_2 exchanges mainly with the carbon in the ocean and in the biosphere (*13, 14*). The CO_2 fluxes out of or into the atmosphere-ocean-biosphere system, such as those due to weathering, sedimentation, or volcanic emissions, are negligible for our purposes, as they total only about 5×10^{-6} of the entire oceanic carbon mass per year (*15*).

For the simulation of systems, it is common to use "well-mixed boxes" (or "compartments"). It is a good approximation to consider a reservoir as a well-mixed box if its internal mixing time is short compared to the time scales of the phenomena discussed, such as exchange with other reservoirs, radioactive decay, or production of matter within the reservoir. In our case, the characteristic times of interest are of the order of a decade to centuries. Therefore, we may represent the atmosphere as a well-mixed box, as its mixing time is a few years.

The ocean, the largest of the three exchanging carbon reservoirs, has a turnover time of the order of 1000 years as determined by means of ^{14}C measurements (*16*). Several CO_2 prognoses have been made on the basis of models in which the ocean consists of two well-mixed boxes (2B models)—namely, a mixed layer or surface ocean, and a "deep sea," which means the whole ocean below the mixed layer (*8–12*). In reality, however, the transition between the mixed layer and the deep sea is continuous, and in view of the long turnover time and the above criterion, the deep sea should not be represented by one well-mixed box. It is therefore not surprising that models with a two-box ocean either fail to give a consistent description of the CO_2 cycle (for instance, giving a wrong natural oceanic ^{14}C distribution when they are fitted to give the correct CO_2 increase) or need additional adjustments to provide satisfactory results. Thus, the mixed layer must be chosen a few hundred meters thick in order to obtain satisfactory results (*8, 10, 17*), while observations yield an average value of about 75 m.

A way to consider the continuous character of oceanic mixing is to describe it by eddy diffusion (*18*), which means that the flux of matter (F) is assumed to be proportional to its concentration gradient ($\partial C/\partial z$)

$$F = -K \frac{\partial C}{\partial z}$$

Broecker *et al.* (*6*) introduced the concept of an eddy-diffusive ocean for estimating future CO_2 levels, but computed the oceanic CO_2 transport only in an approximate way. We present here predic-

U. Siegenthaler is a research associate and H. Oeschger is a professor of physics at the Physics Institute of the University of Bern, 3012 Bern, Switzerland.

tions obtained by means of the box-diffusion (BD) model developed by Oeschger *et al.* (*19*), in which the ocean consists of a mixed-layer box and an eddy-diffusive deep sea below it with a constant eddy diffusivity K; horizontal homogeneity is assumed. All vertical transport processes, including turbulent mixing, formation of deep water in polar seas, and particulate transport of carbon, are thus summarized by eddy diffusion and described by one model parameter, K. The BD model consistently reproduces, besides the observed CO_2 increase, phenomena with very different characteristic time scales—namely, the preindustrial vertical ^{14}C distribution in the ocean, the ^{14}C dilution due to the fossil CO_2 (Suess effect), and finally the ^{14}C increase in the ocean surface after the bomb tests (*19, 20*). Therefore, we can expect that the BD model is suited for reliable forecasts, irrespective of the time constants of future variations in CO_2 production. For comparison, we also present results of a 2B model. Figure 1 shows the BD model schematically.

Carbon dioxide exchange between the atmosphere and the mixed layer is formulated as an exchange between two boxes in the usual way. The exchange coefficient between the atmosphere and the mixed layer and the oceanic transport coefficients—eddy diffusivity in the BD model and the exchange coefficient between the mixed layer and the deep sea in the 2B model—are determined from the preindustrial ^{14}C distribution (95 percent of the atmospheric value for the mixed layer and 84 percent for the total ocean). The gas exchange rate between the atmosphere and the ocean has also been determined by measuring the radon-222 deficiency (difference between the radon-222 concentration and the concentration that would be in equilibrium with its parent isotope, radium-226) in surface waters (*21*), and the results are consistent with the exchange coefficient determined from ^{14}C data. In the 2B model the mixed-layer depth is a free parameter, adjusted so as to predict the observed CO_2 increase, whereas in the BD model the mixed-layer depth is 75 m, independently fixed from oceanographic data.

If the atmospheric CO_2 concentration varies, the amount of dissolved CO_2 gas in the ocean surface layer varies proportionally at chemical equilibrium. Carbon dioxide gas, however, accounts for less than 1 percent of the total inorganic carbon (or "total CO_2") in surface seawater, the main species being bicarbonate (HCO_3^-) and carbonate (CO_3^{2-}) ions. An increase in the CO_2 concentration brings about a shift in the chemical equilibria between dissolved CO_2, HCO_3^-, and CO_3^{2-} such that the relative increase of the total CO_2 concentration is less than that of CO_2 gas alone. This is taken into account by introducing a buffer factor ξ: if the CO_2 pressure is increased by p percent, the total CO_2 concentration of seawater, including dissolved CO_2, HCO_3^-, and CO_3^{2-}, increases by only p/ξ percent (*5, 6, 10*). The factor ξ increases with CO_2 pressure. We will use the following relationship, which we deduced from figure 3 of (*10*)

$$\xi = 9 + 4.9\, p_m/P_m - 0.1\,(p_m/P_m)^2 \qquad (1)$$

where p_m is the excess CO_2 pressure in the mixed layer over the preindustrial pressure P_m, and p_m/P_m is therefore the relative increase. The size of the buffer factor determines in a crucial way the ocean's capacity for taking up additional CO_2. Different authors who computed ξ from chemical data (equilibrium constants and average seawater composition) obtained, for present CO_2 concentrations, values between 9 and 10, but experiments by Rebello and Wagener (*22*) yielded a value of only 7.

The biosphere, comprising in our case living and dead terrestrial organic matter, is the carbon reservoir about which the least information is available. There is not even agreement on its size; the value given by Bolin (*14*), 1.15×10^{18} grams of carbon, is often quoted, but another estimate leads to nearly 10×10^{18} g (*23*). We take 1.5×10^{18} g, including only long-lived plants (*17*). The biosphere is modeled as a well-mixed box exchanging with the atmosphere. [In the original version of the BD model (*19*) we assumed that all plants have the same lifetime, but an exponential distribution of lifetimes, as given by a well-mixed box, seems more appropriate.]

Some plants probably react to a higher CO_2 level by enhanced photosynthetic activity (*24*), so the biosphere might act as a sink for additional CO_2. This is taken into account in the model in the following way: if the atmospheric CO_2 pressure increases by p percent, the CO_2 flux from the atmosphere to the biosphere increases by ϵp percent. The size of the biosphere growth factor ϵ is uncertain, and values between 0 and 0.3 have been used. On the other hand, it has been suggested that there might be a net decrease of the size of the biosphere due to human influence, resulting in a net flux of biospheric CO_2 into the atmosphere (*4, 25, 26*). If this flux were proportional to the global energy production, it could be formally represented by a negative ϵ. Fertilization by CO_2 and deforestation partly compensate each other, and it is not yet known which effect is prevailing. Below, we compare results obtained with $\epsilon = 0.2$, $\epsilon = 0$ (no net biosphere growth), and in some cases with $\epsilon = -0.2$. Attempts are being made to learn more about the role of the biosphere by determining the change in the atmospheric $^{13}C/^{12}C$ ratio since the beginning of industrialization, because the value of this ratio in organic carbon is very different from the values in atmospheric and oceanic CO_2 (*27*). Besides measurements in tree ring series, continuous $^{13}C/^{12}C$ measurements of atmospheric CO_2 or perhaps of CO_2 enclosed in old polar ice might be helpful for this purpose.

The marine biosphere is not considered because it is of negligible influence. The carbon mass of living matter in the sea is only about 1/600 of the atmospheric CO_2. Furthermore, an increased carbon supply cannot lead to increased productivity of the marine biomass because the productivity is limited by phosphorus (*28*), not by carbon, which is abundant anyhow. [In some coastal zones the marine biota are being fertilized by excess phosphorus from human activities (*29*), but for a global budget this is of minor importance.]

The model parameters used and the individual models compared are summarized in Table 1. The models were modified to test the sensitivity of the predictions to changes in the most important modeling features: ocean mixing, buffer factor, and biosphere response. The thickness of the mixed layer and the exchange parameters (k_{am} and K for the BD models and k_{am} and k_{dm} for the 2B model) are chosen as described above.

Validity of the Models

Keeling and co-workers (*30*) have continuously monitored the atmospheric CO_2 level at Mauna Loa, Hawaii, and at the South Pole. In Fig. 2 the observed CO_2 concentrations are compared with the predictions of models 1 to 3 of Table 1, based on the actual CO_2 production rates (*31*).

The preindustrial atmospheric CO_2 level, N_a, is not exactly known; it is probably somewhere between 288 and 295 parts per million (ppm), but even values outside this range cannot be ruled out (*32*). We assumed N_a for each model so that the predicted CO_2 concentration for 1971 was within $\pm$ 0.2 ppm of the observed mean value of 326.5 ppm. The values for N_a thus determined (Table 1) are compatible with the observed values

(*32*). We wish to point out that in the BD models, N_a is the only parameter adjusted to measured CO_2 data; all other parameters are fixed independently. In the 2B model, the mixed-layer depth is also determined from CO_2 observations. Therefore, the 2B results up to 1975 agree closely with the BD curves with corresponding biota growth and buffer factors and are not shown in Fig. 2.

Comparing the world CO_2 production and CO_2 increase between 1958 and 1974, we find that the atmospheric increase was on the average 53 percent of the production. Assuming for a moment that this figure also applies for the cumulative production of 62 ppm until 1974, the original atmospheric level, N_a, is found to be about 296 ppm. The observed concentration curve shows changes in the rate of increase that are not predicted by the models and cannot be attributed to varying CO_2 production. Obviously the exchange parameters between the CO_2 reservoirs are not quite constant with time. We must be aware that relatively small fluctuations in environmental parameters are sufficient to produce CO_2 variations of the observed size, because the total fluxes are much larger than the net fluxes. Thus, the exchange flux between the atmosphere and the ocean is 330 ppm per 7.7 years or 43 ppm/year, while the net flux into the ocean is about 1 ppm/year—that is, about 2 percent of the total flux. Because of the slightly irregular CO_2 increase, the comparison of model computations with observations is not a very decisive model test. However, the BD case with a net decrease of the biosphere ($\epsilon = -0.2$, lowermost curve in Fig. 2) must be ruled out, whereas the case with $\epsilon = 0$ still appears possible. The best fits are achieved with $\epsilon = 0.2$, with either assumption for ξ.

The role of the biosphere needs further discussion, because the biomass may be decreasing because of human use. Woodwell and Houghton (*25*) even supposed that the net biospheric CO_2 flux into the atmosphere might be as high as the fossil flux. The atmospheric CO_2 variations depend, of course, on the time function of CO_2 production; let us assume that the nonfossil flux was always equal to the fossil flux—that is, increased about exponentially with a time constant of ~35 years. For this case, it can easily be seen that such a large total CO_2 production is not compatible with the current ideas about the CO_2 cycle, because the ocean would then have to be a much more effective sink than is suggested by the model calculations. The cumulative production up to 1974 would be twice the fossil production, or 124 ppm, of which about 38 ppm (= 330 ppm − 292 ppm) would reside in the atmosphere, with the rest in the ocean. For a buffer factor of 10 and a 38-ppm atmospheric excess, the ocean could take up, after infinite time (that is, at chemical equilibrium with the atmosphere), about 250 ppm of excess CO_2 (see Eq. 2). This means that if the biosphere had actually released as much CO_2 as stemmed from fossil carbon, the ocean would have reached 34 percent saturation by absorbing 86 ppm of CO_2. This corresponds to an effective depth of saturation of 1300 m, while in case of a constant biomass the ocean is only 10 percent saturated, corresponding to an effective depth of 380 m. A penetration depth of 1300 m for an *e*-folding time of about 35 years is in clear disagreement with the oceanic turnover time of at least 1000 years. Therefore, the high biospheric flux proposed by Woodwell and Houghton cannot be accounted for by faster oceanic mixing than we have assumed. On the other hand, Bolin (*26*), evaluating data of the Food and Agriculture Organization on world forest resources, estimated the present annual nonfossil carbon input to be 10 to 35 percent of the present emission from the use of fossil fuels. Our BD computations for $\epsilon = -0.2$ yield only 10 percent for this ratio, and for 35 percent the biota growth factor ϵ would have to be correspondingly more negative. As we have ruled out even the case $\epsilon = -0.2$ on the basis of Fig. 2, we see that Bolin's estimate, too, contradicts our model results.

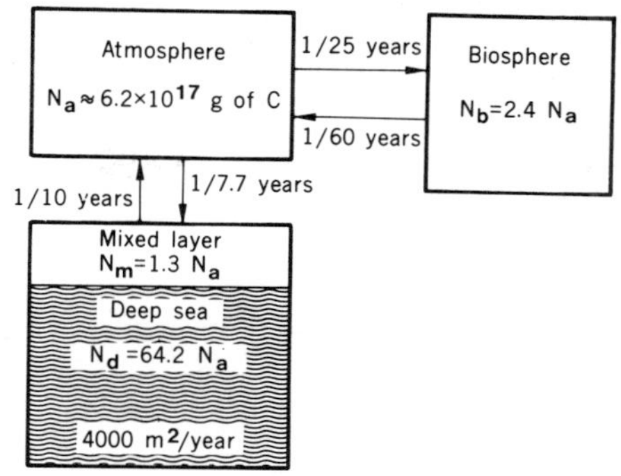

Fig. 1. Global CO_2 model with an eddy-diffusive deep-sea reservoir. The numerical values given are for the reservoir sizes, the exchange coefficients, and, in the deep sea, the eddy diffusivity.

The oceans could, however, take up more excess CO_2 than has been computed if the seawater buffer factor were lower than 9 to 10. But the thermodynamic constants entering into the calculations of the buffer factor are well determined and the average seawater composition is also reasonably well known, so it seems doubtful that the theoretical value is much in error. Another mechanism that might remove excess CO_2 is dissolution of near-surface carbonate sediments due to acidification of the seawater by the higher CO_2 concentration (see below), which would correspond to a lower apparent buffer factor. But this process is probably unimportant on a global scale. We therefore cannot find a sink for the proposed large CO_2 release from the biosphere.

The foregoing discussion is based, however, on the assumption that both fossil and biogenic CO_2 production rates grew in an approximately exponential way. This may be wrong for deforestation, in which case the situation is more

Table 1. General model parameters and specific models used. "Theoretical ξ" means a buffer factor as given by Eq. 1. In model 4, a lower buffer factor is assumed. Models 1 and 4 include enhanced biosphere growth at higher CO_2 levels ($\epsilon = 0.2$), model 3 a net shrinking of the biosphere and models 2 and 5 a constant biospheric mass.

	Parameters common to all models	
N_a	Preindustrial CO_2 in atmosphere	~292 ppm = 6.2×10^{17} g of carbon; see below
N_b	Preindustrial CO_2 in biosphere	$2.4\,N_a$
N_{oc}	Preindustrial CO_2 in ocean	$65.5\,N_a$
N_m	Preindustrial CO_2 in mixed layer	$1.3\,N_a$ (BD); $5.3\,N_a$ (2B)
k_{am}	Exchange coefficient, atmosphere–mixed layer	1/7.7 years
k_{ab}	Exchange coefficient, atmosphere–biosphere	1/25 years
K	Eddy diffusion coefficient, deep sea, for BD	4000 m²/year
k_{dm}	Exchange coefficient, deep sea–mixed layer, for 2B	1/1250 years
ξ	Buffer factor for excess CO_2	See below
ϵ	Biosphere growth factor	See below

Specific models and parameters

Box diffusion (BD)
1) BD with theoretical ξ, $\epsilon = 0.2$, $N_a = 295$ ppm (6.25×10^{17} g of carbon)
2) BD with theoretical ξ, $\epsilon = 0$, $N_a = 291.5$ ppm (6.18×10^{17} g of carbon)
3) BD with theoretical ξ, $\epsilon = -0.2$, $N_a = 287.5$ ppm (6.10×10^{17} g of carbon)
4) BD with $\xi = 7$, $\epsilon = 0.2$, $N_a = 297.5$ ppm (6.31×10^{17} g of carbon)

Two-box ocean (2B)
5) 2B with theoretical ξ, $\epsilon = 0$, $N_a = 291.5$ ppm (6.18×10^{17} g of carbon)

complex. Actually, data for New England (*25*) show that the forest biomass in that region decreased greatly during the last century and has somewhat recovered since 1900. Of course, a large CO_2 input in the past does not contribute to an atmospheric increase today, but rather leads to a slightly declining "baseline." This shows that more data about variations of the terrestrial biomass are urgently needed; also, the oceanic buffering mechanism should be checked again.

On the basis of the discussion above, we prefer the following model characteristics: a diffusive deep sea (BD); a biosphere growth factor of $\epsilon = 0.2$ or 0; and a buffer factor ξ, judging purely from Fig. 2, of 9 to 10 (theoretical value) or somewhat lower. Thus, there remain cases 1, 2, and 4 from Table 1; cases 3 and 5 are not considered good approximations. Nevertheless, results for the 2B model (*5*) are presented below for comparison.

Model Responses

After a long time, any CO_2 input into the system will be distributed among the different reservoirs, the carbon contents of which will then be at new steady-state values. The total input is $n_a + n_{oc} + n_b$, where the n_i denote the excesses over the preindustrial CO_2 masses, N_i, in the atmosphere, ocean, and biosphere. At steady state, according to the definitions of ξ and ϵ, this is equal to

$$n_a + (1/\xi)(n_a/N_a)N_{oc} + \epsilon(n_a/N_a)N_b = (n_a/N_a)(N_a + N_{oc}/\xi + \epsilon N_b)$$

Therefore, the fraction F_a of the total input remaining in the atmosphere forever, according to the model, is given by

$$F_a = \frac{N_a}{N_a + N_{oc}/\xi + \epsilon N_b} \qquad (2)$$

For $\xi = 10$, we get $F_a = 0.125$ for $\epsilon = 0.2$ and 0.132 for $\epsilon = 0$. With higher concentrations, the buffer factor, and therefore also F_a, increases. The exploitable fossil fuel reserves have been estimated as 7.1×10^{12} metric tons of carbon (*9*) or about 11.5 times the preindustrial atmospheric amount of CO_2. If all these reserves were burned, the new steady-state concentration, for $\epsilon = 0$, would be about four times the preindustrial value (with $\xi = 23$ according to Eq. 1), accompanied by a temperature increase of 3° to 6°K (*3*). This state would be reached only after many centuries; before then, much higher CO_2 levels would occur.

The dynamic behavior of a linear model is fully characterized by its response to a pulse input. The solution of any input function can be represented as a superposition of such pulse responses. Our models are linear as long as ξ is constant, which is the case for relatively small CO_2 increases. Nevertheless, the linear pulse responses give a good idea about the general model behavior for larger concentrations too. Figure 3a shows the pulse responses for BD with $\epsilon = 0.2$ and $\epsilon = 0$ and for 2B with $\epsilon = 0$. Obviously, there are considerable differences between the different cases. The mixed layer rapidly reaches chemical equilibrium with the atmosphere, with a time constant (for a 75-m-deep mixed layer) of about 1.5 years (*21*). This is reflected in the quick decrease of the atmospheric CO_2 level during the first years after the pulse input. The mixed layer of the 2B ocean is about four times larger than that of the BD ocean and can take up correspondingly more excess CO_2; therefore, the decline at the beginning is stronger for the 2B than for the BD curves. However, when the CO_2 pressures of the mixed layer and the atmosphere have become approximately equal, CO_2 is transported faster into the eddy-diffusive than into the well-mixed deep sea. There is quite a large difference between the runs with and without biosphere growth ($\epsilon = 0.2$ and 0, respectively), demonstrating the importance of the terrestrial biosphere for the CO_2 cycle.

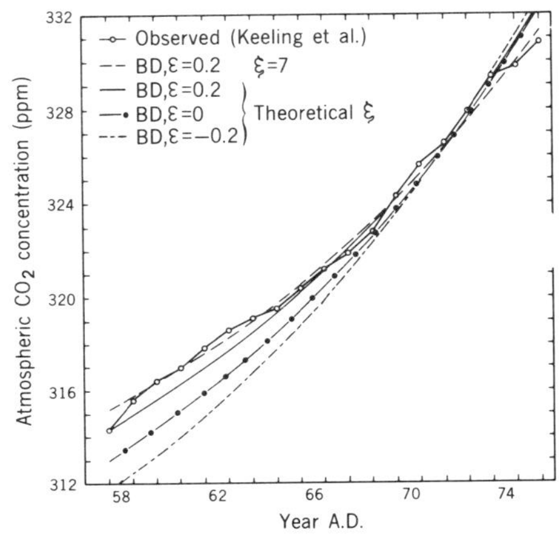

Fig. 2. Observed CO_2 concentrations as measured by Keeling and co-workers, average of South Pole and Mauna Loa data (*30*), and curves computed by use of models 1 to 4 of Table 1. Theoretical ξ means a buffer factor of about 9 and increase with higher CO_2 concentration (see Eq. 1).

The CO_2 decrease is far from exponential (see Fig. 3b), and one can therefore not speak of one relaxation time or time constant for this decline. Thus, in BD with $\epsilon = 0.2$, it takes 21 years until the transient excess (that fraction of the excess above the final equilibrium value), which is plotted in Fig. 3b, has reached half its original value. One-quarter of the original value is only reached after 70 more years, and we have to wait another 240 years, or about 330 years in all, until the value has halved again to one-eighth. After the first decade, the concentration decreases most rapidly for BD with $\epsilon = 0.2$.

Predictions

A detailed prediction of future burning rates of fossil fuel would have to be speculative, but we can indicate upper and lower limits for this function with some confidence. As a lower limit we assume a constant CO_2 production rate at the level of 1975—that is, 1.87×10^{16} g of CO_2, equivalent to 2.4 ppm/year (*31*). Figure 4 shows the results for this case. (Only the predictions for BD, with the theoretical ξ and $\epsilon = 0$ as well as $\epsilon = 0.2$, are shown in Figs. 4 to 6; Tables 2 and 3 show results for cases 4 and 5 as well.) It is noticed that no new equilibrium concentration is reached for a constant production rate, although this would be expected, for instance, for man-made aerosols, for which, because of their short atmospheric residence time, the rate of removal would soon equal the production rate. The fraction of the cumulative CO_2 production residing in the atmosphere is between 51 and 62 percent in 1970, and in 2070, 100 years later, is still between 38 and 59 percent (see Table 2). A doubled CO_2 level would be reached between the years 2160 and 2480.

As an upper limit we assume that all economically exploitable fossil fuels (11.5 times the preindustrial atmospheric CO_2; see above) will be burned, and that the production rate grows by 5.0 percent per year immediately after 1970, as it did from 1960 to 1970. The cumulative production is assumed to increase as a logistic function of time after 1970, so the production rate is given by

$$p(t) = \frac{d}{dt} \frac{3340 \text{ ppm}}{1 + 84 \exp(-t/20 \text{ years})}$$

Figure 5 shows $p(t)$ and the predictions. The CO_2 concentration would have doubled by the year 2020, essentially independent of the model used. The maximum atmospheric CO_2 level, five to ten times the preindustrial value, would be reached 40 to 80 years after the production peak, which is assumed to occur around 2060.

For the upper-limit scenario, the airborne amount of the cumulative production would be between 48 and 83 percent in 2070 (Table 2). For any future CO_2 production function between the two limiting cases discussed, 46 to 80 percent of the total produced CO_2 would still be in the atmosphere 100 years from now, according to the preferred models (cases 1 and 2). Thus, the assumption that the airborne fraction will remain near 50 percent, which is sometimes made (*4, 7*), may lead to a considerable underestimation of future CO_2 levels. It is mainly the dependence of the oceanic buffer mechanism on the CO_2 concentration that is responsible for the high atmospheric load, because at high concentrations the ocean becomes quite ineffective as a CO_2 sink.

Comparing the various predictions, we see that the concentrations obtained with the BD model are lower than those obtained with the 2B model (Table 2). This is because diffusive transport into the deep sea is faster, for short-time disturbances, than transport by exchange between two boxes, as discussed in (*19*). The difference between the BD cases with $\epsilon = 0.2$ and $\epsilon = 0$ (models 1 and 2) demonstrates the importance of any variation of the terrestrial biomass. The assumption of a positive biota growth factor ϵ, constant over centuries, is probably not justifiable. Thus, case 1 predicts in the upper limit 100 percent growth of the biosphere by the year 2200.

With climate models becoming more and more realistic, a maximum permissible atmospheric CO_2 level might be found which should not be exceeded if the atmospheric radiation balance is not to be disturbed in a dangerous way. We do not know such a threshold level at present; let us assume that it would be 50 percent above the preindustrial concentration. This would lead, according to present climate models, to a mean temperature increase of about 1° to 2°K (*3*). The question is, What amounts of fossil fuels may be burned in the future in order not to exceed this 50 percent excess? We have computed these production time histories (*33*) under the assumption that the constant 50 percent excess is reached asymptotically as a logistic function of time (see Fig. 6). The fastest increase of the CO_2 level occurs in the year 2000, and computed CO_2 emission rates are largest between 2000 and 2010—28 to 84 percent above the 1970 value, depending on the model considered. Then they have to level off rapidly, reaching 1970 values between 2030 and 2080. In 150 to 250 years from now, the decrease becomes very slow and the permitted output remains for a long time at 25 to 40 percent of the 1970 rate. The yearly consumption at the end of our century would, of course, be much smaller—by a factor of 2 to 3—than that in our upper-limit case. The integral production would, 100 years from now, be of the order of one preindustrial atmospheric CO_2 mass (Table 3).

This scenario clearly does not allow us to go on burning fossil fuel at the present growth rate for a long time. A modestly growing demand for fossil energy could be met, but around the turn of the century new technologies would have to take over a substantial part of global energy production. Whatever model case is considered, the CO_2 output in the year 2000 permitted by this scenario is much lower than the value of 6 ppm/year rec-

Table 2. Cumulative CO_2 production, P_{cum}, and airborne fraction of P_{cum}, F_a (see Eq. 2). The models are described in Table 1.

Year	P_{cum} (ppm)	F_a (percent) Model 1	Model 2	Model 4	Model 5
1970	53.9	56.3	62.2	51.4	62.4
			*Lower limit**		
2020	173	51.1	58.4	44.2	61.1
2070	293	45.7	54.1	37.7	59.2
2170	532	40.8	49.6	31.1	56.0
			Upper limit†		
2020	438	64.6	71.1	55.5	70.4
2070	2150	69.2	80.4	48.2	82.6
2170	3340	53.6	74.6	27.2	79.8

*Based on a constant production rate after 1975. †Assuming that P_{cum} increases as a logistic function of time after 1970 to 11.5 N_a.

Table 3. Estimates of P_{cum} and F_a for a maximum atmospheric CO_2 level of 438 ppm.

Year	Atmospheric CO_2 (ppm)	Model 1 P_{cum} (ppm)	Model 1 F_a (%)	Model 2 P_{cum} (ppm)	Model 2 F_a (%)	Model 4 P_{cum} (ppm)	Model 4 F_a (%)	Model 5 P_{cum} (ppm)	Model 5 F_a (%)
2020	395	190	52.7	175	59.0	209	46.6	169	61.0
2070	431	304	44.6	268	51.9	357	37.4	245	56.7
2170	438	406	35.2	350	41.8	506	27.8	309	47.3

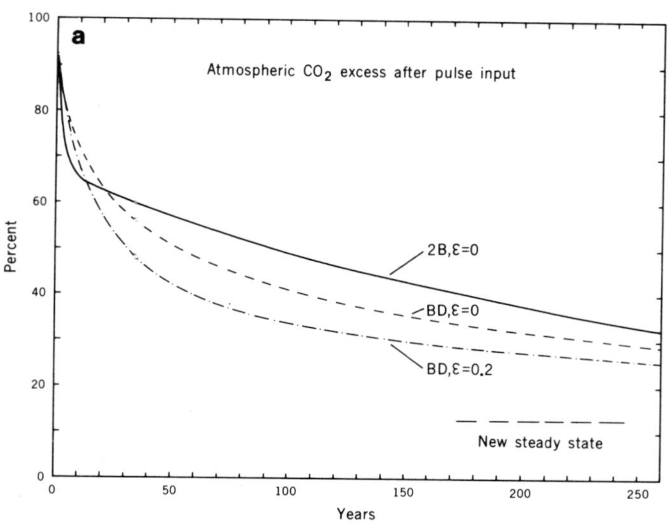

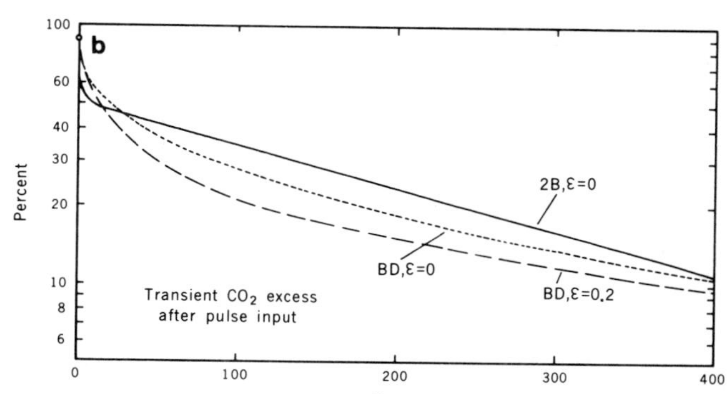

Fig. 3. (a) Responses of models 1, 2, and 5 to a CO_2 pulse input. The models are as described in Table 1, except for a constant buffer factor $\xi = 10$ (linear scale). (b) Same as (a), but only the transient part of the airborne fraction—that is, the part of the atmospheric CO_2 excess that will have vanished into the ocean or the biosphere after infinite time (logarithmic scale).

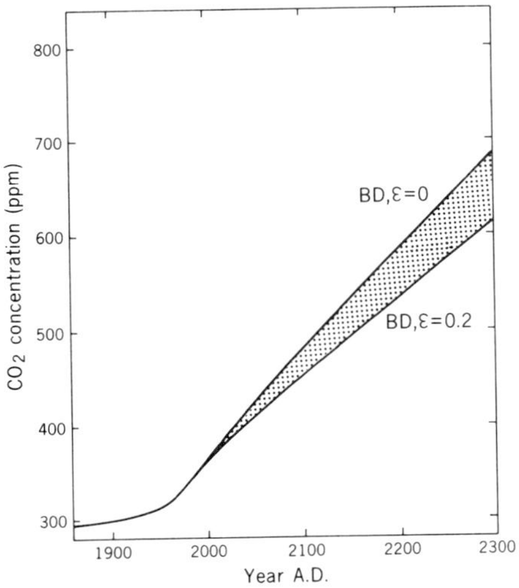

Fig. 4. Predictions of future atmospheric CO_2, assuming a constant production rate of 2.4 ppm/year after 1970. Only the results for a diffusive deep sea and for the theoretical buffer factor ξ (see Eq. 1) are given, for a constant (ϵ = 0) and a growing (ϵ = 0.2) biosphere. The shaded area indicates the concentration range which we consider the most realistic.

ommended for modeling by Rotty (*34*).

It must be emphasized that we have arbitrarily chosen the limit of 50 percent for the allowable CO_2 excess, and it may be too low or too high. It merely serves to illustrate the concept of an energy strategy based on a maximum permissible atmospheric CO_2 increase.

Further Consequences

It has been mentioned that with an increasing CO_2 level the ocean might become undersaturated in CO_3^{2-}, so that the carbonate compensation level might rise and carbonate sediments be dissolved at depth. Once the dissolution of sedimentary carbonate started, the ocean's capacity for taking up industrial CO_2 would increase (*5, 35*). Eventually, even the surface ocean might become undersaturated in CO_3^{2-}, which would threaten the existence of organisms that build calcareous shells; however, surface waters are generally very supersaturated in CO_3^{2-} (*28*), so this danger does not exist in the near future, except perhaps in a few limited areas. We do not want to discuss these problems because (i) the kinetics of carbonate dissolution are not yet fully understood, and (ii) for a study of the oceanic carbonate system, the oceanic circulation and fluxes have to be considered in more detail than they are in global models designed for computing atmospheric CO_2 levels. For instance, the Atlantic and Pacific oceans should be separated because their turnover times and carbonate concentration patterns are significantly different. The formation of deep water at high latitudes by sinking of dense surface water should be considered.

Several authors (*35, 36*) have suggested that higher CO_2 concentrations would lead to enhanced continental weathering of minerals, and thus to a higher calcium input into the oceans, tacitly assuming that the weathering capacity of terrestrial waters is determined by the atmospheric CO_2 partial pressure. However, this is not the case in general, since the equilibrium CO_2 pressure in soil waters and in rivers is usually much higher than the atmospheric level, on the average by about one order of magnitude in the rivers of the world (*37*) This is so because most CO_2 is taken up by the infiltrating water in the top soil zones, where biogenic CO_2 is present at high concentrations. An increase in dissolved CO_2 and, therefore, weathering capacity could thus be caused only indirectly—for instance, through enhanced biospheric activity or a more intensive hydrologic cycle—and would be relatively small compared to the atmospheric CO_2 increase.

There are several feedback mechanisms that have not been mentioned so far. An amplification of the greenhouse effect leads to a warmer ocean surface and thus a lower CO_2 solubility, so that the atmospheric CO_2 pressure grows even more. Eriksson (*38*) computed a 5 percent change of the CO_2 pressure per degree centigrade for equilibrium between atmosphere and ocean. If a doubling of the CO_2 concentration (to ~580 ppm) leads to a warming of 1.5° to 3°K, this would bring about another CO_2 increase of approximately 7.5 to 15 percent, or 44 to 87 ppm. This first-order estimate shows that this feedback effect is quite appreciable. As pointed out by Eriksson, however, it would take a very long time for the whole ocean to reach equilibrium, so that the effect would be smaller on the time scales considered here.

According to Manabe and Wetherald (*2*), not only temperature but also amount of precipitation would rise with a higher CO_2 level. Both changes are favorable for biological productivity (*39*), so that growth of the biosphere is suggested not only on the basis of its dependence on CO_2. Whether these favorable tendencies would prevail over the negative influences on vegetation—extensive human use and air pollution—can only be decided when adequate data are obtained.

Finally, one should keep in mind the question of what factors determine the natural atmospheric CO_2 pressure in the long term, which has not yet been clearly resolved (*15*). Geologic processes that are not discussed here then become important, and the problem is even more complicated than when one considers only variations over centuries.

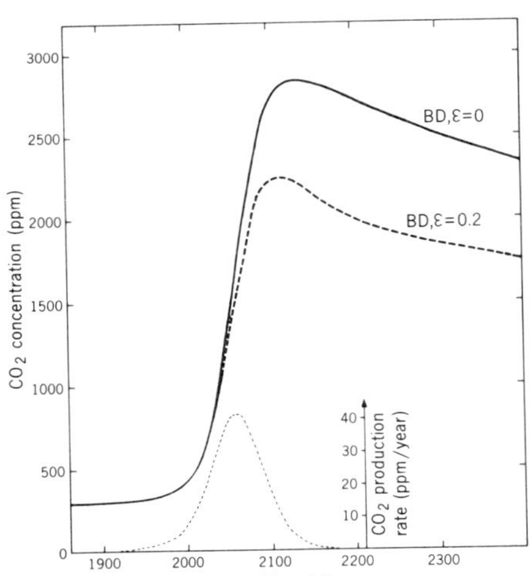

Fig. 5. Predictions for the case in which all recoverable fossil fuel is burned. The lowermost curve shows the assumed CO_2 production rate. The concentration curve for ϵ = 0.2 is dashed because it would predict extensive growth of the biota, which does not seem very probable.

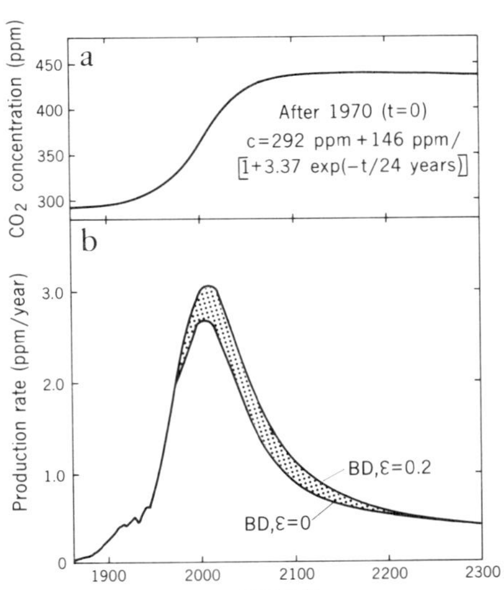

Fig. 6. Carbon dioxide production rates as observed up to 1970, and as permitted after 1970 for an increase of the atmospheric excess in a prescribed way (a) to a maximum of 50 percent.

Conclusions

It has often been noted that about 50 percent of the total CO_2 produced remains in the atmosphere. Bacastow and Keeling (*10*) found, however, that the buffer factor of seawater increases with higher atmospheric CO_2 concentration, which means that the ocean becomes a less effective sink for industrial CO_2. For our upper-limit scenario (all fossil reserves to be burned) the airborne fraction may be as high as 80 percent. After CO_2 production stopped, the atmospheric level would decline very slowly to a new equilibrium state. At equilibrium, the atmosphere would still contain at least one-eighth of the total man-made CO_2. This clearly shows that, once negative effects of the CO_2 increase were observed, it would not be possible to restore the "pre-CO_2" situation simply by stopping CO_2 emissions.

If a maximum CO_2 increase of 50 percent above the preindustrial concentration is tolerable, CO_2 production could still be increased until the beginning of the 21st century, but it would then have to be reduced rapidly. The value of 50 percent for the tolerable increase may be too high or too low; if it is correct, we may burn in total over the next centuries not much more than 10 percent of the known fossil fuel reserves.

Our intention in this article is to present the results that can be obtained by means of relatively simple global models and to indicate the uncertainties still involved. The following are some major problems that are still open.

1) *Biospheric masses and fluxes*. The biomass has probably decreased in some places and increased in others. Is the observed CO_2 increase partly due to a shrinking biosphere, or did the biosphere as a whole take up fossil CO_2?

2) *The oceanic buffer mechanism against CO_2 variations*. There is a discrepancy between thermodynamic computations (*6, 10*), which yield a buffer factor ξ of 9 to 10, and experimental results (*22*), which yield $\xi \approx 7$. Unfortunately, it is difficult to measure the increase of the total CO_2 concentration in the sea, which presently amounts to only about 1 percent in surface waters, while natural variations (for example, with latitude) are much larger.

References and Notes

1. R. M. Goody and J. C. G. Walker, *Atmospheres* (Prentice-Hall, Englewood Cliffs, N.J., 1972).
2. S. Manabe and R. T. Wetherald, *J. Atmos. Sci.* **32**, 3 (1975).
3. S. Schneider, *ibid.*, p. 2060.
4. C. F. Baes, H. E. Goeller, J. S. Olson, R. M. Rotty, *The Global Carbon Dioxide Problem* (ORNL-5194, Oak Ridge National Laboratory, Oak Ridge, Tenn., 1976).
5. S. Bolin and E. Eriksson, in *The Atmosphere and the Sea in Motion*, B. Bolin, Ed. (Rockefeller Institute Press, New York, 1959), pp. 130–142.
6. W. S. Broecker, Y.-H. Li, T.-H. Peng, in *Impingement of Man on the Oceans*, D. W. Hood, Ed. (Wiley-Interscience, New York, 1971), pp. 287–324.
7. W. S. Broecker, *Science* **189**, 460 (1975).
8. L. Machta, in *The Changing Chemistry of the Oceans*, D. Dyrssen and D. Jagner, Eds. (Almqvist & Wiksell, Stockholm, 1972), pp. 121–145.
9. K. E. Zimen and F. K. Altenhein, *Z. Naturforsch. Teil A* **28**, 1747 (1973).
10. R. Bacastow and C. D. Keeling, in (*11*), pp. 86–135.
11. G. M. Woodwell and E. V. Pecan, Eds., *Carbon and the Biosphere* (Atomic Energy Commission, Technical Information Service, Washington, D.C., 1973).
12. C. D. Keeling, in *Energy and Climate: Outer Limits to Growth* (National Academy of Sciences, Washington, D.C., in press).
13. Data and models on the global carbon cycle have been summarized by Baes *et al.* (*4*), and by Bolin (*14*).
14. B. Bolin, *Sci. Am.* **223**, 131 (September 1970); in *The Physical Basis of Climate and Climate Modelling* (GARP Publ. Ser. No. 16, World Meteorological Organization, Geneva, 1975), pp. 225–235.
15. R. M. Garrels and E. A. Perry, in *The Sea*, E. G. Goldberg, Ed. (Wiley, New York, 1974), vol. 5, pp. 303–306.
16. For a summary of work on ^{14}C in the oceans see M. Stuiver, in (*11*), pp. 6–20.
17. C. D. Keeling, in *Chemistry of the Lower Atmosphere*, S. I. Rasool, Ed. (Plenum, New York, 1973), pp. 251–329.
18. K. Wyrtki, *Deep-Sea Res.* **9**, 11 (1962); W. H. Munk, *ibid.* **13**, 707 (1966); J. Craig, *J. Geophys. Res.* **74**, 5491 (1969).
19. H. Oeschger, U. Siegenthaler, U. Schotterer, A. Gugelmann, *Tellus* **27**, 168 (1975).
20. The natural ^{14}C variations, as observed in tree rings, seem to be larger than the ^{14}C variations calculated from models, including the BD model, under the assumption that they are due to varying ^{14}C production due to changing solar activity as reflected in sunspot numbers. However, this is not necessarily due to wrong model results for the attenuation factors, because the relationship between sunspot numbers and changes in ^{14}C production is not firmly established. Furthermore, there may be other reasons for ^{14}C variations, such as fluctuations in ocean mixing.
21. W. S. Broecker and T.-H. Peng, *Tellus* **26**, 21 (1974).
22. A. Rebello and K. Wagener, paper presented at the 4th International Symposium on the Chemistry of the Mediterranean, Rovinj, Yugoslavia, 1976.
23. W. A. Reiners, in (*11*), p. 303. The high carbon content of the terrestrial biosphere is obtained from an estimate of total nitrogen in organic detritus and a C/N ratio of 12.
24. H. Lieth, *J. Geophys. Res.* **68**, 3887 (1963).
25. G. W. Woodwell and R. A. Houghton, in *Dahlem Workshop on Global Chemical Cycles and Their Alternation by Man*, W. Stumm, Ed. (Abakon, Berlin, 1977), pp. 61–72.
26. B. Bolin, *Science* **196**, 613 (1977).
27. H. D. Freyer and L. Wiesberg, in *Isotope Ratios as Pollutant Source and Behaviour Indicators* (International Atomic Energy Agency, Vienna, 1975), p. 49; M. Stuiver, personal communication.
28. W. S. Broecker, *Chemical Oceanography* (Harcourt Brace Jovanovich, New York, 1974).
29. W. Stumm, in *The Changing Chemistry of the Ocean*, D. Dyrssen and D. Jagner, Eds. (Almqvist & Wiksell, Stockholm, 1972), pp. 329–346.
30. C. D. Keeling, R. B. Bacastow, A. E. Bainbridge, C. A. Ekdahl, P. R. Guenther, L. S. Waterman, J. F. S. Chin, *Tellus* **28**, 538 (1976); C. D. Keeling, J. A. Adams, C. A. Ekdahl, P. R. Guenther, *ibid.*, p. 552; C. D. Keeling, private communication. Note that the corrected scale ("1974 manometric mole fraction scale") used in these publications and also in this article is slightly different from the scale used formerly by Keeling and co-workers. The value for the end of 1971 is 326.5 ppm (new scale) compared to 322.5 ppm (old scale).
31. Data up to 1970 are from C. D. Keeling [*Tellus* **25**, 174 (1973)]; data for 1971 to 1974 are from R. M. Rotty (*34*); and the 1975 value (2.40 ppm/year) was estimated by Keeling (*12*).
32. G. S. Callendar, *Tellus* **10**, 243 (1958); J. R. Bray, *ibid.* **11**, 220 (1959).
33. The production functions leading to the prescribed CO_2 excess were computed by leaving the production rate constant for 1 year, then comparing the actual atmospheric concentration with the concentration aimed at (Fig. 6), correcting the production rate accordingly, continuing the procedure for one more year, and so on.
34. R. M. Rotty, *Global Carbon Dioxide Production from Fossil Fuel and Cement, A.D. 1950–A.D. 2000* (Institute for Energy Analysis Research Memorandum [IEA(M)-76-4], Oak Ridge Associated Universities, Oak Ridge, Tenn., 1976).
35. W. S. Broecker, in (*11*), pp. 32–50.
36. P. Möller and P. P. Parekh, *Sci. Total Environ.* **4**, 177 (1975).
37. R. M. Garrels and F. T. Mackenzie, *Evolution of Sedimentary Rocks* (Norton, New York, 1971), chaps. 5 and 6.
38. E. Eriksson, *J. Geophys. Res.* **68**, 3871 (1963).
39. R. H. Whittaker and G. E. Likens, in (*11*), pp. 281–302.

40. Thanks are due to M. Stuiver for stimulating discussions, and to W. S. Broecker and another reviewer, unknown to us, for valuable comments on this article. This work is financially supported by the Swiss National Science Foundation.

II

Conservation and Public Policy

U.S. Energy Demand: Some Low Energy Futures

Demand and Conservation Panel of the Committee on Nuclear and Alternative Energy Systems

This article presents several plausible energy demand scenarios for the United States through the year 2010, each one derived from analytic efforts conducted by the Demand and Conservation Panel (*1*) of the National Research Council's Committee on Nuclear and Alternative Energy Systems (CONAES). While the CONAES study (*2*) covers a range of plausible energy futures—from continued rapid growth in demand to actual reductions in demand—we focus here on futures in which demands are lower, in order to provide insight into how energy demand growth can be reduced, and on the consequences of low energy growth. As a further effort to explore low energy growth, we assume future economic growth to be smaller than it has been in the past or than many think it likely to be in the future. This analysis is not intended to show that low demand futures are the most likely or the most desirable; instead, it is meant to illustrate the opportunities for lower energy demand growth and the public policies required to realize these opportunities.

Summary. The basic features of the U.S. energy supply and utilization system change so slowly that an understanding of the dynamics of major change requires projections that extend several decades. Over such a time most energy-consuming capital stock is replaced, life-styles change, and technology evolves. This study of plausible low energy futures leans heavily on detailed engineering analysis of demand by sector, combined with econometric techniques where appropriate. The results indicate that, given time for the system to respond to prices, regulations, and incentives, U.S. energy demand is very elastic. Consequently, a major slowdown in demand growth can be achieved simultaneously with significant economic growth by substituting technological sophistication for energy consumption.

Low energy futures could result from constraints on supplies that appear as higher prices, import restraints, rationing, taxes, or other public policies (for instance, in response to limitations on oil and gas imports, SO_x and CO_2 production, and nuclear power). They could also result from a national decision to use energy resources more efficiently or from shifts in social priorities (such as less pollution of air and water). Adjustments to these changed conditions, whether in developing new energy supplies, expanding old ones, devising a more efficient utilization system, or changing the mix of goods and services demanded, will require decades of effort. To devise and implement a reasonable plan of action we must gain a better understanding of future supplies of energy (and their cost) and at the same time project plausible levels of demand for energy, identifying factors that can influence the future course of demand (energy price, population and income growth, status of technology, and so forth).

The scenarios presented here suggest that there is much more flexibility toward reducing energy demand than has been assumed in the past. Indeed, it appears that very similar conditions of habitat, transportation, and other amenities could be provided in the year 2010 with primary energy consumption ranging from 60 to 135 $\times 10^{15}$ Btu's (quads). This wide range of plausible future energy demands results from the technically possible responses to various price and policy possibilities.

The CONAES Demand and Conservation Panel consists of John Gibbons, University of Tennessee, Knoxville 37916; Roger Carlsmith, Oak Ridge National Laboratory, Oak Ridge, Tennessee 37830; Macauley Whiting, Dow Chemical Company, Midland, Michigan 48640; Eugene Goodson, Purdue University West Lafayette, Indiana 47907; Clark Bullard, University of Illinois, Urbana 61801; Duane Chapman, Cornell University, Ithaca, New York 14850; Charles Berg, Route 1, Bucksfield, Maine 04220; Norman Bradburn, University of Chicago, Chicago, Illinois 60637; Paul Craig, University of California, Berkeley 94720; Joel Darmstadter, Resources for the Future, Inc., Washington, D.C. 20036; Stephen Rattien, Donovan, Hamester, and Rattien, Inc., Washington, D.C. 20007; Lee Schipper, Lawrence Berkeley Laboratory, Berkeley, California 94720 (on leave at the Swedish Royal Academy of Sciences); and Robert Uhler, Electric Power Research Institute, Palo Alto, California 94303.

Background

For many decades before the 1970's, the price of a unit of energy was a small and steadily decreasing part of almost everyone's budget. Major discoveries of oil and gas, combined with rapidly advancing technology, led to falling energy prices and promoted enormous growth in consumption. Not only did we find ever-increasing uses for energy, but because the relative price was falling there was correspondingly less incentive to be concerned about efficiency of utilization; instead, expansion of production (especially of electricity) was attractive because each new generating unit was more efficient and helped lower the average cost of production. Despite falling energy prices and the consequent lack of concern for conservation, efficiency of energy utilization improved steadily. For example, industry's energy consumption per unit of output fell at a rate of 1 to 1.5 percent per year from 1950 to 1970 (*3*). Indeed, in the past half-century, U.S. energy consumption per capita rose by 50 percent while the productivity of that energy—the ratio of gross national product (GNP) to energy consumed—rose by 70 percent. Part of this improvement came about through conversion from coal to oil and gas (conversion of railroads from coal-fired steam to diesel) but much is attributable to other factors such as new processes and advances in materials (solid state electronics, improved insulation) that made possible more efficient energy conversion and utilization.

Between 1946 and 1973 amenities such as large automobiles, air conditioning, frost-free refrigerators, and home freezers changed from luxuries to necessities and found ready markets. The future, according to many observers, was filled with prospects of ever-cheaper energy that would provide for our many wants, enable resource production from low-grade ores, and turn deserts green with desalted water. Electricity so cheap that it would hardly pay to meter it was even conjectured. Such times of technological Nirvana now seem far more than 15 years in the past.

Several separate events occurred over the past decade that have profoundly influenced both our energy systems and our perception about the future.

• As low-cost domestic oil resources were consumed, the nation began, in the late 1960's, to depend on imported oil, which was then cheap. The hidden costs (vulnerability to cutoffs and discontinuities in price) of that dependence appeared with the oil embargo of 1973.

• Low-cost nuclear power emerged more slowly and turned out to be more costly than anticipated. Costs of other new energy forms (such as synthetic fuels from coal) also escalated rapidly.

• There was increased awareness in the 1960's of external costs of energy production, such as pollution and health and safety costs. External costs contributed to energy price increases in the 1970's as some of them were internalized. Some coal users switched to oil and natural gas, pushing up demand for those rapidly depleting energy forms.

The long-term trend of declining real energy price (that is, corrected for inflation) was reversed in the early 1970's, when major price increases were traumatically introduced by the Organization of the Petroleum Exporting Countries (OPEC) (fall 1973) and other fuel prices subsequently moved upward (Fig. 1). The trend toward higher prices is not likely to be reversed soon. At issue, therefore, is the extent to which society can respond to these higher prices not by sacrifice but through conservation—substitution of activities and increased efficiency of use.

What Is Energy Conservation?

Few concepts are subject to as many different interpretations as energy conservation (*4*, *5*). In recent years, most people have associated conservation almost exclusively with curtailment, as unanticipated interruptions in supplies of oil and natural gas have forced Americans to cut back on energy use. In this article we do not view conservation as curtailment. Instead, conservation includes technological and procedural changes that allow us to reduce demand for energy (or specific scarce fuels) without corresponding reductions in the goods and services we enjoy. That is, conservation is a means of enhancing the energy user's perceived welfare; it leaves society materially better off than it would be otherwise and therefore is an act of enlightened self-interest; it implies that benefits of conservation actions exceed costs. These changes can be accomplished by substituting capital (insulation) or ingenuity (new microprocessor controls on heating systems) for the brute force use of energy to accomplish a particular task. Our definition allows for changes in the market basket of goods and services making up the GNP, as consumers shift to less energy-intensive goods and services to meet their needs (shifting to less energy-intensive modes of travel, buying more durable or repairable goods).

In the United States today the need for economic efficiency alone demands that energy be conserved. Energy is being wasted in buildings and industrial processes because they were designed in an era when today's energy scarcities and price levels were not foreseen; the energy component of operating costs was then too small to warrant much attention. Today we find it cost-effective to upgrade much energy-consuming equipment and to build more energy-efficient vehicles, industrial equipment, and buildings for the future. In the past, government policies led to energy prices far below replacement cost—the price level that would occur under perfect market conditions. As these policies are changed, higher levels of energy conservation will become economic.

Many advocates argue for more energy conservation than even "perfect" market signals would call for. The most commonly cited reason is that we should reduce oil imports to minimize our vulnerability to embargoes and our balance-of-payments deficits. Some of the other reasons have to do with timing; President Carter maintains that energy conservation can give us time to reevaluate our commitment to the plutonium breeder reactor and possibly achieve a significant reduction in the risks of nuclear proliferation. Holdren (*6*) and Lovins (*7*) suggest that too much energy too soon may create greater hazards than too little too late and recommend energy conservation policies. Conservation would provide the nation with the flexibility to pick and choose among long-term energy options as the uncertainties surrounding nuclear waste management, the technology of solar cells, and the climatic effects of fossil fuel combustion are clarified. Kissinger (*8*) points to the possibility that oil exporting nations may accumulate balance-of-payments surpluses large enough to precipitate financial crises and economic chaos in the industrial democracies of the world. Other reasons for conserving energy are based on ethical considerations such as the fact that the United States has only 5 percent of the world's population, yet consumes more than one-third of the world's energy, and the belief held by environmentalists that major external costs remain in the energy system and that much of our finite energy resources should be saved for future generations.

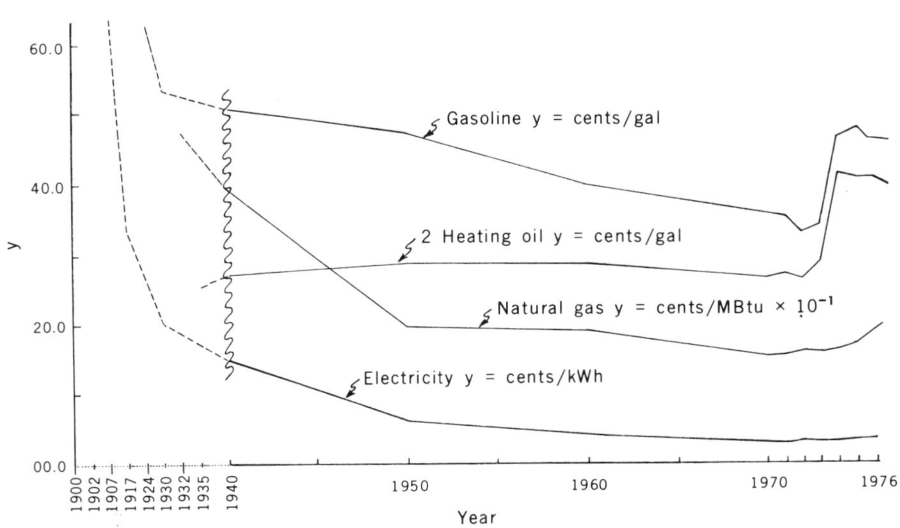

Fig. 1. History of residential energy prices (1976 dollars).

The Demand Study

In this study the Demand and Conservation Panel examined some plausible paths that energy demand might take in the future (through 2010) in response to price and other factors. It was undertaken in support of a more comprehensive CONAES study of the future of energy supply and demand (*2*, *9*). Panel members were chosen to provide a diversity of points of view and disciplinary skills. Additional resources and expertise were brought into the study by several resource groups, which were formed by the Panel to examine specific issues (*1*).

At the outset, the Panel encountered considerable conventional wisdom about energy use that has found extensive, if tacit, acceptance in various segments of American society. The following are some examples of what are now known to be misconceptions about energy use.

• Energy and the production of goods and services are intimately and inextricably linked: energy is a relatively fixed factor in GNP.

• Energy consumption and jobs are inextricably tied together; more energy consumption means more jobs, and vice versa.

• Higher illumination levels generally help productivity; low illumination is injurious to the eyes.

• Reducing the growth of energy consumption implies replacement of machines (bulldozers) with manual labor (men and women wielding picks and shovels).

• Turning down the thermostat at night is counterproductive; the energy used in heating up the house in the morning more than offsets any savings.

This kind of folklore indicated a need to develop a more quantitative technical and economic understanding of the dynamics of energy consumption. The Panel decided to map out a broad range of plausible future levels of consumption (*10*). Since consumption is influenced by many factors, certain things were assumed, such as population growth rate at Census Bureau series II, and others were allowed to vary, notably the price of energy, the economic growth rate, and public policies.

Economic growth was assumed to vary linearly in time, with total (real) GNP doubling by 2010. In a variant analysis GNP was allowed to triple by 2010. The latter case corresponds to an average real GNP growth of 3 percent per year between 1975 and 2010 with higher than average growth in the near term (for example, 4 percent) and lower growth in the long term, reflecting lower population growth. In this article we report results based on the assumption of a 2 percent average growth rate, again with higher than average growth in the near term. It should be emphasized that even in these low-demand futures we assume a doubling of real income over the next 35 years.

Table 1. Detailed price schedules for key fuel types were determined for each scenario. Average consumer prices for distillate oil, natural gas, utility coal, and electricity in 2010 are given in 1975 dollars per million Btu's. Prices are assumed to change slowly and continuously between 1975 and 2010.

Scenario	Consumer price ($/MBtu)			
	Distillate oil	Natural gas	Utility coal	Electricity
		1975		
Actual	2.81	1.29	0.81	7.91
		2010		
I and II	13.49	14.84	3.24	26.37
III	6.74	7.42	1.62	15.82
IV	2.81	3.09	0.81	7.91

For most scenarios, price was used as the primary driving force behind the different outcomes. There were three pricing assumptions, corresponding approximately to prices four times, double, and equivalent to 1975 prices (see Table 1). A very low growth variant, assuming significant changes in life-style, was also studied.

Prices serve many functions in a market economy. They reflect the relative scarcity of resources and serve as signals to producers and consumers in allocating the supply of and demand for energy resources. Prices assumed for the scenarios were allowed to reflect external costs as they affect occupational and public health as well as environmental quality, and the degree to which these costs become internalized in the costs of energy products.

Public policies can serve these functions as well. First, they can and do affect price. Environmental regulations internalize costs, tax policy affects pricing patterns, and regulatory agencies (such as state utility commissions) sometimes set prices. For example, in response to external costs, regulation has set minimum standards on car safety, air pollution, pesticides, strip-mining, and mine safety. Second, conservation education can shift the mix of goods and services demanded as well as their energy efficiency. Third, tax subsidies can and have promoted production. Therefore public policies can be alternatives or surrogates for price changes, seeking to reflect social value and cost and guide producers and consumers in decisions concerning production and consumption.

In analyzing the dynamics of demand over such an extended period of time it was decided to invoke several different techniques.

1) *Near future (the next decade).* Since energy consumption patterns in this period will strongly reflect the characteristics of existing capital stock, the most appropriate approach was felt to be econometric analysis. In applying this technique, recent trends in energy consumption were analyzed and used to project demand as influenced by both price and nonprice factors. Statistical techniques were used to determine the past quantitative relationships of energy

Table 2. Energy demand scenarios.

Scenario	Energy price ratio, 2010/1975*	Energy conservation policy	Energy in 2010 (quads)					
			Buildings	Industry	Transport	Total	Losses†	Primary consumption‡
I	4	Very aggressive, deliberately arrived at reduced demand requiring some life-style changes	6	26	10	42	16	58
II	4	Aggressive; aimed at maximum efficiency plus minor life-style changes	10	28	14	52	22	74
III	2	Slowly incorporates more measures to increase efficiency	13	33	20	66	28	94
IV	1	Unchanged from present policies	20	39	26	85	51	136
1975			16	21	17	54	17	71

*Overall average; assumptions by specific fuel type were made reflecting parity and supply; price increases were assumed to occur linearly over time. The price was assumed to be either that actually charged at the final point of demand or the shadow price reflecting a policy. †Losses include those due to extraction, refining, conversion, transmission, and distribution. Electricity is converted at 10,500 Btu/kWh, coal is converted to synthetic liquids and gases at 68 percent efficiency. ‡These totals include only marketed energy. Active solar systems provide additional energy to the buildings and industrial sectors in each scenario. Total energy consumption values are 63, 77, 96, and 137 quads in scenarios I, II, III, and IV, respectively.

Table 3. Total primary energy consumption in 2010 by fuel type. Each set of numbers represents only one of a wide variety of energy resources that could be used to meet energy demands. Because energy resources are largely interchangeable over the long run, the actual mix in 2010 can be influenced by changes in price, technology, and policy.

Fuel type	Total primary energy consumption (quads)				
	Scenario				1975
	I	II	III	IV	
Liquid fuels*	24	29	38	50	30
Gaseous fuels†	8	9	11	26	17
Coal (direct use)‡	10	11	13	10	4
Electric inputs§	17	26	32	50	20
Total purchased fuels‖	58	74	94	136	71
Active solar¶	5	3	2	1	Negligible
Totals	63	77	96	137	71

*Liquid fuels include petroleum, shale oil, and synthetic liquids derived from coal. †Gaseous fuels include natural gas and gasified coal. ‡Figures do not include coal used for liquid and gaseous fuels (necessary in most of the scenarios) or for electricity production. §Includes coal, nuclear, hydro, geothermal, and oil (for peak demand only). ‖Because of rounding off, totals may not equal the sums for fuel sectors. ¶Estimated use of active solar units in buildings and industry.

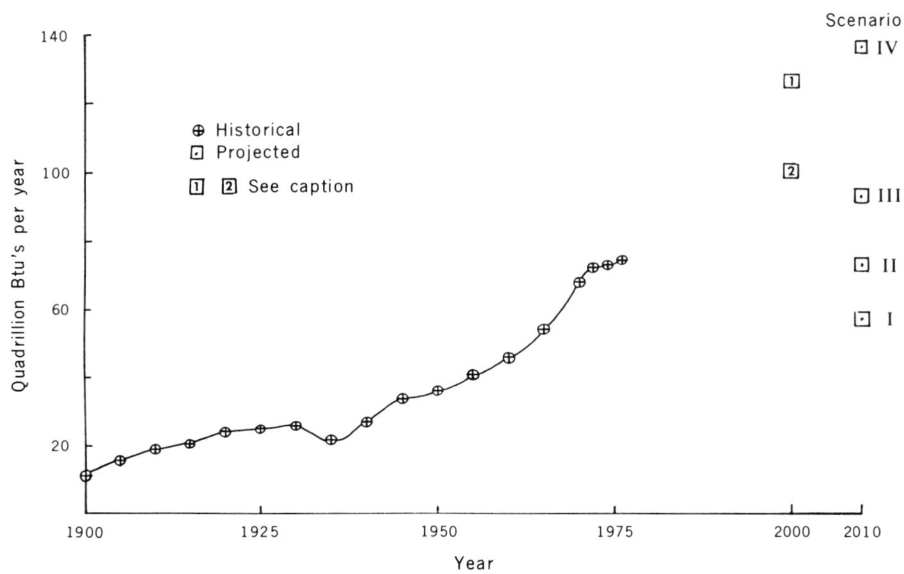

Fig. 2. Summary of total U.S. energy demand scenarios. Some information was obtained on values for 1990, but the detailed evaluations were focused on 2010. Other results shown for comparison are [1] the Energy Policy Project (EPP) (*16*) "tech fix" case and the Institute for Energy Analysis (IEA) (*18*) "high" case, and [2] the EPP "zero energy growth" case and the IEA "low" case.

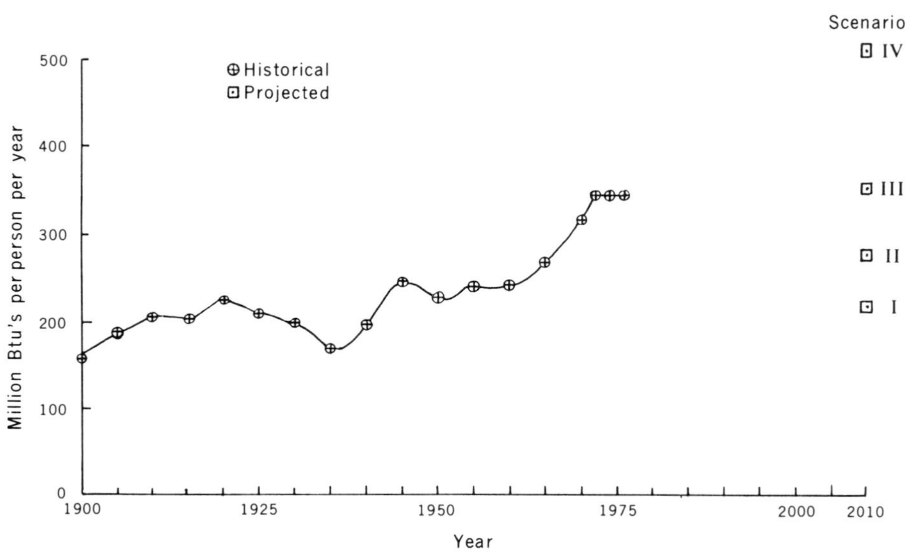

Fig. 3. Per capita energy demand history and projections. Population growth rate was assumed to be that given by Census Bureau series II.

use to energy prices, conservation efforts, incomes, GNP, population, and other demographic and economic factors. Energy demand in the future was defined by the interaction of these observed historical relationships tempered by the assumptions of future economic and demographic variables. Although this technique can be used to examine historical variations in energy use in relation to geographic variations in energy prices and income to obtain near-term insights, it has limited validity for application to the distant future. It cannot logically be applied to periods with major differences in economic structure or energy prices. The contribution of econometrics to the work of the Panel was that it indicated whether the short-term response to assumed prices and incomes in each scenario leads energy demand to follow a path in the near future that is consistent with the Panel's energy projections into the next century.

2) *Longer-term future (to 2010).* Energy consumption characteristics of various amenities (such as transportation and habitat) can change markedly over periods comparable to the average lifetime of energy-consuming capital equipment (cars, air conditioners, and houses). Social conditions and attitudes can also change greatly over a 35-year span. Therefore, even though our analysis will undoubtedly turn out to be somewhat off the mark, it is important because actions taken today will significantly affect the conditions that exist in 2010. A detailed engineering approach was taken to analyze plausible energy trends in each consuming sector over the long term. Energy consumption was partitioned into three sectors—residential and commercial buildings and appliances, industry, and transportation—which were examined independently. The results of the independent sectoral demand analyses were integrated, using energy input-output analysis to make them consistent with a specified market basket of goods and services making up the GNP in each scenario.

For the scenarios presented here, the real GNP growth rate was assumed to slowly decline from 3 percent in the late 1970's to about 1 percent in 2010, corresponding to a linear increase and a doubling of real GNP over the 35-year period. The reduced GNP growth rates in the latter years stem from expected reductions in the rate of growth of population and the labor force. The assumption that GNP will rise while the labor force is leveling off implies a positive rate of growth in productivity, but does not specify the sectors in which it occurs. The scenarios may be taken to reflect various assumptions about relative price changes in energy and nonenergy sectors, or price may be considered simply a surrogate for a variety of government policy options.

All scenarios implicitly account for the development and market penetration of new technologies and products. This is accomplished by specifying the market basket of consumer goods and services consistent with a fixed GNP. Any new product within this limit would have to displace another; only the differential energy requirement would be relevant.

Caveats. Neils Bohr pointed out that "it is very difficult to make an accurate prediction, especially about the future." We feel that for such subjects as energy demand it is even very difficult to make an accurate projection. Thus, we present our major disclaimers about the study.

• It is assumed that the future unfolds smoothly. Energy prices are a primary driving force behind the analyses, and it is assumed that they are known to energy users or that appropriate regulations compensate for the lack of perfect information. If there is extreme uncertainty about energy prices, and reason to believe that they may drop significantly, many persons will choose not to make major capital investments in conserving technologies.

• For most of the outcomes studied it is assumed that personal tastes will not change very much from those of today. Roughly the same (albeit twice as large) market basket of goods is to be provided to consumers in 2010 as today. Changes in values could modify this assumption drastically.

• Because the use of energy in the United States is so inefficient, energy consumed in providing an amenity can be reduced by changes in efficiency over a relatively wide range (*11, 12*). That is, an increase in the real price of energy can make it economically very attractive to use less energy to provide a particular amenity. Recent history shows that the effect of price can just as easily be produced by regulation, standards, incentives, and disincentives; it can also be produced by supply uncertainty, especially in the industrial sector.

• Long-term economic growth is exceedingly difficult to predict because it depends on vagaries of population growth, politics, social attitudes, labor productivity, technological innovation, and resource availability. Therefore any long-term GNP or, for that matter, population assumption is subject to challenge.

• Almost all of our past experience with energy (the basis for econometrically derived price elasticities) has been in an era of falling real energy prices. The lag times associated with price elasticities may not be applicable under conditions of rising real prices.

• Energy consumption models that are primarily based on engineering can only be approximate; this is partly because assumptions must be made about economically rational market behavior. For example, people may choose to buy a slightly cheaper, low-efficiency refrigerator and pay higher operating costs.

• New kinds of unusually energy-intensive consumer goods are not assumed explicitly but neither are new kinds of energy-saving technologies. Either or both could change demand patterns significantly in a decade.

• Energy prices are assumed to reach parity—that is, price is assumed to reflect the quality of energy—by the middle 1980's. (This means that coal will continue to be priced lower than oil or gas on a Btu basis.)

Results of the Analyses

Results are presented here for four demand growth scenarios. Each is believed to represent an internally consistent picture of the U.S. economic system in the

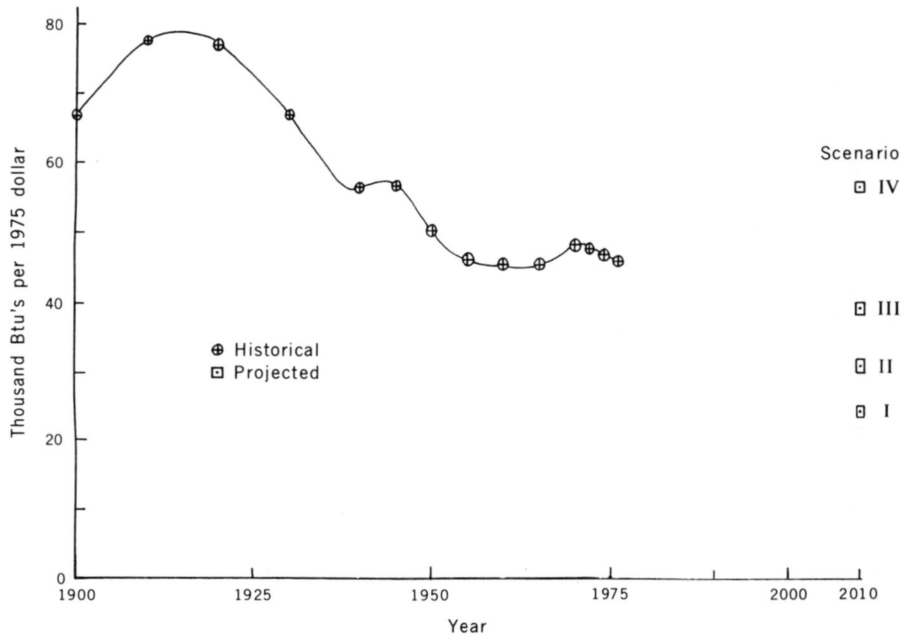

Fig. 4. Time dependence of the energy/GNP ratio. Many factors such as market basket choices and the price of energy can affect this ratio. Long-term trends in the ratio reflect changes in technology as they affect the energy intensity of new capital stock.

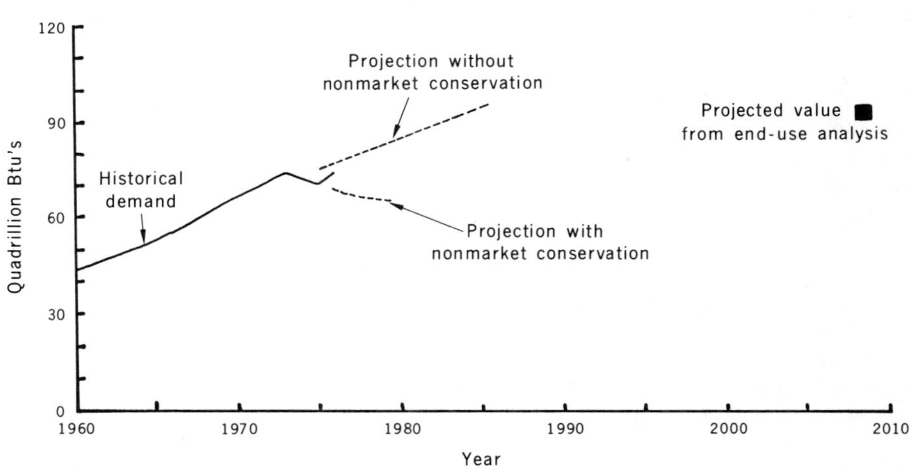

Fig. 5. Energy demand projection for the next 10 years. This example, corresponding to scenario III, indicates the range of total U.S. energy demand that could occur. The upper dashed line corresponds to demand driven by price factors, the lower one to demand influenced by price factors as well as nonmarket factors, including heightened awareness and regulations similar to those existing in 1974 and 1975. The demand value in 2010 obtained through detailed end-use analysis is consistent with extension of the range of econometrically derived demand.

Table 4. Energy efficiencies in 2010, according to scenario IV.

Buildings and appliances		Industry		Transportation	
Type	Intensity*	Type	Intensity†	Type	Intensity‡
Thermal integrity (heating)		Agriculture	0.95	Automobile	20 mpg
Residential	0.76	Aluminum	0.79	Light trucks and vans	16 mpg
Commercial	0.7	Cement	0.75	Air passenger	0.5
Government and education	0.5	Chemicals	0.84§	Truck freight	0.9
Space conditioning		Construction	0.73	Air freight	0.6
Air conditioning	0.94	Food	0.86	Rail freight	1.0
Electric heating	0.9	Glass	0.82		
Gas and oil heating	0.8	Iron and steel	0.83		
Refrigeration and freezing	0.92	Paper	0.76		
Lighting	0.70	Other industry	0.85		

*Energy intensity of new construction and products in 2010 compared with 1975. †Average energy per unit production in 2010 compared with 1975. ‡When figures are not given in designated units they refer to energy intensity in 2010 compared with 1975 (including changes in load factor). §Excluding feedstock.

year 2010. No scenario is presumed to be more probable than the others. They represent a broad range of plausible alternative futures, within the focus of this article on lower-demand futures.

The characteristics of the scenarios are summarized in Table 2. The scenarios highlight the strikingly wide range of energy efficiencies that appear to be technically and economically feasible for a particular level of economic activity. The energy demand paths are illustrated in Fig. 2. The demand levels in 2010 were obtained by using input-output synthesis of the separate sectoral analyses. Trends and projections in terms of per capita energy demand are shown in Fig. 3 and the energy GNP ratio in Fig. 4.

Plausible energy supply mixes corresponding to the scenarios are summarized in Table 3. However, it must be remembered that over a period of decades the mix of fuels (as well as aggregate demand) can change considerably. A discussion of the amount of electricity that could be used is given later in this article. The energy supply mixes shown in Table 3 are quite conventional: resource conservation significantly reduces the imperative for crash programs to develop new technologies and provides the flexibility to eliminate one or more technologies for environmental or safety reasons.

In scenario I we examine a set of fairly extreme changes by considering the combined implications of substantially higher energy prices, significant life-style accommodations, and reduction of energy consumption by equipment to approximately 60 percent of the per capita level today. Some shift of the population to warmer climates and colocation of industry and residences is assumed. For the intermediate scenarios (II and III), we assume substantially higher energy prices and corresponding major market accommodations to obtain amenities less expensively. The results for scenario II indicate the extent to which a society that chooses a high efficiency of utilization, through technical sophistication and innovation of its social institutions, can avoid substantially modifying the nominal market basket of goods and services. The results obtained for scenarios III and IV reflect a future much like today, where efficiency improvements in familiar energy-consuming devices permit production of goods and services with corresponding changes in life-styles. It should be noted, however, that even though the assumed efficiency improvements are cost-effective in terms of assumed energy prices, they are not likely to be achieved in the absence of supportive policies.

Econometric results. Long-run price elasticities were found to be comparable to those reported in other recent studies (*13*). The greatest response observed was the long-run price elasticity of −1.7 (industrial use of natural gas), while the smallest was the elasticity of −0.4 for gasoline use with respect to gasoline price. Historic price responses provide a basis for examining how near-future growth in energy demand may vary with energy price.

Differentiation of conservation, price, and income influences in reducing energy demand in 1974 showed that after the increase in energy prices and the decline in personal incomes throughout the country were taken into account, energy use fell below what would be expected from this price- and recession-induced decline. This nonprice-induced conservation effect accounted for further reductions in use of 6 percent (residential-commercial electricity) to 13 percent (industrial use of coal). It probably reflected lower speed limits, increased energy awareness, and simple improvements in energy management. It is impossible to know whether this nonprice effect will continue in the future; however, it apparently existed in a weaker form in 1975, and essentially disappeared in 1976. One such projection (for scenario III) is reproduced in Fig. 5. It shows the near-future econometric forecast associated with twofold growth in real energy prices from 1975 to 2010.

Opportunities for improved efficiency of use. Analyses of the engineering potential for higher efficiency show that major opportunities exist in all sectors of demand and that the economic incentive to shift to higher efficiency is a strong function of the price of energy. Figures 6, 7, and 8 illustrate this point (*10, 14, 15*). Although the most impressive opportunities for higher efficiency of use occur with *new* capital goods, there are also many good ways to improve the efficiency of the existing system (operation of buildings, industrial processes).

Highlights of the Scenarios

Scenario IV. With the exception of the price of natural gas (which more than doubles by 2010), energy prices remain constant in this scenario over the entire period, possibly because of breakthroughs in the supply sector or public policies that provide subsidies to energy price. Higher gas prices might result from either deregulation of wellhead prices, regulation at increased price levels, or the introduction of synthetic gas produced from coal.

Despite the lack of higher energy prices, the incentive to increase efficiency of use is greater than it was before 1972, because in those years real energy prices were falling. Furthermore, energy prices are higher today than they were when most existing capital investments were made. Therefore there are additional incentives to increase efficiency even without further improvements in the technology of energy utilization. As a consequence, the efficiency of energy use improves modestly over the period. Representative changes are given in Table 4.

In the buildings sector, the lagged effect of postembargo energy prices and public policy actions already under way result in efficiency improvements in new stocks through 1980. Because of continued increases in gas prices, small improvements in most gas appliances occur beyond 1980. The better performance in new products and buildings is felt throughout the period to 2010 as older stocks are replaced. However, improvements occur very slowly; for example, electric space heating continues to be dominated by resistance heaters. In the industrial sector the historic trend to lower energy consumption per unit output continues but at a slower pace, resulting in an overall drop of 18 percent by 2010.

In transportation, per capita energy consumption grows modestly, but technological advances result in small improvements in efficiency. Current policies [such as miles per gallon (MPG) standards for automobiles] continue. No significant shifts occur in freight modes, but the percent of consumer transportation expenditures in air travel increases as automobile travel begins to be saturated (ownership, minutes traveled per day, and so on). Transportation energy intensiveness drops in all sectors. Mass transit travel (per capita) nearly doubles.

Scenario III. In this scenario real energy prices steadily climb, ending in 2010 at twice the 1975 levels. These increases are mostly due to higher production costs as cheap supplies of fossil fuels are exhausted. There are substantial efficiency increases in the buildings, including improvements in thermal integrity of structures and increased efficiency of appliances (see Table 5). Overall building energy use *decreases* at an average annual rate of 0.6 percent, compared to growth of 3 percent per year from 1950 to 1975. The major reason for the decrease in growth rate is the reduction in space heating requirements. Because of differences in assumed energy prices, the relative market share of electricity increases from 21 to 51 percent, while the natural gas share declines from 53 to 21 percent. Electric heat pumps become more efficient and find widespread use. Solar energy becomes increasingly important toward the end of the period for air conditioning, space heating, and water heating. The energy efficiency ratio of new air conditioners in 2010 is close to 10 Btu's per watt-hour (compared with 6 Btu/W-hour in 1975). Because of the increase in energy prices, there are increased expenditures for energy in buildings. However, the percentage of personal income spent for household fuel increases only moderately, from 3.1 in 1975 to 4.3 in 2010. There is a corresponding increase of 3 percent in capital expenditures for buildings and appliances during the period. The technologies to produce the higher efficiencies in this scenario are either currently on the market or are achievable by well-known means. Figures 6 and 7 indicate the kinds of improvements that might be expected in the energy efficiencies of structures and appliances. Similar improvements are likely for other kinds of appliances and buildings.

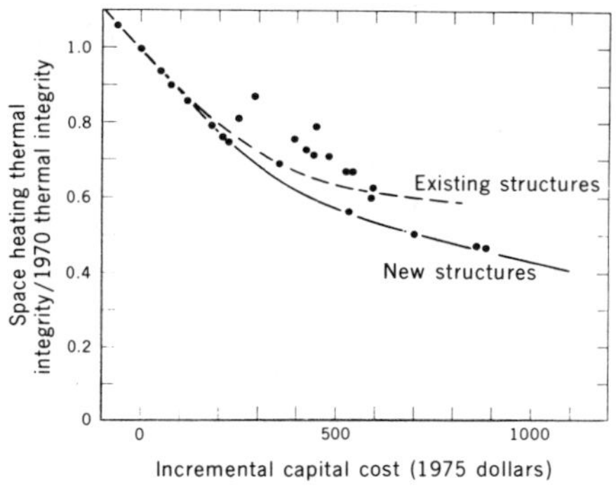

Fig. 7. Energy intensiveness of new and existing residential dwellings as a function of investment in construction (or retrofit). Energy intensiveness falls rapidly for small increases. There are more cost-effective opportunities for new structures because insulation investments cut the required size (and therefore cost) of heating and cooling systems (*15*).

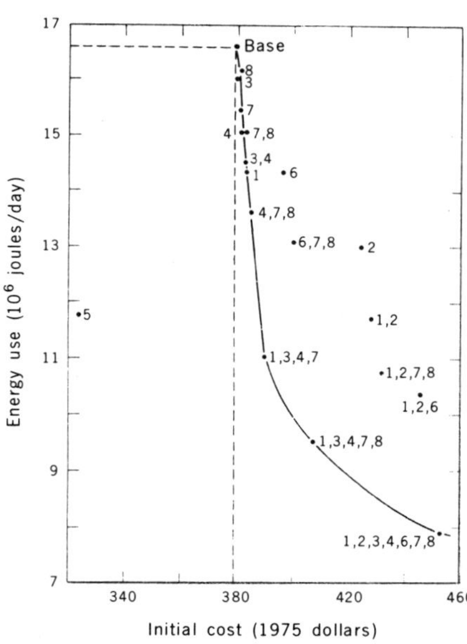

Fig. 6. Energy intensiveness of a typical household refrigerator-freezer combination, plotted against purchase price. The energy intensiveness falls rapidly for relatively small increments in first cost (*14*). Modifications corresponding to the numbered points were: 1, increase insulation thickness; 2, improve insulation thermal conductivity; 3, remove fan from cooled area; 4, add antisweat heater switch; 5, eliminate frost-free and forced air systems; 6, improve compressor efficiency; 7, increase condenser surface area; and 8, increase evaporator surface area.

In the industrial sector the steadily increasing price of energy results in slower production growth in the energy-intensive aluminum and chemicals industries than in industries producing less energy-intensive substitutes. The overall weighted average energy consumed per unit output in 2010 is 26 percent lower than in 1975. Some shifts have occurred between producing sectors (less steel, more aluminum and fiber glass), reflecting changed demands by manufacturers and the construction industry. Industrial cogeneration becomes widely practiced.

In the transportation sector, per capita energy consumption in 2010 remains about the same as in 1975 despite slightly expanded passenger and freight movement. The average energy intensity of automobiles (Btu's per vehicle mile) drops to half of the 1975 value, reflecting large gains in automobile performance (Fig. 8). Airline travel grows at an average 2 percent per year. Mass transit use in 2010 (passenger miles per capita) is 3.5 times that in 1975. Although water freight stays relatively constant, rail freight expands (in ton-miles per capita) by about 30 percent and truck freight by 40 percent, and air freight expands threefold.

Scenario II. In this case consumption in 2010 is about the same as today, after going through a peak around 1990. Energy prices increase substantially for all energy forms over the 35-year period from 1975 to 2010. Real energy prices are assumed to essentially quadruple by

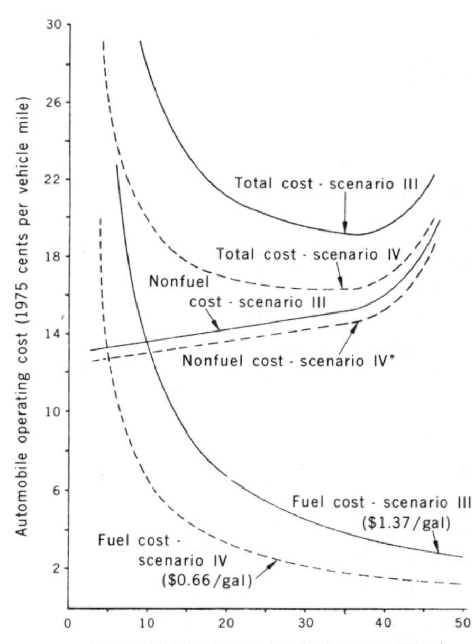

Fig. 8. Total cost of private automobiles plotted against mileage performance achievable by 1985. To provide approximately the same interior space, safety features, and so on, a more energy-efficient car is more expensive, but the added cost is not great for mileages less than 35 to 40 miles per gallon. Balanced against operating costs, the total cost of ownership and operation is remarkably insensitive to mileage. Note that nonfuel cost includes such items as depreciation, maintenance, garaging, parking and tolls, insurance, taxes and fees, and interest lost.

Table 5. Energy efficiencies in 2010, according to scenario III.

Buildings and appliances		Industry		Transportation	
Type	Intensity*	Type	Intensity†	Type	Intensity‡
Thermal integrity (heating)		Agriculture	0.85	Automobile	27 mpg
Residential	0.63	Aluminum	0.63	Light trucks and vans	21 mpg
Commerical	0.6	Cement	0.63	Air passenger	0.45
Government and education	0.45	Chemicals	0.78§	Truck freight	0.8
Space conditioning		Construction	0.65	Air freight	0.6
Air conditioning	0.75	Food	0.76	Rail freight	0.97
Electric heating	0.63	Glass	0.76		
Gas and oil heating	0.75	Iron and steel	0.76		
Refrigeration and freezing	0.68	Paper	0.71		
Lighting	0.70	Other industry	0.75		

*Energy intensity of new construction and products in 2010 compared with 1975. †Average energy per unit production in 2010 compared with 1975. ‡When figures are not given in designated units they refer to energy intensity in 2010 compared with 1975 (including changes in load factor). §Excluding feedstock.

Table 6. Energy efficiencies in 2010, according to scenario II.

Buildings and appliances		Industry		Transportation	
Type	Intensity*	Type	Intensity†	Type	Intensity‡
Thermal integrity (heating)		Agriculture	0.85	Automobile	37 mpg
Residential	0.63	Aluminum	0.55	Light trucks and vans	30 mpg
Commercial	0.42	Cement	0.60	Air passenger	0.42
Government and education	0.35	Chemicals	0.74§	Truck freight	0.6
Space conditioning		Construction	0.58	Air freight	0.6
Air conditioning	0.66	Food	0.66	Rail freight	0.91
Electric heating	0.52	Glass	0.69		
Gas and oil heating	0.72	Iron and steel	0.72		
Refrigeration and freezing	0.58	Paper	0.64		
Lighting	0.60	Other industry	0.57		

*Energy intensity of new construction and products in 2010 compared with 1975. †Average energy per unit production in 2010 compared with 1975. ‡When figures are not given in designated units they refer to energy-intensity in 2010 compared with 1976 (including changes in load factor). §Excluding feedstock.

2010. Public policies provide incentives, taxes, standards and regulations, vigorous research and development (R & D), and public education to help accelerate the United States toward high efficiency of energy utilization.

Because of the higher prices, there are substantial improvements in new appliances, and buildings are carefully designed and constructed for efficient energy use (Table 6). More energy improvements are made in existing buildings than in scenarios III and IV. As a result, overall energy use in buildings declines from 16 quads in 1975 to 11 quads in 2010. Although energy consumption is reduced somewhat, the very high energy prices result in substantial increases in expenditures for fuel. However, increases in income over the period reduce the relative impacts so that residential fuel expenditures increase only from 3.1 to 4.9 percent of personal income. Thermostats are set back and room temperatures are kept lower in winter (68°F) and higher in summer (75°F). Some new appliance technologies are introduced in this scenario. For example, high-efficiency electric and gas heat pumps may be developed and widely adopted with thermal storage (the annual cycle energy system, where heat is stored in water). Decentralized uses of solar energy for space heating, air conditioning, and water heating make a substantial contribution near the end of the period (25 percent of new air conditioners, 50 percent of new space heaters, and 70 percent of new water heaters in 2010). Improved retrofit measures and construction practices are an important aspect of the scenario. Retrofit measures include extensive insulation of previously uninsulated exterior walls. Passive solar house construction, consisting of heavily insulated structures with a large area of windows facing south, become popular after 1990. Typical house construction would include double 2 by 4 inch exterior walls with a full 8 inches of insulation and 12 inches of insulation in ceilings. Some underground or earth-covered construction would further reduce the energy consumption.

Industry responds to the price trends and tight government regulations; it makes large investments in both retrofit and new process development, resulting in an overall decrease in energy consumed per unit output that reaches a value 34 percent lower in 2010 than in 1975. However, the cost of conversion to new, more efficient processes still does not warrant a significantly accelerated write-off of existing plants purely for energy reasons. To achieve this improved efficiency current air pollution standards are not tightened. Throwaway packaging declines. Production of chemical foams and fiber glass for expanded thermal insulation increases; production of materials for strong but lightweight automobiles is expanded. Industrial cogeneration and power plant waste heat utilization become commonplace.

The most efficient automobiles, using new technologies such as the Brayton and Stirling engines, are introduced beginning in the late 1980's. Federal MPG standards set under the Energy Policy and Conservation Act of 1975 (Public Law 94-163) are reached and new standards are set, taking full advantage of the technological potential to increase mileage. People spend about 30 percent more time in cars, on the average, than in 1975. Air passenger miles per capita increase to a level about 60 percent greater than today's, while telecommunications technology improves substantially. Increased demand results in two generations of aircraft by 2010, which provide one-third more seat-miles per gallon. Airplane load factors increase from 55 to 75 percent as a result of government policy. Some modal switching from truck to rail occurs, and improvements in truck fuel efficiency and load factors yield substantial energy savings. Despite expanded per capita use (miles traveled, tons carried) the energy consumed in this sector in 2010 falls by 18 percent. Per capita use of mass transit expands nearly fivefold.

Scenario I. This scenario received less detailed attention than the others; it was derived by making incremental changes in the assumptions for scenario II concerning additional energy-conserving policy actions. As is the case for any level of energy use, many possible configurations of technology and life-style are compatible with total energy use of 63 quads at twice today's GNP. Here we describe only one set of changes relative to scenario II. We emphasize that such a scenario could not occur without aggressive, coordinated long-term policy actions in the areas of land use, transportation, and electric utility regulation. A more restricted set of policies limited to energy use (efficiency standards, energy taxes) probably could not achieve such a large reduction in energy use by 2010.

Energy use in buildings is reduced through improved building materials and construction techniques, including expanded use of passive solar design, and extensive use of such techniques as annual cycle energy systems, as well as active solar units. Continued migration to "sun belt" states occurs, along with acceleration of trends toward multifamily units. These measures combine to reduce end use for space conditioning and water heating to a level about one-third less than obtained in scenario II. Additional savings (about 1 quad) accrue from using steam from cogeneration in commercial and apartment buildings. An additional 0.3 quad of electricity is saved through further increases in the efficiency of appliances.

Substantial shifts occur in industrial production, favoring less energy-intensive goods. Major investments are made in new process technology and in new, highly efficient plants with cascaded use of heat. Since maximum practicable cogeneration (of electricity and industrial heat) was assumed in scenario II, no major additional improvements are assumed here.

The average time spent in automobiles (currently about 53 minutes per person per day) is reduced by 2010 to the 1963 average (43 minutes), in part because of better organized living patterns. This results in a savings of 1 quad of gasoline. In scenario II no improvement in the efficiency of military fuel use was assumed. In scenario I we assumed a 50 percent increase in effectiveness through improved efficiency, saving 1 quad.

No changes are made in the air transport sector beyond those in effect for scenario II. Energy use for freight transportation is reduced by one-third (2.0 quads) by shifts toward a more service-oriented economy and by strong policies promoting shifts from truck to rail transport.

The mix of goods and services constituting the GNP in all other scenarios showed only income effects, such as saturation of food purchases and increases in purchases of durable goods. If present trends continue toward higher prices for all resources (including energy) relative to labor, it may be expected that goods will be made more durable and will be maintained longer. Under such conditions the consumer market basket would shift away from goods toward services (including repair and maintenance).

Scenario I calls for significant changes in energy-consuming technology, and almost all activities become quite energy-efficient. The net result of these changes, in terms of primary fuel use, is a primary energy demand of 58 quads per year, corresponding to 60 percent of today's per capita energy. Scenario I may signal an approach to the level where energy and GNP become tightly coupled even in the longer term because presently anticipated technological improvements are fully utilized in scenario I.

Table 7. Role of purchased (utility) electricity (2010) in terms of total primary energy demand.

Sector	Scenario II*		Scenario III		Scenario IV		1975 (actual)
	Minimum	Maximum	Minimum	Maximum	Minimum	Maximum	
Buildings, quads	10	22	13	25	17	43	12
Industry, quads	8	17	9	20	11	23	8
Transportation, quads†	1	4		4		4	
Total, quads	18	43	22	49	28	70	20
Primary energy, quads	74		94		136		71
Electricity, percent of primary energy‡	25	56	24	49	23	52	28
Average annual electricity growth, percent§	0.2	2.6	0.7	3.2	2.2	4.0	

*Electricity demand growth for scenario I was not estimated here. †A major shift to electricity for ground transportation is not assumed here. However, if about half of transportation were shifted to electricity by 2010 the additional electrical demand, beginning in 1990, would be $\sim$5 GW/year. ‡These numbers represent upper and lower bounds, based on corresponding policies. The likely consumption in each scenario, barring such policies, is given in Table 2. §Over the period 1975 to 2010.

Discussion of Results

Demand for electricity. The demand for energy as electricity has grown substantially faster than the demand for total energy (7.1 percent per year in the decade 1960 to 1970, compared to 4.2 percent per year for all energy). Analyses based on a broad range of assumptions suggest that the demand for electricity will grow substantially more slowly in the future. Ranges of average electricity demand growth are given in Table 7. These ranges are rather broad because of the ease with which electricity can be interchanged with other energy forms in new construction. Thus, an electric heat pump and a gas heating system are almost directly interchangeable. Similarly, for many industrial heat applications and ground-based transportation either electricity or gaseous or liquid fuels can be used. Only in a few applications does electricity appear to serve uniquely (lighting, specialized industrial applications, computers).

For the maximum electricity use analyses virtually all electric heating of buildings was assumed. In the industrial sector, the maximum (purchased) electricity scenario describes a situation where there is almost no self-generation of electricity, and where many processes are electrified. At the same time, the transportation sector was still assumed to operate primarily on liquid fuels. Major advances in electric energy storage (such as high-efficiency batteries) could change this situation significantly, however, and lead to even greater demands for electric energy than those included in the analyses given here. In the minimum electricity analyses, gas and oil provide virtually all heating needs, and electricity use is restricted to electric drive and other critical processes. Since the figures pertain only to electricity purchased from utilities, they understate the actual extent to which the industrial sector relies on (internally generated) electricity.

Economic effects. The scenarios imply a large potential for long-term change in the ratio of energy use to GNP. As long as the transition to energy-efficient capital stocks occurs smoothly and over a significantly long period of time, there is no reason to expect major adverse effects on the GNP. There appear to be no major differences in labor requirements between scenarios I and IV. Since only 16 percent of the labor force is employed in the sectors responsible for 70 percent of industrial energy consumption, the potential for appreciable impacts in that area is small. Labor in construction and maintenance of buildings, appliances, and automobiles could increase, offsetting reduced employment in energy- producing sectors.

Essentially all capital goods require energy (power plants, industrial equipment, buildings). Today about 25 percent of new capital available in the private sector goes for energy supply and conversion facilities (*16*). This is expected to increase in the future as more capital-intensive energy supply technologies are introduced (solar electric systems, synthetic fuel, nuclear plants). We assume for the purpose of analysis that this would increase to 30 percent in 2010 for scenario IV (energy prices about the same as today's). We will now consider the differences between this "nominal" case and scenario II.

In scenario II energy consumption is about 40 percent lower than in scenario IV, so we assume that the capital needed for energy supply facilities is likewise reduced by 40 percent to only 18 percent of total capital. This leaves 82 percent of new private capital available for energy-consuming stocks, representing about a 12 percent increase over the nominal 70 percent level. Each of the sectoral analyses indicated that the incremental capital cost of achieving scenario II efficiency levels is less than 10 percent of total non-energy capital investment. Therefore, the capital not used for energy production could match or exceed the additional capital needed to improve energy efficiency of energy-consuming equipment.

To check sensitivity to assumptions about GNP growth, the Panel considered another scenario in which conditions were assumed identical to those in scenario III except that the GNP is 40 percent higher in 2010 (nearly tripling instead of doubling the 1975 levels). Energy use increases by only 35 percent. This might be interpreted as implying an income elasticity of about 0.90, which is not inconsistent with the results of Herendeen and Tanaka (*17*) for the U.S. economic system in the early 1960's. This less than proportional increase in energy use is due entirely to shifts in the market basket of goods and services.

Conclusions

The purpose of this article has been to report on plausible energy futures corresponding to the low range of possible energy supplies. The Panel concluded that whatever driving forces might be involved, it will be technically feasible in 2010 to use roughly a total amount of energy as low as that used today and still provide a higher level of amenities, even with total population increasing 35 percent. It should be noted, however, that even under such low-energy scenarios there would still be enormous requirements for developing new energy supplies to replace depleted resources. The results of the analysis can be discussed in terms of responses to some frequently raised questions about the U.S. energy system.

Is a lower-energy future a bleak one for the United States? Probably the most important single finding of the study is that our national well-being can improve while energy growth is constrained to varying degrees. For this to occur requires higher energy prices, more regulation, or both. However, in every sector of the economy major increases in energy efficiency can be made by using presently available technology, and even greater improvements can be made with technology now under development. The large discrepancies between present energy efficiencies and those that are thermodynamically obtainable, together with economic analysis, show that improvements in efficiency of 1 percent per year or more are sustainable over a number of decades in the United States. If slower population growth trends continue, present amenity levels could be maintained and might be slowly expanded (at about half of past economic growth rates) without major increases in energy use.

Is a low-energy future a low-technology future? Probably just the reverse is true. A low-energy future offers strong incentives for technological innovation. In many recent inventions information has been substituted for energy. Modern computers and communication equipment perform better than past units but use only a small fraction of the energy. The techniques used to bring about energy reductions reported in this study in almost every case rely on the use of advanced technology.

If energy demand growth was so rapid in the past, how can it be so slow in the future without major effects? The most significant reasons for rapid energy growth since World War II have been the following.

- Steadily declining energy prices relative to the prices of other goods and services.
- Expanding consumer population.
- Introduction and rapid market penetration of energy-intensive consumer products that provide basic amenities (such as individualized transportation, comfortable habitat, and household conveniences).
- Increasing cost of labor, which increased the substitution of capital and energy for labor.
- Major increases in disposable income, enabling rapid expansion of consumer buying power.

None of these factors is likely to continue at its historic pace. Many past projections of energy futures simply did not take into account inevitable changes in these trends. Where they are taken into account, future demand growth is projected to be slower even without active conservation policies. In all the scenarios discussed in this article we assumed virtually complete saturation of the use of air conditioning and other major appliances by 2010, but did not explicitly account for new "phantom" appliances as energy-intensive as air conditioning. But neither did we count such new technologies as increasing energy efficiency. Considering that real energy prices will probably increase in the future, the latter type of technological development is more likely than the former.

The long-term substitutability of labor and capital for energy indicates that goods and services can continue to be provided in the future with less energy input. However, to do this requires other human and material resources. Therefore, unemployment in power plant construction is compensated by greater employment in constructing more energy-efficient commercial buildings and residences; lower requirements for steel in automobile construction are offset by an increased demand for aluminum and fiber glass; and control system designing for nuclear power plants is replaced by control system designing for space conditioning in commercial buildings. As long as these changes occur slowly, their impacts are minimized. It is essential to have time for adjustment.

What constrains our progress toward more efficient energy utilization? We have pointed out that existing products and technological know-how, if fully utilized, will permit major improvements in energy efficiency with little change in life-style. We can also be confident that improved technologies can be developed through R & D investments in this relatively unexplored field. Because there are already many technical opportunities for improved energy efficiency, we must conclude that the most important near-term constraints or impediments are not technological ones. They may be categorized as follows.

1) Price signals: energy consumers tend to weigh their conservation investment decisions against current average energy prices, whereas energy producers (such as utilities) weigh investment decisions against long-run incremental prices. Since energy prices at the margin are now usually considerably higher than average prices, producer and consumer investment decisions are not made on equivalent bases. Thus investments are biased toward increasing supply rather than moderating demand. This situation would be largely eliminated if prices were adjusted (as through an energy surcharge tax) to approximate long-run incremental costs.

2) Time: given time, changes that would be highly traumatic if made quick-

ly can be relatively easy. Unfortunately, Americans have a tendency to want to do things instantly. (As Adlai Stevenson said, "We Americans seem never to see the handwriting on the wall . . . until our back is up against it.") Retrofitting is important and economical, but since the greatest technical opportunities to save energy can be achieved in new capital stock, it is clear that attendant efficiencies can be achieved only as that stock is replaced. The time needed for such stock turnover ranges from about a decade (air conditioners, automobiles) to a half-century (industrial processes). Time is also required to introduce new ways of designing and building energy-consuming items.

3) Standards and regulations: public policies that were developed when energy was cheap and plentiful frequently have the effect of constraining energy-conserving actions. These policies include natural gas price controls, freight transportation regulations, building codes, procurement procedures, and tax policy. Most of these are federal policies but there are also many state and local ones. Progress in energy efficiency that is rapid enough to meet national needs will require policies that directly influence energy-consuming activities.

If relative energy prices stay at about their present level, what is the likely future of demand? If we utilize available technology and make wise economic decisions (minimum total cost), future energy demand growth will be considerably slower than most past projections have indicated. Even a small increase in real energy prices provides a significant opportunity for cost saving through higher efficiency of energy use. Depending on the details of future events, U.S. energy consumption could actually peak within the next 20 years even though well-being continues to improve. But it seems more likely that demand growth will simply slow down. Much depends on public policies (especially those that affect investment), population growth, labor cost, consumer choices, and economic growth. Of this, however, we are certain: many of the principal factors that drove the high energy demand growth in the past are no longer with us.

References and Notes

1. The members of the CONAES Demand and Conservation Panel are listed in the author note. Its chairman is J. Gibbons. The Panel's Resource Groups and their chairmen are: Buildings, R. Carlsmith; Industry, M. Whiting; Transportation, E. Goodson; Integration, E. Bullard; and Economics, D. Chapman. Extensive technical assistance was provided by R. Barnes, E. Berndt, L. Colquitt, K. Friedman, E. Hirst, K. Johnson, T. Mount, D. Pilati, and T. Tyrrell. In addition in each Resource Group a number of experts from universities, business, and government contributed to the study. Although the Panel reached a consensus about the main points brought out in this study, there were individual differences of opinion about some details. These were small enough that no dissenting or minority opinions were recorded.
2. Report of the National Research Council Committee on Nuclear and Alternative Energy Systems, in press. This article summarizes some of the information supplied to CONAES for their consideration but does not necessarily represent their opinions or conclusions.
3. R. Barnes, "The evaluation of new energy sources for process heat," prepared for the National Science Foundation by Dow Chemical Company, September 1975.
4. See, for example, J. Gibbons and R. Sant, *The Changing Economics of Energy*, B. Abrahamson, Ed. (Westview, Boulder, Colo., 1975), pp. 105–140; W. Tavoulareas and C. Kaysen, *A Debate on "A Time to Choose"* (Ballinger, Cambridge, Mass., 1977).
5. L. Schipper and J. Darmstadter, *Technol. Rev.* **80**, 40 (January 1978).
6. J. Holdren, testimony before the Subcommittee on Energy and Power, Committee on Interstate and Foreign Commerce, U.S. House of Representatives, 22 March 1977.
7. A. Lovins, *Soft Energy Paths: Toward a Durable Peace* (Ballantine, New York, 1977).
8. H. Kissinger, address before the Alliance to Conserve Energy, Chicago, 11 June 1977.
9. Interim Report of the National Research Council Committee on Nuclear and Alternative Energy Systems, National Academy of Sciences, January, 1977. See also P. M. Boffey, *Science* **195**, 380 (1977).
10. CONAES Panel on Demand and Conservation, *Outlook for Energy Demand and Conservation* (National Research Council, Washington, D.C., in press).
11. K. W. Ford, G. I. Rochlin, M. H. Ross, R. H. Socolow *et al.*, Eds., *Efficient Use of Energy: A Physics Perspective* (American Institute of Physics, New York, 1975).
12. E. P. Gyftopolous, L. J. Lazaridas, T. F. Widmer, *Potential Fuel Effectiveness in Industry* (Ballinger, Cambridge, Mass. 1975).
13. See, for example, T. Mount and T. Tyrrell, in (*10*), appendix.
14. R. Hoskins and E. Hirst, *Energy and Cost Analysis of Residential Refrigerators* (Report ORNL/CON-6, Oak Ridge National Laboratory, Oak Ridge, Tenn., 1977).
15. E. Hirst and J. Jackson, *Energy* **2**, 131 (1977).
16. See, for example, Ford Foundation Energy Policy Project staff, *A Time To Choose: America's Energy Future* (Ballinger, Cambridge, Mass., 1974).
17. R. Herendeen and J. Tanaka, *Energy* **1**, 165 (1976).
18. C. Whittle *et al.*, *Economic and Environmental Implications of a U.S. Nuclear Moratorium, 1985–2000* (Institute for Energy Analysis Report ORAU/IEA 76-4, Oak Ridge Associated Universities, Oak Ridge, Tenn., 1976).
19. The CONAES study has been conducted under the auspices of the National Research Council and supported in part by the Department of Energy.

Efficient Energy Use and Well-Being: The Swedish Example

Swedes use less than two-thirds as much energy per capita as Americans, at the same standard of living.

Lee Schipper and Allan J. Lichtenberg

It is often said that there is a direct relationship between per capita energy use and standard of living as measured by gross national product (GNP) (*1*). However, examination of the energy and GNP statistics for the most industrialized countries indicates a large spread in the ratio of energy use per unit of GNP (see Fig. 1). This article compares energy use in the United States, one of the countries with a high energy/GNP ratio, with that in Sweden, a country which in 1971 used approximately 60 percent as much energy as the United States to generate each dollar of GNP. Sweden was chosen not only because of its low energy/GNP ratio, but also because the GNP per capita is essentially the same in both countries. Moreover, much of the economic activity and many of the demographic features in Sweden are similar to those in the United States. Thus, evaluating the differences in energy utilization between these two countries may illuminate strategies for saving energy (*2*).

Studies of energy conservation in the United States indicate that the more important of these strategies, taken together, could reduce energy consumption 25 to 40 percent (*3–5*), while lowering pollution, reducing capital requirements for energy production, and generally raising

Lee Schipper is a research assistant at the Lawrence Berkeley Laboratory and an energy specialist with the Energy and Resources Group, University of California, Berkeley 94720. Allan J. Lichtenberg is a professor of electrical engineering and computer sciences, and chairman of the Energy and Resources Group, University of California, Berkeley.

Fig. 1. The energy/GNP ratio for several countries over time, with hydroelectric power counted at 3 kwht/kwhe. From Linden (*1*). For a discussion of units, see (*2*).

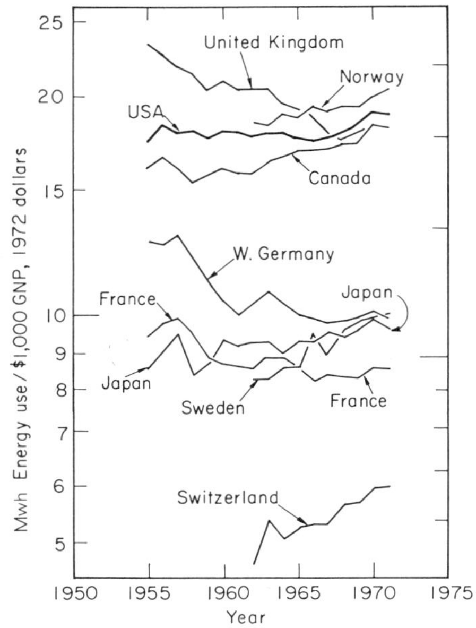

employment. But the interrelationships among economic inputs (including energy) within an economy are complex. Thus, examination of an economy that is similar to ours but requires substantially less energy than our own may provide guidance in understanding the total effect of energy conservation.

Interest in energy use and conservation has stimulated a number of international comparisons (*6, 7*), as well as new evaluations of data from single countries (*8, 9*). A preliminary study concerned with a number of countries showed some of the differences reported here, but no conclusions were drawn (*10*). In a study of the United States and West Germany (*11*) comparisons were developed further, and methods for conserving energy in the United States were discussed; the conclusions reached are in qualitative agreement with those in this article. Two other comparisons of U.S. and Swedish energy consumption have been undertaken (*6, 7*), and we have been able to compare our data with theirs. Although there are many small discrepancies in data from different sources, in no cases are these discrepancies large enough to change our general conclusions.

Factors Entering into International Energy-Use Comparisons

Many factors enter into the determination of the energy/GNP ratio. Among these are energy costs relative to other costs, government policies including taxes, subsidies, and regulations, and demographic and cultural variables. One meaningful measure of the effect of energy prices on consumption is the price elasticity of demand, defined as the ratio of the percentage change in demand to the percentage change in price, other factors being held constant. A study of the long-term elasticity for electricity in the United States, for example, gave values of –1.2 for residential use, –1.8 for industrial use, and –1.4 for commercial use (*12*). Recent studies indicate that the long-term elasticity for gasoline may be as great as –0.75 (*13*). The long-run effects of energy prices can be seen qualitatively in Fig. 1. The high energy/GNP countries are those that historically have had cheap energy (relative to other goods and services); the United States, Canada, Great Britain, and Norway are examples. The countries with lower energy/GNP ratios are those that have been relatively fuel-poor, especially since World War II. Although Sweden, for example, has had abundant hydropower, the country has been increasingly dependent on imported petroleum, particularly for nonelectric uses. Consequently, electricity has been inexpensive relative to fuel, with both price and per capita consumption very similar to that in the United States. Motor fuels, on the other hand, have been taxed heavily in Sweden, and per capita consumption of these refined petroleum products has been far below U.S. consumption. Similar taxes have been the rule in other oil-poor countries. Although oil for home heating has been relatively inexpensive in Sweden (comparable to U.S. oil prices) the large amounts demanded for long winter heating seasons acted in place of higher prices to stimulate conservation efforts (*14–18*).

Sweden's energy/GNP ratio was rising during the 1960's, probably because of changes in life-style similar to those which had taken place in the United States a decade or two earlier. These changes include greater living space and gasoline use per capita. The ratio has since stabilized and then fallen in the period of high energy prices after the oil

Table 1. Basic economic and social indicators for the United States and Sweden (1971). Data for the United States are from (*24*), data for Sweden from (*22, 23*) and fact sheets distributed by the Swedish Institute, New York.

Indicator	United States	Sweden
Physical characteristics		
Population (million)	207	8.1
People per square mile	57	47
Climate-heating [degree-days per year (68°F)]*	5,500	9,200
Economic activity		
GNP (dollars per capita)	5,051	4,438
Energy consumption (kilowatt-hours per capita)	96,000	52,450
Steel (kilograms per capita)	620	680
Cement (kilograms per capita)	342	430
Fertilizer (kilograms per capita)	105	67
Paper (kilograms per capita)	224	540
Food (per day)		
Energy (kilocalorie per capita)	3,300	2,850
Protein (grams per capita)	99	80
Cereals (grams per capita)	176	168
Meat (grams per capita)	310	142
Health and education		
Doctors per 1000 persons	1.5	1.35
Dentists per 1000 persons	0.49	0.72
Hospital beds per 1000 persons	7.8	15
Infant deaths per 1000 births	19	11.1
Teachers per 1000 students	34	60
Newspaper copies per 1000 persons	301	534
Books published per 1000 persons	0.39	0.94
Conveniences		
Telephones per capita	0.59	0.56
Television sets per capita	0.45	0.32
Automobiles per capita	0.45	0.3
Automobile passenger miles per capita (1970)	7,900	5,050
Refrigerators (percent of households)	100	93
Freezers (percent of households)	28	46
Clothes washers (percent of households)	76	41
Vacuum cleaners (percent of households)	88	89

*We use 68°F for the United States and Sweden and have adjusted figures for the United States accordingly (*27*).

embargo, as has the ratio in most other countries. The United Kingdom and West Germany had a falling energy/GNP ratio during the 1950's and early 1960's, probably due to a shift away from coal, which was their main cheap source of fuel, to more expensive substitutes.

One factor that can be important in determining the energy/GNP relationship is the relative industrialization or type of industry in a country. Certain products are particularly energy-intensive, including steel, aluminum, cement, paper, and plastics. The effect of changing the output mix is most noticeable in comparing Luxembourg, where the steel industry plays a dominant role in the economic structure, with Switzerland, where banking, insurance, timepieces, and other items of high value added per unit of energy expended predominate. Luxembourg has an energy/GNP ratio of 30 kilowatt-hours total per dollar compared to Switzerland's 6 kwht per dollar (*10*). The energy/GNP ratios of Great Britain and New Zealand were found to differ by a factor of 2 (*5*), which may be partly attributable to the degree of wealth based on agriculture in New Zealand. However, effects due to the agricultural sector are usually small among industrialized nations. For the countries in Fig. 1, agricultural sectors comprise between 3 and 5 percent of the total GNP; if any correlation exists, it is between energy use and the size of the services sector, which will be explored further below.

The effects of cultural and life-style differences on energy consumption are very difficult to quantify but are clearly very important. Cultural patterns, although not wholly controlled by the marketplace, may be tempered over long periods of time by prices and fuel availability. Some of the current pattern of intensive energy use in the United States and Canada can be traced to the availability of fuel wood during the 19th century (*19*). In 1850, for example, with a per capita energy consumption of 30.8 × 10^3 kwht, including wood, the United States used as much energy per capita as Switzerland does today.

We take the years 1970 to 1972 as our comparison period, because complete data are available and energy prices and use trends were relatively stable. Where appropriate, we use data from other years. In 1971, there was a mild recession in Sweden; total energy use was slightly higher in 1970, and our Swedish industry statistics were taken from that year. Because of variations in exchange rates, relative GNP's can change independent of changes in real wealth. Unless otherwise noted we employ the exchange rate of 5.18 Swedish crowns (Skr) per dollar, which applied until late 1971. The rate was as low as 3.9 Skr per dollar in 1973 and has since stabilized at 4.3 Skr per dollar in 1975 and 1976. It is generally believed that the old exchange rate undervalued the Swedish crown relative to the dollar.

In comparing energy/GNP ratios, addi-

Table 2. Per capita energy consumption in the United States and Sweden in 1971. Data for the United States are from (*28–30*); we included 1,000 kwh per capita in wood wastes (*30a*); the totals in the kwh and kwht columns do not agree because of differences in counting hydropower. The Swedish data are from (*16, 21, 31*), with feedstocks estimated from (*32–34*); we included 4,000 kwh per capita in wood wastes; hydropower was counted at 3,413 Btu/kwhe in the kwh column. All kwht values were calculated by distributing utility losses to end consumers; consumption of electricity within electrical sectors was counted in "Industry." The kwht column for the United States includes hydropower at 10,460 Btu/kwhe; that for Sweden counts all electricity at 10,400 Btu/kwhe. The actual "heat rate" for thermal and back-pressure plants in Sweden is 8,870 Btu/kwhe, including distribution losses; the rate for production only is 7,780 Btu/kwhe. Cogenerated electricity in the paper industry is excluded from the kwht columns.

Consumption	United States			Sweden		
	kwh	kwhe	kwht	kwh	kwhe	kwht
Transportation	24,025	25	24,075	7,350	200	7,775
Commercial	9,600	2,150	14,250	7,375	1,500	10,625
Residential	13,500	2,300	18,450	11,125	1,400	14,150
Industry	28,900	3,300	36,000	20,400	4,200	29,450
Feedstocks	5,600		5,600	2,500		2,500
Utility losses (actual)*	14,200			3,700		
Actual consumption†	95,825	7,775	98,375	52,450	7,300	64,500
Energy embodied in foreign trade‡	1,800§		1,800	−4,600‖		−4,600
Net consumption¶	97,625	7,775	100,175	48,150	7,300	59,900

*Hydropower was counted at 3,413 Btu/kwhe. Other losses are according to actual consumption. †Actual consumption refers to fuels and electricity, including petroleum refining losses and other captive fuels. ‡Embodied energy includes the process energy of refined fuels but not the energy available when the fuel is burned. §The import-export energy balance for the United States is from (*35*). ‖The import-export energy balance for Sweden is from (*16*). ¶Exports of coal, crude oil, or refined products are excluded from this balance.

Table 3. Passenger transportation data for the United States (*36–38*) and Sweden (*33, 40, 41*). Division of modes into urban (within areas of population 30,000) and intercity [from (*41*)] does not exactly correspond to our classification by local and intercity; PM, passenger mile.

Passenger mode	United States (1972)			Sweden (1970)		
	D_J' (PM per capita)	E_J' (kwh/PM)	I_J (kwh per capita)	D_J' (PM per capita)	E_J' (kwh/PM)	T_J (kwh per capita)
Automobile*						
< 30 miles	4,850	1.72	8,330	1,825		
> 30 miles	4,200	1.02	4,300	3,225		
Totals	9,050	1.41	12,630	5,050	0.74	3,760
(Totals for 1970)	(7,900)	(1.41)	(11,200)			
Bus†						
Local (< 30 miles)	112	0.50	56	460	0.41	200
Intercity (> 30 miles)	122	0.30	42	25		
Rail‡						
Local (< 30 miles)	64	0.21 (0.63)§	13.7	85	0.16 (0.48)§	15
Intercity (> 30 miles)	21.3	0.87	18.6	356	0.25 (0.75)§	90
Total land	9,370	1.36	12,760	5,975	0.68	4,065
Air domestic	490	3	1,500	46	1.12(?)	275(?)
Air international‖	243(?)	1.38(?)	335	200		
Other passenger and military			1,500		?	200
Total passenger	10,103		16,095	6,221		4,540

*Hirst (*36*) gives 1969 load factors that imply an overall load factor (ratio of passenger miles to vehicle miles) for automobiles of 1.7, which seems unreasonably low. A load factor of 2.2 is implied in (*37, 38*), and a load factor of 1.9, which we adopt, is assumed in (*42*). There was a similar discrepancy in the Swedish data, most references giving an implied overall load factor of 2, with one giving 1.7. We adopt 2, since the driving in Sweden is dominated by family driving to a greater degree than that in the United States. †The U.S. bus fleet is 75 percent diesel; the Swedish bus fleet is 10 percent electric, the remainder either diesel or gasoline. ‡The U.S. local rail service is electric; intercity rail service is 75 percent diesel, the rest electric. In Sweden 90 percent of rail service is electric, the rest diesel. §Electricity figures are net values, and the E_J's in parentheses reflect a theoretical 3 kwht/kwhe. ‖The figures for international fuel and passenger miles are uncertain.

tional problems arise. Comparing the size and content of the GNP has received considerable attention (*20*). In our study we give indications of the structure of the economy in Sweden and in the United States, highlighting the differences and similarities, comparing various physical standards of well-being. Accounting for differences in climate, geographic factors, population distribution, and so on, is also important; we have made comments on this problem where applicable. The method of counting the contributions of hydropower and of combined electricity and heat generation can be significant in international comparisons, and is discussed below. We find that no matter how one counts hydropower the difference in energy use between Sweden and the United States is large, especially since the largest contrasts appear in transportation, space heating, and process heat applications. The use of noncommercial sources of energy, usually considered only when discussing less-developed countries, is important to our comparison. The paper industry in Sweden, which accounts for fully 15 percent of the total consumption of energy there, actually generates 60 percent of its fuel internally from waste forest products. Together with other waste products, including urban wastes, these noncommercial fuels account for 9 percent of Sweden's total fuel use in 1971 (*21*) compared to 1 percent for the United States. Finally, a troublesome statistical problem is inconsistency between different information sources. For example, the fuel used by agricultural and construction equipment could be counted in transportation or industry, depending on how figures are kept. Similarly, self-generated electricity, district heating, by-product fuels (such as coke gas), noncommercial fuels, consumption of energy by energy producers, and so forth must be carefully sorted out. We believe we have resolved these various problems to the point that the remaining errors are only a few percent.

Table 4. Automobile data for the United States and Sweden (1970). (Conversions used: 1 U.S. gallon = 33.75 kwh; 1 mile = 1.6 km.)

Parameter	United States	References	Sweden	References
Persons per vehicle	2.25	(*42*)	3.4	(*16*)
Licensed drivers per capita	0.5	(*38*)	0.4	(*16*)
Passenger miles per capita	7,900	(*42*)	5,050	(*40*)
Vehicle miles per capita	4,160	(*37*)	2,560	(*46*)
Miles per vehicle*	9,360	(*37*)	8,900	(*46*)
Load factor	1.9	(*42*)	2.0	(*40*)
Average weight (kg)	1,700	(*43*)	1,100	(*44*)
Miles per gallon†				
Actual	13.7	(*37, 43*)	24	(*43–45*)
Theoretical	12.5		20	
Kilowatt-hours per passenger mile	1.4		0.73	
Kilowatt-hours per capita	11,200		3,710	

*The surprising similarity between the U.S. and Swedish values suggests that in Sweden second cars are replaced by mass transit, and a significant number of families have no car at all. †The U.S. theoretical value is estimated from the weight-fuel economic statistics of the Environmental Protection Agency (*43*); the actual value is determined by dividing actual miles driven by fuel consumed. The Swedish theoretical value from Ullén (*44b*) matches the actual value for Sweden.

Table 5. Goods transportation data for the United States and Sweden. The U.S. data are from (*30, 30b, 39*). Swedish data are from (*16, 40*), with the breakdown for truck by distance based on the 1973 distribution.

Transport	United States (1972)			Sweden (1970)		
	D_J (ton-mile per capita)	E_J (kwh per ton-mile)	T_J (kwh per capita)	D_J (ton-mile per capita)	E_J (kwh per ton-mile)	T_J (kwh per capita)
Truck						
Local (0 to 30 miles)	360	1.95	700	339	0.58	200
Intercity (>30 miles)	2069	0.63	1430	1284	0.86	1100
Total truck	2429	0.88	2130	1623	0.8	1300
Rail	4132	0.19	800	1350	0.06 (0.18)*	80
Domestic air	20	7.5	150			
Water						
Domestic			420	704	0.3	190
International†			480			1600
Total goods	6585		3980	3670		3170
Nonrevenue goods transport (agriculture, forestry, construction, and so forth)			1850			470
Pipeline			200‡			
Other			120‡			930
Totals	6585		6150			4570

*The value in parentheses reflects a conversion factor of 3 kwht/kwhe. †Excluded from totals in Table 2. ‡These are 1971 data.

Sweden and the United States: Physical and Economic Comparison

In Table 1 we compare physical characteristics, economic activity and various measures of well-being in the United States and Sweden (*22–24*). Although the populations differ by a factor of 25, the population densities are similar, as is the distribution into densely populated urban centers and sparsely populated rural regions. Movement to the suburbs, fostered by the automobile, started earlier and is more advanced in the United States, although there are signs of such a trend in Sweden (*25, 26*). The natural distances over which goods must move are larger in the United States, although in Sweden much of the lumber, iron ore, and electric power flows from the sparsely populated far north to the more crowded south. The climate in Sweden is more severe; the number of degree-days, weighted by population distribution, is close to 9200 in Sweden, comparable to the value in North Dakota, whereas the weighted U.S. average is approximately 5500 degree-days (*27*).

Table 1 indicates that in 1971 the United States had a GNP per capita 10 percent higher than Sweden's at the then current exchange rates. However, for each dollar of GNP Sweden required only 68 percent as much energy as the United States. Correcting for the energy embodied in foreign trade (see below) reduces the 1971 Swedish figure to 61 percent. Despite the difference in energy use, the total per capita production of basic industrial commodities is comparable in Sweden and the United States.

Basic well-being is difficult to compare quantitatively. As seen in Table 1, food intake is similar, with Americans eating considerably more meat (about twice the Swedish per capita consumption), which, per gram of protein, is more energy-intensive than most other foods. In health and education, Sweden leads the United States in almost all categories. When the comprehensive health and social security system in Sweden is examined this difference is even more striking.

The large number of automobiles and TV's in the United States is accounted for mainly by multi-unit ownership by families. Transportation convenience is, in fact, comparable, because public transportation is more readily available in Sweden, and domestic distances are generally smaller (*25, 26*). Swedes have more second homes (500,000 in all) per capita than Americans, and most of the population enjoys 4 weeks of paid vacation each year. Thus we conclude that the living standards are comparable quantitatively in Sweden and the United States, but the mix is substantially different, with somewhat less energy-intensive economic activities and life-styles in Sweden.

Table 6. Per capita residential and commercial energy use in the United States (*3, 30b, 49, 49a*) and Sweden (*16, 41, 53, 54*) in 1972.

Use	United States	Sweden
Residential		
Direct fuel (kwh)		
Heating	9,660	8,200
Water heating	1,950	3,300
Gas appliances	630	125
Second homes		300
Electricity (kwhe)		
Refrigerator and stove	610	530
Lighting	335	105
Air conditioning	300	
Other appliances	590	475
Heating	280	400*
Water heating	500	
District heating saving		−1,300†
Total net use (kwh)	14,855	12,135
Electric conversion loss at U.S. rate‡	5,230	3,020
Total gross use (kwht) (with actual losses)‡	20,085	15,135 (12,820)
Commercial		
Floor space (m²)	10	13§
Direct fuel (kwh)		
Space heat	5,625	4,800
Water heat	790	
Air conditioning	200	
Electricity (kwhe)		
Air conditioning	205	
Lighting	1,250	625
Electric heat and other	310	1,075
Total net use (kwh)	8,380	6,500
Electric conversion loss‡	3,530	3,200
Total gross use (kwh) (with actual losses)‡		(7,280)

*Includes hot water in all-electric homes, and second homes. †Assigned to residential sector for convenience. ‡Losses are counted at the U.S. rate of 2 kwht/kwhe for uniform comparison, as in Table 2. Actual Swedish losses (0.46 kwht/kwhe) are reflected in the totals in parentheses. §This value is obtained from the volume of commercial office space by assuming a 4-m room height.

Comparison of Energy Use

In Table 2 we compare energy use in the United States and Sweden (*16, 21, 28–35*). Sweden uses less energy per capita in all sectors, the largest difference being in the transportation sector. There are considerable differences in basic materials processing in the industrial sector and electricity use in the residential and commercial sectors.

A useful formula that summarizes the uses of energy (T_J's) is: energy use $= \Sigma_J E_J D_J = \Sigma_J E'_J D'_J = \Sigma T_J$, where the D_J's are the dollar demands for goods and services and the E_J's are the energy intensities of those demands; or, in physical terms, the D'_J's are the quantities of goods and services and the E'_J's the energy intensities associated with those quantities.

When data are disaggregated in this way, both the relative mix of modes (D_J or D'_J) and the efficiency of those modes $(E_J)^{-1}$ or $(E'_J)^{-1}$ can be compared among countries. Energy use in the economy can be lowered by shifting economic activity to less energy-intensive sectors (D_J) or by increasing the efficiency (lower E_J) of production of a given D_J. We use this formalism in the specific comparison of U.S.-Swedish energy use. The differences in E_J's between countries indicate possibilities for energy conservation through technical change, without requiring changes in life-style.

Transportation. Table 3 shows basic passenger transportation data for Sweden and the United States (*33, 36–42*). Major differences exist in all modes. In addition to the striking differences in automobile D', E', and T, we note that Swedish passenger transportation is more heavily concentrated in rail (including subway) and bus modes, at the expense of the automobile and the airplane. All Swedish E'_J's are lower than the corresponding U.S. values. This is due in part to higher load factors and the extensive use of air and bus charters.

In Table 4 we consider the automobile in more detail (*37, 38, 40, 42–46*). We see that the Swedish D' is only 62 percent of the U.S. figure, and E' is only 60 percent of the U.S. figure. The biggest contributor to efficiency is the lower weight of Swedish cars (1100 compared to 1700 kilograms). The weight distributions are given in Fig. 2 (*43–45*). Interpolation of Environmental Protection Agency (EPA) measurements of fuel consumption as a function of inertial weight suggests that weight alone accounts for a 30 percent difference in energy consumption per mile (*43*). The lack of power extras, automatic transmissions, and air conditioners in Swedish cars reduces fuel demand further, as does the lower ratio of engine displacement to car weight.

In addition to these technical differences there are differences in automobile utilization that have significant consequences. For trips of 10 kilometers or less, for which automobile fuel consumption per mile is nearly double average fuel consumption (*47*), the Swedes use private cars and public transit in the ratio 55/45 (percentage of trips) (*45*). In the United States, by contrast, the ratio is

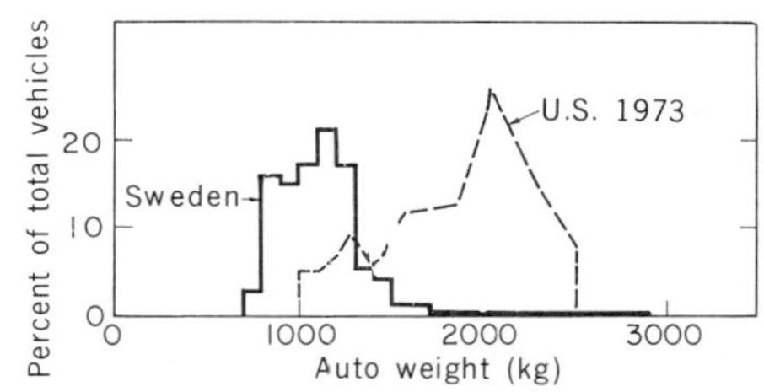

Fig. 2. Distribution of automobiles by weight. 1974. Swedish data are for 1974 and are from (*44*); U.S. data are for 1973 and were estimated from (*43a*).

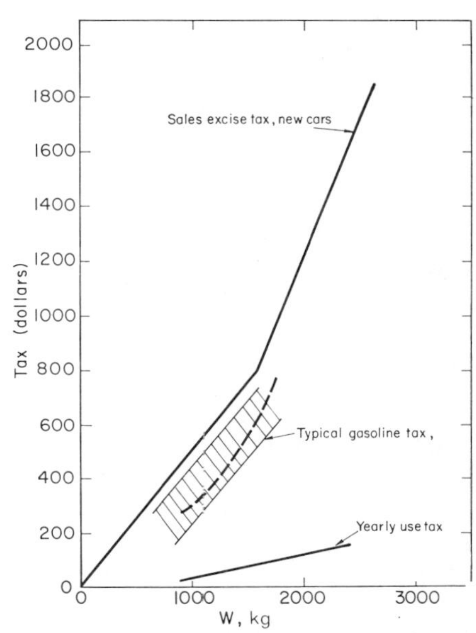

Fig. 3. Taxes on automobiles and gasoline in Sweden in 1974 (based on the old exchange rate of 5.18 Skr/$). Sales excise tax was computed from $45 per 100 kg to 1600 kg and $58 per 50-kg additional weight. The shaded area and dashed line give an estimate of gasoline tax as a function of weight, using 14¢/liter for a typical 20,000 km/year. The yearly use tax was computed from $32 + $8.40 [weight (kg)-900]/100. Data are from (*23, 44a, 48*).

Table 7. Residential space energy consumption (fossil fuels only). Data for the United States are from (*14, 15, 24, 51*); values are for single-family dwellings, except the kwh per capita value, which includes all dwellings. Swedish data are from (*16, 17*); MFD = multiple-family dwelling, SFD = single-family dwelling.

Parameter	United States	Sweden		
		MFD	SFD	Average
Persons per housing unit	3.3	2.1	3	
Rooms per housing unit	5.1	3.2	4.5	
Persons per room	0.66			0.66
Average area (m^2)	115	70	110	
Degree-days (68°F)	5,500			9,200
Kilowatt-hours per housing unit	34,000	16,350	28,750	
Kilowatt-hours per square meter	300	235	260	
Kilowatt-hours per degree-day	6.2	1.77	3.10	
Kilowatt-hours per square meter per degree-day	0.054	0.025	0.028	
Kilowatt-hours per capita	9,150			8,200

Table 8. Heating intensities by climatic regions for the United States (*14, 15*) and Sweden (*17*). In the Swedish data (*17*), curves for electrically heated homes were adjusted upward to reflect oil furnace efficiencies and construction.

Parameter	United States			Sweden		
	California	Pennsylvania	Minnesota	Malmö	Stockholm	Norrbotten
Degree-days (68°F)	1,900	5,500	8,500	7,700	9,200	13,000
Kilowatt-hours per square meter per degree-day	0.11	0.063	0.049	0.028	0.027	0.026

90/10 (*38*). This traffic accounts for 65 percent of all automobile trips in the United States, resulting in lower average driving cycle efficiencies. Thus it becomes apparent why actual miles per gallon in Sweden are higher than predicted by interpolation of EPA measurements (*43*); the driving cycle demands less energy. Surprisingly, load factors in both countries average approximately 2. This is probably because the smaller size of families in Sweden compensates for that country's higher ratio of family to commuter use.

Speed limits further reduce Swedish automobile energy use. The Swedish speed limit during our period of comparison was as high as 110 km/hour (68 mile/hour) on only about 10 percent of the largest highways. In the United States highway speed limits were commonly 65 mile/hour or more in 1971.

It has been noted (*25, 26*) that the use of alternatives to the automobile in Sweden has gradually eroded. Nevertheless, the automobile's share of all passenger miles has stabilized at 82 percent in Sweden (*33*); the U.S. figure is 92 percent (*38*). The availability and use of mass transportation in local and long-distance travel is an important factor in the optimization of the use of the auto discussed above (*25, 26*). In Stockholm, Gothenburg, and Malmö, where more than 25 percent of Sweden's population resides, mass transit, motor bikes, and pedal bikes account for 75 percent of all commuting (*41*). The figure for the entire country is 46 percent. Mass transit provides half of this, mostly in the cities named above. Most of the cities of more than 50,000 people in Sweden have bus systems and economic incentives, including subsidies, to encourage their use by riders going into the city center. In Gothenburg, for example, one can obtain a round-trip ticket for the price of a single fare by using the streetcars and buses at off-peak daytime hours; in Stockholm and other cities a 70-Skr pass allows unlimited transportation on all rail and bus lines. Buses are often as close as 4 minutes apart during peak hours, and rapid rail and buses provide direct service to places as much as 40 km from the city centers. Thus, to the city or suburban dweller in Sweden, mass transit presents a viable and economic alternative to the use of an automobile, and development of suburbs and new towns around rail and bus stations reflects the popularity of mass transportation. For longer trips, alternatives to automobile transport in Sweden are also available. Intercity buses, semicharter buses, and trains carry 20 percent of the passenger miles in trips over 50 km. Swedish Railways offers hourly departures between Malmö, Gothenburg, and Stockholm during day and early evening hours, traveling at average speeds of 80 to 100 km/hour.

The tax system has strongly affected use of the automobile in Sweden. In 1971, the gasoline tax of 50 cents per gallon raised the price by 250 percent to 70 cents per gallon (*23, 48*). Automobile excise taxes and yearly fees rise in proportion to vehicle weight according to the formula shown in Fig. 3. These fuel and weight taxes influence owners to purchase light cars, as the small proportion of cars heavier than 1700 kg (the U.S. average) shows (Fig. 2). In addition, the excise taxes raise the cost of a new car in Sweden (compared to the United States), providing an incentive to keep an older car in running condition. The average car in Sweden has a lifetime of about 14 years compared to a U.S. average of less than 10 years.

Nontax disincentives have also been employed to discourage use of automobile transit in Sweden. In Stockholm there is no 24-hour free street parking in the greater downtown area, and parking fines begin at $12.50. Both Stockholm and Gothenburg have set up systems of barriers, one-way streets, mass transit–only lanes or passageways, and pedestrian-only streets that further discourage use of cars.

For freight transport, as shown in Table 5, the largest difference in per capita energy use is associated with distances through which goods are moved (*16, 30, 30b, 39, 40*). A lesser, though still important, factor is the energy intensity of freight movement. Although a complete study of efficiency is yet to be made, some important factors can be identified. Long-haul trucks are more energy-intensive in Sweden than in the United States, but short-haul freight is much less energy-intensive. Small station wagons and four-cylinder microbuses or diesel minitrucks are used extensively for short hauls in Sweden, in contrast to the heavier pickup or panel trucks used in the United States, so that mode and vehicle are more closely matched to the demands of the task. Much of the difference in freight miles would be accounted for by shipments of Swedish exports of raw materials through other countries, exports that far outweigh (literally) imports. However, these freight miles are not distinguished in our study. Also, coal and other fuels are transported over much greater distances in the United States than in Sweden.

Energy used in foreign passenger travel, particularly in European countries, where this constitutes a significant fraction of what corresponds to domestic travel in the United States, may distort comparative energy use analysis. This is particularly true of air travel. Nearly every passenger flight connecting Sweden with anywhere stops in Copenhagen, where most of the fuel for the trip is put aboard. Thus, Danish fuel intensity per air passenger mile is abnormally high (*8*), while that for Sweden is low (*16*). It is also difficult to credit passenger miles when foreign visitors travel to or within a country. Because of these uncertainties, we have refrained from drawing conclusions from the great differences in E' for air passenger travel in Table 3.

Residential and commercial energy use. A comparison of energy use in the residential and commercial sectors is given in Table 6 (*3, 16, 30b, 41, 49–51b*). Although the per capita consumption is significantly lower in most categories, a full appreciation of the differences is only obtained by examining the D'_j's and E'_j's separately.

Space heating, consuming more than one-half of the total residential energy (Table 7), shows very large differences in efficiency when account is taken of the differing climates and the actual energy use per unit area of residential or commercial space. The larger number of degree-days in Sweden is compensated for by considerably lower heating intensity (kilowatt-hours per square meter per degree-day). A study of insulation in Swedish homes and apartments showed that heat loss through walls declined steadily to a typical U value of 0.06 British thermal unit per hour per square foot per degree Fahrenheit (*16*). One can almost guess the year of construction of a residence in Sweden by the U values, the scatter from the average value for any year of building being very low (*16*). This indicates that factors such as building codes have acted to permit only energy-efficient (and economic) construction in housing (*52*). In contrast, U values in the United States have been set mainly by a weak Federal Housing Administration (FHA) minimum property standard, which before 1971 was 0.12 Btu hour^{-1} ft^{-2} $°\text{F}^{-1}$ for ceilings and 0.19 for walls (*50*). The factor of 2 difference between U.S. and Swedish U values is nearly equal to the average ratio of heating intensities. Swedish houses also have correspondingly less infiltration and heat loss through glass because of the use of double glazing (*17*).

Although the lower heat loss in Swedish houses is in part a response to the more severe climate, this is not the primary reason, as seen by comparing heating intensities in terms of degree-days in various regions in the United States and Sweden (Table 8). Although there is little overlap between the U.S. and Swedish degree-day values, the plots of intensity (kilowatt-hours per square meter per degree-day) against degree-days would clearly lie on different curves for Sweden and the United States. Also, the Swedish values are nearly independent of degree-days, reflecting the use of similar standards in their four climate zones (*52*).

In Sweden, the mix of single-family dwellings and apartments (multiple-family dwellings), 42 and 58 percent, respectively, is considerably different from that in the United States, where in 1970 the corresponding figures were 71 and 29 percent. However, this does not account for much of the difference in heating efficiency, as the kilowatt-hours per square meter was only slightly lower in Swedish apartments than in single-family dwellings, and the kilowatt-hours per capita was also very similar, because of the higher number of people per house in single-family dwellings. In apartments, common metering of all units in a building removed the incentive to conserve, raising both space heat and hot water use (*41*).

Use of electric resistance heating in Sweden was increasing rapidly (*53*), as was U.S. use (*49*), until the embargo of 1973 caused a reevaluation of the overall effectiveness of such systems. In 1972, 7 percent of Swedish homes (15 percent of single-family dwellings) were heated electrically, similar to the 8 percent in the United States, but much less than the approximately 20 percent in Norway, where hydroelectricity is the largest

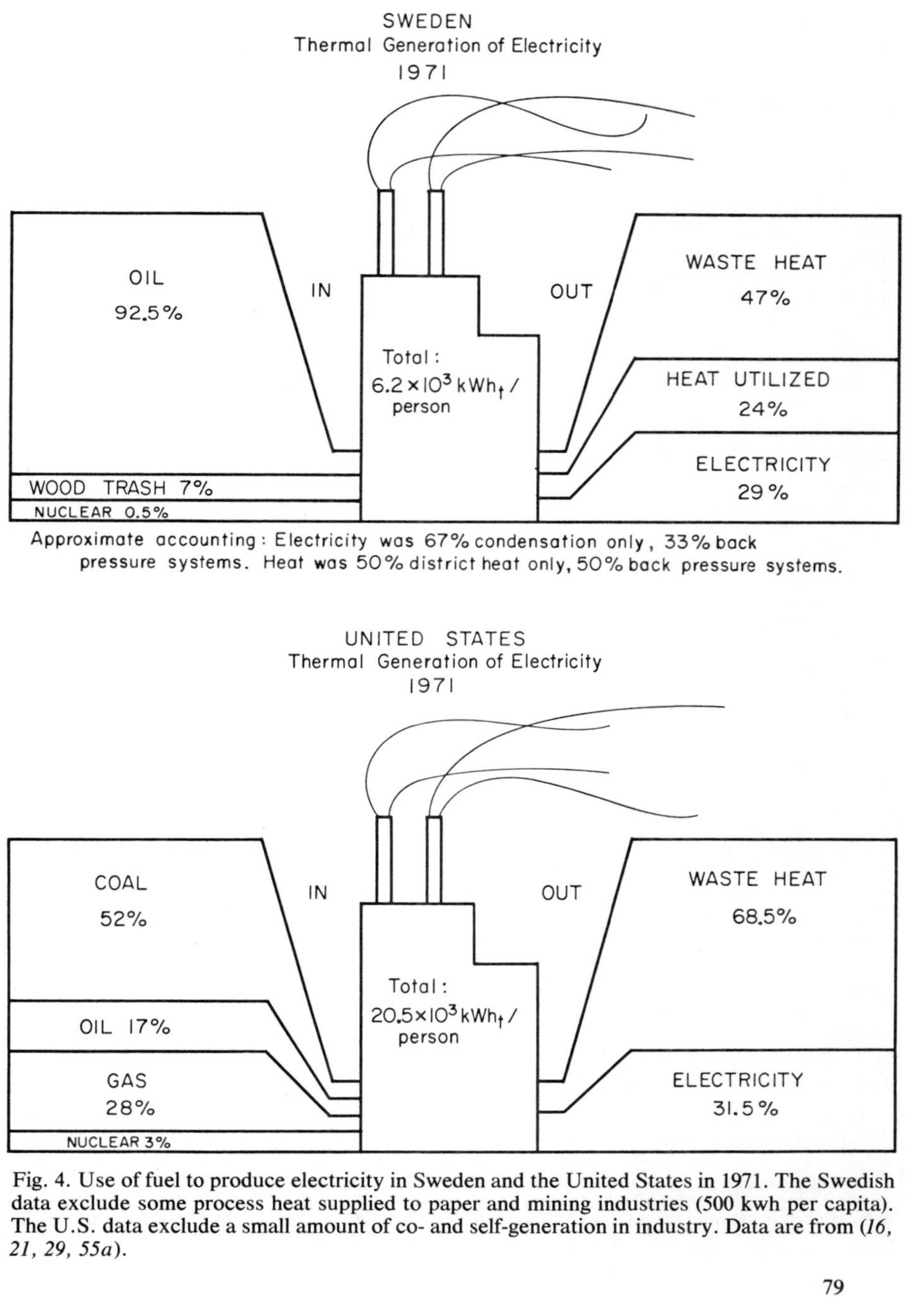

Fig. 4. Use of fuel to produce electricity in Sweden and the United States in 1971. The Swedish data exclude some process heat supplied to paper and mining industries (500 kwh per capita). The U.S. data exclude a small amount of co- and self-generation in industry. Data are from (*16, 21, 29, 55a*).

single contributor to the total energy supply. The average heat losses in Swedish all-electric homes are two-thirds of those in oil-heated homes (*17*).

In the commercial sector, overall energy use per square meter of space may be as much as 30 percent lower in Sweden than in the United States (*16*), even before the difference in heating degree-days is considered. The heating intensity, when measured in kilowatt-hours per square meter per degree-day, is approximately 2.5 times lower in Sweden. We attribute this mainly to the same differences in insulation, ventilation, and construction standards that applied to the residential sector, but the energy consumed in the commercial sector is reduced further by more realistic lighting standards, which also lower the need for cooling. (Unlike many large buildings in the United States, Swedish office buildings do not generally require air conditioning in winter to remove the heat produced by high lighting levels.)

In Table 6 the important residential and commercial uses of electricity are compared. Higher U.S. energy use arises primarily from a combination of factors: significantly more use of large appliances like dryers and large "frost-free" refrigerators, excess lighting, and more small appliances (*53, 54*). Air conditioning is conspicuously absent from Swedish electricity use, but accounts in the United States for only 12 percent of electricity used in the residential and the commercial sectors and only 3 percent of total energy use. The per capita total energy use for space cooling in the United States is roughly equal to the total consumption for space heating of factories in Sweden, a factor unimportant in the United States (*30a, 54*).

Water heating, another major energy user, requires typically 6,200 kwht per household in apartments (central water heating) and 10,500 kwht per household for single-family dwellings in Sweden, while the corresponding U.S. figures are 9,600 and 11,500 kwht. Much of the hot water in Sweden is prepared in centralized systems, eliminating some of the losses typical of American single-unit water heaters. On the other hand, the larger systems are not easily metered individually; in studies of energy use in apartments in Sweden (*16, 41*) it was noted that occupants paying individually for heat, hot water, and electricity use at least 15 percent less than those paying indirectly by sharing the cost in the rent.

An important mechanism for supplying space heat in Sweden is district heating, in which central stations either produce heat alone, or cogenerate heat and electricity. District heating supplies 19 percent of the total residential heat needs in Sweden (*16*). The energy balance for Swedish thermal power plants shows that 24 percent of the kilowatt-hour input appears as warm water or steam, primarily for heating of homes and buildings, and 29 percent as electricity (*21, 55a*). Figure 4 illustrates the combined electricity-heat balance for the United States and Sweden for 1971. In Malmö, a city of 250,000, combined elec-

Table 9. Energy use in industry—an economic overview. Data for the United States are from (*30, 56, 57*), data for Sweden from (*16, 33–35*). Values of kwh include kwhe. The E_J and T_J values are net; for gross kwht, multiply by [(2 kwhe/kwh) + 1], where the kwh and kwhe are from T_J. The U.S. figures for kwhe include self-generation, but these are not included in the E_J's. The U.S. E_J's are for 1971; value added from (*56*) was inflated to 1971 dollars.

Industrial sectors	D_J		E_J (kwh/$)		T_J			
	Value added ($/capita)		U.S. (1971)	Sweden (1970)	U.S. (1971)		Sweden (1970)	
	U.S.	Sweden			kwh	kwhe	kwh	kwhe
Manufacturing								
Paper*	62	101	44	75	3,200	290	7,625	1,300
Market pulp	2	34			125	25	3,680	500
Paper mills	24	60			2,500	230	3,895	800
Percent of sector†	40%	84%			82%	88%	99%	100%
Chemicals	156	62	25	18	3,930	575	1,135	540
Organic	16	7			1,575	110	80	250
Inorganic		9			1,220	250	110	100
Plastics and fibers	24	18			630	80	305	80
Agricultural	8	7			115	15	120	55
Percent of sector	31%	49%			90%	79%	54%	90%
(Feedstocks consumed)‡					(4,600)		(1,600)	
Petroleum products	30	11.5	142.9	81.7	4,000	145	940	30
Refining§	25	8	152.0	112.5	3,800	135	900	23
Stone, glass, and clay	51	50	36.3	32.5	1,850	120	1,625	150
Primary metals	110	103	51.8	37.7	5,700	710	3,880	910
Basic steel	46	74			4,390	190	3,065	500
Alloys	10	3			80	35	280	160
Nonferrous	8	6			640	300	370	225
Percent of sector	57%	81%			90%	74%	96%	97%
Total energy-intensive‖	421	328	44.4	46.4	18,680	1,840	15,205	2,930
Other manufacturing	1,320	808	3.4	2.0	4,525	1,050	1,600	710
Total manufacturing	1,741	1,137	13.3	14.8	23,205	2,750	16,805	3,640
Energy harvest (excluding refining and electrical utilities)					2,500	230	500	280
Mining					570	100	570	180
Agriculture and forestry					1,825	55	510	200
Construction (excluding vehicles)					900	16	650	85
Total industry (excluding feed stocks)					29,020	3,290	19,035	4,385

*Includes wood wastes. †The value added is given for groups that are more energy-intensive than the average. Percent of sector gives the percentage of the sector contained therein. It can be seen that Sweden's value added is more concentrated in these sectors. ‡Feedstocks for Sweden are estimated from (*31–33*). These are excluded from totals. §Includes captive consumption not counted in most Swedish studies, but found in (*31, 32*). Feedstocks are subtracted from refining losses in (*31*). Here 500 kwh per capita of nonfuel petroleum (lubricants and so forth) is omitted, but it is counted in Table 2. The Swedish refining *T* could be as low as 700. ‖Excludes self-generation for electricity totals, except 400 kwh of self-generation in paper and pulp industries in Sweden.

tricity-heat stations provide heat for more than 50 percent of the homes (*55b*). The overall effect of these systems, after the slightly lowered production of electricity is taken into account, is a net saving of fuel of 1300 kwh per capita, which is 2 percent of the total energy consumption in Sweden. These savings are somewhat offset by the high demand for heat and hot water in unmetered apartments in Sweden compared to single-family dwellings.

Industrial energy use. In both Sweden and the United States the largest use of energy in industry is for basic materials processing. In Sweden this energy use is highly concentrated, five sectors accounting for 85 percent of the net use (*16, 21, 30–35, 56–64*).

In Table 9 we see that larger fractions of Sweden's manufacturing value added and energy use are concentrated in the five energy-intensive sectors. Additionally, the energy use in each sector and the value added are more concentrated toward materials processing—organic chemicals versus drugs, paper mills versus paper products, and so forth. Thus the mix of output in Swedish industry is more energy-intensive than in U.S. industry. This is reflected in the E_J for the total of the energy-intensive industries, which is slightly higher in Sweden than in the United States, as is the aggregate E_J for all manufacturing. While some energy-intensive products, such as plastics, chemicals, and aluminum, are made in greater per capita quantities in the United States, steel, cement, paper, and pulp are made in greater per capita amounts in Sweden. Much of Sweden's energy-intensive raw output is exported.

However, these measures of intensity can be misleading. As Table 10 shows, the process energy intensity (E') is significantly lower in Sweden for virtually every product, usually because of reduced process heat requirements. We note that similar differences in process energy intensities were found in the study of West Germany (*11*). These findings suggest that Sweden's industry is more energy-efficient than our own. More important, though, these findings stress the inaccuracy of measuring energy use, or efficiency, by aggregate ratios of energy use to value added or GNP, as done in (*1*).

Swedish industries use more electricity as a fraction of industrial energy consumption, or as a fraction of all electricity used in the whole economy, than their American counterparts. This effect

Table 10. Materials and energy consumption data for the United States and Sweden. Data are for 1970 and 1971; U.S. data are from (*5, 7, 11, 24, 30a, 56, 59*), Swedish data from (*7, 16, 23, 32, 33, 60*). Electricity was included (net) in E_J. The T_J value for Sweden reflects 3 kwht/kwhe.

Material	D_J' (kg/capita)		E_J' (kwh/kg)		E_J' (kwhe/kg)		T_J (kwh/capita)		T_J* (kwht/capita)	
	United States	Sweden	United States	Sweden	United States	Sweden	United States	Sweden	United States	Sweden
Basic steel*	580	650	7	4.8	0.5	1.0	4000	3100	4640	4420
Aluminum†	17	9	17.7	17.7	17.0	17.0	300	160	880	465
Oil refined‡	2900	1400	1.4	0.7	0.05	0.05	4060	900	4350	940
Market pulp‡	~ 1	550	9	6.7	1	1		3685		4900
Paper, including pulping§	260	550	9.5	6.6	1.5	1.5	2470	3630	2860	4730
Cement	342	460	2.0	1.6	0.1	0.1	685	735	755	830
Organic chemicals‖	234	89	6.7	4.0			1575	355	1800	855
Inorganic chemicals‖	100	87	12.2	4.4			1220	390	1720	600
Plastics and fibers‖	51	43	12.3	5.0			630	215	790	375
Fertilizer‖	105	67	1.0	1.8			115	115	145	230
Feedstocks (energy)‖	480	215	11.63	11.63			5600	2500		

*We did not include the energy content of scrap, estimated at an average of 500 kwh/ton for the United States and 1000 kwh/ton for Sweden, averaged over all steel. †Counts only the smelting of Al_2O_3 to Al. Refining of bauxite takes place in the United States but not in Sweden. ‡The U.S. oil refining E_J is from (*56, 59*). Swedish losses are estimated from (*31*) and (*33*). The last source gives a very low figure of 0.65 kwh/kg, but estimates based on the known flow of oil through refineries indicate 1.0 kwh/kg. We do not account for differences in refinery output mix. §Pulp and paper values include the energy in wood wastes and liquors. This amounts to 1000 kwh per capita for the United States (*30a, 59*), and about 4000 kwh per capita for Sweden (*16, 21*). Sweden uses more wood waste for fuel per ton of output, and uses fewer external fuels as well. Swedish electricity was one-third cogenerated, and U.S. electricity about half that. ‖Values of E for chemicals are difficult to obtain and to compare. Feedstocks used, including road oils, were converted to kilograms by using the approximate relation 1 kg (oil equivalent) = 11.63 kwh. Energy consumed by uranium enrichment in the United States is counted in industrial chemicals (*57*).

Table 11. Energy intensities and costs in industry. The U.S. data (for 1971) are from (*56, 57, 60*), Swedish data (for 1970) from (*16*). Price figures are for purchased fuels only. Electricity data are for purchased electricity, except for the Swedish paper industry. Subscripts e and f mean electricity and fuel.

Industry	T_J (kwh/capita)		E (kwh/\$)		P (¢/kwh)		P_e/P_f	
	United States	Sweden	United States	Sweden	United States	Sweden	United States	Sweden
Five energy-intensive industries (excluding feedstocks)*								
Fuel	16,840	12,275	40	37.4	0.15	0.20	5.4	3.75
Electricity	1,840	2,930	4.4	8.9	0.81	0.75		
Other manufacturing†								
Fuel	3,475	900	2.6	1.1	0.19	0.36	6.3	3.1
Electricity	1,050	710	0.8	0.9	1.2	1.1		
Total manufacturing								
Fuel	20,315	13,175	11.7	11.6	0.16	0.22	6.25	3.7
Electricity	2,890	3,640	1.7	3.2	1.0	0.82		

*Energy-intensive industries include paper [Standard Industrial Classification (SIC) 26, Svensk Näringsgrenindelning (SNI) 341], chemicals (SIC 29, SNI 353 and 354), stone glass and clay (SIC 32, SNI 36), and primary metals (SIC 33, SNI 37). †"Other" industries include SIC 20 to 25, 27, 30, 31, and 34 to 39, and SNI 31 to 33, 342, 355, 356, 38, and 39.

Table 12. Typical energy prices in the United States and Sweden. The exchange rate used is $1 = 5.18 Skr (1960 to 1970) and 4.30 Skr (1974). The U.S. data are from (*15, 24, 57*); Chern (*58*) gives the following prices (¢/kwh) for U.S. industry as a whole in 1971; gas, 0.13; coal, 0.12; oil, 0.23; electricity, 0.98; and other, 0.25. Compare with Swedish prices. The Swedish data are from (*16, 23, 48*) and a 1975 press release from the Swedish embassy.

Energy type	United States				Sweden			
	1960	1970	1974	1970 (¢/kwh)	1960	1970	1974	(1970) (¢/kwh)
Oil products (¢/gallon)								
Gasoline*	30	35	45	1.04	53	61	116	1.82
Diesel	23	28	35	0.83	42	48.8	90	1.45
Heating oil								
Small customers	15	18	35	0.50	13.3	13.2	40.6	0.37
Large customers	10.5	12	25	0.33				
Heavy oil	7	8	23	0.23	7	8.5	22.5	0.24
Gas (¢/MM Btu)								
Residential	82	87	113	0.29		550	680†	1.9
Industrial								
Firm Service	51	50		0.17				
Interruptable service	33	34		0.11				
Coal, industrial ($/ton)‡	10	13	25	0.14		18		0.2
Electricity (¢/kwh)								
Base	2.75	2.75		2.75	3.14‖	2.12‖	2.3§‖	
Base and space heating	1.75	2.0		1.5		~1.5	2.0§	
Industrial	1	1	1.5	0.4–2.1		0.93	1.8§	0.6–2.2

*Swedish gasoline taxes: 42¢ per gallon in 1970, about 68¢ per gallon in 1974. The U.S. prices include a tax of 10 to 13¢ per gallon. †Data for 1973. ‡Coal price excludes captive and utility coal. §Data for 1975. ‖Swedish figures are based on 1700 kwh/year (1960), 3000 kwh/year (1970), and 2000 kwh/year (1974).

can be understood by noting that historically nearly all of Sweden's electricity has been generated from hydropower, the predominant domestic energy resource; industries could be expected to utilize this resource, which has been less costly than steam-based electricity. Since electricity prices are similar, we attribute the higher electric intensity of Swedish industry to the lower ratio of the price of electricity to the price of fuels, as compared to that in the United States, where most electricity is steam-based. These costs and the quantities used are summarized in Table 11.

Other factors in Sweden tend to reduce specific industrial energy consumption compared to that in the United States. Sixty percent of all fuel used in the paper industry (which consumed 15 percent of all energy in Sweden) is provided internally by barks and liquors, as opposed to 35 percent in the United States (*21*); but a third of the electricity used by that industry (and smaller fractions elsewhere) is cogenerated with steam production (*16, 21, 55a*), thus reducing fuel needs. Cogeneration has also achieved considerable energy savings in Germany (*11*), and is considered to be economic for the United States (*65*), where half of the electricity consumed in the paper industry is self-generated, but only a small amount is cogenerated.

Like process industries, assembly industries in Sweden tend to show a lower use of fuel per unit of product (or value added) than those in the United States. This is in spite of greater space heating requirements in Sweden, which in some cases surpass requirements for electric drive and lights. A total of 20 percent of Sweden's industrial energy use is for space heating. In the entire Volvo concern, encompassing several large assembly plants, 1974 energy use was estimated at 0.6×10^9 kwh for space heat and hot water, a similar amount for process heat, and an equal amount for electricity, of which one-third went for lighting and office use. [Volvo was able to cut its total energy use 25 percent after the oil embargo through "leak plugging" (*62a*).] If Swedish industrial fuel use were adjusted to take into account the difference in climate, it would be 10 percent lower.

The relatively more modern equipment in Swedish industry—Sweden's capital stock has grown significantly faster than ours as the Swedish GNP approached ours—certainly contributes to the higher efficiency. Both U.S. and Swedish industry have improved energy efficiency through technological change since World War II, in spite of generally falling energy prices (*56*). A comparison of data collected by Meyers *et al.* (*56*) with Swedish data (kilowatt-hours per ton or per dollar) suggests that the energy intensity of Swedish industry today would lie 10 to 15 years hence on the projected curves of Meyers *et al.* for U.S. industry. Significantly absent from Swedish industrial energy use were (and are) "interruptable" gas at bargain prices and cheap coal, two fuels that have been important to many U.S. industries and whose low price and availability fostered higher energy use in the past. The importance of relatively higher prices for energy in Sweden as a major factor in its relative energy efficiency was emphasized in a series of studies by Carlsson (*61*); see also (*16*), vol. 2, appendix 3.

Both official Swedish government forecasts (*16*) and the views of individuals in industry (*61–63*) reflect the belief that optimization to counter ever-increasing fuel prices will further reduce specific energy requirements of Swedish industry toward the end of the century, as many have also predicted for the United States (*3, 4, 56*). Since Sweden traditionally has paid a high industrial wage, the saving of energy has come about not by direct substitution of labor for energy, but through the substitution of energy management (*62*) and capital [Carlsson in (*16*), vol. 2, appendix 3] for energy.

Other factors in resource use in Sweden contribute to both lower demand per product and lower demand for energy-intensive products themselves. It was noted above that Swedish cars outlast American cars, weigh less, and use materials that themselves require less energy than their American counterparts. Furthermore, Swedish consumers have maintained the widespread use of returnable bottles. Other utilization patterns (relative sizes of D_J) are interesting; in the late 1960's plastic bags became popular, only to be replaced by paper again as the cost of plastic, made from imported petroleum, rose compared to the cost of paper, made largely from domestic sources. We can generally conclude that cultural and institutional factors combine with economic and technical factors to effect energy savings in the industrial sector in Sweden relative to the United States. This is mainly done by increasing efficiency (lowering the E_J's); but changing the mix of products (mix of D_J's) actually consumed in Sweden toward lower energy intensity is also significant in some areas.

Imports and exports of goods. Since imports and exports comprise an important part of economic activity it is important to evaluate the energy embodied in nonenergy trade, as well as the process energy embodied in refined fuels, such as gasoline. We consider direct energy to be that applied by the producer of a good or service, and indirect energy that required to produce the materials and services used by the producer. For the United States, Herendeen and Bullard (*35*) found that while nonenergy imports and exports contained equal amounts of energy, the imports of refined oil embodied more energy than exports of coal and refined oil products (excluding the heat of combustion of these fuels). A similar balance for Sweden was evaluated by the Energi Prognos Kommitéen (*16*). The results, summarized in Table 2, show that Sweden's energy use per capita is overestimated due to an export surplus of embodied energy; the United States has a small import surplus. In contrast, Denmark's per capita energy use is considerably underestimated, as shown by Elbaek (*8b*), who found that the energy balance of trade amounted to an import of 20 percent of the energy consumed in Denmark. By contrast, West Germany has a large export surplus (*11*). Note that in every case the energy embodied in import of fuel is much larger than any of these figures. We conclude that an accounting of the energy embodied in foreign trade widens the difference in energy use between Sweden and the United States.

Electricity production and district heating. Sweden is rich in hydropower, an energy source that accounted for approximately 14 percent of all energy and 75 percent of all electricity produced in 1972. In Swedish statistics the electricity is counted at 85 percent First Law efficiency. Since most of Sweden's hydroelectric resources are in the far north, transmission line losses are greater than in the United States, per net kilowatt-hour sold.

About 35 percent of Sweden's fuel-based electricity came from combined heat-electricity systems, which produce more useful kilowatt-hours per kilowatt-hour total consumed than do purely electrical-thermal plants (*16, 55a*). In 1971 Sweden consumed fuel amounting to about 4.11 Mwh per capita to produce 1.77 Mwh per capita of electricity, for a "heat" rate of 2.32 kwht/kwhe. This is illustrated in Fig. 4, in which 0.8 Mwh per capita heat-only production is included. If the heat and electricity had been generated separately, about 1.3 Mwh per capita additional fuel would have been required if the heat was then produced in central plants, and about 1.5 Mwh per capita more if it was produced in smaller boilers. Only the half of the combined heat-electricity and heat-only systems that are in or near cities, to supply residential and commercial heat, are included in Fig. 4. The other half, located in industries, primarily paper, doubles the energy savings given above. Central heat-only plants provide heat for 600,000 dwellings, at 85 percent fuel to home (First Law) efficiency, compared to 65 percent for boilers in apartments. This saves 5100 kwh per dwelling or 375 kwh per capita. Another 25 kwh per capita is saved by district heating of buildings, for a total savings of about 400 kwh per capita from heat centrals. These savings must be added to those obtained by use of district heat from combined generation. The use of back pressure and heat-only centrals, in Sweden, leads to a heat rate of about 2.3 kwht/kwhe. Applying this thermal heat rate to hydropower in Sweden leads to a value of kwht which is 12 percent greater than the net kwh. This is in contrast to the 20 percent increase in Table 2 that was obtained by applying the U.S. heat rate to Swedish hydropower.

An additional factor that should be taken into account when comparing electricity use is that, stimulated by the low ratio of the price of electricity to the price of fuel, Swedish industry tended to use electricity for a wider range of tasks. Had electricity been 85 percent thermally generated, as in the United States, it would have been more expensive, and therefore used more sparingly (*66*).

Analysis of Differences in Energy Use; Conclusions

In Table 12 we show explicitly some important energy prices for Sweden and the United States. The largest price differences occur in road fuels, even before the higher taxes on automobiles in Sweden are considered. Electricity, on the other hand, has been relatively inexpensive (compared to fuel) in Sweden, because in the past a large share of electricity has been hydropower (*66*). In 1971 electricity use in Sweden (7400 kwh per capita) was close to that in the United States (7700 kwh per capita), but more of this total was used in the industrial sector in Sweden and more in the residential and commercial sector in the United States. Other fuels in Sweden lie between these two extremes, being slightly more expensive in Sweden (before 1973) and used more efficiently there as well. Since the price of oil used for home heating in Sweden was comparable to U.S. values (until 1973), the length of the heating season, as well as institutional factors mentioned above, must account for the efficient use of that fuel for space comfort. Significantly, Sweden had no natural gas or domestic coal, two fuels whose low prices certainly encouraged intensive use in the United States.

Higher energy prices alone, however, do not account for the more efficient energy use in Sweden. In this article and elsewhere (*4*) it has been stressed that, while a particular set of energy prices determines a mix of energy and other economic factors that allows production for the least cost, institutional and social factors determine how close individual consumers, firms, and society as a whole come to this most economic energy use. In the United States, for example, mortgage policies and market considerations constrain developers to minimize first costs, rather than life-cycle costs, constraints that do not appear to be applicable to construction in Sweden. Also, building codes have imposed energy-conserving construction more uniformly in Sweden, and the Swedish government has given priority to energy conservation in housing loans (*52*). Energy conservation in passenger transport in Sweden has also been strongly influenced by government policy, in this case mainly through the market mechanism, by various taxes and incentives. These factors also have important synergistic effects. Good intracity transport, and high costs of operating an automobile, tend to keep the population more concentrated. In addition to maintaining the viability of the public transport system itself, this situation also affects housing and living patterns in energy-saving ways. With increased population densities apartment living is more common, potentially effecting energy savings through fewer external walls, better insulation, and more efficient heating systems. Shopping also becomes easier, with more neighborhood stores; trips are shorter, often on foot; and smaller storage facilities are required, resulting in smaller refrigerators with consequent electricity savings.

In a recent study of energy use in the United States, Hannon (*67*) suggested that lowering the energy requirement for an economy by changing life-styles and the mix of consumer goods (the D_J's) would be difficult, because consumer expenditures would generate energy requirements no matter how they were directed. We have shown here that in Sweden the D_J's are shifted toward less energy-intensive activities, and the E_J's toward higher efficiency. For both effects, dollars saved by saving energy in one activity and respent on another do not, on the average, generate as much energy use as expenditures for a more energy-intensive mix of D_J's, or activities with less efficient E_J's, would have done. All energy intensities are reduced through higher efficiencies (conservation), and shifts from more to less energy-intensive activities are made at the same dollar level. Sweden, like other European countries, developed these energy economies to offset its higher energy prices and balance of payments problem resulting from importing energy. This resulted in a higher standard of living for a particular level of energy consumption. This suggests the answer to the dilemma posed by Hannon: in the face of energy scarcity and consequent rising energy prices, consumers in the United States would seek to maintain their standard of living by optimizing energy use both through increased energy efficiency and through shifting to less energy-intensive activity.

Future work should further elucidate both the underlying causes of and the mechanisms for achieving higher energy use efficiency in Sweden. At this point we offer some tentative conclusions about energy use based on our comparison of Sweden and the United States.

1) For a particular GNP, the efficiency of energy use, demographic factors, and the mix of goods and services share in determining the energy requirements of an economy.

2) Some of the main factors that have accounted for the reduced energy use in Sweden are smaller automobiles, more use of mass transit, more insulation and tighter construction, more efficient industrial processes, and the use of cogeneration and district heating.

3) Higher efficiency (lower E values) accounts for the largest portion of the lower energy use in Sweden arising from these factors.

4) Counting hydropower at 3 kwht/kwhe, as is done in U.S. accounting, results in a ratio of Swedish to U.S. energy consumption of 0.6, while using a one-to-one accounting results in a ratio of 0.5. The difference, although significant, does not account for the dramatically lower Swedish energy use.

5) The most important variable affecting energy use and energy efficiency is the relative price of energy with respect to other resources. However, institutional and social factors are also important.

6) It is necessary to consider individual energy intensities (E_J's) as well as levels of activity (D_J's) in order to understand energy uses and needs. Consideration of only total energy use (ΣT_J) or the energy/GNP ratio (*1, 28, 68*) obscures dramatic differences in intensity (or efficiency) and economic structure. Similarly, forecasts of energy "needs" in which the aggregated quantities are used also overlook vital details and trends in the components of E and D, components that may be more or less sensitive to changes in energy prices.

Although we have seen that energy use in Sweden is generally more efficient than that in the United States, both countries can improve energy use effectiveness by optimizing to the higher energy prices that have developed since the period we examined. Our comparison indicates that many energy conservation measures are available to the United States, as energy prices continue to rise. The Swedish economy performs well as a (relatively) energy-efficient economy, suggesting that more efficient energy use will not interfere with the function of the American economy. While we hesitate to give an exact figure, we suggest that Swedish methods of energy conservation, including smaller cars, better structures, and more efficient use of process heat, would result in a savings of 30 percent of the total energy used in the

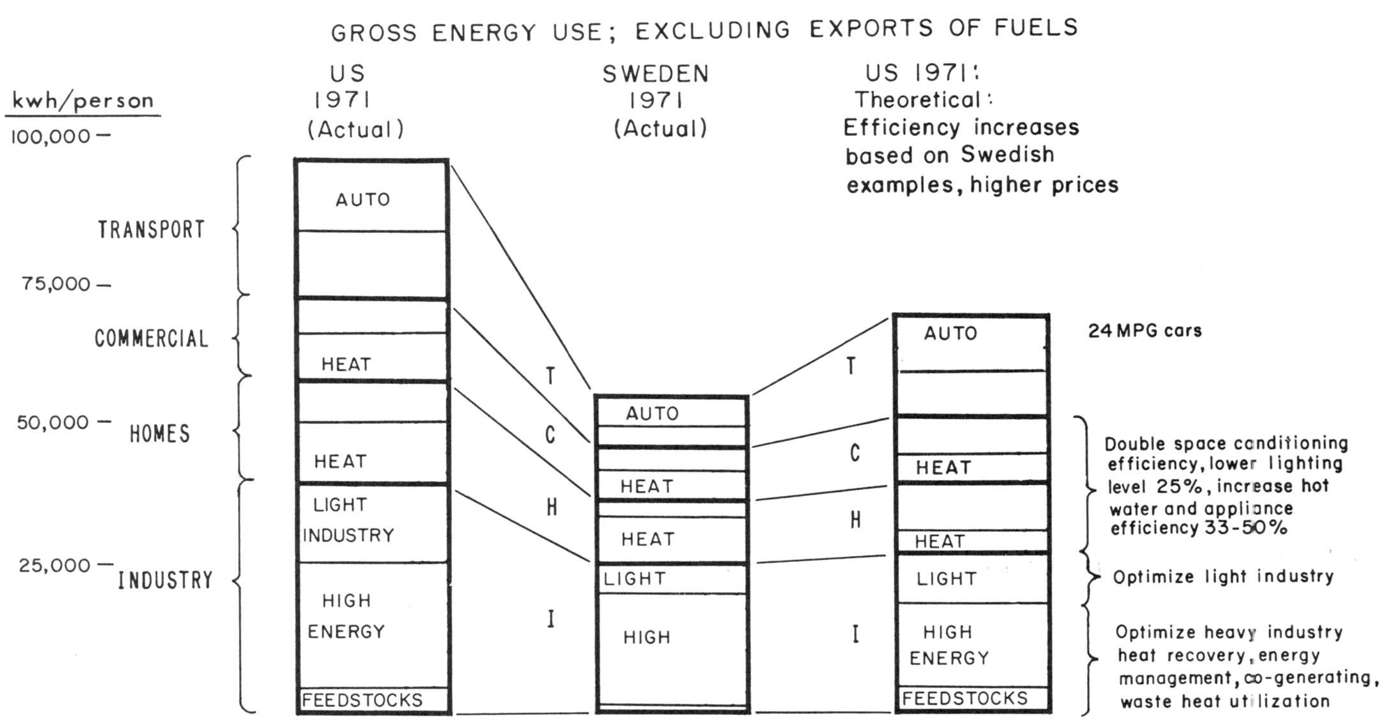

Fig. 5. Summary of U.S. and Swedish energy use in 1971, and theoretical U.S. energy use based on Swedish intensities in industry, space conditioning, and automobiles (miles per gallon). For theoretical U.S. use it is also assumed that appliance use and lighting levels decrease by 33 percent. Freight, airlines, and energy extraction are ignored, but higher air-conditioning and lighting efficiency are factored in. Life-style factors (such as numbers of appliances and passenger miles) were not considered. Values in the column at the left are kwh per capita. Compare with (*3—5*).

United States (Fig. 5). Thus, international energy use comparisons, far from suggesting an inevitable coupling between level of economic activity and energy use, actually suggest ways in which more well-being can be wrought from every Btu of fuel and kilowatt-hour of electricity consumed in a given country (*69*).

References and Notes

1. H. R. Linden, *Testimony Before the President's Energy Resources Council* (10 December 1974), or *Resource, Economic and Historic Considerations Related to Issue of Energy Policy* (1975). Both are available from the Institute for Gas Technology, Chicago, Ill.
2. Except where popular use dictates American units (for example, miles per gallon or degree-days Fahrenheit), we use kilowatt-hours as our standard unit: kwh means fuel or electricity used in end consumption; kwhe refers to electricity only; and kwht refers to total consumption of energy with electric conversion losses included at the rate of approximately 2 kwh lost per kwhe produced. All tons are metric.
3. M. Ross and R. Williams, *Assessing the Potential for Fuel Conservation* (Institute for Public Policy Analysis Report 75-02, State University of New York, Albany, 1975); available from M. Ross, Department of Physics, University of Michigan, Ann Arbor; *Technol. Rev.*, in press.
4. L. Schipper, *Energy Commun.* **2** (No. 4), 333 (1976); in *Annual Reviews of Energy*, J. Hollander and M. Simmons, Eds. (Annual Reviews, Palo Alto, Calif., 1976), vol. 1, pp. 455–517.
5. A. Makhijani and A. Lichtenberg, *Environment* **14** (No. 5), 10 (1971).
6. "World alternative energy strategies project" (S. Carhart and P. Craig, personal communication) and "Energy consumption and economic activity and comparative patterns in nine industrial countries," Resources for the Future, Inc., Washington, D.C. (J. Darmstadter, J. Dunkerley, J. Allerman, in preparation).
7. A. Doernberg, "Comparative analysis of energy use in Sweden and the United States," *Brookhaven Natl. Lab. Rep. BNL-20539* (1975).
8a. Industri Department, *Danmarks Energi Forsyning i Fremtiden* (Denmark's Energy Supply in the Future) (Copenhagen, 1974).
8b. See also B. Elbaek, *Energi—Energi—Energi Kris* (Munksgaard, Copenhagen, 1975).
9. Ministry of Industry, *Energiforsyningen in Norge på Fremtiden* (Energy in Norway in the Future) (Oslo, 1974). See also H. Parr, P. Hofseth, K. Byrne, J. Riekeles, *Energi Miljø og Samfunn* (Energy, Environment, and Society) (Norges Naturvernforbund, Oslo, 1974).
10. J. D. Parent, *Some Comments on Energy Consumption and GNP* (Institute for Gas Technology, Chicago, Ill, May 1974; revised October 1974).
11. R. Goen and R. White, *Comparison of Energy Consumption Between West Germany and the United States* (Government Printing Office, Washington, D.C., 1975).
12. T. D. Mount, L. D. Chapman, T. J. Tyrrell, "Electricity demand in the United States: An economic analysis," *Oak Ridge Natl. Lab. Rep. ORNL-NSF-EP-49* (1973).
13. W. J. Mead, private communication.
14. Foster Associates, *Energy Prices 1960–1970* (Ballinger, Cambridge, Mass., 1974).
15. Federal Energy Administration and American Gas Association, *A Pilot Project in Homeowner Energy Conservation* (Government Printing Office, Washington, D.C., 1974).
16. Energi Prognos Kommitéen, Ministry of Industry, Stockholm, *Energi, 1985–2000 (Energi Prognos Utredningen [EPU])* (Allmänna Förlaget, Stockholm, 1974).
17. K. Munther, *Energi-Förbrukning i Småhus* (Energy Use in Single Family Dwellings) (Statens Institut för Byggnadsforskning Report 58, Svensk Byggtjänst, Stockholm, 1974).
18. In this and other examples Swedish usage is adjusted to reflect 60 percent fossil-fueled heating (First Law) efficiency.
19. S. H. Schurr and B. C. Netschert, *Energy in the American Economy, 1850–1970* [Johns Hopkins Press (for Resources for the Future), Baltimore, Md., 1960].
20. I. Kravis, *A System of International Comparisons of Gross Product and Purchasing Power* (Johns Hopkins Press, Baltimore, Maryland, 1975).
21. *Effektivare Energi Använding* (More Efficient Energy Use) (Report 181, Ingenjörs Vetenskaps Akademi, Stockholm, 1975).
22. *Some Facts About Sweden* (Skandinaviska Enskilda Banken, Stockholm, 1974–1975). We use 5.18 Skr = $1. In 1974 the applicable rate would be approximately 4.30 Skr = $1.
23. *Statistiska Årsboken* (Statistical Abstracts of Sweden) (Statistiska Centralbyrån, Stockholm, 1972–1975), with English summaries.
24. Department of Commerce, *1975 U.S. Statistical Abstract* (Government Printing Office, Washington, D.C., 1975).
25. K. Hultgren, *Människan och Bilsamhället* (People and the Automobile Society) (Norstedt, Stockholm, 1974).
26. L. Anell, A. Hedborg, M. Lönnroth, L. Ingelstam, *Skall vi Asfaltera Sverige?* (Should We Pave Over Sweden?) (Norstedt, Stockholm, 1971).
27. We use 68°F as a base temperature for both countries, and we ignore appliance heat loads and so forth. This understates heating differences because appliance use is somewhat greater in the United States, while insolation in Sweden is small compared to nearly all of the United States in the winter. Peak cold temperatures are not as low in Sweden as in our North Central, Rocky Mountain, and New England states, but average temperatures are comparable, and the Swedish winter is longer.
28. W. Dupree and J. West, *United States Energy Through the Year 2000* (Government Printing Office, Washington, D.C., 1972, updated 1975).
29. E. Cook, *Study of Energy Futures* (Environmental Design Research Associates, Chapel Hill, N.C., 1975).
30. R. Knecht and C. Bullard, *Direct Use of Energy in the U.S. Economy* (Technical Memo No. 43, Center for Advanced Computation, University of Illinois, Urbana, 1975).
30a. Stanford Research Institute, *Patterns of Energy Consumption in the U.S.—1968* (Government Printing Office, Washington, D.C., 1972).
30b. M. Beeler, *Source Book for Energy Assessment* (Brookhaven National Laboratory, Upton, N.Y., 1975).
31. *Verksamheten* (Activities) (Ångpanneföreningen, Stockholm, 1971–1976).
32. *SOS Industri 1970–73* (Sweden's Official Statistics—Industry 1970–73) (Statistiska Central Byrån, Stockholm, 1972–1975).
33. Energi Program Kommittéen (EPK), *Energiforskning*: vol. A. *Energin och Näringslivet* (Energy R & D: vol. A, Energy and the Economy) (Allmänna Förlaget, Stockholm, 1974).
34. The latest energy use data for Sweden are published as part of the Energy Conservation Bill in the Parliament: *Energy-Hushållning* (Energy Husbandry) (Ministry of Industry, Stockholm, February 1975). See also *Energiförsörjningen 1975–1980*, Rapport av Statens Industriverk (Allmänna Förlaget, Stockholm, 1976).
35. R. Herendeen and C. Bullard, *Energy Syst.*, in press.
36. E. Hirst, "Energy intensiveness of passenger and freight traffic, 1950–1970," *Oak Ridge Natl. Lab. Rep. ORNL-NSF-EP-41* (1973).
37. Federal Highway Administration, *Estimated Motor Vehicle Travel in the United States* (Washington, D.C., 1970–1975).
38. Federal Highway Administration, *Nationwide Personal Transportation Survey* (Government Printing Office, Washington, D.C., 1970).
39. Federal Energy Administration, *Project Independence: Energy Conservation: Transportation* (Government Printing Office, Washington, D.C, 1974).
40. Ministry of Communications, *Regionala Trafikplanering* (Regional Traffic Plan) (Stockholm, 1972); *Transporter i Sverige* (Transportation in Sweden) (Report Dsk 1975: 4, Stockholm, 1974); Swedish Institute, *Fact Sheet of Swedish Transportation* (Swedish Information Service, New York, 1973).
41. Energi Beredskaps Utredningen (EBU), *Energi: Beredskap i Kristid* (Energy: Preparedness During Crisis) (Allmänna Förlaget, Stockholm, 1975).
42. Motor Vehicle Manufacturers Association, *Automobile Facts and Figures* (Detroit, 1975); *Motor Vehicles and Energy* (Detroit, 1974).
43. K. Austin and K. Hellman, *Passenger Car Fuel Economy* (SAE Paper 730790, Society of Automotive Engineers, New York, 1973).
44a. *Motor Traffic in Sweden* (AB Bilstatistik, Stockholm, 1974–1975).
44b. See also J. Ullén, *Bilfakta* (Auto Facts) (Jan Ullén AB, Stockholm, 1975), vol. 28.
45. Energi Program Kommittéen, *Energiforskning*: vol. C, *Transport och Samfärdsel* (Energy R & D: vol. C, Transportation) (Allmänna Förlaget, Stockholm, 1974).
46. Statens Vågverk, *Personbilarnas Årliga Körlängd* (Yearly Private Car Vehicle-Miles) (Stockholm, 1974).
47. K. Austin and K. Hellman, *Passenger Car Fuel Economy as Influenced by Trip Length* (SAE Paper 750004, Society of Automotive Engineers, New York, 1975).
48. Svenska Esso, *Oljeåret 1974* (Oil Year 1974) (Stockholm, 1975).
49. A. Makhijani and A. J. Lichtenberg, *An Assessment of Residential Energy Use in the U.S.A.* (ERL M-310, Univ. of California Engineering Research Laboratory, Berkeley, 1973).
50. J. Moyers, "The value of thermal insulation in residential construction," *Oak Ridge Natl. Lab. Rep. ORNL-NSF-EP-9* (1972).
51a. Federal Energy Administration, *Residential and Commercial Energy Use Patterns, 1970–90* (Government Printing Office, Washington, D.C., 1974).
51b. See also S. Dole, *Energy Use and Conservation in the Residential Sector: A Regional Analysis* (R-1641-NSF, Rand Corporation, Santa Monica, Calif., 1976).
52. Indeed, the mortgage law of 1957 gave priority to home builders or buyers who intended to insulate beyond the building code minimums. A newer program (1975) provides greater subsidies, direct grants, and local testing programs, with total outlays of over $350 million over 3 years.
53. Centrala Driftledningen, *Sveriges Elkonsumtion, 1975–1985* (Sweden's Electricity Use, 1975–1985) (Stockholm, 1972).
54. Energi Program Kommittéen, *Energiforskning*: vol. D., *Uppvärmning och Lokal Komfort* (Energy R & D: vol. D, Heating and Space Comfort) (Almänna Förlaget, Stockholm, 1974).
55a. Swedish District Heating Association, *Combined Heat and Power, 1975* (Värmeverksförening, Stockholm, 1976).
55b. R. Johnson and L. Bell, *Malmö's Powerplants* (Industriverket, Malmö, Sweden, 1973).
56. J. Meyers *et al.*, *Energy Consumption in Manufacturing* (Ballinger, Cambridge, Mass., 1975).
57. Department of Commerce, *Census of Manufacturers* (Energy Data for 1971) (Government Printing Office, Washington, D.C., 1972).
58. W. Chern, "Electricity demand by manufacturing industries in the U.S.," *Oak Ridge Natl. Lab. Rep. ORNL-NSF-EP-87* (1975).
59. E. Gyftopolous, L. Lazaridis, T. Widmer, *Potential Fuel Effectiveness in Industry* (Ballinger, Cambridge, Mass., 1975).
60. S. I. Kaplan, "Energy demand patterns of eleven major industries," *Oak Ridge Natl. Lab. Rep. ORNL-TM-4610* (1974).
61. B. Carlsson, in *IUIs Långtidsbedömning 1976* (Industrins Utrednings Institut, Stockholm, 1976), p. 270; lecture at Industrial Energy Day, Stockholm, May 1976.
62a. Swedish Federation of Industries, *Energy Conservation in Swedish Industry* (Industriförbundets Förlag, Stockholm, 1974).
62b. A. Iveroth and B. Helmerson, *Industrin och Energi* (Industry and Energy) (Industriförbundets Förlag, Stockholm, 1974).
63. Statens Industriverk, *Tätorternas och den Tunga Industrins Energi Försörjning* (Energy Supply to Industry and Built-Up Areas) (Liber Förlag, Stockholm, 1976).
64. U. Norhammer *et al.*, *Samlat Energikunnande* (Collected Energy Skills) (Ångpanneföreningen, Stockholm, 1975).
65. Dow Chemical Co.; Environmental Research Institute; Greenspann-Townsend, Inc.; Cravath, Swaine, and Moore, Inc., *Energy Industrial Center* (National Science Foundation, Washington, D.C., 1975).
66. In Sweden, government-owned power stations, including hydroelectric facilities, produced 50 percent of the total electricity. These stations were financed at commercial interest rates, in contrast to most government-owned power in the United States.
67. B. Hannon, *Science* **189**, 95 (1975).
68. F. Felix, *Electr. World* **184** (No. 9), 64 (1975).
69. J. Holdren, *Bull. At. Sci.* **31**, 26 (December 1975).

70. This research was supported in part by the Lawrence Berkeley Laboratory, University of California, under ERDA contract W-7405-Eng-48. We also acknowledge the support of the Energy and Resources Group, University of California. One of us (L.S.) has also received grants from the Society for the Advancement of Scandinavian Studies and the Swedish Information Service, San Francisco. We acknowledge the helpful suggestions of J. Holdren, R. Herendeen, C. Bullard, and B. Carlsson. Måns Lönnroth (of the Secretariat for Future Studies, Stockholm) provided invaluable help in coordinating our research efforts in Sweden. This work is based on report LBL-4430 from the Lawrence Berkeley Laboratory.

Energy Conservation in New Housing Design

Jack E. Snell, Paul R. Achenbach,
Stephen R. Petersen

The total annual energy use, including space heating and cooling, in a typical detached house built before 1970 in Baltimore, Maryland, with a floor area of 140 m^2, amounts to an average 187 gigajoules (10^9 joules) (*1*) of heating fuels and 44 gigajoules of electrical energy (*2*). At 1975 average prices for natural gas and electricity (*3*), the corresponding annual cost is about $840. If the house has electric resistance heating, its 1975 annual energy costs could be as much as $1400. These household expenditures are nearly double the expenditures for the same energy requirements in 1970. Thus a primary incentive for increasing the thermal efficiency design of new housing is the increasing cost of energy. However, this is not the only factor stimulating energy consciousness in housing design. The more or less regular recurrence of brownouts in some large cities since 1970, the subsequent unavailability of natural gas to new residential customers in many parts of the country, and finally, of course, the Arab oil embargo of November 1973 all served to place in focus both the vulnerability of the United States to an energy shortage and changing and uncertain patterns of future fuels availability (*4*). Hence the more efficient use of energy has become an urgent national goal which is providing a further impetus to the consideration of energy conservation design features in housing.

The potential impact of these strong economic and national policy thrusts on both existing and new housing is significant. Existing dwellings that have low thermal efficiency and cannot be improved economically will become increasingly obsolete. This will result in decreased property values (relative to more thermally efficient dwellings) and, in some cases, outright abandonment. To a large extent, this obsolescence is due to a general failure to anticipate the dramatic increase in energy costs that has occurred since 1970 and now promises to continue, although at a lesser rate, at least through the end of the century. If such premature obsolescence is to be avoided in new housing, its energy efficiency must be increased substantially over that of most of the existing stock of housing.

In this article, we examine the potential implications of energy cost and availability for new housing design over the rest of the century. Our discussion is based on the premise that the rise in the price of fuels relative to that of labor and materials is focusing the attention of home builders and home buyers on total life-cycle costs. This shift in emphasis from first-cost, short-term, and speculative factors to the energy conservation technologies stimulated by life-cycle cost analyses portends a number of major physical changes in new residences over the next few decades.

This thesis is developed by looking first at the housing market, the historical influences of certain building practices, and fuel price changes in recent years. Attention is directed primarily to single-family detached dwellings including mobile homes, single-family attached homes, townhouses, garden apartments, and low-rise housing units. Next, major design parameters related to energy are examined in the context of a simplified life-cycle cost model, with numerical examples. The constraints imposed by existing technology, building and occupancy practices, and energy costs on total life-cycle costs are examined for their implications to future housing design practices. Then, the technological opportunities for energy conservation as a design consideration are discussed, with an emphasis on identifying the extent to which many design parameters related to energy will change in the next few decades. Finally, we focus on the question of the implications of energy conservation on new housing design, ownership, and operation and on the identification of several specific research and policy issues.

Dr. Snell is Chief, Office of Energy Conservation, Mr. Achenbach is Chief, Building Environment Division, and Mr. Petersen is a member of the Building Economics Section, Technical Evaluation and Application Division; all of the Center for Building Technology, Institute for Applied Technology, National Bureau of Standards, Washington, D.C. 20234.

Major Factors of Change

Existing housing and new construction data presented by the Federal Energy Administration (FEA) in its Project Independence report (*5*, table 2.6) show several important trends (Table 1). Without explicit consideration of future energy price increases and anticipated energy conservation measures, FEA predicted major increases in mobile home construction and essentially constant shares of the national housing inventory for single-family detached, low-density, low-rise, and high-rise units over the 1970–1990 time period. Thus, FEA estimated that, even in 1990, 70 percent of all housing would be single-family detached units or mobile homes, 25 percent would be low-density and low-rise units, and only 5 percent would be high-rise units. The FEA also suggests that new mobile home construction in 1990 will nearly equal the 1970 inventory of these units. Our analysis identifies influences that could significantly alter these expectations.

About 60 percent of all the energy used in a typical house is devoted to environmental conditioning and the remaining 40 percent of use is directly related to the activities of the occupants, such as water heating, cooking, lighting, laundry, food preservation, dishwashing, waste disposal, personal hygiene, entertainment, and communication (*5*, figure 1.2). These occupant services are provided by a wide variety of appliances, most of which use electricity. In the 1960's, energy use in housing was increasing at an annual rate of 4 to 6 percent, largely due to the ingenuity of industry in providing a large variety of attractive and convenient appliances for the home (*2*, p. II-1). This proliferation of appliances was viewed as an increase in the standard of living, and their energy use was of little concern prior to 1970 because energy was considered relatively plentiful and inexpensive.

The growth in energy consumption per capita was altered significantly with the oil embargo of 1973. The immediate response was to institute austerity in living habits to reduce the demand for foreign oil. Austerity took the form of lowered house temperature, reduced use of appliances, lowered lighting levels, decreased mobility in transportation, and similar measures for conserving energy. It was demonstrated that the energy use of almost any building can be reduced by 20 percent by more careful operational practices and by maintaining a consciousness of energy conservation (*6*). Some homeowners also made improvements in the thermal properties of their homes by the use of insulation, storm windows, and caulking. Both austerity and building improvement practices have continued since the oil embargo ended, partly because of dedication to the national goal of self-sufficiency, but mostly because of the very sharp rise in the price of all forms of energy since 1973.

Few, if any, of the gains made in the environmental quality of housing over the last few decades need to be sacrificed in order to reach the desired goal of energy self-sufficiency in this country. Improved design of housing, increased efficiency in equipment and appliances, and better control and operation of equipment and systems can reduce the energy requirements of new housing from 30 to 50 percent as compared to practices prevailing before 1970 (*6*). One answer to the question of what influences there are to motivate home buyers or builders to seek these changes can be found in examining the economic factors facing these decision-makers. The extent to which energy conservation features can best be utilized in new housing can be determined objectively in a life-cycle cost context where both first costs and recurring costs are considered.

While life-cycle cost analysis has been a valid means of investment analysis for many years, it has only recently been used explicitly as an aid to energy conservation in the design of buildings (*7, 8*). Historically, the construction cost or sales price of housing has dominated investment decisions in the housing market. As a result, energy conservation features that were incorporated into the building design were often based on comfort considerations or on a reduction in the costs of heating and cooling equipment rather than on a reduction of long-term operational costs.

An examination of U.S. prices related to energy and construction between 1950 and 1970 (Table 2) will help to explain this first-cost syndrome. During this period, natural gas and fuel oil prices rose at an annual rate of 2 percent in current dollars and actually declined in real terms. During the same time period, the price per kilowatt-hour of electricity (at given usage levels) showed almost no increase in current dollars and, in fact, declined substantially in real terms. At the same time, construction costs increased at an annual rate of approximately 3.5 percent in current dollars and increased in real terms as well, while mortgage interest rates more than doubled. Anticipation of higher inflation rates at the end of this period also resulted in higher discount rates for the evaluation of future energy savings. Thus, while first costs and mortgage costs were increasing at a relatively rapid rate, energy costs appeared to be becoming less and less significant, even in a life-cycle cost context.

Since 1970, however, this situation has reversed. Energy costs have increased dramatically. Not only have they outpaced construction cost increases, they have, more importantly, reversed direction from declining real costs to increasing real costs. While it is unlikely that the sharp rate of growth in energy costs experienced between 1970 and 1975 will continue through the end of the 20th century, homebuyers' expectations as to the rate of future energy price increases must be considerably higher than just a few years ago.

Moreover, two important factors are occurring that greatly influence life-cycle heating and cooling costs. (i) There has been an accelerating switch from less expensive natural gas and fuel oil to more expensive electric heat, especially of the resistance type. (ii) The percentage of new residences with central air conditioning has nearly doubled since 1966 (Table 3).

Table 4 shows the effective cost per gigajoule of space heating output (adjusted for equipment conversion efficiency), based on average billed energy prices between 1950 and 1975. The present value of the annual cost to provide a gigajoule of heat to the conditioned space of a residence over its expected lifetime, N, can be calculated in a straightforward fashion with the use of the present worth factor (PWF). The PWF is calculated with the use of N, the annual discount rate, and the expected annual rate of increase in unit energy costs (*9*).

In a typical house built in 1970, natural gas was used as a heating fuel (Table 4) at an average cost of $1.41 per gigajoule of output, with an expected price increase (calculated on the basis of historical experience) of about 2 percent. A discount rate based on a 1970 mortgage interest rate of 9 percent and a conservative lifetime of 30 years results in a present-value life-cycle cost of $17.70 per gigajoule of output. The typical house built in 1975 uses electric resistance heat at $6.94 per gigajoule of output. Assuming that an expected annual price increase of at least 5 percent is reasonable, and again using a 9 percent discount rate, we calculate the present-value cost of gigajoule output annually for 30 years as $121.31, nearly seven times greater than that for a gas-heated house in 1970. Moreover, if air conditioning is added, the total expected present value of providing heating and cooling require-

Table 1. Representative housing data for the United States. Data from (*5*).

Year	Number of housing units (× 10^6)					
	Mobile home	Single-family detached	Low density (duplex)	Low-rise, townhouse, garden apartment	High rise	Total
1970	2.1	44.6	11.0	6.5	3.3	67.7
1990	7.4	59.4	14.8	8.8	4.5	94.9
Construction, 1990	1.9	6.6	2.2	1.2	0.6	12.5

Table 2. Price indexes for heating fuels (on the basis of billed costs), construction costs, and consumer prices. Data from (*3, 21–25*).

Year	Natural gas	Fuel oil	Electricity* (1 January)	Residential construction	Consumer price index
1950	69.9	72.6	97.5	63.0	72.1
1955	82.9	86.0	99.3	72.5	80.2
1960	100.1	89.0	102.4	81.8	88.7
1965	99.9	94.4	102.9	90.4	94.5
1970	107.4	109.3	99.7	122.4	116.3
1975	182.1	230.6	187.0	184.3	162.3

*On the basis of 20,000 kwh annual usage in cities of more than 50,000 (1975 index extrapolated from 1974 index on the basis of 12,000 kwh annual usage rates; 1950–1960 indexes extrapolated on the basis of 6,000 kwh annual usage rates).

Table 3. Space heating and cooling equipment installed in new single-family housing (percent). Data from (*26*).

Year	Gas	Oil	Electric	Central air conditioning
1965	64	13	20	25
1970	62	8	28	34
1975*	35	9	55	48

*Extrapolated from 1974 data.

ments over a 30-year period rises even further relative to the expected cost, just 5 years ago, of gas heating alone.

Decision Model for Optimization of New Housing Design

Energy conservation features offer little appeal when their only visible impact is either increased first cost or substitution for convenience or esthetic design features. The picture changes dramatically when life-cycle costs are considered. These costs can be computed on the basis of annual costs over the useful life of the building or the present value of all such costs. If properly considered, each of these approaches gives identical results. The first, often referred to as "Pity-em" in the trade, or *P*rincipal-*I*nterest, *T*axes, *E*nergy, and *M*aintenance, provides a convenient assessment of the relative affordability of various housing types compared to the householder's salary or take-home pay. The second approach is more closely related to standard financial analysis procedures and is therefore used in the following examples.

Since space heating accounts for the principal share of residential energy use, a simplified model of how these parameters affecting this energy use contribute to life-cycle cost is examined. The analysis may be readily extended to space cooling, lighting, and other energy uses. In this context, the major elements of the life-cycle cost of a house are the first costs of the building shell and the heating system, and the present value of heating system operations over the useful life of the building.

The first cost of the building shell is directly related to its surface area, A, and its overall thermal resistance, R_o (expressed as reciprocal watts per square meter per degree Kelvin), by the expression $(a_1 + p_1R_o)A$. The parameters a_1 and p_1 represent costs per unit area and, in general, may take on different values for increasing ranges of values of R_o to account for piecewise linear shell costs. The first cost of the heating system is directly related to its design capacity, $(\Delta t\ A)/R_0$ (where Δt is the design temperature differential between indoors and outdoors), with a fixed cost a_2 and a unit capacity cost, p_2. Then

$$\text{First cost} = (a_1 + p_1R_o)A + a_2 + p_2(\Delta t\ A)/R_o \quad (1)$$

Recurring annual energy costs related to heat transmission through the building shell are a function of the average annual climatic heating load relative to occupant indoor comfort requirements, H (often represented by heating degree days × 24 hours per day); the heating equipment conversion efficiency, η; the cost per gigajoule of energy used for heating, p_3; and A and R_o. Present-value life-cycle energy costs (*2*) can be found by applying the PWF to the average annual energy costs at current energy prices, as follows:

$$\text{Present-value energy costs} = (p_3H/\eta) \times (A/R_o) \times \text{PWF} \quad (2)$$

so that the simplified, life-cycle cost of space heating for a house is obtained by combining Eqs. 1 and 2

$$\text{Life-cycle cost} = (a_1 + p_1R_o)A + a_2 + [p_2\Delta t + p_3(H/\eta) \times \text{PWF}](A/R_o) \quad (3)$$

This is obviously an oversimplified equation. It does not explicitly account for a number of factors—such as wind speed, solar insolation, landscaping, building and window orientation, architectural treatment, air infiltration or controlled ventilation, occupant behavior, and the like—each of which, in many cases, could significantly influence actual residential energy use. However, this equation does identify the major determinants of life-cycle energy costs in housing design—that is, size, thermal resistance, climatic location, energy prices, PWF, and equipment efficiency. The optimal value of thermal resistance R_o^* is that which minimizes life-cycle space heating costs (Eq. 3). In the simplest case, where a_1 and p_1 are constant, R_o^* is obtained by differentiating Eq. 3 with respect to R_o

$$R_o^* = \left[\frac{p_2\Delta t + p_3\dfrac{H}{\eta} \times \text{PWF}}{p_1}\right]^{1/2} \quad (4)$$

Equation 4 is generally valid for R_o^* in cases where insulation can be increased with no structural modification to the building shell. Otherwise, a_1 and p_1 may take on different values for different intervals of R_o. In such cases, the optimal value of thermal resistance is obtained by selecting, from those values of R_o which define intervals of p_1 and the values of R_o^* defined by Eq. 4, the one for which Eq. 3 is minimized. R_o^* is primarily a function of incremental heating system capacity costs, energy prices, the incremental cost of thermal resistance, equipment efficiency, climate factors, and the PWF. However, R_o^* is unrelated to the total area of the building shell.

Equations 3 and 4 can be used in the context of an example to analyze the potential design implications of various energy conservation measures that influence life-cycle cost. For this purpose, consider a hypothetical single-family residence in New York City, drawn from a recent study (*10*). The major design parameters of this and similar dwellings are given in Table 5.

A 150-m^2, one-floor residence is typical of much recent U.S. construction. The major elements of first and operating costs are shown in Table 6. The major energy-related components (shell and heating and cooling system) account for 30 percent of the first cost. Annual energy costs (fuel oil at 10 cents per liter and electricity at 4 cents per kilowatt-hour) are $1342. Space heating accounts for 77 percent of the total energy used but less than half the annual energy costs, which consist primarily of lighting, hot water,

Table 4. The U.S. average cost per 10^9 joule output to conditioned space by fuel type (on the basis of billed rates). Data from (*3, 21, 22*). The assumed seasonal conversion factors are as follows: oil and gas, 0.6; electric resistance, 1.0; and heat pump, 1.7.

Year	Cost (dollars)						
	Natural gas		Fuel oil		Electricity (1 January)		
						Per 10^9 joule	
	Per therm	Per 10^9 joule	Per gallon	Per 10^9 joule	Per kwh*	Resistance	Heat pump
1950	0.058	0.92	0.123	1.39	0.013	3.61	2.12
1955	0.069	1.09	0.145	1.64	0.013	3.61	2.12
1960	0.083	1.31	0.150	1.69	0.014	3.89	2.29
1965	0.083	1.31	0.160	1.81	0.014	3.89	2.29
1970	0.089	1.41	0.185	2.09	0.013	3.61	2.12
1975	0.151	2.39	0.390	4.40	0.025	6.94	4.08

*On the basis of the billed cost of annual consumption in the 15,000- to 20,000-kwh range.

and appliances. Since cooling makes up such a small portion of the energy load in this example, it is excluded from the following calculations.

Optimal shell resistance. Equation 4 can be solved for the optimal value of attic resistance, on the basis of current insulation costs, a New York climate, a discount rate of 9 percent, a fuel price increase of 5 percent, and a lifetime of 40 years (*11*). Such a solution gives R_o* values of 5.3 for oil heat, 8.3 for electric resistance heating, and 6.3 for electric heat pump operation. From Table 5, for the roof, a design U_o of 0.42 ($R_o = 1/U_o = 2.4$) is considered typical in much existing housing. These R_o* values are two to three times greater than conventional usage of residential attic insulation provides. Likewise, although a U_o of 0.57 ($R_o = 1.75$) is considered typical for opaque wall areas in existing housing, an R_o of 3.2 to 4 has been shown to be economically feasible in most new housing (*9*). Multiple glazing and significant increases in floor insulation are also called for in many cases. As a result, heating (and cooling) energy requirements related to the shell of a house can be halved in comparison to past construction practices by increasing the overall thermal resistance to a more economical level.

Energy price increase. If energy prices were assumed to increase at an annual rate of 10 percent, R_o* in attics would increase from 5.3 to 7.9 in the above example for oil heat. Not only is optimal design highly sensitive to energy price, but uncertainty about future energy prices can lead to major diseconomies in residential operations. The net effect of this on economic useful life is difficult to estimate. Even now it is usually impractical to retrofit existing housing with enough wall insulation to meet presently optimal insulation levels for new housing. If energy prices do continue to increase significantly there is a possibility of rapidly accelerating economic obsolescence in a significant fraction of existing, particularly single-family detached, housing.

Housing type and size. Shell area (*A*) is a major parameter of both housing design and energy usage, as shown by Eq. 3. Comparison of a single-family detached unit, for which shell area is 270 m² (not including floor), with a similarly insulated, two-story, inside townhouse unit of the same floor area but a shell area of only 135 m² reveals that heating and cooling expenditures are approximately halved, in addition to a significant reduction in construction costs. Further reductions in unit shell area are found in mid- and high-rise residential complexes.

Table 5. Representative housing data: Parameters of prototype designs in northeastern United States (*5*, figure 1.2).

Parameter	Mobile home	Single-family detached	Low density (duplex)	Low-rise, townhouse, garden apartment
Dwellings per building	1	1	1	20
Floor area (m²)	68	150	100	85
Total floor area (m²)	68	150	200	1,700
Plan area (m²)	4 × 17	10 × 15	13.3 × 15	15 × 56.6
Building height (m)	2.2	3	3	5
Shell area (m²)	160	270	344	1,690
Shell area per dwelling (m²)	160	270	172	85
Degree days (K)		2,900		
Summer design temperature (°C)		30.5 db/ 24.4 wb		
Glass (percent)		15		15
Wall, U_o (watt/m²K)		1.36		1.93
Roof, U_o (watt/m²K)		0.42		0.40
Annual electricity consumption (kwh)		18,600		404,564
Oil (liter)		5,980		85,500
Lighting (watt/m²)		11		11
Infiltration (air changes per hour)		1.0		0.5
Cost (construction) ($)	9,000*	33,500	50,000*	351,000
Mechanical, electrical systems ($)	2,000*	7,000	12,000*	80,700

*Estimates of values not found in the cited references.

The opposite is true for mobile homes, which have the highest surface-to-volume ratio of the types of dwellings considered here. A mobile home may have as little as half the floor area of a small detached house but its unit energy use is greater than half of that for the single-family house.

Equipment efficiency. As heating system efficiency increases, the cost per gigajoule output from the system to the conditioned space decreases, reducing both life-cycle energy expenditures and the optimal R_o. Heat pumps, which may be 50 to 100 percent more efficient than electric resistance heating equipment, can have a substantial impact on optimal house design and total life-cycle costs where electricity is the most practical heating source. In large building complexes, total energy systems (such as the Modular Integrated Utility System) may provide greater efficiencies and use lower cost fuels than systems available for use in low-density communities.

Climate. Another major factor influ-

Table 6. Example of a single-family detached house in New York City (*5*, figure 1.2).

Item	Dollars	Percent
First cost		
Energy related		
Shell	6,464	19.3
Heating and cooling system	3,520	10.5
Lighting	672	2.0
Electrical	784	2.3
Hot water	128	0.4
Appliances	1,000	3.0
Total	12,568	37.5
Other		
Interior finish, plumbing, design, and the like	20,932	62.5
Total first cost	33,500	100.0

Energy source	10^6 joule/m²	10^9 joule	Fuel oil (liter)	Electricity (kwh)	Cost (dollars)
*Operating costs**					
Oil for heating	1,555	233.3	5,980		598
Electricity for cooling	41	6.2		1,700	68
for hot water	178	26.7		7,300	292
for lighting and power	233	34.9		9,600	384
	2,007	301.1	5,980	18,600	1,342

*Annual energy use and cost (floor area = 150 m²).

encing life-cycle energy cost is climate. This factor directly influences the operating cost portion of life-cycle cost and has a somewhat smaller impact on optimal design resistance, as reflected by the square-root relationship of Eq. 4. Life-cycle cost economy may require highly differentiated building practices in various sections of the country in order to fully reflect heating and cooling load differentials as well as regional differences in energy and construction costs.

Other parameters. This life-cycle cost model and optimization process can be expanded to include other variables related to design. First costs can reflect changes in infiltration characteristics, the installation of cooling equipment, increased heating and cooling equipment efficiencies, solar shading coefficients, window size, and other design characteristics related to energy use. Annual operating costs can be expanded to include the reduced energy requirements and possible increased maintenance costs associated with these improvements. As a result, optimal values of R_0, infiltration rates, size and efficiency of heating, ventilating, and air-conditioning (HVAC) systems, shading coefficients, and other design parameters related to energy can be determined simultaneously.

Energy Conservation as a Design Parameter for Housing

Having established, on the basis of life-cycle cost, a solid rationale for major energy-related changes in future housing design, we can discuss the technological realities relative to such change. In the decades preceding 1970, the building construction industry, architects, engineers, and manufacturers of building materials and equipment invested a great deal of research and development effort to make housing more responsive to the users' needs and to improve the livability of houses. The environmental characteristics of houses that received a great deal of attention in these developments were: (i) efficient use of enclosed space; (ii) quality of the thermal environment (temperature, humidity, air circulation); (iii) indoor air quality (control of solid, liquid, and gaseous contaminants); (iv) use of daylight and sunlight; (v) effective visual communication with the out-of-doors; (vi) acoustical privacy; and (vii) indoor and outdoor esthetics and landscaping. In addition to providing a satisfactory level of each of these environmental characteristics, it was also necessary to provide adjustability of many of them in response to variations in weather, time of day, or activities of the occupants.

Attention among researchers in the building community is directed now toward various means of responding to the influences of fuel price increases and national energy conservation policies. There is considerable optimism that major energy savings are attainable without a deterioration in essential performance. In fact, there is good reason to believe that a more careful study of user needs in illumination, thermal environment, ventilation air, and equipment control may yield information leading to both energy conservation and improved building. Available options for the housing designers include the following.

1) *Insulation of walls, ceilings, and floors*. A frame wall made of studs 38 by 140 mm spaced 610 mm apart (2- by 6-inch studs on 24-inch centers) has as much strength as the more conventional wall made of 38- by 89-mm studs spaced 406 mm apart; the former uses no more wood than the latter and allows 50 percent more insulation to be used. The use of 38- by 140-mm studs is becoming common practice in some parts of the country (*12*). New truss designs allow insulation up to 305 mm thick to be installed across the entire attic area.

2) *Air infiltration*. Few houses are now built with an air leakage rate of less than one-half building volume per hour (0.5 air change per hour). More typical construction has an air leakage of 1 to 1.5 air changes per hour. Heating the leakage air often accounts for a third or more of all the heating energy requirements of the house. Only 9 to 17 m^3 per hour per person of fresh air is required to meet the oxygen and odor control requirements of people (*13*). Of course, alternative schemes for residential ventilation are available to provide the added fresh air needed during the hour or so of daily cooking time and the occasional periods when groups of people are being entertained. The application of caulking and sealing compounds at all building joints and at all penetrations of the exterior walls during construction is very cost effective and can reduce air leakage to 0.5 change per hour or less in most housing and simultaneously improve building comfort by reducing draft. This is approximately twice the fresh air required for oxygen and odor control. Use of available new compounds, although more expensive in first cost, can eliminate costly periodic replacement.

3) *Windows*. Windows provide daylight in a house and visual communication with the out-of-doors. In addition, when properly designed and located, they can in many cases provide net heating gains from solar energy during the heating season. However, a window not properly exposed to sunshine, even if double-glazed, can lose ten times as much heat as an equal area of well-insulated wall, while 20 times as much heat is lost if it is single-glazed. For this reason, the reduction of window area to the minimum needed for satisfactory daylighting and view to the outside is a wise approach in many situations. Double-glazed windows are cost effective over large areas of the United States; where fuel prices are relatively high and the winter weather is severe, triple-glazing can be cost effective as well.

4) *Landscaping and shading*. Shrubbery planted close to the house and evergreen trees (*14*) located about a building height away from the house on the windward side can significantly reduce the effective wind velocity on the building surfaces. This reduces air leakage and also the transmission loss, especially for windows. Deciduous trees of appreciable height can provide beneficial shade on the house during summer and allow sunshine to impinge on windows and walls in winter.

There are a number of opportunities to improve the energy efficiency of the heating and air-conditioning equipment in houses by 5 to 10 percent, and several additional techniques with even greater potential. These are as follows.

1) *Heat pump*. A heat pump for space heating use will result in a seasonal heating efficiency (relative to fossil fuels consumed) about equal to that attained with oil or gas heating equipment, and will save one-third to one-half of the electrical energy that would be required for an electrical resistance heating system. The heat pump is best adapted to the central and south central latitudes of the United States, where the magnitude of the heating requirements is intermediate. The heat pump is less efficient in very cold climates. Also, in the far south, the air-conditioning load predominates, and a system designed only for cooling can be made more efficient than a heat pump.

2) *Solar heating system*. A solar heating system can easily be designed to provide one-half to two-thirds of all the heating energy required in a house. However, the first cost of such systems is sufficiently high that they are now competitive only with electrical resistance heating, and then only in areas where electric energy is relatively expensive (*15*).

3) *Ventilation*. Numerous means are available to provide fresh air in any room in a house when the outdoor air condi-

tions are favorable relative to the indoor conditions. If the occupants are willing to allow the indoor temperatures to fluctuate from 21° to 27°C and to use an attic fan for ventilation of the occupied spaces in lieu of air conditioning whenever the outdoor air is below indoor temperature, a very high saving in electrical energy for air conditioning can be obtained in most sections of the United States. Such a system is usually controlled manually but could be automated. Small exhaust fans, manually operated, are often used to remove odorous or moist air from cooking areas and bathrooms. The same principle could be used in conjunction with solid-state automatic or programmable controls to augment house ventilation during periods of high occupancy, thus eliminating the need for continuous high infiltration of outdoor air into houses.

Options for designs of equipment and appliances that are more conserving of energy include better matching of demand and capacity, modularization of heating and cooling equipment to increase off-peak efficiency, separate supply of outdoor air for combustion, automatic ignition systems, automatic (fail-safe) combustion air dampers, and combustion- and cooling-cycle designs with improved efficiency. Research has provided some quantification of benefits, such as the modularization of heating equipment (*16*), the automatic night set-back of thermostats (*17*), the supply of combustion air from outdoors (*18*), the penalties of oversizing heating equipment (*16*), and duct air leakage in attics (*19*). Other options mentioned above require better technical documentation of potential energy savings, and nearly all of the options require economic analysis to correlate cost effectiveness with climate, operational requirements, energy prices, PWF's, and other factors related to life-cycle costs.

It is highly probable that technologies not in significant use now, such as wind power, fuel cells, energy storage systems, and total energy systems, will become candidates for energy conservation in individual houses or communities in the future. The large number of options available for energy conservation in the building envelope and in equipment and appliances indicates that many trade-offs are possible, and that there may be several combinations of options that are substantially equivalent in energy conservation and economic value. Commercially available computer programs make it possible to evaluate many alternatives and, therefore, will undoubtedly find increased use in residential building design.

Implications for Future Housing Design

The preceding analyses demonstrate that rising energy prices and the potentials for increased energy conservation are powerful forces influencing change in housing design practices to follow the dictates of life-cycle performance optimization. The precise nature of these influences is still unclear; present data and analytical capabilities for detailed life-cycle analyses are imperfect. Nonetheless, major influences may be perceived. These are listed in Table 7. All of the indicated changes should improve housing quality while increasing its life-cycle affordability.

Design influences related to energy will, of necessity, stimulate major changes of practice in the building industry that affect architects, engineers, materials and equipment manufacturers, and regulating officials. Two cloudy and potentially troublesome issues are the impacts of energy conservation on mobile home economics and the prospect of accelerated obsolescence of much existing single-family housing that is thermally inefficient.

Private industry and government leaders who believe that new housing should incorporate energy conservation features must consider what the best techniques are for making it happen. Education, persuasion, financial incentives, and regulatory procedures have all been proposed and advocated by various groups for stimulating more energy-efficient housing designs in the near future. There are also significant political, national security, international trade, and foreign policy considerations in adequately motivating a national energy conservation program. A variety of actions has been taken in each of these areas. A few states have adopted energy conservation regulations. The American Society of Heating, Refrigerating, and Air-Conditioning Engineers has developed a voluntary standard (*20*), and several agencies of the federal government have been seeking to develop building performance standards for energy conservation.

There is not now a consensus of opinion in the United States on whether measures for stimulating increased energy conservation in buildings should be pressed on a voluntary or mandatory basis. Both processes are being advanced with considerable vigor. The future directions followed in housing design in order to attain a higher level of energy efficiency will probably not differ much under voluntary or mandatory procedures. The rate of progress might be greater and the level of conservation attained might be higher under mandatory procedures, but the directions will be dictated by presently known technologies and those now being researched and developed. When significant advances have been made in improving the efficiency of energy utilization in the resi-

Table 7. Anticipated impacts of energy conservation on new housing design.

Housing type
Smaller, higher density, fewer detached houses
Increased shift to townhouse and low-rise from single-family detached
Diminished relative attractiveness of mobile homes in life-cycle cost terms
Improved designs with lower unit demands will help keep fossil fuel economically competitive in many areas for years to come
Architectural features
Thicker wall (cavity and sandwich) and roof construction for more insulation
Fewer picture windows, more double- and triple-glazed windows, some specially coated glass
More functional windows—designed as passive solar collectors
Tighter, better sealed joints, higher performance sealants, better workmanship
Better control of moisture to protect insulation
Attention to shape, orientation, landscaping in design
Control of air movement between floors—fewer open stairwells and split-level designs
Insulated foundation walls in cold climates
Greater thermal resistance for more expensive fuels, such as electric heating
Better thermal comfort and greater acoustical privacy
Mechanical systems
Smaller, more efficient HVAC equipment, better load matching
Customized ventilation to provide outdoor air when and where needed
Widespread use of heat pumps in moderate climates—integrated heat pump and solar heating
Solar space heating in selected climates where gas and oil are in short supply
More zoning and multipoint control systems in larger residences
Electrical load management control options for hot water and appliances
Solar water heating in the south
Institutional
Life-cycle cost-based performance standards (voluntary or mandatory) for new housing design
Labeling of houses, equipment, and appliances for energy use, cost, and performance
Householders knowledgeable of how to operate homes efficiently

dential sector and energy conservation ceases to be a priority national concern, the quality of man's habitat will have been unmistakably improved.

References and Notes

1. 1 Btu = 1055 joules.
2. D. G. Harvey, *Residential Energy Consumption–Phase I* [Department of Housing and Urban Development, Report HAI-1 (Contract H-1654 to Hittman Associates, Columbia, Md., March 1974)].
3. Bureau of Labor Statistics, *Retail Prices and Indexes of Fuels and Utilities* (Department of Labor, Washington, D.C., December 1975), tables 1 and 2.
4. Federal Energy Administration, *Project Independence Report* (Washington, D.C., November 1974).
5. *Final Task Report—Residential and Commercial Energy Use Patterns 1970–1990*, prepared by the Interagency Task Force on Energy Conservation under direction of Council on Environmental Quality (conducted by Arthur D. Little, Inc.), (Federal Energy Administration Project Independence Blueprint, Washington, D.C., November 1974), table 2.6.
6. National Bureau of Standards, *Technical Options for Energy Conservation in Buildings, Technical Note 789* (Department of Commerce, Washington, D.C., 1973); Guidelines for Saving Energy in Existing Buildings, *Building Owners and Operators Manual ECM-1* (Federal Energy Administration, Washington, D.C., 1975).
7. S. R. Peterson, *ASHRAE Trans.*, in press.
8. J. C. Moyers, *The Value of Thermal Insulation on Residential Construction: Economics and the Conservation of Energy* (Atomic Energy Commission, Oak Ridge National Laboratory, Oak Ridge, Tenn., 1971).
9. $$\text{PWF} = \sum_{t=1}^{N} \left(\frac{1+P}{1+D}\right)^{t}$$ where P is the average annual rate of fuel price increase, D is the discount rate, and N is the building lifetime.
10. Arthur D. Little, Inc., *Report to Federal Energy Administration, An Impact Assessment of ASHRAE Standard 90-75* (Federal Energy Administration contract C-78309, Cambridge, Mass., December 1975), pp. 17, 20, 43, 44, and 77.
11. Parameters assumed for calculation in text; H = (24 hour/day) × (5219 degree days); p_4 = \$0.80/R m²; PWF = 20.4 (N = 40 years; 9 percent discount rate, 5 percent rate of fuel price increase); oil = \$0.10/liter, η = 0.6; electric resistance = \$0.04/kwh, η = 1.0; and the heat pump = \$0.04/kwh, η = 1.7.
12. Owens/Corning Fiberglas, *The Arkansas Story* (Toledo, Ohio, 1975).
13. *The American Society of Heating, Refrigerating, and Air-Conditioning Engineers Standard 62-73, Natural and Mechanical Ventilation* (United Engineering Center, New York, 1973).
14. G. E. Mattingly and E. F. Peters, *Wind and Trees: Air Infiltration Effects of Energy in Housing* [Center for Environmental Studies Report No. 20 (NSF Undergraduate Research Grant 11237), Princeton University, Princeton, N.J., 1975].
15. J. A. Duffie and W. A. Beckman, *Science* **191**, 143 (1976).
16. G. E. Kelly and D. A. Didion, *Energy Conservation Potential of Modular Gas-Fired Boiler Systems, NBS BSS 79* (Department of Commerce, National Bureau of Standards, Washington, D.C., December 1975).
17. D. A. Pilati, *The Energy Conservation Potential of Winter Thermostat Reductions and Night Setback* (Oak Ridge National Laboratory, ORNL-NSF-EP-80, Oak Ridge, Tenn., February 1975).
18. G. Peoples, Addendum to Proceedings of Conference on Improving Efficiency in HVAC Equipment and Components for Residential and Small Commercial Buildings (Joint sponsors: Federal Energy Administration American Society of Heating, Refrigeration, and Air-Conditioning Engineers, Department of Commerce, National Bureau of Standards, and Ray W. Herrick Laboratories of Purdue University, Lafayette, Ind., October 1974), p. A66.
19. P. R. Achenbach, C. W. Phillips, W. T. Smith, *ASHRAE Trans.* **70**, 390 (1964).
20. *The American Society of Heating, Refrigerating, and Air-Conditioning Engineers Standard 90-75, Energy Conservation in New Building Design* (United Engineering Center, New York, 1975).
21. Federal Power Commission, *Typical Electric Bills 1975* (FPC-86, Washington, D.C., 1975), table 1.
22. Federal Power Commission, *All Electric Homes in the United States* (FPC R-85, Washington, D.C., 1975), table 1.
23. Bureau of Economic Analysis, *Business Statistics* (Department of Commerce, Washington, D.C., 1973), p. 55.
24. Bureau of Economic Analysis, *Survey of Current Business* (Department of Commerce, Washington, D.C., 1975), vol. 55, p. S-8.
25. American Gas Association, *1972 Gas Facts* (Arlington, Va., 1973), table 138.
26. Bureau of the Census, *Construction Reports–Series C25, Characteristics of New One-Family Homes: 1974* (Department of Commerce, Washington, D.C., 1975), table 16.

Transportation Energy Conservation Policies

Higher gasoline taxes and new car fuel economy standards are effective energy saving policies.

Eric Hirst

Between the end of World War II and 1972, transportation fuel use grew steadily and rapidly because of increases in both passenger and freight traffic, shifts toward the use of less energy-efficient modes, and declines in energy efficiency for individual modes (*1*). However, since 1972 a number of forces have emerged that may significantly alter these trends in the future.

These forces include the Arab oil embargo and subsequent sharply higher prices for gasoline. After nearly two decades of falling "real" prices, the price of gasoline increased 26 percent between 1972 and 1974; since then prices have risen even higher (*2*). Because of these higher gasoline prices, personal consumption expenditures devoted to gasoline increased 23 percent between 1972 and 1974 (to $36 billion in 1974). Prices of new automobiles are also rising rapidly—up from an average of $3700 in 1972 to $4000 in 1973 (*3*)—although they are rising no faster than the overall Consumer Price Index.

In addition to these economic forces, a number of institutional changes are under way or under serious consideration. The National Mass Transportation Assistance Act of 1974 authorized the expenditure of nearly $12 billion during the period 1975 to 1980. Unlike previous federal programs for mass transit, the 1974 legislation authorized operating, as well as capital, grants for transit systems. The automobile industry, under pressure from the federal government, agreed to improve new car fuel economy 40 percent between 1974 and 1980 [from about 14 to 20 miles per gallon (mpg)]; recent federal legislation (*4*) requires 20 mpg in 1980 and 27.5 mpg in 1985. Modifications to the Federal Highway Trust Fund allow funds to be used for mass transit improvements and to encourage carpooling; several communities are beginning to institute significant carpool programs.

The extent to which these and other new forces will operate on traditional patterns of personal travel and land use to change the energy intensiveness and energy use of our transportation system is the subject of this article. I examine the period 1950 to 1972 with respect to personal travel and energy use for it, review the relative energy efficiencies of different urban and intercity passenger systems, discuss several policies for reducing transportation fuel use, and compare the energy savings likely with each of these policies in 1980 and 1985.

Four policies were selected for discussion here: (i) improving mass transit, (ii) increasing carpooling, (iii) raising gasoline prices, and (iv) imposing new car fuel economy standards. I chose these

The author is a research engineer in the Energy Division, Oak Ridge National Laboratory, Oak Ridge, Tennessee 37830. The work described in this article was performed while he was director of the Office of Transportation Research, Federal Energy Administration, Washington, D.C. 20461.

Table 1. Energy impacts of transportation conservation measures. Energy savings are calculated relative to a baseline in which auto travel is 1.2×10^{12} vehicle-miles (VM) in 1980 and 1.4×10^{12} VM in 1985, urban travel accounts for 55 percent of this total, and average auto fuel economy is 14 mpg for both years and 12 mpg in urban areas. Average automobile occupancy is 2.2 PM/VM and urban occupancy is 1.6 PM/VM (PM = passenger-miles).

Policy	Estimated energy savings (thousand barrels per day)	
	1980	1985
Increase percentage of urban travel carried by mass transit from 2.5 percent in 1973 to 5.0 percent in 1980 and 7.5 percent in 1985	52	122
Increase carpooling sufficiently to reduce work-trip auto travel by 10 percent in 1980 and 1985	69	105
Increase gasoline prices by 20 percent starting in 1975	484	700
Increase new car fuel economy from 14 mpg in 1974 to 20 mpg in 1980 and 22 mpg in 1985	568	1327

because they are important and widely discussed, and because there are analyses with which to evaluate their effectiveness. However, there are several other options, such as stricter enforcement of a speed limit of 55 miles per hour, wider adoption of right-turn-on-red, better urban traffic control systems, and a host of changes related to air traffic and freight traffic.

I conclude that, during the next decade at least, significant passenger transportation energy savings can be achieved only by improving new car fuel economy; such improvements can either be required by legislation or induced by increases in gasoline prices. Behavioral changes (greater use of mass transit and carpooling) are surprisingly insensitive to purely economic forces unless they are so strong as to be politically infeasible. Table 1 summarizes major results of the analyses discussed in this article. These results suggest the need for more and better programs to encourage people to change their attitudes toward energy use and personal transportation.

Historical Trends in Passenger Travel and Energy Use

Total transportation fuel use grew from 8.9 QBtu (8.9×10^{15} Btu) in 1950 to 18.3 QBtu in 1974 (*5, 6*) with an average annual growth rate of 3.0 percent. Between the mid-1960's and 1972 the growth rate was much higher at 4.7 percent a year. However, transportation fuel use increased only 3.3 percent between 1972 and 1973 and actually declined 3.2 percent between 1973 and 1974 (*5*). The 1974 decline was due to a combination of sharply higher fuel prices, spot shortages during the summer, the Arab oil embargo that winter, and the 2 percent decline in real gross national product (GNP) between 1973 and 1974.

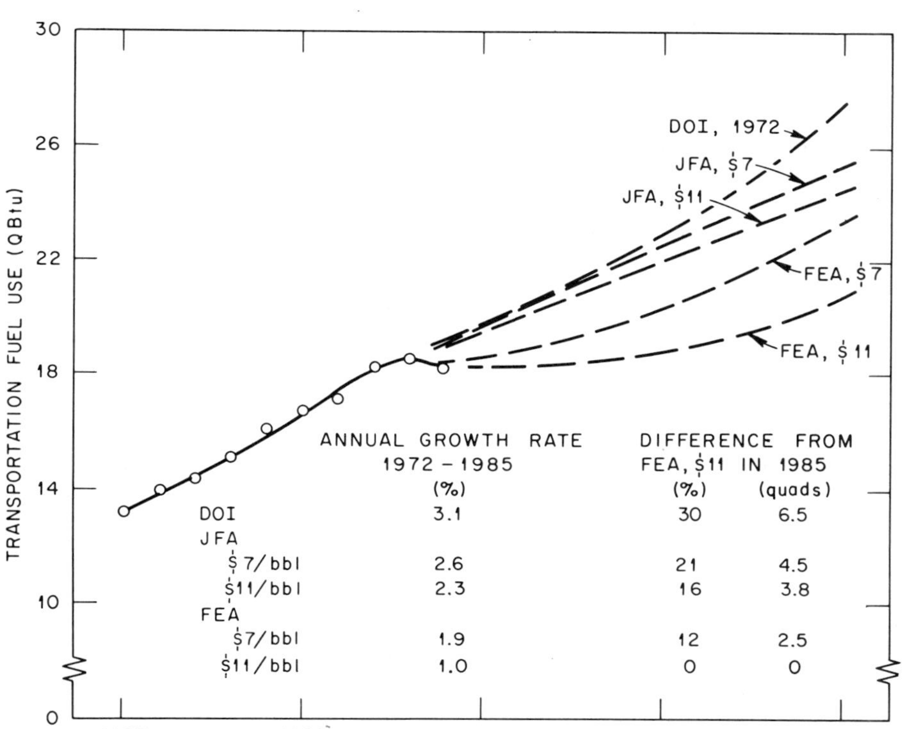

Fig. 1. Total transportation fuel use and forecasts.

Table 2. Energy intensiveness of passenger modes, 1972. Values were updated using data sources and methodologies discussed in (*1*). These national averages—ratios of total fuel used to total traffic carried by each mode—depend strongly on each mode's operating characteristics (load factor, trip length, speed, equipment type, and so forth). Approximate load factors (fraction of seats occupied) for these modes were: intercity airline, 0.5; automobile, 2.4 PM/VM; rail, 0.4; bus, 0.5; urban automobile, 1.6 PM/VM; rail transit, 0.4; and bus transit, 0.2.

Mode	EI (Btu/PM)	Mode	EI (Btu/PM)
Intercity		Urban	
Airline	7700		
Automobile	3100	Automobile	6700
Rail	2700	Rail transit	2600
Bus	1500	Bus transit	3000

Figure 1 shows actual transportation fuel use from 1965 through 1974 and projections to 1985 from three different sources. The Department of the Interior (DOI) projection (*7*) was prepared in 1972—long before recent oil price increases. The other two sets of forecasts, by Jack Faucett Associates (*8*) and the Federal Energy Administration (*6*), were prepared during the summer of 1974 as part of the Project Independence effort. These forecasts used crude oil prices (in 1973 dollars) of $7 and $11 a barrel. The variation among forecasts is considerable. The DOI forecast is much higher than the others, presumably because it assumes the low oil prices of the 1960's; its growth is equal to the long-run growth rate over the past two decades. The other forecasts show growth rates far below the historical trend. If these latter forecasts prove correct, considerable fuel savings will be achieved in the transportation sector because of fuel price increases alone (due almost entirely to increases in new car fuel economy).

Intercity passenger traffic is carried primarily by automobile and, to a lesser extent, by airplane, bus, and train (*1*). The variation in energy intensiveness (EI) among these modes (Table 2) is considerable. Buses and trains are the most efficient modes, followed by autos and airplanes. In 1972, EI for airplanes was five times higher than for buses. However, airplanes are the fastest mode and automobiles are the most convenient.

Between 1950 and 1972, the fraction of intercity passenger traffic carried by airplane climbed rapidly at the expense of

Table 3. Energy conservation impacts of transit improvements. Numbers in parentheses are percentages of regional transportation fuel use saved. Data are from (*13, 14*).

Strategy	Estimated savings (gallons per day)
Regional bus—Atlanta Fare reduced from 40 to 15 cents Service increased from 19 million to 22 million bus-miles per year Ridership increased 28 percent	9,000 (0.5)
Corridor service, bus—Washington, D.C. Shirley Highway (Route I-95), 11-mile busway in median Ridership increased from 1,900 to 11,500 per day in 5 years 40 percent of riders were auto drivers 30 percent of riders get to bus by car	3,000 (0.1)
Corridor service, rail—Philadelphia Lindenwold line Line carries 30,000 riders per day 28 percent of riders were auto drivers 90 percent of riders get to line by car	− 450 (0)

trains and buses. Energy consumption rose 192 percent as a result of a 156 percent growth in traffic and a 14 percent increase in overall EI (*1, 9, 10*). This increase in EI was due to increases in EI for individual modes and to the shift from buses and trains to airplanes.

Urban passenger traffic is carried almost exclusively by car, with only a small and declining fraction carried by mass transit (buses and electric) (*1*). As Table 2 shows, mass transit is two to three times as energy-efficient as cars (*10–12*). Urban EI values are double comparable intercity values because of poorer vehicle performance (fewer miles per gallon) and poorer utilization (fewer passengers per vehicle) in cities.

Between 1950 and 1972, the fraction of urban passenger traffic carried by cars steadily increased. Energy use grew 219 percent, caused by a 161 percent rise in traffic and a 22 percent increase in EI (*1, 10–12*). Increased EI was due to higher individual modal EI and to the shift from mass transit to automobiles.

Figure 2 shows how EI for urban modes increased between 1950 and 1973. Similar increases in energy intensiveness occurred for the intercity modes, except for railroads.

Improved Mass Transit

Although mass transit is considerably more energy-efficient than are automobiles, it now accounts for such a small fraction of total urban passenger travel (2.5 percent in 1973) that its short-term potential contribution to energy conservation is slight. The data shown in Table 3, from three recent mass transit programs (*13, 14*), suggest that the energy impacts of transit fare reductions and service improvements [such as expanded area coverage and reduced time between vehicles (headway)] are almost negligible.

There are several reasons for the slight energy impacts shown in Table 3. First, mass transit accounts for a tiny fraction of urban travel and an even smaller fraction of the urban travel energy budget. Thus sizable increases in mass transit traffic will have only slight impacts on total urban traffic and energy use. Second, while reduced fares and improved service will increase ridership, the experience cited above suggests that less than half the increase comes from former automobile drivers. The remainder are auto passengers, walkers, users of other transit systems, and people who formerly stayed home. Only shifts of auto drivers to mass transit reduce overall energy use. Third, expanded route coverage and reduced headways lower system load factors; this increases EI and energy use. Fourth, automobiles are often used to gain access to transit systems; the energy used for this must be subtracted from the energy savings due to the shift from auto to transit.

Figure 3 shows the sources of increased bus ridership (*14, 15*) due to the improvements in the bus system of Atlanta, Georgia, summarized in Table 3. In 1972, fares were reduced from 40 cents to 15 cents and a number of service improvements were instituted: some lines were extended, some were revised, new routes were established, and headways were reduced overall. The net impact of these changes was an increase in annual coverage from 19 to 22 million bus-miles.

Bus patronage increased 28 percent after the fare reduction and service improvements. Reducing fares increased load factors by increasing ridership with no increase in bus-miles. Service improvements, on the other hand, lowered load factors because ridership, in Atlanta at least, increased more slowly than did bus-miles. Overall, the combination of reduced fares and increased service raised load factors slightly. Because more than half of the new riders were not formerly auto drivers, the fuel saving—9300 gallons a day—is slight.

The major conclusion that can be drawn from Table 3 and Fig. 3 is that transit improvements alone offer little hope of large energy savings. Improving mass transit can save energy only if the increased transit ridership comes primarily from automobile drivers. Increasing transit patronage by attracting people from nonauto modes (such as other transit systems, walking, and bicycling) will probably increase urban passenger energy use. Thus saving energy through increased transit requires both the carrot and the stick. The carrot is to induce people to travel by mass transit and the stick is to force people out of their cars.

Even if transit improvements and auto disincentives are effective, mass transit is unlikely to provide substantial energy savings during the next decade. The potential energy savings are limited by the small size of the present transit plant and the small fraction of urban travel moved by transit. Doubling the percentage of urban travel carried by transit from 2.5 percent in 1973 to 5.0 percent in 1980 would require 100,000 new buses during this 7-year period, compared with the 1973 national fleet of 46,000 buses (*11*).

Assuming that funds can be found to finance the purchase of these buses; that drivers, mechanics, and managers can be trained during this period; that ridership will increase; and that the new riders will come from automobiles; the energy savings for 1980 are equivalent to 52,000 barrels of crude oil a day (*10–12*). If the percentage of urban passenger travel carried by transit increases to 7.5 percent in 1985 (a tripling of its 1973 share), the national savings would be equivalent to 122,000 barrels of crude a day. As shown later, these savings are small relative to those possible with other transportation measures.

Although the short-term energy conservation potential of increased mass transit is slight, this does not mean that transit improvement programs should be abandoned. Changes in urban travel patterns are likely to require at least a decade because of long lags associated with changes in land use patterns, auto ownership, and individual attitudes toward public transportation. Thus, unless improvement projects are undertaken now, the long-term potential benefits of mass transit will never be realized. Also, mass transit offers other benefits besides reduced energy use: less congestion during peak periods, fewer traffic fatalities, and increased mobility for those with limited access to autos. Finally, combining transit improvements with auto disincentives provides a transportation alternative to those dislodged from their automobiles.

Carpooling

The average automobile load factor for urban work trips is presently only 1.2 passenger-miles per vehicle-mile (PM/VM) (*12*). Thus there is an enormous energy conservation potential in the empty seats in automobiles traveling during the daily peak hours. Increasing this load factor from 1.2 to 1.6 PM/VM would save 440,000 barrels of crude oil a day in 1980. However, there are serious questions concerning the methods available to induce greater auto occupancy and the effectiveness of these methods.

Cambridge Systematics, Inc. (CSI), has analyzed a variety of carpool promotion options (*16*), using a modification of their disaggregate travel demand models to estimate auto ownership and changes in work and nonwork travel in response to these policies. Their model has been applied to Washington, D.C., using data collected during the D.C. 1968 Home Interview Travel Survey. They have examined a number of carpool alternatives related to the cost and supply of parking, costs of auto travel (tolls, gasoline taxes, carpool subsidies), direct regulation of urban travel (auto-free zones, gasoline rationing), and employer incentives.

The preliminary outputs of the CSI model for a number of such alternatives are shown in Table 4. These results are short-run impacts; that is, they do not include long-run changes in auto ownership that will occur because of these policies.

Table 4 shows that parking incentives for carpools and major increases in parking costs substantially reduce the amount of solo driving and increase both carpooling and mass transit use. However, these reductions in commuter travel are partially offset by increases in nonwork auto travel, which arise because more automobiles are available for nonwork travel. In the long run many of these "extra" cars will probably not be replaced; thus the long-run energy savings are likely to be larger than the short-run savings.

As an example, consider the predicted

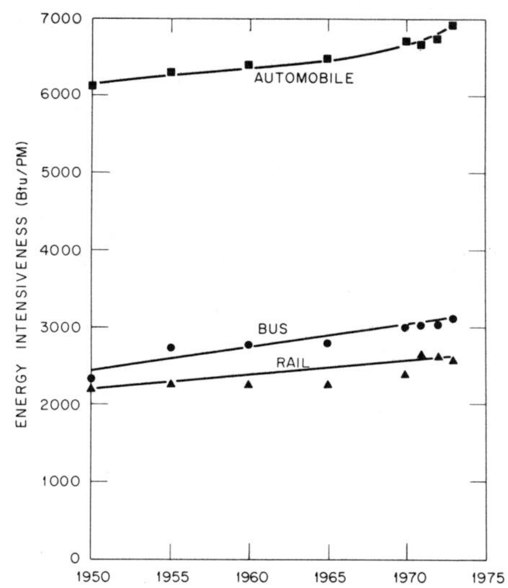

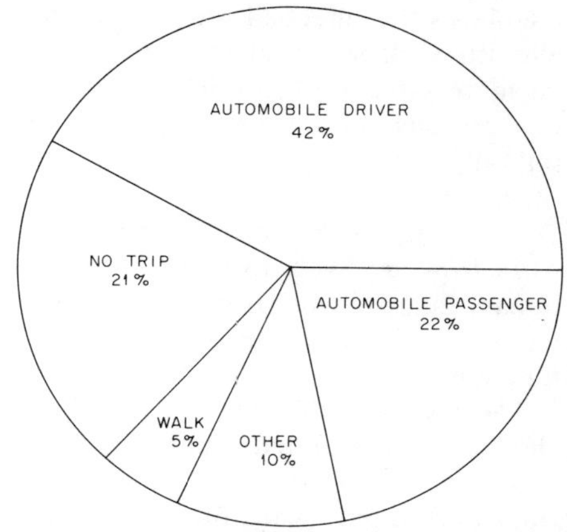

Fig. 2 (left). Energy intensiveness of urban travel modes, 1950 to 1973. Fig. 3 (right). Previous travel modes of new bus riders in Atlanta.

impacts of an increase in areawide parking costs of $3 per day. The parking surcharge reduces work-trip miles traveled by 10 percent, a 3.9 percent reduction in overall urban travel. However, the induced nonwork travel amounts to 1.4 percent of the urban total. Thus the net impact of the surcharge is only a 2.5 percent reduction in urban auto travel.

The percentage reduction in fuel use is less than the percentage reduction in travel because of fuel penalities associated with the increased trip circuity and extra weight in carpooling and increased number of cold-starts in nonwork trips. Thus, urban auto fuel use is cut by only 2.1 percent compared with the 2.5 percent reduction in travel due to the parking surcharge. This is equivalent to a national fuel saving of 69,000 barrels of crude oil a day in 1980.

Let us assume that, in the long run, only half the induced short-run nonwork auto travel occurs. Then, if long-run equilibrium is achieved in 1985, the 3.9 percent reduction in urban travel due to increased carpooling, coupled with the 0.7 percent increase in nonwork travel, reduces urban travel by 3.2 percent. This amounts to a 2.7 percent reduction in urban auto fuel use, equivalent to 105,000 barrels of crude oil a day in 1985.

These results—fuel savings of 2.1 and 2.7 percent in 1980 and 1985, respectively—show a remarkable insensitivity of commuters to changes in travel costs. Adding $3 a day to the base cost of parking increases the overall cost of commuting by 100 to 200 percent. The response to this enormous change in cost is only a 10 percent reduction in work-trip travel.

Table 4. Energy impacts of carpooling policies in Washington, D.C. "Parking incentives" refer to the restriction that close-in parking spaces are reserved for carpoolers. "Parking incentives and parking costs" add a $2-per-day parking charge for solo drivers. Data are from (*16*). Base values (row 1) are those used to compute the percentage changes given below them.

Policy	Work-trip mode share (%)			Vehicle-miles traveled			Fuel consumption (gallons per day)
	Drive alone	Carpool	Mass transit	Work	Nonwork	Total	
Base values (excluding weekend travel)	52.9	25.4	14.5	10.4	16.7	27.1	2.58
			Percentage change from base value				
Parking incentives	−10.69	22.05	0.40	−3.37	1.02	−0.64	−0.55
Parking incentives parking costs	−22.27	43.82	4.55	− 9.82	2.50	−2.24	− 1.83
Base parking cost + $1 (areawide)	−5.07	4.50	10.61	− 3.27	0.71	−0.81	− 0.68
Base parking cost + $3 (areawide)	−15.61	13.93	32.57	− 10.20	2.29	−2.49	− 2.07
Base parking cost + $3 (central business district only)	−6.52	2.97	17.83	− 4.04	1.04	−0.92	− 0.78
Carpool subsidy, 5 cents/PM	−4.04	11.31	− 5.06	−2.47	0.43	− 0.68	−0.50

Policies that affect all automobile travel (both work and nonwork) are likely to be more effective in saving energy for two reasons. First, work trips account for only a third of all auto travel (although they account for about 40 percent of all auto gasoline use). Second, nonwork trips are likely to be more discretionary than are work trips and therefore more sensitive to changes in dollar and time cost.

The results of this ongoing CSI study should be interpreted cautiously. The data used to construct the model are from 1968, a time when fuel prices were low, Washington's transit service was poor and deteriorating, Americans loved their cars, and incomes were rising steadily. Thus the data are from an era in which our attitudes and behavior strongly favored automobile ownership and use. Also, models such as that developed by CSI capture the major demographic and economic variables affecting travel decisions, but they cannot explicitly model changes in attitudes. To the extent that attitudes toward energy use, carpooling, and automobiles have changed since 1968, the model's results are in error. Finally, CSI's model evaluates only the short-run impacts of policy changes. As households adjust their auto ownership and home location in response to these policies, the impacts of the policies may increase.

Gasoline Taxes

One policy that is often discussed but rarely embraced is an increase in the federal tax on gasoline, currently at 4 cents a gallon. Those favoring an increase argue that it would reduce gasoline use, allow maximum consumer choice (in terms of changes in both vehicle use and vehicle ownership), and be simple and inexpensive to administer. Opponents argue that its economic burden on low-income families would be intolerable, that consumer demand for gasoline is insensitive to price changes, and that further increases in gasoline prices would adversely affect economic recovery.

The Office of Energy Systems in the Federal Energy Administration (FEA) developed a simple econometric model for estimating changes in automobile travel, fuel economy, and ownership in response to changes in gasoline prices (and also to exogenous changes in new car fuel economy) (*17*). The model contains three behavioral equations that estimate annual demands for automobile travel, new car sales, and new car fuel economy as functions of income, unemployment, gasoline price, and the average age of the automobile stock.

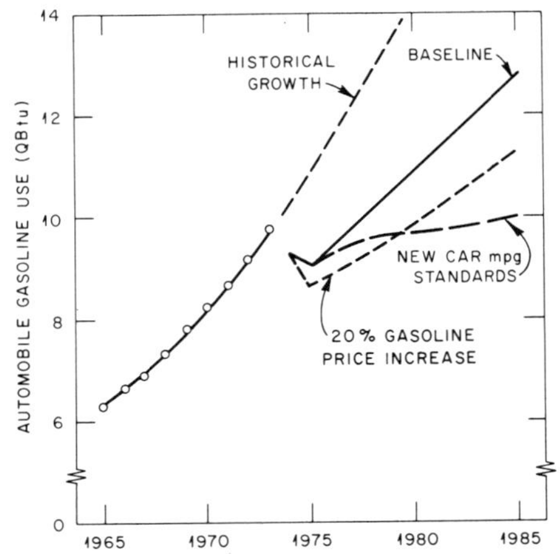

Fig. 4. Automobile gasoline use forecasts to 1985.

Automobile gasoline use from 1965 to 1973 and several projections to 1985 (using the FEA model) are shown in Fig. 4. The gasoline tax in these simulations is expressed as a constant percentage of the base price to avoid problems with inflation (in which a current dollar tax becomes weaker each succeeding year); the tax (20 percent) is imposed in 1975 and remains in effect during the 10-year simulation.

Between 1965 and 1973, automobile gasoline use increased at an average annual rate of 5.6 percent. The top curve in Fig. 4 shows the continuation of this historical trend. The curve marked baseline represents the output from the FEA model, assuming that the "real" price of gasoline remains constant at its assumed 1975 level (55 cents a gallon) and that no new federal policies are imposed that would affect auto ownership and use. The baseline curve shows that recent increases in gasoline prices are expected to substantially reduce growth in gasoline use: from 5.6 to 3.5 percent annually. Increasing gasoline prices by an additional 20 percent further slows the annual growth rate to 2.7 percent. Relative to the baseline fuel use, the 20 percent tax saves the equivalent of 484,000 barrels of crude oil a day in 1980 and 700,000 barrels a day in 1985. These fuel savings are nearly an order of magnitude larger than those due to programs that increase carpooling or mass transit use.

The short-run price elasticity of demand for gasoline implied by the FEA model is −0.21; the long-run elasticity is −0.72 (*17*). These elasticities are in good agreement with those derived in other studies (*18*).

Figure 5 shows changes in new car fuel economy predicted by the model. In the baseline, fuel economy improves from 14 mpg in 1974 to 15 mpg in 1978 and then drops back to about 14.3 mpg during the next 7 years. This decline in fuel economy is due to the assumed rise in personal incomes: as incomes grow, the price of gasoline becomes a less important determinant of new car purchase decisions. The 20 percent gasoline tax simulated here causes a sharp increase in new car fuel economy to 17 mpg in 1978. Then fuel economy drops to an average of 16.2 mpg during the next 7 years.

The curves of Figs. 4 and 5 show that the major response to higher gasoline prices is an increase in new car fuel economy rather than a decline in auto travel. During the first year, however, approximately three-fourths of the change in gasoline use is due to reduced driving and only one-fourth is due to improvements in fuel economy (*18*). Auto travel is reduced 4 percent in 1975 and 1976, 2 percent in 1980, and 1 percent in 1985 because of the 20 percent increase in gasoline price (*17*). Thus, as more and more new cars (purchased after the gasoline price increase) enter the fleet, changes in auto travel become negligible.

New Car Fuel Economy Standards

The FEA model (*17*) can also be used to evaluate the impacts of mandated improvements in new car fuel economy. Figure 4 shows the model's predictions of automobile gasoline use when standards are imposed that require average new car fuel economy to increase from 14 mpg in 1974 to 20 mpg in 1980 and to 22 mpg in 1985. Figure 5 shows the imposed fuel economy schedule.

The gasoline savings due to these improvements in new car fuel economy are substantial—even higher than those due to the 20 percent gasoline price increase. However, because only about 10 percent of the automobile stock is "rolled over" each year, it takes a few years before the standards have a major impact on fuel use. The schedule simulated here saves the equivalent of 568,000 barrels of crude oil a day in 1980 and 1,327,000 barrels a day in 1985.

Because improvements in new car fuel economy reduce the cost of driving per mile, one impact of the standards is a slight increase in the amount of auto travel; 3 percent in 1980 and 5 percent in 1985. Figures 4 and 5 show that the 20 percent increase in gasoline price initially has a stronger effect on both new car fuel economy and auto fuel use than do the new car fuel economy standards considered here. However, as a larger and larger fraction of the automobile stock is influenced by a stricter and stricter fuel economy standard, the savings due to the standard equal and then surpass those due to the fuel price increase: the crossover point for the policies simulated here is in 1979. These results suggest that a gasoline tax increase is a potent short-run measure, while new car fuel economy standards are powerful in the long run.

Although the FEA model is an interesting and useful tool with which to analyze automobile gasoline use policies, it contains several weaknesses that must be considered in interpreting its outputs. The equations estimating demands for new cars and new car fuel economy are independent of automobile prices. This implicitly assumes that improvements in fuel economy will not affect vehicle prices and that automobile prices will change in the future (relative to the other explanatory variables in the model) as they have in the past. The model also assumes that a prespecified fraction of the existing automobile fleet is scrapped each year. Thus the dynamics of the new and used car markets and auto ownership are quite restricted. Finally, the model treats all automobiles as homogeneous entities; there is no market class distinction between, say, compacts and luxury cars. In spite of these deficiencies, the FEA model is probably the most useful tool available for analyzing these policy alternatives.

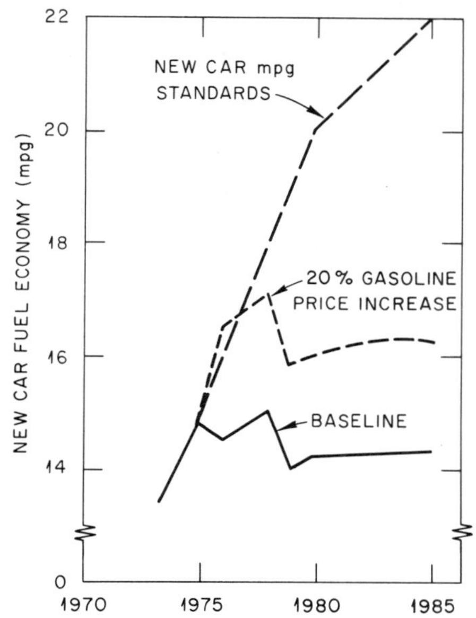

Fig. 5. New car fuel economy forecasts to 1985.

Conclusions

Table 1 summarizes the estimated energy savings for 1980 and 1985 due to the four transportation policies discussed here. Clearly, policies that directly affect automobile ownership and use (gasoline taxes and fuel economy standards) are much more effective in saving energy than are policies designed to shift travelers to energy-efficient alternatives (mass transit and carpools).

This suggests that, during the next decade at least, attention should be focused on technological means to reduce transportation fuel use. If the numbers in Table 1 are approximately correct, they show that behavioral changes (with respect to personal travel) are quite difficult to effect: people are extremely resistant to purely economic forces that seek to change their travel modes and extent of travel.

These results also suggest (implicitly) the need for much greater emphasis on informing people about energy problems and the need for conservation and demonstrating the attractiveness and viability of energy-efficient practices. These activities will encourage consumers to change their tastes and attitudes; changes in consumer behavior will then follow. In other words, social norms need to be modified so that people will want to carpool, will want to use mass transit, and will want to own small cars.

The analyses on which these conclusions rest involve techniques and data that are far from satisfactory. The models generally capture the important economic variables (such as prices and incomes). However, intangibles such as comfort, convenience, reliability, safety, what the neighbors think, and whatever else goes into individual decisions on how, when, and where to travel are not captured by these models. Therefore the models can predict behavior only when these intangibles (the variables not included in the models) do not change.

The models are generally based on data from the 1960's. To a surprising, perhaps frightening, degree, the United States today is quite different from what it was 10 years ago. This is true for fuel prices and also true with respect to individual expectations for the future and attitudes toward the environment, energy conservation, and automobiles. Thus decade-old data (even when used with very good models) may yield inaccurate estimates of behavioral changes. Also, the FEA model (*17*) is estimated by using time-series state-level data. CSI, on the other hand, uses detailed cross-sectional data for a single city for 1 year (*15*). Because of these differences in data, the results of these two models may not be consistent with each other.

Finally, a comprehensive model with which to evaluate these policies in combination does not now exist. Therefore, it is not possible to evaluate potential synergisms among the four policies considered individually in this article. Combinations of auto disincentives (such as an increased gasoline tax) and improvement of alternatives (carpooling, mass transit) are, intuitively at least, quite appealing. Such combinations are likely to maximize energy savings while minimizing adverse impacts of auto disincentives on mobility.

References and Notes

1. E. Hirst, *Energy Intensiveness of Passenger and Freight Transport Modes: 1950–1970* (Report ORNL-NSF-EP-44, Oak Ridge National Laboratory, Oak Ridge, Tenn., 1973).
2. "Retail prices and indexes of fuels and utilities," Bureau of Labor Statistics monthly press releases, U.S. Department of Labor, Washington, D.C.
3. National Automobile Dealers Association, Research Department, Washington, D.C., unpublished data.
4. Energy Policy and Conservation Act (PL 94-163, 94th Congress, 1975).
5. "U.S. energy use down in 1974 after two decades of increases," Bureau of the Interior, Washington, D.C., 3 April 1975; see also earlier press releases.
6. Federal Energy Administration, *Project Independence Report* (Government Printing Office, Washington, D.C., 1974).
7. W. G. Dupree, Jr., and J. A. West, *United States Energy Through the Year 2000* (Department of the Interior, Washington, D.C., 1972).
8. Jack Faucett Associates, *Project Independence and Energy Conservation: Transportation Sectors*, Project Independence Blueprint Final Task Force Report (Federal Energy Administration, Washington, D.C., 1974).
9. E. Hirst, *Total Energy Use for Commercial Aviation* (Report ORNL-NSF-EP-68, Oak Ridge National Laboratory, Oak Ridge, Tenn., 1974).
10. Federal Highway Administration, *1973 Highway Statistics* (Department of Transportation, Washington, D.C., 1974); see also earlier annual issues.
11. *'74–'75 Transit Fact Book* (American Public Transit Association, Washington, D.C., 1975).
12. *1974 National Transportation Report* (Department of Transportation, Washington, D.C., 1974).
13. D. E. Boyce, K. Nguyes, T. Noyelle, V. R. Vuchik, "Impact of rapid transit on fuel consumption and cost for the journey to work," University of Pennsylvania, draft final report for the Federal Energy Administration, June 1975.
14. J. P. Curry, "Case studies of transit energy and air pollution impacts," DeLeuw Cather and Co., draft final report to the Environmental Protection Agency, March 1975.
15. *Analysis of Transit Passenger Data* (Metropolitan Atlanta Regional Transportation Administration, Atlanta, Ga., 1973).
16. W. A Jessiman, J. H. Suhrbier, T. J. Atherton, Cambridge Systematics, Inc., Cambridge, Mass., paper presented at the 55th annual meeting of the Transportation Research Board, Washington, D.C., January 1976.
17. J. Sweeney, "Passenger car use of gasoline: An analysis of policy options," draft report, Office of Energy Systems, Federal Energy Administration, Washington, D.C., February 1975.
18. S. Wildhorn, B. K. Burright, J. H. Enns, T. F. Kirkwood, *How to Save Gasoline: Public Policy Alternatives for the Automobile* (Report R-1560-NSF, Rand Corporation, Santa Monica, Calif., 1974); B. K. Burright and J. Enns, *Econometric Models of the Demand for Motor Fuels* (Report R-1561-NSF, Rand Corporation, Santa Monica, Calif., 1975).

19. I thank M. Rodekohr, C. Rappaport, J. Gibbons, and J. Suhrbier for their assistance in preparing and critically reviewing the manuscript for this article.

Energy Conservation and the Consumer

A tax placed on energy and adjusted to wage levels would ease a change to a more labor intensive economy.

Bruce Hannon

In this article I seek to explain what is known about the connections among consumer activities, the direct and indirect demands of these activities for energy and employment, and the consumer's options for energy conservation. When the consumer's activities demand employment, they require the consumer's own services, but when they demand energy they require the stored energy resources of the earth. This fact underpins the three dilemmas of energy conservation that I will discuss in detail. The consumer's control over the energy he uses to heat, cool, and light his residence or fuel his auto will not be discussed; the methods for conservation in these areas are well understood.

Individual control over energy use is viewed as an alternative to governmental assumption of that control. Energy conservation is necessary because of (i) the environmental effects of unbridled consumption; (ii) the long lead time and massive capital allocations necessary for future increases in energy supply; (iii) the instabilities associated with a large dependence on foreign energy supplies; (iv) the need for an enduring national goal that unifies the nation and does not require massive new economic growth; and (v) because energy is a fundamental ingredient in any economic system. This last point indicates the necessity to focus on the conservation of energy rather than steel, for example. A proper plan to conserve energy would conserve steel. In general, capital can be recycled but, of course, energy can not. Labor can not only be recycled, it can be multiplied.

During the industrial revolution in the United States there has been an unprecedented thrust to substitute the energy from fossil fuels for human labor to accomplish material transformations (toolmaking) and to achieve mechanization. From the worker's and consumer's view, this substitution has generally meant less drudgery in the work place, greater buying power, and more leisure. To the industrialist, the substitution of tools and energy for labor has provided a means to control and predict production costs and, consequently, profits.

Material and energy costs have been low, relative to labor, as well as predictable for almost 50 years. Machines and fuels are incapable of striking for higher wages. Economic growth seems to have absorbed those workers displaced by mechanization. Thus technology, material and energy use, and employment and economic growth appear to be intimately linked. But now the finiteness of resources comes into view as this nation's striving for a better life spreads internationally. We are beginning to understand the links among heavy consumption, environmental damage, and degradation of the quality of life. We are beginning to see the need for one more adaptive act of humankind, the need for a group morality.

It should be within the reach of each individual consumer to perceive that he must forgo certain forms of energy consumption now in order to ensure their availability to his offspring or to his own generation in the future. As the human body eventually reaches a condition of zero growth, then so must that population as a whole. Such an achievement, however, would require our casting out such mathematically impossible maxims as "the greatest good for the greatest number" and substituting, perhaps, "life-styles of elegant frugality." Then we would act to preserve the knowledge of how to live healthfully, limit our numbers, and minimize our per capita consumption of resources, particularly energy. We would recognize that, because people's wants are infinite and because resources are finite, only relative wants are important; and that economic growth increases the disparity in relative wants, eases our severe interpersonal and interindustry competition, and increases the uncertainty of future events.

Economic growth and development will probably continue until the assured available energy is almost entirely demanded for maintenance of the society. Stabilization of economic growth now is often argued as being a means of reducing the inevitable future trauma of living on an energy budget. There would obviously be problems, however, with the egalitarian national life-style that would be required to achieve stabilization now. We would have to recognize that the rising world population will result in the rising collision of individual and national freedoms as well as in the accelerated depletion of stored energy; and that the concept of individual and corporate ownership, apparently an adaptation to resource shortages, will probably give way to public or societal ownership, as the shortages worsen. It is unclear whether a totally democratic society could exist with severe shortages. If it could not, smaller and smaller fractions of the population would participate in resource allocation decisions. How to control the decision-makers would then become a problem. As long as the population at large could perceive an equitable distribution of the dwindling resources then, presumably, the society could remain stable. Plato's concept of leaders or "guardians" deprived of all physical wealth and steeped in leadership seems an appropriate model for controllers of a stable society with limited resources.

Before we plan to conserve energy or to increase our energy supplies, we must understand the total energy and employment cost of every good and service. Only then will we be able to determine the energy conserved by consuming one good or service instead of another, or the energy cost of substituting a new technology, a new manufacturing process, for example, for another. The changes that may have to be made in order to reduce our energy consumption can be thought of as efficiency improvements in our current life-style. As I shall show, many such changes may be resisted and, unless they are fully understood, they may not provide significant net energy savings.

To calculate the energy cost of one unit of an item, one must determine the direct inputs of goods and services required to produce that item. Because each of these inputs has its own set of inputs, and so on, one obtains such a multitude of small inputs that leaving off the next round does not significantly change the total energy requirements of the original item. I have

The author is director of the energy research group at the Center for Advanced Computation, University of Illinois at Urbana-Champaign, Urbana 61801.

used the economic input-output data published by the U.S. Department of Commerce (*1*), and the data of Herendeen (*2*), to calculate the direct and indirect energy and employment demands of a number of personal consumption activities and to analyze trade-offs between human and mechanical energy.

Substituting Energy for Labor: The First Dilemma

If conserving energy or any material resource is judged to be the main path to our survival, the impact of such a path on employment must be known. Since we, as laborers, decide our future, and since we have repeatedly increased our wages and controlled the cost of the resource base, the potential redistribution of labor resulting from a conservation program is of primary importance.

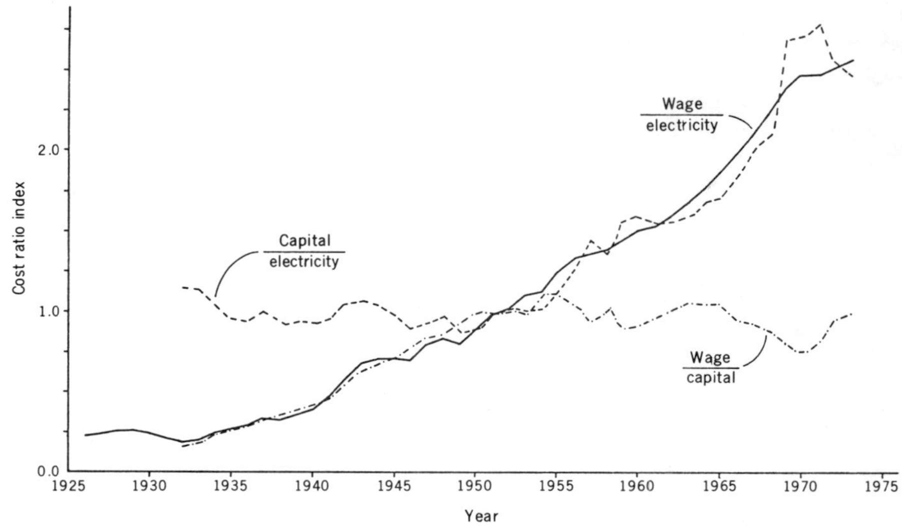

Fig. 1. Industrial labor, electricity, and capital cost ratios in the United States, 1926 to 1973, in current dollars. Based on manufacturing worker's hourly wages, the industrial price of a kilowatt-hour of electricity, and the yields on AAA corporate bonds (1951 = 1.0). Data from the Department of Commerce (*1*), Edison Electric Institute (*21*), and Moody's *Industrial Manual* (*22*).

Let us observe the problem from the point of view of an industrial decision-maker. He compiles his current capital, materials, labor, and energy costs and projects them on the basis of his expectations of the future. Trend analyses would have shown that his raw materials and energy costs are either stable, unless his product is derived from agriculture, or declining, as in the case of electricity, his principal energy substitute for labor. In terms of dollars, tool costs would have been rising primarily because of their labor content, but the most rapidly rising, and the most unpredictable, costs would have been those for manual labor. An important factor in the industrialist's decision-making process would be his anticipated relative unit costs of energy and of labor. The ratio of these two costs are formed into an index (wage divided by price of electricity) and plotted for the years 1926 to 1973 in Fig. 1. In that period the ratio grew by a factor of 10. It grew by 230 percent between 1950 and 1970, during which time electrical machinery costs increased by about 54 percent (current dollars). The curve in Fig. 1 shows a leveling to a declining trend during recession periods, possibly indicating an attempt of the economic system to equilibrate resource and labor costs.

Total costs of electricity relative to total product costs are small. In 1963, one average dollar of industrial sales caused only about 2.5 cents of electricity sales (*3*), and about 65 cents went for wages, directly and indirectly (*4*). As shown in Fig. 1, the rapidly rising cost of capital since 1950 has probably deterred a more rapid switch to a more electric, less labor intensive production system. Since 1950, the marginal productivities of labor and capital have been rising relative to that of electricity, while the ratio of the marginal productivities of labor and capital has been relatively neutral. This capital cost, however, contains the cost of building and heating machinery which cannot be considered a substitute for labor. The generally higher cost of capital should therefore aid in the return to a more labor, less electric intensive system. The point is twofold: the cost of electricity has failed to represent its importance in the marketplace, and the situation has grown worse with time.

If we take energy, labor, and capital to be factors required for production, then, as the cost of energy is increased relative to the cost of labor, the economic system will adjust, in theory, such that the marginal productivity (a change in gross national product resulting from a unit change in factor use) of energy will increase relative to that of labor. If the wage levels and labor productivity are constant, this change means lower energy use. In other words, Fig. 1 indicates that the productivity of labor has, historically, increased more rapidly than, and at the expense of, electrical energy productivity. To reverse this process, the price of electricity must rise relative to wages. This same phenomenon appears to be true for overall energy and labor use (*5, 6*).

The overall effects of economic growth on energy and labor demand were suggested by an investigation (*7*) in which it was demonstrated that if, in 1963, the economic sectors had been allowed to grow at the proportional expense of all the remaining sectors, most industries would have become more energy demanding and less labor demanding. In a special case study (*8, 9*) it was shown that the system that produces beverages would be less energy demanding and more labor demanding if the beverages were delivered in refillable rather than throwaway containers. Most important was the type of labor that would be involved in the change. Jobs would be lost in the highly organized, high-wage paying, can-maker's plants and would be gained in the low-wage paying, relatively nonorganized, retail sector. The lack of acceptance of this plan was the direct result of opposition from organized labor and intensive lobbying by can- and bottle-making interests.

Another example of increasing energy and decreasing labor utilization is found in a comparison of the energy and employment costs of protein production (*3*). The production of a unit of milk protein requires about 2.7 times as much energy as a unit of cheese protein, and yet requires only about one-fifteenth as many jobs. This result derives from the nature of the different processes: milk contains mostly water and it must be homogenized, pasteurized, and trucked, each of which operations is an energy but not labor intensive process.

Herein lies the first dilemma. Energy conserving policies would increase overall employment in general by decreasing the number of high-wage jobs and increasing the number of low-wage jobs, but the people holding the former belong to the most highly organized unions. Thus organized labor, after struggling to achieve reasonable wages and personal dignity in the marketplace, and having slowly relinquished much dignity because of the inevitable relationship between high wages and assembly lines, for example, would now have to be asked to forgo their higher wages in order to avoid the general unemployment induced by declining energy supplies. Such a situation would be tenable only if prices and profits were thoroughly controlled. It also follows that technological developments have, historically, resulted in jobs being denied to marginally employed workers.

Income and Energy Use: The Second Dilemma

The relation between personal income and energy use can best be introduced by asking three questions:

1) What fraction of total energy use does the U.S. consumer control (directly and indirectly) through his purchase of goods and services?

2) How does this direct and indirect fraction vary with income?

3) How might consumers proceed to change their purchasing habits so as to reduce their energy dependence?

The first question is answered in Table 1, in which the data indicate the importance of personal consumption expenditures. Such expenditures demanded two-thirds [one-third of this amount for the automobile (*10*)] of all U.S. energy and 62 percent of all employment in 1963. Government activities accounted for about 13 percent of all energy consumption and employed about 21 percent of the total labor force. Capital formation demanded about 10 percent of all energy and 12 percent of all employment. Exports accounted for most of the remaining energy (coal) and labor consumed. The fractions of the various energy types allocable to personal consumption is surprisingly constant, except for coal. Most coal was consumed in steelmaking and electrical generation and about 16 percent was exported (directly and indirectly) in 1963. In the near future it will be possible to compare the data of Table 1 with similar data from 1967.

Figure 2 provides the answer to the second question. Family income after taxes is plotted against the direct, indirect, and total energy demand associated with the spending of that income (*10*). For example, the poor generally traveled by bus (when they traveled) and the rich usually traveled by plane. Both are indirect energy uses. Middle income people traveled by auto, which represents a direct energy use. The rate of direct energy use slowed as income increased, but the rate of indirect energy use did not. Together they yielded a total energy demand that was nearly linear with income. Here we find the second dilemma: spending money in any way demanding energy, and the extra dollar spent required almost the same energy when spent by a poor family as it did when spent by a rich one. It appears that the only way to save energy is to reduce income, unless the variation of energy consumption about the "total" line is very large. However, as I shall show in the next section, saving energy means saving dollars and that leads to additional energy use.

From the supporting data for Fig. 2 (*10*), it is evident that the average family used about 55 percent of its total energy directly. It is also evident that two-thirds of the lowest income group's total energy use was direct. Just the opposite is true of the upper income group: two-thirds of its total energy use was indirect. A further examination of these data shows that modification of the behavior of the upper middle income class would be the most fertile area for energy conservation. Herendeen is now investigating the energy-income redistribution effects in view of the slight nonlinearity of the total energy line in Fig. 2.

The third question is more difficult to answer. Consumer spending patterns would appear to be a function of income level. That is, as family income increases, buying patterns change. For example, meat becomes a larger part of the diet; travel is more frequent and by faster modes; and more investments are made. As income rises, more goods and services are consumed, resulting in rising indirect energy demands.

As an example of the utility of Fig. 2, let us consider a person living in New York. It was reported in 1973 (*11*) that the average New Yorker is much less energy demanding than the average U.S. citizen, even though his personal income is 25 percent higher than average. This, it was assumed, was because New Yorkers live in apartments and use mass transit heavily. But the data of Fig. 2 suggest that if a person has a 25 percent higher than average income, then his energy use is also nearly 25 percent higher than average. It is apparently difficult and expensive in both dollars and energy to service people living in a city. Or, perhaps living in a city induces a more frenetic, consequently more costly, pattern of behavior.

Figure 3 with the enlargement of its central portion shows the relative (direct and indirect) energy and labor cost per 1963 dollar of personal consumption on each of the 83 different consumption categories of the U.S. Department of Commerce. In general, the services, except transportation, seem to be distributed along a roughly horizontal (constant energy intensity) line with high-wage services on the left and low-wage services to the right (for example, physicians on the left, hospitals on the right). Goods and transportation services are distributed along a roughly vertical line which joins the service line at about its lower-third point. The energy

Table 1. Percentages of energy and labor (direct and indirect) for final demand activities, during 1963. The data are from Hannon and Abbott (*15*). The definitions of the activities are from (*1*).

Activity	Energy type: Coal	Crude oil	Refined petroleum	Electricity	Gas	Total primary energy	Labor*
Personal consumption expenditures	53.2	71.3	70.8	69.2	72.2	66.6	62.2
Gross private fixed capital formation	14.9	8.6	7.3	10.5	10.5	10.3	12.1
Net inventory change	1.2	1.3	1.5	0.9	1.0	1.3	0.9
Exports†	15.9	6.6	7.7	4.9	5.0	9.0	3.6
Federal government purchases, defense	6.6	5.6	6.1	6.5	4.7	5.9	8.4
Federal government purchases, other	1.0	0.9	0.9	0.4	0.9	0.9	2.5
State and local government purchases, education	2.4	1.8	1.6	3.0	2.0	2.0	4.6
State and local government purchases, health, welfare, and sanitation	1.0	0.8	0.7	1.1	0.9	0.8	1.5
State and local government purchases, safety	0.2	0.3	0.3	0.2	0.2	0.2	0.9
State and local government purchases, other	3.6	2.9	3.1	3.3	2.5	3.1	3.3
Total final demand	100.0	100.0	100.0	100.0	100.0	100.0	100.0

*Man-year basis: includes household and government workers but not military personnel. †Dollar value of imports not used.

purchases are, logically, highest on this line.

The possible lack of independence among some of the personal consumption items is demonstrated by the averaged point "owning and operating an auto" in the upper-center portion of Fig. 3, and by comparing this point with the points "gasoline and oil," "new and used cars," "auto repair," "auto insurance," and "tires." Knowledge of the individual auto expenditures could guide the consumer on the energy and employment effects of his marginal dollar expenditures, but the overall effect of owning an auto is given by the "average" point.

Figure 3 gives the total energy and employment effects produced by the spending of a consumer dollar on the 83 different consumption categories. The figure does not show the secondary effects. Suppose, for example, that you purchase a new car. The energy and employment effects shown in the figure represent only those effects which occur in the web of industries which supply the auto industry (including the fabricating, transporting, and retailing). These effects do not include the energy and employment demands created by the personal consumption of the workers who receive your new car money from the manufacturer, nor the energy and labor demanded by the workers of the producers of these goods as they too spend their wages. As money passes through each new set of hands, some is saved and taxed and eventually the effect dies out. If one assumes that for every dollar spent on a new car, two more are spent via the multiplier effect on average personal consumption, then each new car dollar spent, which represents one dollar of growth in the gross national product, demands 60,000 British thermal units (new car) and 172,000 Btu (two average 1963 personal consumption expenditure dollars) for a total of 232,000 Btu (1 Btu is equivalent to 1055 joules). The expansion from 60,000 to 232,000 Btu demonstrates the effects of the growth dollar on energy demand. It has a similar effect on employment demand.

Respending Dollars Saved: The Third Dilemma

Let us now look more deeply into the energy and labor needed to supply an entire service. Herendeen (*2*) has shown that all transportation consumed (directly and indirectly) about 42 percent of the U.S. energy in 1963, with the personal auto accounting for half of this demand. Such knowledge prompted us at this laboratory to investigate energy use and transportation options. Table 2 shows the dollar, energy, and employment costs of ten alternative passenger modes. The intercity modes were developed for an approximate 300-mile (482.7-kilometer) intercity distance in order to allow the five modes to be considered competitive. All costs are very sensitive to load factors. In terms of dollars and energy, the plane is easily the most expensive, and the train is the most employment intensive. The intercity bus costs the least in dollars and in energy.

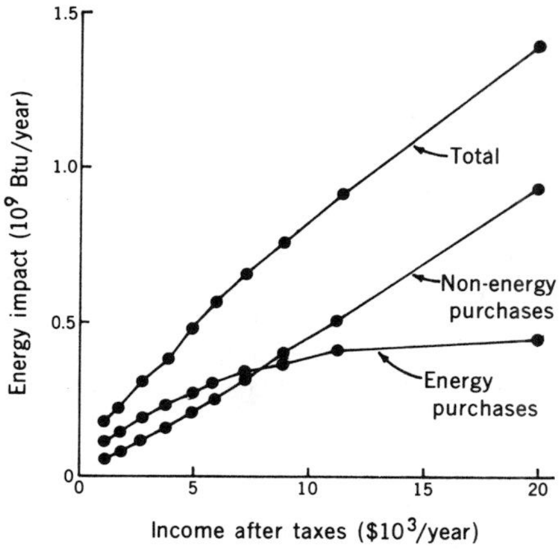

Fig. 2. Direct, indirect, and total family energy impact plotted against family income, 1960 to 1963. From Herendeen (*10*).

The urban passenger has a variety of modes to choose from, as shown in Table 2. These modes are increasingly unpopular but decreasingly energy expensive as one moves down the list. Energy consumed in walking was that used to supply the food consumed by the average person for the energy used in excess of that used by the body in the resting position.

It has been pointed out by Bullard (*12*) that the important factor in energy conservation is the rate at which energy is saved on the transfer from one activity to another. Tables 3 and 4 show rates of energy savings (British thermal units saved per dollar saved) for shifts from each transportation mode to the other. The rates vary from about 17,000 to 633,000 Btu per dollar. In all cases except in the urban transportation shifts from car to bus, dollars are saved if energy is saved, and vice versa. For example, the traveler who switched from urban bus to bicycle would save energy (and dollars) at the rate of 51,000 Btu per dollar. If he were not careful to spend his dollar savings on an item of personal consumption which had an energy intensity greater than 51,000 Btu per dollar then his shift to the bicycle would have been in vain.

It is easy to notice that in every instance, a change in transportation mode that would conserve energy would also save

Table 2. Selected results on the total dollar, energy, and labor impacts of consumer options in transportation during 1971. Data are expressed in terms of the requirements to move a million passengers 1 mile (1.6 km) (NA, not applicable).

Transportation mode	Load factor	Thousands of dollars (1971)	Energy* 10^6 Btu	Energy* Direct (%)	Jobs
Intercity transportation					
Car	2.9 people	55	5900	51	3.7
Plane	53% full	58	9800	73	3.8
Bus	47% full	39	2700	51	3.1
Train	37% full	44	4000	58	7.2
Electric commuter†	31% full	128	9900	11	8.5
Urban transportation					
Car‡	1.9 people	69	8900	58	4.2
Bus‡	12.0 people	105	5300	57	8.3
Motorcycle	1.1 people	57	4200	49	1.55
Bicycle	1.0 people§	26§	1300	59§	1.7‖
Walking	1.0 people	NA	710	51	0.7

*Vehicle transportation fuel only. †The "PATH" commuter system, New York–New Jersey, 1971 [see Penner (*16*)]. Other intercity data from Sebald and Herendeen (*17*). ‡Data from Hannon and Puleo (*18*). §From Hirst (*19*). ‖Center for Advanced Computation estimate based on Hirst (*19*).

Table 3. Intercity transportation. The energy that would be saved by shifting from each transportation mode to another. Plus or minus signs preceding numbers indicate, respectively, an increase or a decrease in energy use; plus or minus signs after the numbers indicate, respectively, an increase or decrease in dollar cost. Calculated from Table 2.

Shifting from	Shifting to: Car	Plane	Bus	Train
Car		+355,000+	–200,000–	–633,000–
Plane	–355,000–		–374,000–	–414,000–
Bus	+200,000+	+374,000+		+260,000+
Train	+633,000+	+414,000+	–260,000–	

Table 4. Urban transportation. The energy that would be saved by shifting from each transportation mode to another. Plus or minus signs preceding numbers indicate, respectively, an increase or a decrease in energy use; plus or minus signs after the numbers indicate, respectively, an increase or decrease in dollar cost. Calculated from Table 2.

Shifting from	Shifting to: Car	Bus	Motorcycle	Bicycle	Electric commuter
Car		−100,000+	−392,000−	−177,000−	+ 17,000+
Bus	+100,000−		0	− 51,000−	+200,000+
Motorcycle	+392,000+	0		− 61,000−	+ 80,000+
Bicycle	+177,000+	+ 51,000+	+ 61,000+		+ 84,000+
Electric commuter	− 17,000−	−200,000−	− 80,000−	− 84,000−	

dollars (except in the case of changing from urban car to bus transportation). This provides us with the third dilemma: that is, what does the consumer do with the dollars he saves after he has shifted to a cheaper mode of transportation? He can spend it or save it. In either case, energy will be required to provide for this freed expenditure. The second dilemma indicates that the consumer will adjust to increasing wealth through a change in purchasing habits, and his total energy use will rise proportionately to his income. However, armed with certain information, the consumer can purposefully direct his income so that he does save energy on balance. It should be clear, however, that he can never save more energy by redirecting certain portions of his income than he can by becoming that much poorer.

Table 5 is a rank order listing by energy intensity of the top 20 (in terms of dollars) personal consumption activities. Data on the activities of owning an auto and owning a home are listed so that they can be compared with the energy intensity of the average personal consumption dollar. Labor intensities are also given so that the interested reader may determine the impact of his own respending on the labor situation.

In the above example, the bus passenger turned bicycle rider (see Table 4) could spend his dollar savings on anything below and including "private investment" in the activity list of Table 5, and save some energy. The net energy is found by subtracting the number in Table 5 from the transfer savings of Table 4. If, for example, his savings were spent on "restaurants," the net energy savings would be 51,000 minus 32,400, which equals 18,600 Btu per dollar saved. Also from Tables 2 and 5,

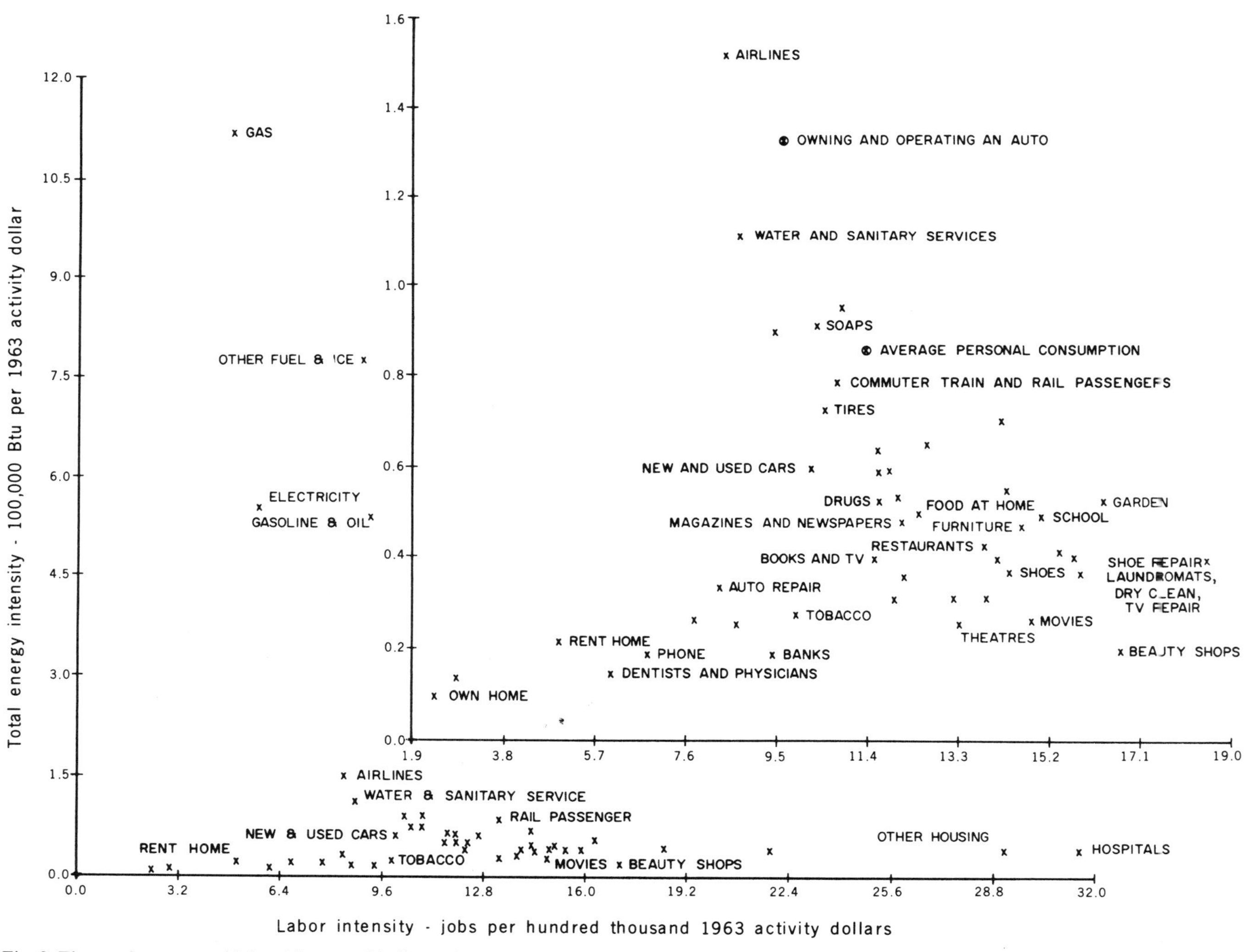

Fig. 3. The total energy and labor (direct and indirect) intensity of various personal consumption activities, 1963. From the Center for Advanced Computation's energy-employment model, August 1974. The inset is an enlargement of the central portion of the data.

Table 5. The energy and labor intensity of the 20 activities in which personal consumption expenditures (PCE) are highest in terms of dollars. Ranked in order of decreasing energy intensity during 1971. From Hannon and Puleo (*18*).

PCE sector description	Energy intensity (Btu per dollar)	Labor intensity (jobs per $1000)	PCE sector description	Energy intensity (Btu per dollar)	Labor intensity (jobs per $1000)
Electricity	502,500	0.0436	(Federal taxes)	(36,300)	(0.082)
Gasoline and oil	480,700	0.0729	Women's and children's clothing	38,100	0.1000
(Housing)	(144,000)	(NA)	Restaurants	32,400	0.0875
(Auto ownership)	(111,500)	(.081)	Men's and boys' clothing	31,400	0.0984
Cleaning preparations	78,100	0.0733	Religious and welfare activity	27,800	0.0863
(Average PCE)	(70,000)	(0.080)	Private hospitals	26,100	0.1718
Kitchen and household appliances	58,700	0.0551	Automobile repair and maintenance	23,500	0.0483
New and used cars	55,600	0.0775	Financial interests except insurance	21,500	0.0784
Other durable house furniture	54,600	0.0894	Tobacco products	19,800	0.0585
(Private investment)	(45,600)	(0.066)	Telephone and telegraph	19,000	0.0585
Food purchases	41,100	0.0852	Rented home	18,300	0.0350
Furniture	36,700	0.0917	Physicians	10,700	0.0325
			Own home	8,300	0.0167

one can calculate the net impact on labor at four jobs gained for every million dollars saved on the bus to bicycle transfer.

Most of the savings rates are so large in Tables 3 and 4 that they could be overcome only if the consumer spent his savings directly on electricity or gasoline and oil. However, the consumer must be aware that the purchase of an extra car, for example, means greater purchases of gasoline and oil as well.

An interesting case arises in the shift from urban car to bus. Here the consumer saves 100,000 Btu per extra dollar spent on bus transportation. This extra dollar must be produced by reducing expenditure on other things, since urban bus transportation is more expensive than car (Table 2). Suppose the extra dollars came from a reduction in average personal consumption spending; then from Table 5, the total energy savings rate would be 170,000 Btu per dollar.

Table 6 shows the same phenomena as Tables 2 to 4. But in this case, home appliance alternatives are compared. The annual total dollar, energy, and employment costs were compiled for manufacturing, operating, and maintaining each appliance. From these data the energy savings rate can be calculated and compared to the data of Table 5 to determine the net energy savings.

Significant energy savings rates are obtainable if fewer appliances are used and if gas is used instead of electricity in stoves, dryers, and water heaters. This savings rate is reduced, if not reversed, by the energy demands of the alternative consumption activity.

Table 5 shows that the federal tax dollar and the private investment dollar are relatively low in energy intensity. This fact indicates that shifts from average personal consumption to federal spending or investment would reduce energy demands. However, many of these activities, for example, building highways or power plants, foster accelerated energy use and lower employment demands than personal consumption. Heavy construction programs can therefore be viewed as energy-wasteful disemployment programs. It is tempting for the energy conservation policy-maker to devise methods to divert the personal consumption dollars into federal spending or private capital formation. If such a procedure were followed for the purpose of conservation, it would be truly deceptive not to focus these dollars on processes which lead to a lowering of the average national energy intensity [see (*13*) for example].

In general, if all consumers were to make such carefully precalculated consumption shifts as outlined above, national income would be preserved, the energy intensity of personal consumption would be lowered, and employment would be maintained if not increased—and it would certainly be redistributed. However, the amount of net energy savings might be small because of the respending effect, and because of the fact that people will not want or need a sufficient quantity of the lower energy goods and services, and be-

Table 6. The dollar, energy, and labor impacts of consumer activities in the home during 1971 (including manufacturing, operating, and maintaining the home). Herendeen and Sebald (*20*).

Type or number of appliances	Requirement per year: Dollars (1971)	10^6 Btu	Hundredths of a job	Energy savings rate* (Btu per dollar)
Kitchen appliance sets†				
Spartan (4 appliances)	152.50	64.9	1.09	+304,000
Moderate (11 appliances)	252.50	95.3	1.85	+254,000
Plush (20 appliances)	337.90	117.0	2.47	
Clothes dryer				
Gas	27.6	8.1	0.23	+500,000
Electric	39.4	14.0	0.27	−383,000
Outdoor	3.4	0.2	0.03	
Stove				
Gas	33.5	13.4	0.26	+198,000
Electric	50.7	16.8	0.36	
Refrigerator (12 cubic feet‡)				
Conventional	45.0	11.1	0.35	+608,000
Frost-free	55.7	17.6	0.40	
Water heating				
Gas	46.0	37.7	0.22	+325,000
Electric	104.0	56.6	0.50	
Electronic media				
Black and white television	21.7	5.4	0.18	+ 79,000
Color television	67.2	9.0	0.63	−120,000
Radio	4.1	1.2	0.03	

*When the activity is compared with the next one down the list; plus and minus signs mean, respectively, increases or decreases in energy use and dollar cost. †Including a stove, clothes dryer, and water heater all run on gas. ‡Twelve cubic feet are equivalent to 0.33 cubic meter.

cause of the likelihood of too few people actually being aware of their options. In any event, there is a limit to the savings that can be realized by such life-style changes which preserve national income.

Conclusion: Resolving the Three Dilemmas

The problem of energy conservation by the consumer has been examined in terms of three dilemmas: First, when wages increase relative to costs, then energy use increases through the process of mechanization. Second, energy use and income are linearly connected such that the spending of an average additional dollar of income demands nearly the same amount of energy, regardless of one's income level. Thus, doubling one's income doubles one's energy use so that the phenomenon of "rising expectations," so ingrained in the American worker, is limited by the amount of available energy. Third, saving energy usually means saving money, the respending of which reduces, if not eliminates, the energy first thought saved. Not respending the savings means a reduction in national income (gross national product) and money saved too quickly means a reduction in employment. But this last effect is somewhat abated by the general increase in labor intensity that is associated with the lower average energy intensity. This effect could also be thwarted by each worker agreeing to work fewer hours. Both solutions, that is, not respending one's savings and working fewer hours, would eventually result in decreased per capita wealth. It should also be pointed out that convenience demands energy, for example, high speed autos and airplanes, and one-way packaging. Giving up convenience is the equivalent of the consumer's accepting a lower implicit wage, and saving energy.

There probably are no popularly acceptable solutions to these three dilemmas. If wages, as well as proprietor's incomes, were reduced relative to the cost of energy, or if the population decreased, or both, net energy consumption would be reduced without creating widespread unemployment. However, strong labor unions attest to the irreversibility of wage rates.

Perhaps an initial energy resource tax, geared proportionately to the wage level, could be placed on fuels, particularly on those used to make electricity. Such a tax would thrust our most important resource into a position in the marketplace competitive with its chief substitute, manpower. The tax revenue could be used to reward those who reduced their energy use or those who were severely disadvantaged by the tax. In effect, this money could be used to cover the initial costs of converting to a life-style with lower energy demands. With energy costing more, the severance tax revenue would generally be spent in a manner compatible with reducing our energy requirements. Also, in the long run, the tax would imply lower real wages because of its demanding a lower labor productivity through the substitution of labor for energy.

It is appropriate that an already energy intensive nation take steps to increase its productivity per unit of energy expended. Raising the cost of energy relative to the cost of its substitute appears to be the best way for us to achieve such an increase in that it induces through the market mechanism the change to a more labor intensive technology. The tax approach might also give us time and revenue to redirect technology into improving productivity from available energy and into easing the transition to lower labor productivity. While this relative cost change would probably be produced by an economic recession, through a reduction in the average real wage, the uncontrolled wage reduction that might result under such conditions could also cause much unemployment unless the technology of production could very rapidly become more labor intensive. The ratio of the energy tax to wages should be increased at a rate which maximally increases productivity from energy without creating an unacceptable level of unemployment.

The impact on prices of the two basic types of energy taxes (one based on the unit of energy, the other on the dollar value of energy) were estimated by Bullard (*6*). The tax based on the British thermal unit is preferred by those favoring energy conservation because it does not discriminate against small energy users who normally pay a higher unit price for energy than do large users.

Another method that would conserve energy, and that is more fundamental than taxation, would be energy rationing through the use of coupons. Energy storage would be guaranteed to the consumer by his being given coupons that could be used for different types of energy, such as gasoline or electricity, at different exchange rates. Thus the consumer would have the option of conserving energy through the process of ownership of a stored versatile fuel. This concept implies the existence of an energy interest rate that depends on consumer decisions to forgo present energy consumption in order to ensure a future supply. The energy coupons would be purchased from the federal government which consequently would need to "own" the energy resources. Such an ownership would allow the society to jointly set a national energy interest rate which would dictate how rapidly the proved energy reserves of the society could be consumed.

In the long run, we must adopt energy as a standard of value (*7*) and perhaps even afford it legal rights (*14*). For energy, like no other entity, cannot be recycled or multiplied. It flows from the sun and from the energy inventories of the earth. Shall we live off the flux or the storage? How fast and under what circumstances shall we draw it from storage? We would answer these fundamental questions correctly if we set energy as the basis of value and structured our economic and legal system accordingly.

As a final point, we must consider the major problem related to the veracity of the energy crisis. Is the crisis contrived or real? The conservative approach is to believe that it is real and act accordingly. If the crisis is real, we shall be viewed by future generations as terribly wise; if the crisis is false, we shall have had an instructive relief from the dancing mirage of our manifest destiny.

References and Notes

1. U.S. Department of Commerce, *Input-Output Structure of the U.S. Economy, 1963* (Government Printing Office, Washington, D.C., 1969).
2. R. A. Herendeen, *Use of Input-Output Analysis to Determine the Energy Cost of Goods and Services* (Center for Advanced Computation, Document No. 69, Univ. of Illinois, Urbana, 1973).
3. B. Hannon, *Technol. Rev. (M.I.T.)* **76** (No. 4), p. xx (February 1974).
4. A. J. Walderhaug, "The composition of value added in 1963," *Survey of Current Business* (Government Printing Office, Washington, D.C., 1973), pp. 34–44.
5. U. S. Department of Commerce, *Handbook of Labor Statistics* (Government Printing Office, Washington, D.C., 1972), p. 220.
6. C. W. Bullard, *Energy Conservation Through Taxation* (Center for Advanced Computation, Document No. 95, Univ. of Illinois, Urbana, 1973).
7. B. Hannon, *Ann. Am. Acad. Polit. Soc. Sci.* **410**, 139 (November 1973).
8. ———, *Environment (St. Louis)* **14** (No. 2), 14 (March 1972).
9. H. Folk, *Two Papers on the Employment Effects of Mandatory Deposits on Beverage Containers* (Center for Advanced Computation, Document No. 73, Univ. of Illinois, Urbana, 1972).
10. R. A. Herendeen, *Am. Soc. Mech. Eng. Tech. Digest* 73-WA/ENER-8 (1973).
11. Resources for the Future, *Patterns of Energy Consumption in the Greater New York City Area* (Washington, D.C., July 1973).
12. C. W. Bullard, Energy Research Group, Center for Advanced Computation, Univ. of Illinois, Urbana.
13. R. Bezdek and B. Hannon, *Science* **185**, 669 (1974).
14. C. D. Stone, *Should Trees Have Standing?* (Kaufmann, Los Altos, Calif., 1974), p. 12. In this classic piece, Stone argues for legal rights for natural objects in the same sense that corporations have them (neither are "persons"). It is not too difficult therefore to wish to extend legal rights to such nonphysical entities as energy, the main purpose being to provide legal controls over the rate of its consumption and depletion and even over the purposes for which it is used.
15. B. Hannon and N. Abbott, *Dollar, Energy and Employment Impacts of Personal Consumption Activities* (Center for Advanced Computation, Document No. 128, Univ. of Illinois, Urbana, 1974).
16. P. Penner, *Dollar, Energy and Labor Cost: Electric Commuter, Intercity Truck, Intercity Bus, Motorcycle, Walking* [Technical Memos. 24 19B, 31, 30, 29, respectively (Center for Advanced Computation, Univ. of Illinois, Urbana, 1974)].
17. A. Sebald and R. Herendeen, *The Dollar, Energy and Employment Impacts of Air, Rail and Automobile Passenger Transportation* (Center for Advanced Computation, Document No. 96, Univ. of Illinois, Urbana, 1974).

18. B. Hannon and F. Puleo, *Transferring from Urban Cars to Buses: The Energy and Employment Impacts* (Center for Advanced Computation, Document No. 98, Univ. of Illinois, Urbana, 1974).
19. E. Hirst, *Energy Use for Bicycling* [Report ORNL-NSF-EP-65 (Oak Ridge National Laboratory, Oak Ridge, Tenn., 1973)].
20. R. Herendeen and A. Sebald, *The Dollar, Energy and Employment Impacts of Certain Consumer Options* (Center for Advanced Computation, Document No. 97, Univ. of Illinois, Urbana, 1974).
21. Edison Electric Institute, *Historical Statistics of the Electric Utility Industry Through 1970* (Edison Electric Institute, New York, 1970), table 45S and supplements.
22. *Moody's Industrial Manual 1974* (Moody's Investor Service, New York, 1974), vol. 1, pp. 247, 248.
23. I thank Hugh Folk, Clark Bullard, Robert Herendeen, and Lee Schipper for their helpful suggestions. Supported by the RANN Program of the National Science Foundation and the Ford Foundation's Energy Policy Project.

Effects of Federal Residential Energy Conservation Programs

Conservation programs can substantially cut energy use and save money for households.

Eric Hirst and Janet Carney

This article addresses two key issues concerning residential energy conservation strategies, their effects on energy use and on household economics. Two conclusions can be drawn. (i) Future levels of residential energy use are subject to considerable control; the analyses reported here show a range in energy use in the year 2000 of 16 to 28 quads (1 quad = 10^{15} British thermal units) (*1*). (ii) Implementing conservation programs generally saves money for consumers; for example, we estimate that the programs proposed by President Carter (*2*) will save households $27 billion between now and the year 2000.

The basis for these conclusions is a set of analyses prepared with a detailed engineering-economic model of residential energy use developed at Oak Ridge National Laboratory (ORNL) (*3, 4*). This model simulates household energy use at the national level for four fuels (electricity, gas, oil, and other), eight end uses (space heating, water heating, refrigerators, freezers, cooking, air conditioning, lighting, and other), and three housing types (single-family, multifamily, and mobile homes). Each of these 96 fuel-use components is calculated for each year of the simulation as functions of the stocks of occupied housing units and new construction, equipment ownership by fuel and end use, the thermal integrity of housing units, the average unit energy requirements for each type of equipment, and usage factors that reflect household behavior. The model also calculates annual fuel expenditures, equipment costs, and capital costs for improving the thermal integrity of new and existing structures at the same level of detail. These cost figures allow us to develop simple benefit/cost ratios for each program evaluated. Figure 1 shows the major elements of the simulation model; the structure is such that the model is sensitive to the major demographic, economic, and technological determinants of residential energy use.

Nine different residential energy "futures" are evaluated with the energy model. The high projection (case 1) assumes that "real" fuel prices remain constant from 1976 through the end of the century and that no federal conservation programs are adopted. In case 2, fuel prices are allowed to rise between 1976 and 2000. As with case 1, no government conservation programs are implemented. Changes in energy use are entirely voluntary and come about as a result of normal market forces only. Case 2 is considered our "base line," against which all the other cases are compared.

Cases 3 through 6 consider the residential conservation programs authorized by the 94th Congress (*5, 6*) and expanded upon in President Carter's April 1977 energy message (*2*). These include appliance efficiency standards, thermal performance standards for new construction, a retrofit program to affect 90 percent of the nation's housing stock, and a combination of these three programs. Cases 7 through 9 consider the energy and economic effects of additional conservation programs, including much larger fuel price increases, elimination of all market imperfections affecting residential energy use, and both of these factors.

Each program and policy is evaluated for its effects on residential energy use (by fuel, end use, and in aggregate) and on household economics (fuel bills and capital costs for equipment and structures) between 1977 and 2000. Tables 1 and 2 summarize the energy and economic effects of each of the nine cases discussed here (*7*).

High Energy-Use Projection

Inputs to the ORNL energy-use model required to develop a projection include population, fuel prices, per capita income, and specifications for government conservation programs (for example, appliance efficiency standards, tax incentives for retrofitting homes, and fuel price increases). Table 3 shows the values of population (*8*), households (*3*), housing distribution (*3*), and per capita income (*9*) used in all the projections discussed in this article. The bases for these inputs are discussed in (*4*).

For case 1, we assume that real fuel prices remain unchanged between 1976 and 2000 at their 1976 values (*10*): $2.81 per 10^6 Btu's for electricity, $1.88 per 10^6 Btu's for gas, and $2.83 per 10^6 Btu's for oil. Finally, we assume that there are no government programs that encourage households to reduce energy use. In other words, we ignore recent legislation [the Energy Policy and Conservation Act (*5*) and the Energy Conservation and Production Act (*6*)] and the Carter Administration's proposed programs.

However, the model estimates changes in new equipment and structure efficiencies and changes in usage (that is, the intensity with which households use

The authors are affiliated with the Energy Division, Oak Ridge National Laboratory, Oak Ridge, Tennessee 37830.

existing stocks of equipment) because of the fuel price increases during the early 1970's. Based on information obtained from Owens-Corning Fiberglas Corporation (*4*), we assume that 14.3 million single-family and 2.0 million multifamily units will be retrofitted during the period from 1974 through 1980. We assume that these actions (additional attic insulation, clock thermostats, caulking and weather stripping, furnace tune-up) will cut energy use for space heating 20 percent and will cost $225 for each single-family and $130 for each multifamily unit (*4*).

Outputs from the energy model, given these inputs, show residential energy use growing from 16.3 quads in 1976 to 18.3 quads in 1980, 23.6 quads in 1990, and 28.1 quads in 2000. The average growth rate during this 24-year period is 2.3 percent per year as compared with 3.6 percent per year between 1950 and 1975 (*11*).

The contribution of different fuels to the total changes during the projection period. Because of the sharp increase in petroleum prices during the early 1970's and consumer preference for gas and electricity, the fraction of household energy use accounted for by oil is expected to decline from 17 percent in 1976 to only 7 percent in 2000. Electricity's share will probably increase from 45 to 61 percent for the reasons given above and also because of growing ownership of electric air-conditioners and electric food freezers. The contribution of gas to the total is expected to decline from 34 to 31 percent during this period; the use of other fuels also will probably decline from 4 to 1 percent.

Energy use grows more slowly in case 1 than historically for several reasons. First, fuel prices are assumed constant in case 1 while historically fuel prices have declined. Second, the effects of the fuel

Table 1. Alternative residential energy-use projections: energy use.

Case	Description	Energy use (quads)				Cumulative (1977–2000)	Average annual growth rate, 1976–2000 * (%)
		1980	1985	1990	2000		
	No government program						
1	High: constant (1976) fuel prices, no government conservation programs	18.3	21.0	23.6	28.1	543.6	2.3
2	Base line: same as case 1 with rising fuel prices	17.8	19.5	21.1	24.2	493.6	1.7
	Federal conservation programs						
3	Base line plus appliance efficiency targets	17.5	19.0	20.5	23.6	482.3	1.5
4	Base line plus new construction standards	17.6	19.1	20.6	23.4	482.8	1.5
5	Base line plus retrofit program	17.1	18.3	20.0	23.1	468.9	1.4
6	Base line plus combined federal program	16.7	17.5	18.9	21.6	447.2	1.2
	Additional conservation program						
7	Combined federal program plus 50 percent fuel price increases	16.7	15.2	15.4	17.0	384.2	0.2
8	Combined federal program plus no market imperfections	16.7	17.1	18.0	19.9	428.9	0.8
9	Combined federal program, 50 percent fuel price increases, no market imperfections	16.7	14.9	14.8	15.8	371.4	−0.1

*The model's estimate of residential energy use in 1976 was 16.3 quads.

Table 2. Alternative residential energy-use projections: direct economic effects. Totals in column 5 have been rounded off.

Case	Description	Present worth of cumulative (1977–2000) expenditures at 8% real interest rate (10^9 1975 dollars)				Energy-related residential expenses in 2000 as a percentage of personal income†
		Fuels	Equipment*	Structure thermal integrity*	Total	
	No government program					
1	High: constant 1976 fuel prices, no government programs	596.2	0	0	596.2	2.5
2	Base line: same as case 1 with rising fuel prices	659.6	2.1	1.0	662.7	3.0
	Federal conservation programs					
3	Base line plus appliance efficiency targets	646.3	10.7	0.7	657.7	2.9
4	Base line plus new construction standards	647.8	2.2	5.0	655.0	2.9
5	Base line plus retrofit program	628.0	2.7	16.7	647.3	2.9
6	Base line plus combined federal program	603.4	11.2	20.7	635.3	2.7
	Additional conservation programs					
7	Combined federal program plus 50 percent fuel price increases	698.8	6.1	20.8	725.8	3.2
8	Combined federal program plus no market imperfections	585.8	20.0	22.9	628.7	2.5
9	Combined federal program, 50 percent fuel price increases, no market imperfections	680.1	15.8	23.4	719.3	3.1

*The incremental capital cost figures for equipment and structures are relative to those for the high case. For equipment, the increments include changes in both ownership and efficiences. For structures, the increments include only thermal integrity changes. †The comparable figure for 1976 was 3.1 percent.

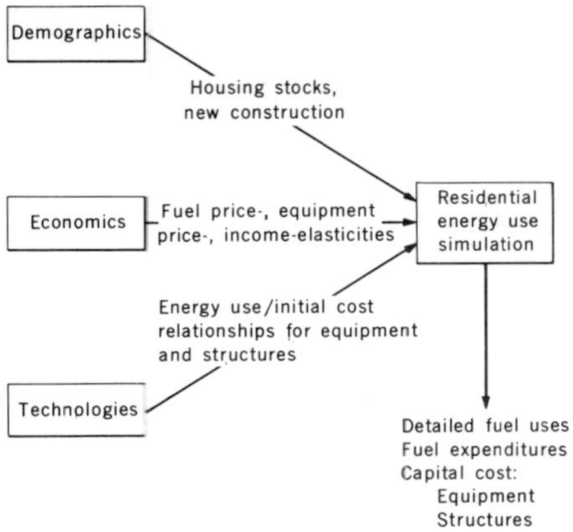

Fig. 1. Schematic of ORNL residential energy-use model.

price increases in the early 1970's are felt slowly over time and dampen energy growth in the projection period as households replace equipment and structures with systems that are more energy-efficient. A third reason relates to "saturation." Between 1950 and 1975, household ownership of air conditioners, refrigerators, freezers, heating systems, and water-heating systems increased dramatically. By 1975, almost all households had heating and water-heating equipment; more than half of all households had air-conditioning systems. Thus, the potential for increasing ownership of known energy-using systems is slight (*12*).

Base-Line Projection

Case 2 differs from case 1 only with respect to fuel prices. The fuel price trajectories used as inputs in our model for this and succeeding projections are obtained from the Federal Energy Administration (*9*) and the Brookhaven National Laboratory (*13*) energy models. These projections indicate a substantial increase in real gas prices (average annual growth of 2.3 percent between 1976 and 2000) and moderate increases in electricity (0.9 percent per year) and oil (1.2 percent per year) prices (Fig. 2). In the base line, residential energy use grows from 16.3 quads in 1976 to 17.8 quads in 1980, 21.1 quads in 1990, and 24.2 quads in 2000 with an average annual growth rate of 1.7 percent (see Fig. 3).

Changes in the distribution of energy use by fuel (case 2) are similar to those in case 1: electricity is expected to increase its share of the total and all other fuels will probably decline in importance. Because of rapidly rising gas prices, the share accounted for by gas is expected to drop to 24 percent in 2000 (compared with 31 percent in case 1) (Fig. 3).

The economic penalty associated with rising fuel prices is surprisingly mild. In 1976, fuel costs amounted to 3.1 percent of total personal income. In case 2, fuel costs amount to 3.0 percent of personal income in the year 2000. Thus, even though the average fuel bill per household increases from $570 in 1976 to $730 in 2000, growth in personal income more than compensates for fuel cost increases.

The mild economic penalty associated with the increase in the average fuel price of almost 40 percent between 1976 and 2000 is due primarily to the voluntary response of households to price increases. This response takes two forms. In the short run, households are expected to reduce their usage of existing stocks of equipment. For example, thermostats are set back during the winter. Thus, in the year 2000, households in case 2 use 10 percent less fuel for heating than they do in case 1. In the long run, as existing stocks of equipment and structures wear out, households will probably replace them with more efficient systems. For example, the average efficiency of new heating equipment in the year 2000 is about 15 percent higher in case 2 than in case 1.

Federal Conservation Programs

Here we evaluate the energy and economic effects of residential conservation programs in four elements: (i) appliance efficiency targets, (ii) thermal performance standards for new construction, (iii) weatherization (retrofit) of existing housing units, and (iv) all of the above. The time between congressional authorization and full implementation of a program can be several years. The programs discussed here were all authorized by the 94th Congress (*5*, *6*); President Carter has proposed stronger programs in each area (*2*, *9*). However, none of the programs has yet been implemented.

Each of these programs is compared to case 2 in terms of energy use and household costs. The inputs discussed earlier

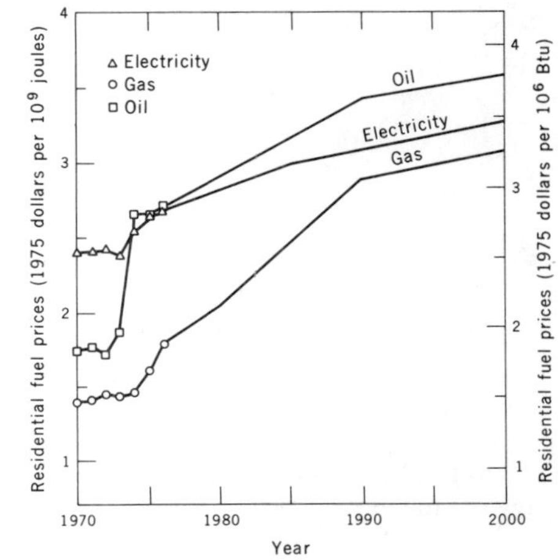

Fig. 2. Assumed fuel prices to the year 2000.

(Table 3 and Fig. 2) remain unchanged for these evaluations.

Appliance efficiency targets (case 3). The Federal Energy Administration (FEA) (now part of the Department of Energy) administers the program to develop and implement a set of appliance efficiency targets such that the average efficiency of new appliances sold in 1980 is at least 20 percent higher than the 1972 average (*5*, *6*). President Carter has proposed that the existing voluntary program be made mandatory (*2*). The FEA targets are shown in Table 4 (*9*).

Consider the efficiency targets for water heaters as an example (Figs. 4 and 5). The target calls for a 15 percent reduction in energy use for electric water heaters. Our analysis (*14*) suggests that this target could be met by replacing existing jacket insulation (2 inches of fiberglass) with 4 inches of urethane foam and adding 1 inch of fiberglass insulation to the distribution line (1 inch = 2.5 centimeters).

These measures would increase the purchase price of the water heater $42. The reduction in annual fuel bills would be $36 (at the 1976 electricity price). The extra cost of the improved electric water heater is repaid in a year.

The target for gas water heaters (20 percent reduction in energy use) could

Table 3. Inputs assumed for all projections of residential energy use (*3*, *8*, *9*).

Year	Population (10^6)	Households (10^6)	Distribution of occupied housing units (%)			Per capita income (1975 dollars)
			Single-family	Multi-family	Mobile home	
1970	205	63	69	27	3	5,420
1975	213	71	67	29	4	5,850
1976	215	72	67	29	4	6,050
1980	223	81	65	31	5	7,150
1985	234	91	63	32	5	7,970
1990	245	99	62	32	6	8,890
2000	262	114	61	33	6	10,570

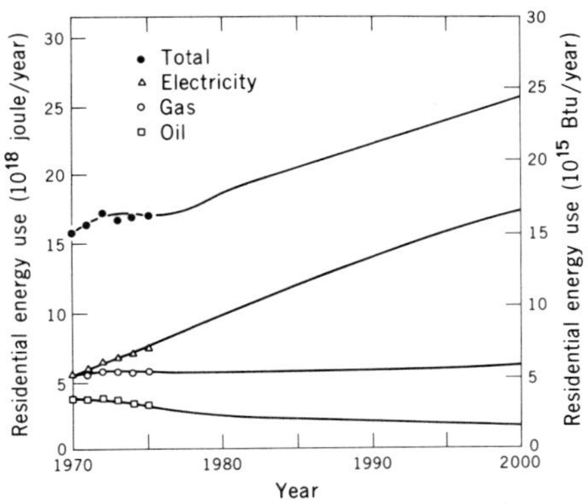

Fig. 3. Base-line projection of residential energy use to the year 2000.

be met by replacing the 1 inch of fiber-glass insulation on the jacket with 2 inches of urethane foam, adding 1 inch of insulation to the distribution line, and reducing the pilot rate. This would increase the cost of the water heater \$39. The reduction in annual gas bills would be \$13. This investment is repaid within 3 years. These examples (and our analyses of other appliances) suggest that the FEA appliance efficiency targets provide good investment opportunities for households.

Tables 1 and 2 summarize the energy and economic effects of adopting these efficiency targets (Table 4) in 1980. In evaluating this case, the model chooses either the given appliance efficiency target or the voluntary response to fuel price changes, whichever yields more efficient equipment. Thus the standards affect equipment choices only when the marketplace does not.

Energy savings increase from 0.2 quad in 1980 to 0.6 quad in 1990 and 0.7 quad in 2000. The cumulative energy saving (from 1977 through 2000) is 11.3 quads. About 85 percent of the energy savings is electricity; this is so because of shifts to electricity for heating and water heating and because the other end uses are powered only by electricity.

The reduction in fuel bills exceeds the increase in capital costs by \$5 billion. The benefit/cost ratio for the appliance program (at the assumed real interest rate of 8 percent) is 1.6. This result suggests that the proposed appliance program would save both energy for the nation and money for households.

New construction standards (case 4). The Energy Conservation and Production Act (ECPA) (*6*) requires the Department of Housing and Urban Development (HUD) to develop thermal standards for the construction of new buildings within 3 years (by 1979). These standards must then be implemented by the states, but only if Congress first approves them. The President's energy program proposed to "advance by one year, from 1981 to 1980, the effective date of the mandatory standards required for new residential and commercial buildings" (*2*, p. 8). Table 5 summarizes the likely improvements in space-heating and air-conditioning loads resulting from these standards (*15*). These standards provide larger percentage savings in multifamily units than in single-family units. This result is consistent with the changes likely from implementing the American Society of Heating, Refrigerating, and Air-Conditioning Engineers (ASHRAE) 90-75 standards or the June 1974 HUD standards (*16*). We also assume that all mobile homes constructed between 1976 and 2000 will meet the recent HUD standards (*17*).

The incremental capital cost of constructing a gas-heated single-family home in accordance with the standards shown in Table 5 is about \$500 (*3*). This figure includes the cost of labor and materials for additional insulation, storm doors, and storm windows minus the savings for smaller heating and air-conditioning equipment. The reduction in the annual heating bills would be \$90 (at the 1976 gas price), and the reduction in an-

Table 4. Assumed reductions in 1980 energy requirements for new equipment from FEA appliance efficiency targets (1970 = 1.0) (*9*).

Space heating	
Electric	1.0
Gas	0.81
Oil	0.93
Water heating	
Electric	0.85
Gas	0.80
Oil	0.81
Refrigerators	0.68
Freezers	0.77
Air conditioners	
Room	0.65
Central	0.80
Other appliances	0.90

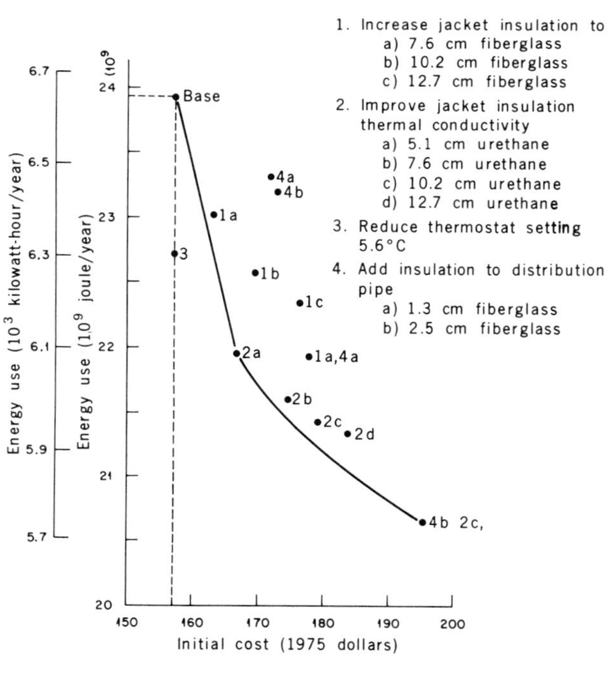

Fig. 4 (left). Energy use versus retail price for a typical electric water heater.

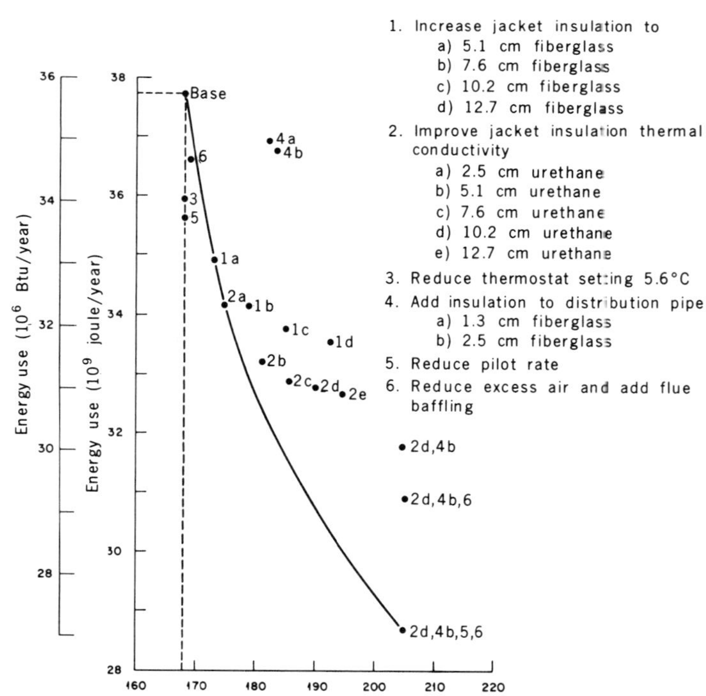

Fig. 5 (right). Energy use versus retail price for a typical gas water heater.

nual cooling bills would be \$35. Thus, the investment in tighter building construction is paid back in 4 years.

Tables 1 and 2 summarize the energy and economic effects of implementing the new construction standards of Table 5. The energy savings increase from 0.1 quad in 1980 to 0.5 quad in 1990 and 0.8 quad in 2000. The cumulative energy savings total 10.8 quads, almost as much as the savings due to the appliance efficiency program.

About 67 percent of the cumulative energy savings is electricity, 21 percent gas, and the remainder oil. Electricity accounts for so much of the savings because all air-conditioning savings are electricity and because more than 40 percent of new housing units are heated with electricity.

The economic benefits of the new construction standards are 50 percent larger than those for the appliance efficiency targets. Fuel bill reductions exceed additional construction costs by almost \$8 billion. The benefit/cost ratio for these standards is 2.9.

Retrofit program (case 5). A number of provisions in the Energy Policy and Conservation Act (EPCA) (*5*) and ECPA (*6*) encourage weatherization of existing structures. For example, ECPA authorizes the FEA to provide financial assistance to low-income households to weatherize their structures; it also authorizes HUD to conduct demonstration programs to provide financial assistance for improving the energy performance of existing buildings. The April 1977 energy message proposes a number of measures to meet the goal of "insulating 90 percent of all residences" (*2*, p. 3). These measures include tax credits for retrofits, requirements that electric and gas utilities assist their customers in weatherizing structures, increased funding for the weatherization program of low-income households, and implementation of a rural home weatherization program.

Table 5. Assumed improvements in thermal integrities for residential structures (1970 = 1.0).

Housing unit	1980 HUD	Retrofit actions*	
		Voluntary	Federal
Single-family units			
Space heating	0.70/0.60†	0.80	0.65
Air conditioning	0.71	0.84	0.70
Multifamily units			
Space heating	0.48	0.80	0.72
Air conditioning	0.58	0.84	0.78
Mobile homes‡			
Space heating	0.80		
Air conditioning	0.84		

*Voluntary retrofits are assumed to apply to 14.3 million single-family and 2.0 million multifamily units between 1974 and 1980. The federal retrofit program is assumed to apply to 42.3 million single-family and 7.3 million multifamily units between 1974 and 1984; this includes the voluntary retrofits. †The first number applies to electrically heated homes; the second number applies to homes heated with gas and oil. ‡Mobile home standards were published by HUD in December 1975 and went into effect in June 1976. We assume that these standards will remain in force unchanged through the year 2000.

Based on conversations with FEA staff (*15*), we assume the parameters for the retrofit program shown in Table 5. The retrofit costs per housing unit are \$580 for a single-family unit and \$240 for a multifamily unit. We assume that these reductions in heating and cooling demands will be implemented in 42 million single-family homes and 7 million multifamily homes by 1985.

Tables 1 and 2 summarize the energy and economic effects of retrofitting these housing units. The energy savings increase from 0.7 quad in 1980 to 1.2 quads in 1985; the savings then decline slowly through the end of the century (1.1 quads in 2000). The cumulative savings of 25 quads is double the savings from either the appliance efficiency targets or the new construction standards. The retrofit savings are large both because so many housing units are affected and because the assumed improvements are large.

The economics of the retrofit program are quite attractive. Reductions in fuel bills exceed increased capital costs by \$15 billion. The benefit/cost ratio for this program is 1.9.

Combined federal program (case 6). Tables 1 and 2 summarize the energy and economic benefits of adopting all three of the programs discussed above (see also Figs. 6 and 7). The energy savings increase from 1.1 quads in 1980 to 2.2 quads in 1990 and 2.6 quads in 2000. The cumulative saving of 46 quads is 9 percent of the base line. Energy growth between 1976 and 2000 is reduced from 1.7 to 1.2 percent per year.

The combined program reduces energy-related costs to consumers by \$27 billion. The overall benefit/cost ratio for the program is 2.0. Energy-related expenditures are reduced from 3.0 to 2.7 percent of personal income in the year 2000. Figure 7 shows that the incremental capital costs of improved equipment and structures exceed fuel bill reduction until 1982. After 1982, however, the annual fuel bill reductions exceed extra capital costs.

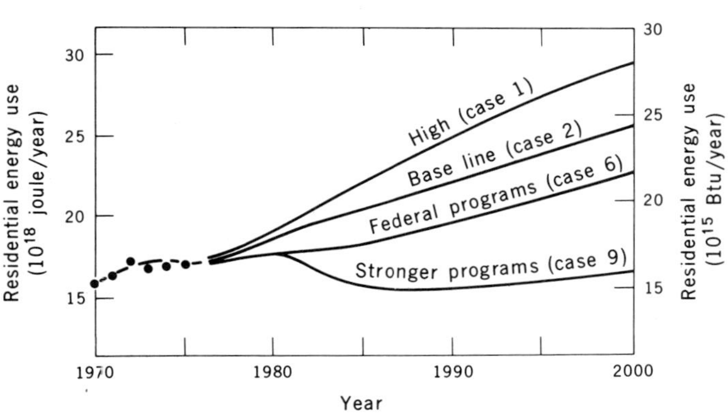

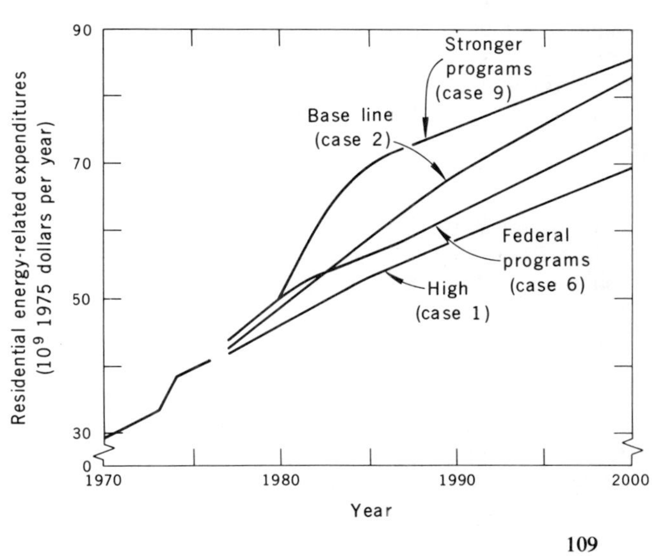

Fig. 6 (left). Alternative projections of residential energy use to the year 2000. Fig. 7 (right). Alternative projections of residential energy-related expenditures to the year 2000.

Additional Programs

Results in the preceding section suggest that the conservation programs proposed and planned by the federal government are likely to save large amounts of energy (46 quads) and money ($27 billion). Now let us examine two additional (stronger) programs for their effects on energy use and household economics.

The first program (case 7) involves large increases in fuel prices. Case 7 is the same as case 6 (combined federal program) except that fuel prices are increased by 10 percent in 1981, 20 percent in 1982, and so on until fuel prices are higher by 50 percent for the years 1985 through 2000. Such fuel price increases might occur because of increasingly strict environmental standards (for example, sulfur removal from power plants, strip-mine reclamation, increased costs for nuclear fuel reprocessing and storage), because of the increasing scarcity of gas and oil and the consequent higher costs of extraction (and the higher opportunity costs for these fuels), and because of the social costs associated with large energy production facilities. Alternatively, governments might choose to tax fuels (*18*).

Increasing the fuel prices shown in Fig. 2 reduces energy use by 2.3 quads in 1985, 3.5 quads in 1990, and 4.6 quads in 2000 relative to the "future" with the combined federal program (case 6). Growth in energy use is cut from 1.2 to 0.2 percent per year.

These higher fuel prices increase consumer expenditures on fuels and also on efficient equipment and structures. Energy-related expenditures in the year 2000 amount to 3.2 percent of personal income, compared with 3.0 percent in case 2 and 2.7 percent in case 6. The cumulative increase in expenditures (at 8 percent) is $90 billion relative to case 6 and $63 billion compared with case 2. Although the relative economic costs of the increased fuel prices are small compared with personal income, the absolute costs are large. If the additional fuel expenditures are returned to consumers through tax rebates, there will be no economic cost of the higher fuel prices. There will, however, be an income transfer from households that use large amounts of fuel to those that use small amounts.

The second program (case 8) involves greater efficiency improvements in new equipment and structures than those shown in Tables 4 and 5. The federal government's proposed standards reduce life cycle costs to consumers but they do not minimize life cycle costs. In case 8, the model selects the "optimal" [in the sense of minimum life cycle costs at consumers' implicit rates of return; see appendix tables of (*4*)] combination of equipment and structures beginning in 1980. Such changes could come about through stronger government regulatory programs, or they might occur through increased awareness and motivation on the part of consumers. At the present time, it is difficult for consumers to collect and process the information they need to make "rational" decisions concerning equipment and structure efficiency. However, government education programs, energy efficiency labels, and other information programs could provide those data and thereby encourage consumers to choose more efficient equipment and structures.

In case 8 we allow the model to select the mix of equipment and structures that minimizes life cycle costs to consumers beginning in 1980. Relative to case 6, energy savings increase from 0.4 quad in 1985 to 0.9 quad in 1990 and 1.8 quads in 2000. The average growth in energy use is cut from 1.2 to 0.8 percent per year. The energy benefits of eliminating market imperfections after 1980 are much greater when compared with case 2 (see Tables 1 and 2).

The economic benefits of either forcing or encouraging consumers to make purchase decisions at their implicit interest rates are significant. The reduction in fuel bills relative to the base line exceeds the increase in capital costs by $34 billion.

In case 9 we combine the two preceding changes. Energy use drops 10 percent between 1980 and 1985 and then increases slowly to the end of the century. Energy use in 2000 is 3 percent less than it was in 1976. Thus, case 9 yields zero energy growth in the residential sector (see Fig. 6).

Energy-related household expenses in 2000 (case 9) total $86 billion, compared with $75 billion in case 6 and $83 billion in case 2 (see Fig. 7). As a portion of personal income, case 9 costs amount to 3.1 percent in 2000, compared with 2.7 percent for case 6 and 3.0 percent for case 2. The cumulative increase in energy-related costs amounts to $57 billion compared with the base line (case 2).

Summary

We used a detailed engineering-economic simulation model of residential energy use to evaluate the effects of nine different residential energy-use "futures." These "futures" are described in terms of annual and cumulative energy use from 1977 through 2000 by fuel, end use, and in aggregate. Outputs from the model also include economic information on the costs to households of fuels, equipment, and thermal improvements to new and existing structures. The major outputs from these nine cases are shown in Tables 1 and 2 and Figs. 6 and 7.

Our conclusions are as follows:

1) Residential energy use will almost surely grow more slowly during the remainder of this century than it did in the past. Energy use grows at 2.3 percent per year in our high projection (case 1), compared with an average growth of 4.0 percent per year between 1950 and 1972. If residential energy use grew at 4.0 percent per year from 1976 through 2000, it would reach 42 quads in 2000—almost 50 percent higher than the estimate from case 1. This significant reduction in energy growth is due to the long-term effects of fuel price increases from 1970 through 1976, reductions in population growth, the absence of new energy-intensive household activities, and the near-saturation of existing household activities.

2) In case 1 we assume that real fuel prices will remain constant between 1976 and 2000. However, all projections we have seen show increases in fuel prices to the year 2000. In case 2, the base line, we use fuel prices that are roughly 40 percent higher in 2000 than in 1976. The effect of these price increases is to cut energy use 14 percent in 2000. In case 2 we assume no government programs. Changes in the energy efficiencies of equipment and structures and changes in household behavior occur only because of voluntary responses to fuel price changes. Thus, the "business as usual" response to the assumed higher fuel prices is to cut energy growth to 1.7 percent per year; energy use in 2000 is only about half of what it would be if historical trends (4.0 percent per year) continued through the end of the century.

3) Implementation of conservation programs that encourage or force manufacturers to produce and consumers to purchase more efficient equipment and structures saves energy and money. These programs (cases 3 through 6)—appliance efficiency standards, thermal performance standards for new construction, and a program to retrofit existing housing units—reduce energy growth from 1.7 percent per year (case 2) to 1.2 percent per year. The cumulative energy savings total 46 quads (69 percent electricity, 22 percent gas, 9 percent oil). In addition, these programs reduce life cycle costs to consumers of owning and

operating households. The present worth of the net benefits (at 8 percent) amounts to $27 billion.

Of the three programs, retrofitting existing homes provides the largest energy and economic benefits. This is so because we assume that all the retrofits are accomplished by 1984; most of these housing units are assumed to remain in the housing stock through the year 2000. However, many of the new units affected by the HUD standards will not be built until the 1990's; they will provide much smaller cumulative energy savings.

The energy savings due to the appliance program would be larger if standards for electric space-heating systems were imposed. Significant improvements in the efficiencies of electric systems are possible through the use of electric heat pumps. Presently available heat pumps require about 55 percent as much electricity for heating as resistance heating systems; more advanced heat pumps might require only 40 percent as much electricity.

4) Because the federal programs examined here offer benefits to society in terms of reduced energy consumption and benefits to households in terms of reduced costs and less frugal usage patterns, we examined the potential for stronger conservation programs (cases 7 through 9). Specifically, we evaluated the effects of large fuel price increases and the elimination of market imperfections in the production and consumption of residential appliances, equipment, and structures after 1980. The model results show that zero energy growth in the residential sector can be achieved with higher fuel prices. However, these fuel price increases raise costs to consumers. For example, case 9 (a combination of two additional, stronger programs) shows an increase in the energy-related expenditures for the year 2000 from 3.0 percent in the base line to 3.1 percent of personal income. The cumulative direct cost to consumers of higher fuel prices is $63 billion.

The ORNL energy model used to develop these projections contains many assumptions and limitations (*3, 4*). Some of these are particularly important with respect to interpretation of the results presented here:

1) The model contains a simple algorithm to determine the extent and pace with which manufacturers and households improve equipment and structure efficiencies in response to changes in fuel prices. Lack of both theory and data prevent us from adequately validating this portion of the energy model. If the model overestimates the market response to fuel price changes, then our estimates of the benefits of government conservation programs are too low. On the other hand, if the model underestimates market response, then the need for government intervention is less than indicated.

2) The model uses relationships between the annual energy use and capital cost of equipment (or structures). These relationships are based on engineering analysis and cost estimates from manufacturers and other sources. If we underestimate the costs of improving efficiencies (for example, Figs. 4 and 5), then we overestimate the economic benefits of government regulation, and vice versa.

3) Inputs concerning the government programs obviously affect our estimates of their effectiveness. These inputs are based on extensive discussions with FEA staff and analyses of the technical and economic feasibility of their programs. We assume that the programs are implemented in a timely, orderly, and logical fashion. If the programs are delayed, are weaker than specified in Tables 4 and 5, or both, then the energy and economic benefits will be less than estimated.

4) Many government policies and programs in effect, under development, and under consideration were not analyzed. These include consumer information programs, energy-efficiency labels for household equipment and structures, elimination of master-metering in apartment houses, changes in electricity and gas rate structures, and government R & D programs to develop new energy-efficient systems. Some of these may be even more effective at saving energy and money than the programs examined here.

Despite these caveats, it seems clear (to us at least) that government residential energy conservation programs can substantially reduce energy use and household expenditures. Programs stronger than those in the National Energy Plan (*2*)—especially programs to improve equipment and structure efficiencies in the last 1980's and 1990's—are likely to yield even larger energy and economic benefits.

References and Notes

1. Quantities are given in British units: 1 quad = 10^{15} Btu's; 1 Btu = 1055 joules. Electricity use figures are in terms of primary energy (11,500 Btu's per kilowatt-hour); that is, they include losses in generation, transmission, and distribution. Figures for gas and oil do not include losses associated with refining and transportation.
2. The White House, *The President's Energy Program* (20 April 1977); Executive Office of the President, *The National Energy Plan*, 29 April 1977 (Government Printing Office, Washington, D.C., 1977).
3. E. Hirst, J. Cope, S. Cohn, W. Lin, R. Hoskins, *An Improved Engineering-Economic Model of Residential Energy Use* (Publ. ORNL/CON-8, Oak Ridge National Laboratory, Oak Ridge, Tenn., April 1977); E. Hirst, W. Lin, J. Cope, *Q. Rev. Econ. Bus.* **17** (No. 2), 7 (1977).
4. E. Hirst and J. Carney, *Residential Energy Use to the Year 2000: Conservation and Economics* (Publ. ORNL/CON-13, Oak Ridge National Laboratory, Oak Ridge, Tenn., September 1977).
5. 94th Congress, *Energy Policy and Conservation Act*, PL 94-163, 22 December 1975.
6. ———, *Energy Conservation and Production Act*, PL 94-385, 14 August 1976.
7. Although the results of our projections are given here as certainties, there is considerable uncertainty about what will happen by the year 2000. Part of the uncertainty relates to the exogenous variables such as fuel prices and population. The rest relates to the model itself, both its structural form and the particular coefficients used in the model equations.
8. Bureau of the Census, "Projections of the population of the United States: 1975–2050" (Current Population Reports, Series P-25, No. 601, U.S. Department of Commerce, Washington, D.C., October 1975).
9. J. Carlin, personal communication.
10. All prices and incomes are in terms of constant 1975 dollars. Electricity is in terms of primary energy (*1*).
11. E. Hirst and J. Jackson, *Energy* **2** (No. 2), 131 (June 1977); Bureau of Mines, "Annual U.S. energy use up in 1976" (press release, U.S. Department of the Interior, Washington, D.C., 14 March 1977).
12. Our energy model does not explicitly allow for the introduction of new end uses (for example, sidewalk deicing, inside air filtration, swimming-pool heating). However, the model does include an "other" end use, and this is allowed to grow as incomes rise (depending on the growth of fuel prices). In the high projection, energy use for other purposes increases from 1.4 to 2.6 quads between 1976 and 2000.
13. S. Carhart, personal communication.
14. R. Hoskins and E. Hirst, *Energy and Cost Analysis of Residential Water Heaters* (Publ. ORNL/CON-10, Oak Ridge National Laboratory, Oak Ridge, Tenn., June 1977).
15. D. Quigley, personal communication.
16. *An Impact Assessment of ASHRAE 90-75, Energy Conservation in New Building Design* (A. D. Little, Inc., Cambridge, Mass., December 1975); *An Energy and Economic Assessment of HUD's Minimum Property Standards* (A. D. Little, Inc., Cambridge, Mass., October 1976).
17. *Fed. Regist.* **40** (No. 244), 58752 (18 December 1975).
18. The National Energy Plan (*2*) proposes to price fuels at their "replacement cost" so that fuel prices reflect the costs of providing additional units of that fuel. Under this proposal, natural gas, for example, might be priced at the cost of importing liquefied natural gas or at the cost of producing synthetic gas from coal.

19. Work supported by the Department of Energy under contract with the Union Carbide Corporation.

III

Future Supply in Evolution

Ocean Boundaries and Petroleum Resources

National-international ocean boundaries are discussed in relation to petroleum resources.

Hollis D. Hedberg

During the last decades, the focus of world interest and activity in exploration for petroleum has been moving more and more into the oceans. In the highly competitive petroleum industry, rewards have traditionally gone to those who were willing and able to be the first to brave the hazards of strange and forbidding environments in order to preempt the choice petroleum areas they might contain. Thus, exploration has advanced through the mountains, the deserts, the jungles, the swamps, and the arctic tundras on land, and is already widespread in the coastal waters of the seas and oceans.

The entry to each new environment has been made possible only as specialized equipment and technology could be developed to meet the special problems of each. The complexities of resource extraction offshore once seemed completely prohibitive, but the development of marine geophysical prospecting and various types of marine drilling installations suited to various water depths has now made exploration and production in shallow and moderately deep offshore waters of the continental shelf entirely feasible and, in some respects, less difficult than in some land areas. Currently, there are hundreds of offshore fields producing oil and gas in the world; 17 percent of U.S. production and 18 percent of world production in 1973 came from these offshore sources (*1*).

Progress into deeper waters of the outer shelf, the slope, and beyond still presents formidable obstacles. However, exploration holes for petroleum now have been drilled in water depths of as much as 655 meters (2150 feet), and production has been established in water depths of more than 125 meters (about 400 feet). The latest record of the Joint Oceanographic Institutions for Deep Earth Sampling Deep Sea Drilling Program is a hole penetrating 1411 meters (4656 feet) below the ocean floor in a water depth of 4975 meters (16,417 feet).

Costs of offshore petroleum drilling and production operations are extremely high, and in waters as deep as 1000 meters or more would be fantastically so. However, there is now little doubt that drilling and production at these and greater depths can and will be accomplished eventually if the prospective returns are great enough.

Political-Economic Setting

Apart from matters of technology, two factors during the last few years have drastically altered the situation with respect to petroleum from the deep offshore waters. (i) The recent multifold increase in the price of crude oil on world markets and the recent emphasis on an inevitable future shortage of supplies of hydrocarbon energy have escalated the values attributed to potential offshore petroleum properties. (ii) The wave of nationalization of the petroleum industry, together with the increased prices of crude oil and the energy crisis, has stimulated a keener sense of national proprietorship in the ocean beds by those countries with ocean coastlines. In contrast with this latter point has been the appeal so eloquently voiced by U.N. Ambassador Pardo from Malta in 1967 and endorsed in word if not in spirit by many others since—the doctrine that the oceans are the heritage of all mankind and that the resources of the deep-ocean floor belong to the community of nations rather than simply to those nations who happen to control coastlines adjacent to them.

National-International Boundary

The result of the interplay of these factors has been the impasse at meetings of the U.N. Committee on the Peaceful Uses of the Seabed and the Ocean Floor Beyond the Limits of National Jurisdiction, starting in New York in 1968 and most recently held in Geneva during the spring of 1975, on the question of to whom the ocean floors belong. Another meeting is scheduled for March 1976 in New York but a solution satisfactory to the various parties concerned is dubious.

Nearly everyone recognizes that there should be an outer limit to the areal extent of a coastal state's jurisdiction over the ocean floor extending outward from its shores. Also, it is generally agreed that the great central regions of the deep ocean floor should be under international jurisdiction. The big problem is just how far out the boundary between national and international domains should be drawn.

It is critically important that the position of this boundary be decided promptly. It is equally important that it be decided wisely. Once petroleum exploration drilling has progressed out to water depths which may bring it close to this boundary, and the position of particularly attractive areas for petroleum development begins to be known, the problem of attaining any boundary agreement based on principle rather than on self-interest, greed, or military power will have become infinitely more difficult.

Various boundaries based on depth of water or distance from shore, or a combination of the two, have been proposed but none have been adopted. At the moment, the proposal which is said to have the most support is that of a national-inter-

The author is professor emeritus of geology at Princeton University, Princeton, New Jersey 08540.

national boundary 200 nautical miles (1 nautical mile = 1852 meters) from shore. However, for subocean resources this proposal lacks any basis in nature or logic; it is largely an inheritance from the fishing claims of certain South American countries. As applied to petroleum (and other mineral resources) such a boundary would be unacceptable to many countries who would lose much potentially valuable territory naturally pertaining to them, whereas others would needlessly be given huge tracts of ultra-deep ocean bed which would more appropriately have been assigned to an international oceanic regime.

Moreover, although the 200-mile boundary superficially has a simple and definite sound, there appears to be little realization of the many problems involved in its practical implementation. One of these is the lack of any uniform rule among nations for determining the coastal base lines from which the 200-mile distance would be measured. Another is the unsettled question of whether the 200 miles should be measured from the mainland coast only, or whether any or all of the myriad reefs, islets, and islands of the continental shelves, far from shore but belonging to coastal nations, might also be used to extend their 200-mile limits farther seaward. Furthermore, probably few have considered how difficult accurate definition on the ocean floor of 200-mile limits measured off from an irregular coastline might be, although the time might come when such precise and detailed definition of the national-international boundary would be of the greatest economic and political consequence.

Requirements for Boundary

Any plan for a boundary between national and international jurisdiction that could, in the long run, prove acceptable to the nations of the world would be so in proportion as it satisfied the following conditions.

1) It should be a plan with a logical or natural reason behind it rather than one that is purely artificial.

2) As a minimum, it should give to each nation what appears to be the natural prolongation of its land mass—continental or insular—beneath the oceans.

3) It should allow each coastal nation a substantial maritime zone of ocean bottom adjacent to its shores, regardless of the narrowness of its shelf or the steepness of its slope.

4) It should be a plan which it would be equitable to impose uniformly on all nations without exception or special modification for any.

5) It should be a plan that provides for the boundary problems of islands, whether on the shelf or in the deep sea and whether isolated or in archipelagos.

6) It is desirable that it allow each coastal nation to be represented in the fixing of its own precise boundary, which should be defined by coordinates of latitude and longitude within internationally agreed guidelines.

7) It should be a plan which can accomplish uniformly for all nations whatever turns out to be the agreed wishes of the nations with respect to a national-international division of the ocean bottom out beyond the minimum coastal state limits.

8) Finally, it should be a plan which, while recognizing the natural oceanward extent of the coastal state domain, at the same time leaves some substantial part of the potentially valuable resources of the ocean bottom to the international domain.

Continental Margin Boundary Guide

In contrast to the many artificial distance boundaries and the ever more artificial and impractical water-depth boundaries that have been proposed, there is one obvious, natural, and logical guide to a division between national and international jurisdiction over the ocean floor. This is based on the difference between the topographically high-standing continental and island blocks of the earth's surface—the traditional domain of mankind and the nations—and the topographically low-lying deep-ocean floor which has been conventionally international territory (*2*).

The line of division between the two is at the continental margin (or insular margin), and the boundary problem should have been approached from the very beginning with the recognition that the margin is the natural reference base or guide for any boundary between coastal state and international jurisdiction.

The term "continental margin," as used here, is simply the now-submerged *edge* of the continent, as the word margin itself implies. Its position and breadth depend entirely on how sharply one defines that edge.

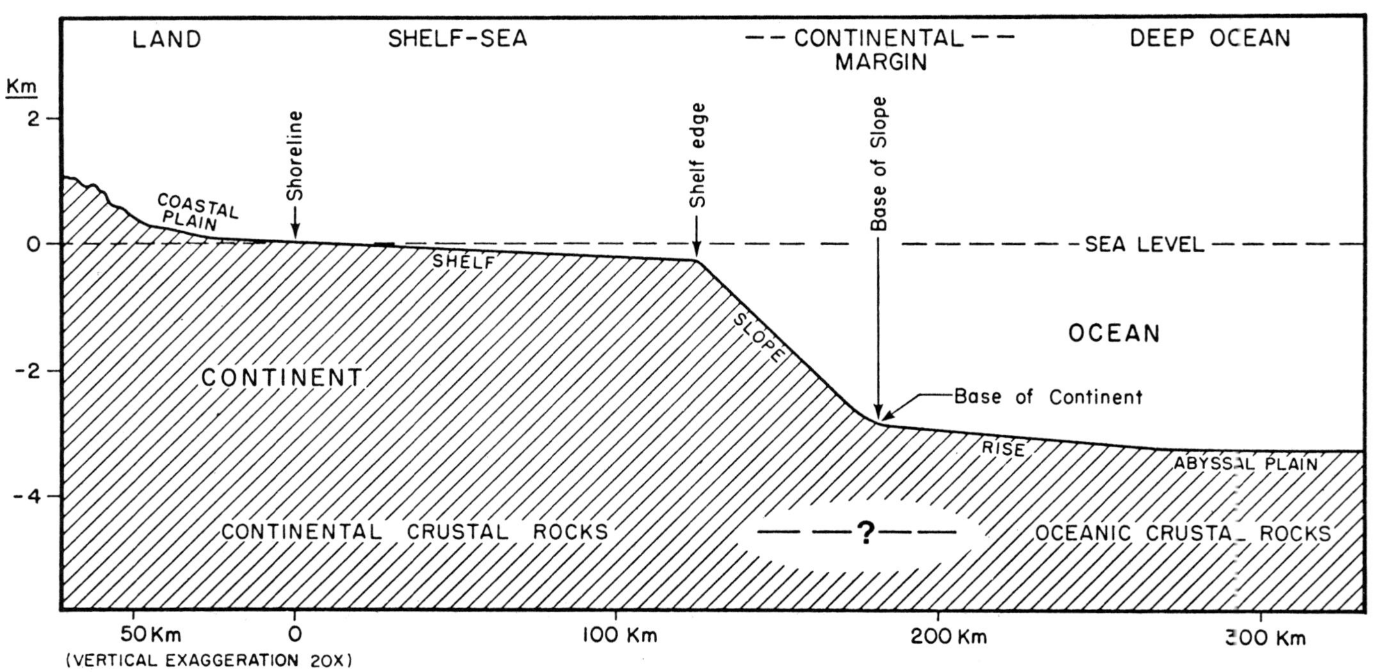

Fig. 1. Geomorphic features of the continental margin. [Drawn originally for (*1*)]

The edge may be considered to be the frontal face of the submerged continental mass (or insular mass) as it faces the deep oceans; and as such it is the continental slope (or the insular slope). Most exactly, however, it is the *base* of that slope that marks the outer limit or base of the continent (or island). (The true inclination of the margin features is far more gentle than the exaggerated scale in Fig. 1 might suggest.)

The continental margin as thus defined coincides in a very general way with a geological change from a relatively light but thick continental crust to a relatively dense but thin oceanic crust—a change that is probably fundamentally responsible for the difference in elevation between the continents and the deep-ocean floors. The continental masses of relatively low-density rock tend to float high in the heavier oceanic crust. However, this geologic change, although scientifically important, is too ill-defined and uncertain in position to be of much help in drawing a political boundary.

The most practical basis for the definition of the outer limits of continent and island masses is not geologic but geomorphic. It is based on the simply and directly observable surface form of these earth features and is readily identifiable by anyone at least in a general way. This outer geomorphic limit of the continents and islands—the boundary separating these high-standing masses from the intervening vast and deep ocean basins—is the most natural and most logical guide to where the political boundary between national and international jurisdiction over the resources beneath the ocean floor should lie.

Base-of-Slope Boundary Zone Plan

The only boundary scheme that appears to meet closely the eight requirements listed above is one which calls for the base of the continental slope (or the insular slope) to serve as a general guide to an internationally prescribed boundary zone, within which the exact position of the boundary between national and international jurisdiction over mineral resources beneath the ocean floor would be designated by the coastal state itself.

The continental slope is probably the single most impressive and most extensive feature of the earth's surface, with a linear extent of 300,000 kilometers on the ocean floor and a height of as much as several kilometers. The slopes of islands are almost equally impressive. However, although the base of the slope is a remarkably widespread and distinctive feature well suited to serve as a general guide to a boundary, it is not definable sharply enough to serve as the boundary itself. Hence an essential feature of any proposed plan that uses the base of slope as a guide is the creation of a boundary zone that would extend oceanward from the approximate position of the base of the slope for an internationally agreed distance within which the precise boundary would be drawn by the coastal state itself (Figs. 2 and 3). It would seem impracticable for the standard width of the boundary zone to be less than 100 kilometers (54 nautical miles), although it might be as much wider as the consensus of nations considered desirable.

The principal reasons for the boundary zone are to take care of uncertainties in precise identification of the base of the slope, as well as to guarantee to all coastal nations as a minimum the submerged part of the continental or island mass adjacent to their shores and naturally pertaining to them and to allow them uniformly such additional area as the nations in joint consultation might agree to be desirable. An additional advantage of the boundary zone scheme is that it would allow the final designation of the precise boundary to be made by the coastal country itself, although within internationally agreed

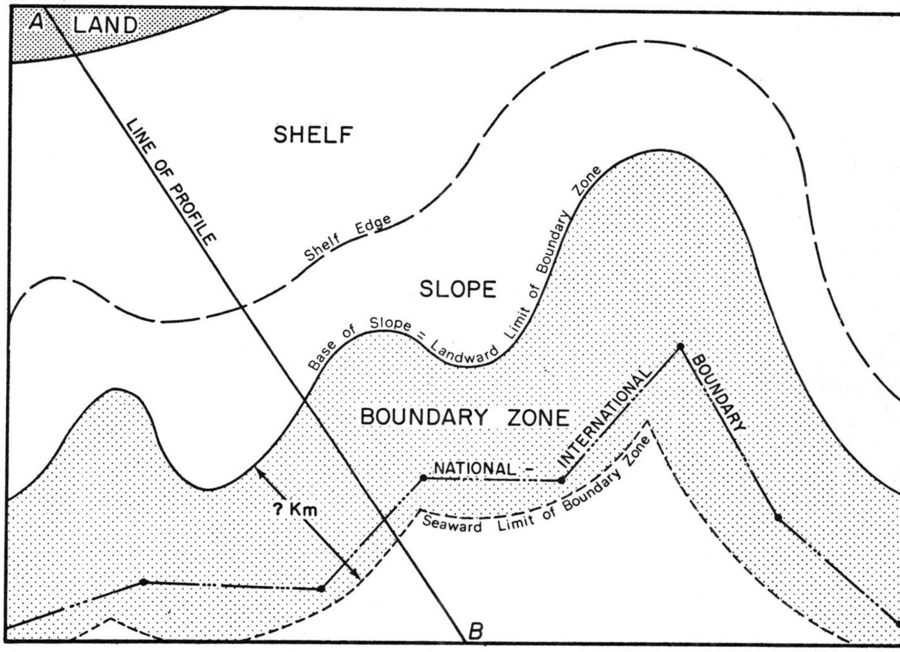

Fig. 2. Base of slope, boundary zone, and national-international boundary. [Drawn originally for (*1*)]

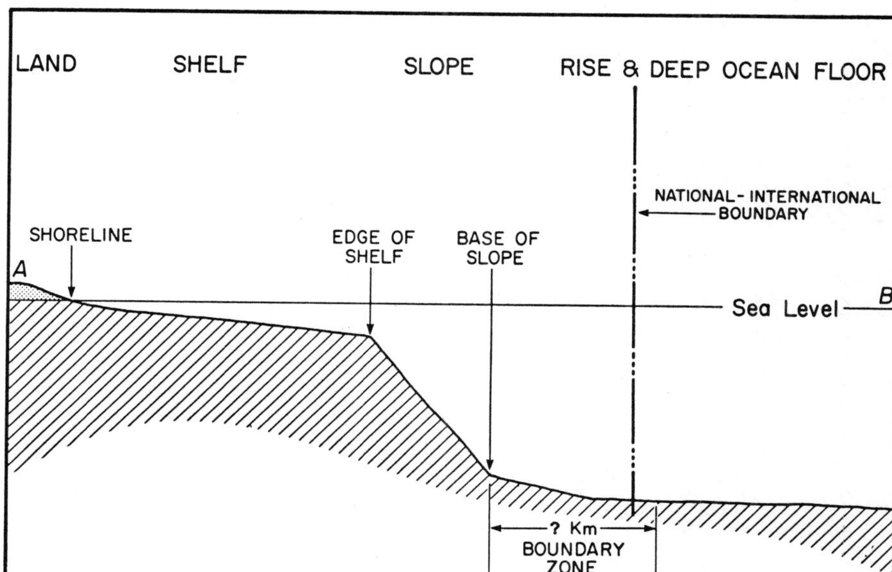

Fig. 3. Profile across area of Fig. 2. [Drawn originally for (*1*)]

boundary zone limits. Finally, the boundary zone concept carries the added advantage of allowing the final boundary to be drawn by simple straight lines connecting fixed points of latitude and longitude, which makes it a practical, precise, and readily determined boundary—a feature quite lacking in a 200-mile-from-shore boundary.

As must be the case with respect to the application of any boundary formula, the establishment of an international boundary commission with qualified technical personnel would be necessary to make certain that the precise boundaries drawn by the coastal states fell within the internationally prescribed limits. This body, working in conjunction with individual advisory national oceanographic committees, would determine in advance the approximate position of the base of the slope (or its reasonable projection through areas of uncertainty) for each coastal state. Based on this approximate base-of-slope line and the standard distance adopted by the nations-in-concert for the width of the boundary zone, the coastal state would itself proceed to draw its own precise boundary. This would then be submitted to the international boundary commission for approval, for linking with the boundaries proposed by adjacent countries, and for recommendation for acceptance by the nations-in-concert.

This, in essence, is what seems to be the simplest, most equitable, and most practical procedure for choosing and defining boundaries between coastal state and international jurisdiction over the mineral resources beneath the ocean floor. Details of the scheme have been explained elsewhere (*2*). Some years ago, using the best bathymetric maps available to me, I tentatively tried out the feasibility of drawing an approximate base-of-slope line around most of the continental and insular areas of the world, and believe that it could be done quite satisfactorily by a group of experienced oceanographers working for the international boundary commission. Some examples showing the approximate position of the base-of-slope line in various parts of the world, boundary zones of alternative widths of 100, 200, and 300 kilometers, and supporting profiles were published in 1973 in order to illustrate the method and its feasibility (*2*, attachments 1–8). Further experience and information indicate that some of these initial attempts require modification.

It should be stressed that the proposed plan is recommended only for a national-international division of jurisdiction for *mineral resources* beneath the oceans. It is not a plan for a division of jurisdiction over oceanic waters for navigational purposes, nor is it necessarily a plan for division of fishing rights. These purposes involve other factors and should be settled accordingly. It might not even be a plan for minerals *on* the ocean floor (such as manganese nodules), although it could be. It is rather a plan that concerns only mineral resources *beneath* the ocean floor requiring drilling or mining and fixed ocean floor installations for their recovery—principally petroleum resources.

Defects of 200-Mile Boundary Scheme

Some of the practical difficulties in application of a 200-mile-from-shore boundary formula have been mentioned above. However, in view of the publicity that has been given to the 200-mile proposal, it may be well to consider also some of its more fundamental defects.

Just as there is no logical or natural basis for an arbitrary 200-mile-from-shore limit, so also the distribution of mineral resources is much less related to the very superficial feature of the present shoreline than to the more fundamental base of slope, or base of continent. The world's great thicknesses of sediments with promising petroleum prospects are more closely related to the base of the slope than to some fixed distance from shoreline.

Adoption of a 200-mile-from-shore formula would cause many "wide-margin" countries to lose large potential petroleum-bearing areas situated on the natural seaward projection of their lands. These would be countries whose base-of-slope line in places is more than 200 miles from shore, such as Argentina, Australia, Canada, People's Republic of China, Ireland, Great Britain, the United States, and the Soviet Union. At the same time, many countries with narrow shelves and slopes would by the 200-mile formula be given vast tracts of deep-ocean bottom of little or no value to them and in no way an apparent prolongation of their lands beneath the sea. Some examples are Chile, Peru, Malagasy.

In contrast, the base-of-slope formula would assure to all countries all adjacent ocean floor naturally pertaining to them, and only so much more as the nations agreed to allot uniformly to all coastal states. If the extent given by the 200-mile formula were considered a desirable minimum, this could be attained by making the standard width of the boundary zone great enough (for example, 300 kilometers, or 162 nautical miles) to always allow the drawing of a boundary at least 200 miles from shore. Moreover, this could be done without robbing the wide-margin countries of their natural rights or without the complication of having to establish different formulas for wide-margin countries than for narrow-margin countries. Conversely, the international oceanic domain could be adjusted readily to whatever size the nations-in-concert might wish, simply through the standard width assigned to the boundary zone.

The base-of-slope formula, as compared with the 200-mile formula, provides a much readier means of expressing the wishes of the nations-in-concert and relating them to essential issues, foremost of which is the wish of each nation to be assured of the sub-ocean bottom minerals that it feels naturally belong to it.

Petroleum Prospects Beneath Oceans

Critical to any wise decision on jurisdiction over the mineral resources beneath the ocean floor is a proper understanding of the nature of their occurrence. Aside from manganese nodules, phosphate deposits, and other surface deposits on the ocean floor, it seems clear that petroleum is by far the most important sub-ocean bottom resource to be considered. It also seems clear that thoughts of petroleum resources have been a dominating factor in the attitude of the nations with respect to proposed boundaries between national and international jurisdiction.

We know from the geology of existing oil-field areas that marine geological environments of the type that we can associate with the continental margins constitute some of the world's most favorable settings for the genesis and accumulation of petroleum. This is in part because of the abundant supply of organic matter there, which is derived from both terrestrial and marine sources by drainage from the land and from marine life fed by upwelling, nutrient-rich waters from the ocean depths. In part, it is because of the restricted bottom circulation and rapid sedimentation rate of many of the margin seas which have been conducive to the preservation of organic matter. And in part, it is because the extraordinary crustal mobility of these margin belts adjacent to the continents has allowed the thick accumulation of a variety of sediments, the gradual cooking of the organic matter under sedimentary overburden, the collection of the generated hydrocarbons in reservoir beds, and the development of abundant trapping features through folding, faulting, unconformities, and stratigraphic changes.

Although a very substantial part of the world's petroleum is already coming from accumulations beneath the ocean floor, almost all is from the landward, relatively shallow waters near shore on the continental shelves. We still know little or nothing directly from actual drilling for petroleum about the prospects of strata beneath the deeper waters adjacent to the margin—those of the continental slope and beyond the base of the slope.

Such information as we have suggests that much of this deeper water area associated with the margins may also have a good petroleum potential. Geophysical surveys and the very limited amount of deep-sea drilling and sampling done for scientific purposes indicate that under much of this deep water lie sequences of sediments deposited, not under bathyal conditions as might offhand have been expected, but under all the favorable shallow-water sedimentary and structural environments of the sediments of the known oil-field areas of the present coastal plains and continental shelves.

However, about as far as we can justifiably go in appraising the prospects of this deep-ocean floor is to recognize that in general it has petroleum potential and that much of it is worthy of drilling exploration for petroleum when the need and incentive are great enough to compensate for the tremendous costs involved in operations in deep, open oceanic waters far from shore bases. Furthermore, in considering the prospects of entirely undrilled ocean areas, it should be recognized that they are truly unknowns. At this stage we can only speculate as to how well they may meet the geological requirements that experience and the progress of the science have shown are necessary for the creation of commercial petroleum accumulations.

The principal requirements for the development of a geological setting favorable to petroleum prospects may be summarized briefly as follows.

1) There must be a rich source of organic matter of the right kind.

2) Conditions must be favorable for the preservation of the organic matter until it can be buried by sediments.

3) Conditions must be favorable for the thermochemical conversion of the organic matter in the sediments to fluid petroleum (oil or gas). This requires an adequate thickness of blanketing sediments (1000 meters or more) so that, with the local geothermal gradient, favorable generating temperatures (50° to 150°C) may have been attained in the buried sediments, but not exceeded to such point as would have destroyed the fluid petroleum generated.

4) Conditions must be favorable for the movement of petroleum from source rocks to carrier beds. This presupposes the presence, in or adjacent to the fine-grained, organic-rich source rock sediments, of porous and permeable rock layers into which the newly generated petroleum can migrate in response to compaction pressure and other expulsive forces, and along which it can move to accumulation centers.

5) Accumulation traps must be present. These are structural or stratigraphic features where petroleum moving along carrier beds might be concentrated to form commercial accumulations.

6) There must be impermeable cover rocks to prevent the escape of petroleum and its dissipation at the surface. Thick shales and evaporites in the section are commonly effective cover rocks.

7) Proper timing in the development of the various conditions mentioned above is necessary; for example, traps formed after the petroleum has departed are of course worthless.

The more nearly an area promises to qualify with respect to the factors listed above, the better are its prospects. However, failure in any one factor may nullify promise in all the others, and until extensive drilling is done in an area it is rarely possible to evaluate all of the critical factors with assurance.

In some ways it is much easier to eliminate areas of little or no prospects than it is to determine how good are those with some prospects. One factor stands out as essential to the fulfillment of almost every one of the requirements listed, and that is the presence of an adequate thickness of sediments. Of course, mere thickness of sediments does not necessarily mean favorable petroleum prospects, but the lack of an adequate thickness to give temperatures sufficient to convert organic matter into petroleum, or to give pressures necessary to have caused its migration to accumulation centers, is sufficient in itself to condemn the prospects of many areas.

Geophysical surveys have already effectively eliminated vast areas of the ocean floor from serious consideration for petroleum by showing that they have only a few hundred meters of sediment. They have also shown that particularly thick sections of sediments commonly exist along the margins of the continents (and in associated seas and small ocean basins) and in some regions extend far out beyond the base of the slope. A thickness of about 1000 meters has commonly been used to separate areas of no promise from those that may have some promise, depending on the character of the rocks. Figure 4 shows estimated thicknesses of sediments above basement in the Atlantic Ocean as interpreted from seismic surveys (*3*).

Unfortunately, there is no reliable answer to the demand for quantitative estimates of the magnitude of petroleum resources in unknown and undrilled areas such as those beneath the deep oceans. Many persons have yielded to the pleas of the public, governments, or the United Nations and have come out with figures. This is perhaps good, because there is a variety in these estimates which tells its own story and because, taken as a whole and in application to large enough regions, such estimates probably do give a worthwhile order of magnitude answer. However, the truth is that the estimators do not know, nor does anyone know before drilling. Quantitative estimates of petroleum resources in new undrilled areas should probably be prefaced by a zero so that a true impression of the estimator's justified assurance may be given (*4*).

Relation of Boundary to Petroleum

The margins of continents and islands appear to have constituted a generally favorable environment for petroleum accumulation. Abundant petroleum production already comes from the landward edge of the margin—the near-shore continental shelf—and even most of the production from petroleum fields on land comes from uplifted sediments which were once deposited in close association with the continental margins of past geologic time.

A dividing line between national and international jurisdiction drawn at today's base of continental slope would leave a large share of this promising belt of potential petroleum-bearing sediments on either side. The landward side of course has the better-known potential, is in shallower water, and is more accessible. However, some of the thickest sedimentary deposits known lie just seaward of the base-of-slope line and constitute the so-called continental rises—huge sediment-filled sumps which, although somewhat questionable as regards adequate reservoirs and traps, must certainly have been good generators of petroleum. The oceanward side of the base of slope may thus have a high petroleum potential although it has the disadvantages of its remoteness from land and the great depths of water in which any operations would have to be carried out.

However, it is not the base of the slope itself but a boundary zone adjacent on its oceanward side that is recommended as the site for the national-international boundary. Thus, the width adopted as standard for this boundary zone would be of critical importance. A boundary drawn near the outer edge of a boundary zone of minimum width (100 kilometers) would still leave much of the areas of thick sediment in the international zone, although in very deep water. On the other hand, the nations-in-concert might decide that a wider boundary zone was desirable—200 or 300 kilometers, perhaps—expressly in order to give the coastal states more of the possible petroleum resources of the so-called rises, or perhaps in order to give complete continuity to the sea floor areas between the islands of archipelagic countries like Indonesia, Philippines, or Mauritius, or to completely divide up the very promising seas and small ocean basins, such as the Gulf of Mexico, the Bering Sea, or the Black Sea, among the adjacent bordering states.

Obviously, any increases in the width of the boundary zone would leave less and less to the international regime and give more and more to the coastal states. However, even with a boundary 300 kilometers oceanward from the base of the slope there would still be some large ocean floor areas with more than 1 kilometer of sediment thickness left in the international zone (for example, off of Argentina, Canada, India, the United States, and the Soviet Union,

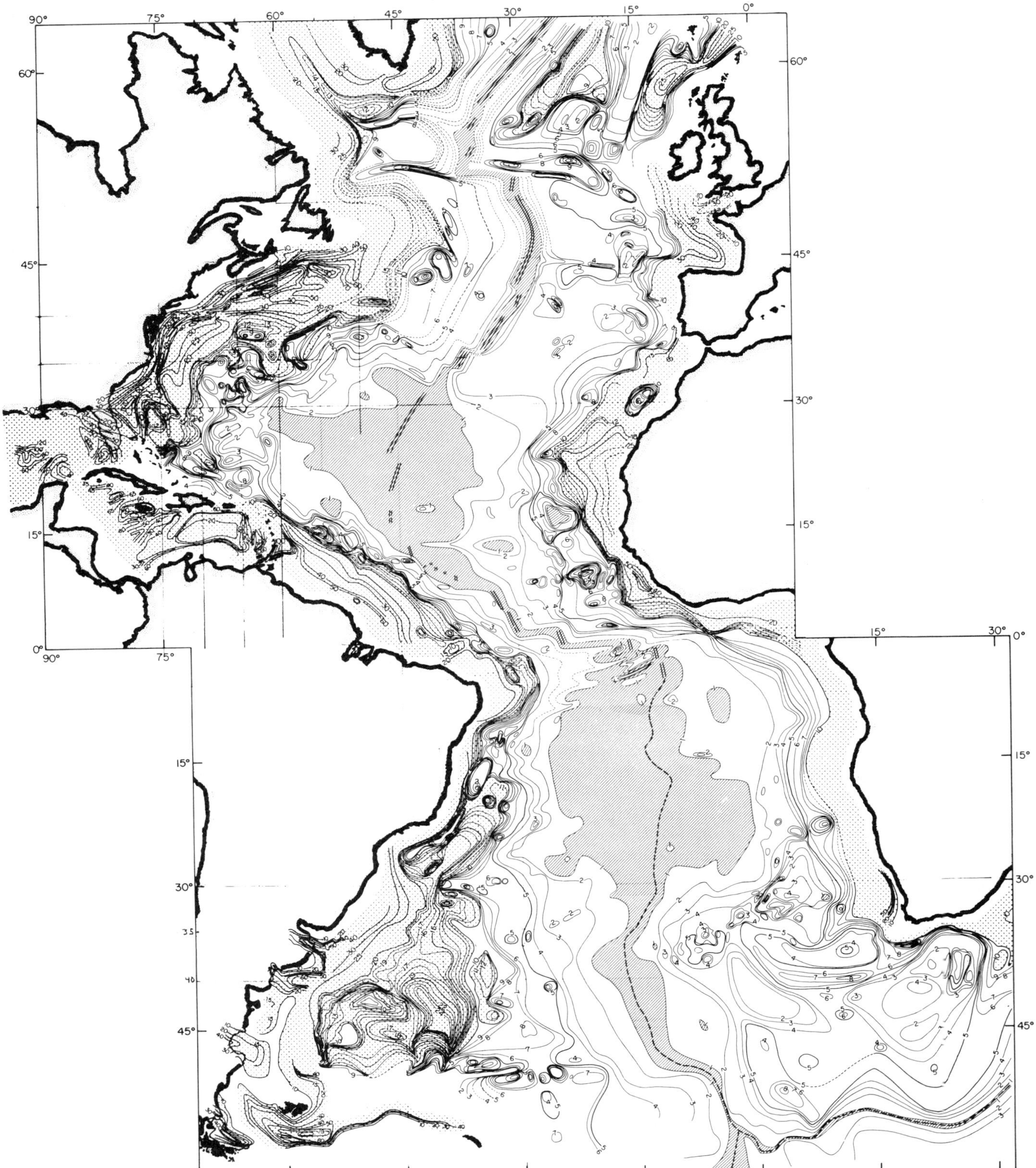

Fig. 4. Thicknesses of sediment above basement in Atlantic Ocean (*3*). Diagonal hachuring indicates areas of less than 100 meters of sediment. Dotted pattern indicates areas of more than 1000 meters of sediment. Numbered contour lines show thicknesses of sediment in hundreds of meters.

among others). The boundary zone concept leaves a flexibility in decision to the nations-in-concert as to how much of the world's prospective petroleum territory they wish to make international, while at the same time guaranteeing by the base-of-slope guideline that the coastal nations shall have at a minimum all that naturally belongs to them.

I am optimistic about the eventual petroleum production prospects of the outer continental margins, but certainly with a boundary beyond the base of slope the ocean territory that would go to the international regime would have little prospect of early petroleum development regardless of how good its ultimate potential might be.

The problems of drilling and development in the extreme water depths and remote locations of the outer margin may be so great and the costs so overwhelming that many of these areas may never even be given a chance to yield actual production. Certainly, only extremely lush reservoirs in this environment could be economic in the near future. And certainly, any areas beyond the base of the slope could not be a great source of international revenue in the next few years, as many seem to have hoped.

Moreover, the plan, once envisioned by some, to immediately benefit the developing countries through the proceeds of petroleum production beyond a national-international ocean floor boundary can no longer practically be carried out, because firm coastal state claims have already preempted the sea bottom out beyond any water depths that will be attractive for petroleum development for many years to come.

If it is the serious will of the nations to assign some part of the world's sea floor resources for the benefit of developing nations, a much more expeditious and practical procedure would be for the nations-in-concert to agree to dedicate to this purpose a certain percentage of coastal state government revenue from all ocean installations actually producing petroleum from water depths greater than a specified number of meters or at distances from shore greater than a specified number of kilometers. Accurate determination of such depths or distances could readily be carried out for each producing installation. International revenue from such sources, however, should of course come entirely from the coastal government's share and not from private entrepreneurs, because the development of even moderately deep and distant water areas will require maximum financial incentives, not additional tax burdens, in order to stimulate interest and compensate for the huge financial risks involved.

Proposed Boundary off Eastern United States

Figure 5 is a bathymetric map of the ocean floor off the eastern (Atlantic) coast of the United States, including adjacent offshores of Bahamas, Bermuda, and Canada (*5*). The base-of-slope boundary guide and alternative zones, as well as the 200-mile-from-shore line, are shown. The base-of-slope line does not coincide with the conventional boundary between slope and rise provinces off eastern United States but has been drawn farther offshore to make it correspond approximately with the outermost principal flattening of the seaward gradient of the ocean floor at the continental edge. More detailed or revised bathymetry might change its position somewhat, but to have used the conventional slope-rise boundary would have put the base-of-slope line at a water depth of about 2000 meters, high up on the edge of the geomorphic continent rather than at its base.

Thicknesses of sediments (in kilometers) from the isopach map of Fig. 4 (*3*) have been transposed roughly to Fig. 5 where they are shown as numbers. However, more recent investigations have indicated that maximum thicknesses of sediment near the axis of the sediment-filled trough adjacent to the continent are somewhat greater than previously supposed and in several areas exceed 10 kilometers.

Assuming that a sediment column of a minimum thickness of 1 kilometer is necessary to make petroleum prospects interesting, it is evident that all of the offshore area out to the base-of-slope line qualifies in that respect. The 100-kilometer boundary zone beyond the base of slope also appears to have an adequate sedimentary thickness for petroleum prospects. Boundary zones of 200 and 300 kilometers would include successively less interesting thicknesses of sediment but even a 300-kilometer boundary zone would leave some sediment thicknesses greater than 1 kilometer on its oceanward side.

Figure 5 illustrates differences that would result from the proposed 200-mile-from-shore boundary between national and international jurisdiction as compared with a boundary related to the base of slope.

North of latitude 38°, the 200-mile line would coincide quite closely with the base of slope if the 200 nautical miles were measured from the mainland of Nova Scotia. However, if the position of the 200-mile line were also influenced by Sable Island (a small Canadian island at the edge of the shelf southeast of Nova Scotia), a quite different boundary would result. Both possibilities are shown in Fig. 5. This is an example of one of the problems with the 200-mile scheme—the question of how small shelf islands should be treated in determining a base from which the 200 nautical miles should be measured. In contrast, using the base-of-slope guide there would be no such ambiguity.

Between latitudes 34° and 38°N, there is a large divergence between the 200-mile line and the base-of-slope line. The 200-mile line fails to include for the United States most of a broad terrace area which is underlain by a substantial sediment thickness. Likewise, between 29° and 32°N, the 200-mile line fails to include in national territory the interesting Blake Ridge, which has thick sediments and has already shown evidence of gas hydrates (*6*). South of latitude 30°N the 200-mile line is not shown because it is uncertain how it might be measured in this area of both United States and Bahama islands shorelines.

Petroleum Prospects off Eastern United States

In evaluating the significance of proposed national-international boundaries to ownership of petroleum resources off the East Coast of the United States, three cautions need to be emphasized. They are: (i) the unreliability of attempts at quantitative estimates of the petroleum resources of this area; (ii) the tremendous costs of any drilling and production operations in the remote deep waters beyond the edge of the continental shelf; and (iii) the remoteness in time when technology and economics might unite to make such areas practically commercial.

There have been many estimates of the petroleum resources off the Atlantic Coast of the United States during the last few years. Some have been very sanguine, others less so; all have one thing in common—they are necessarily based on little or no drilling knowledge and so are of little quantitative reliability. Such estimates, with the false impressions of accuracy which they too often give, may actually be a hindrance to good planning and may impede realization of our true need—the need to find out.

Probably the best (and most conservative) job of appraisal (*7*) is that which, however, applies to the offshore only out to a water depth of 200 meters. This estimate is based on the coordinated studies of a large group of specialists and is given as a range from 0 to 9 billion barrels plotted against a scale of probabilities (for example, 75 percent probability that there will be more than 1.5 billion barrels, 25 percent probability that there will be more than 4 billion barrels, and so on). However, even with this estimate, perhaps another probability factor needs to be applied—the probability that any estimate in a previously undrilled region will not be even close to the truth. What group of specialists, however skilled, would have come close to estimating the petroleum resources of the North Slope of Alaska before Prudhoe Bay, even with the most thorough study and after several wells had been drilled?

The important initial task is not a quantitative estimate but a careful and thorough appraisal of all geological, geochemical, geophysical, and economic data available, with the purpose of answering the

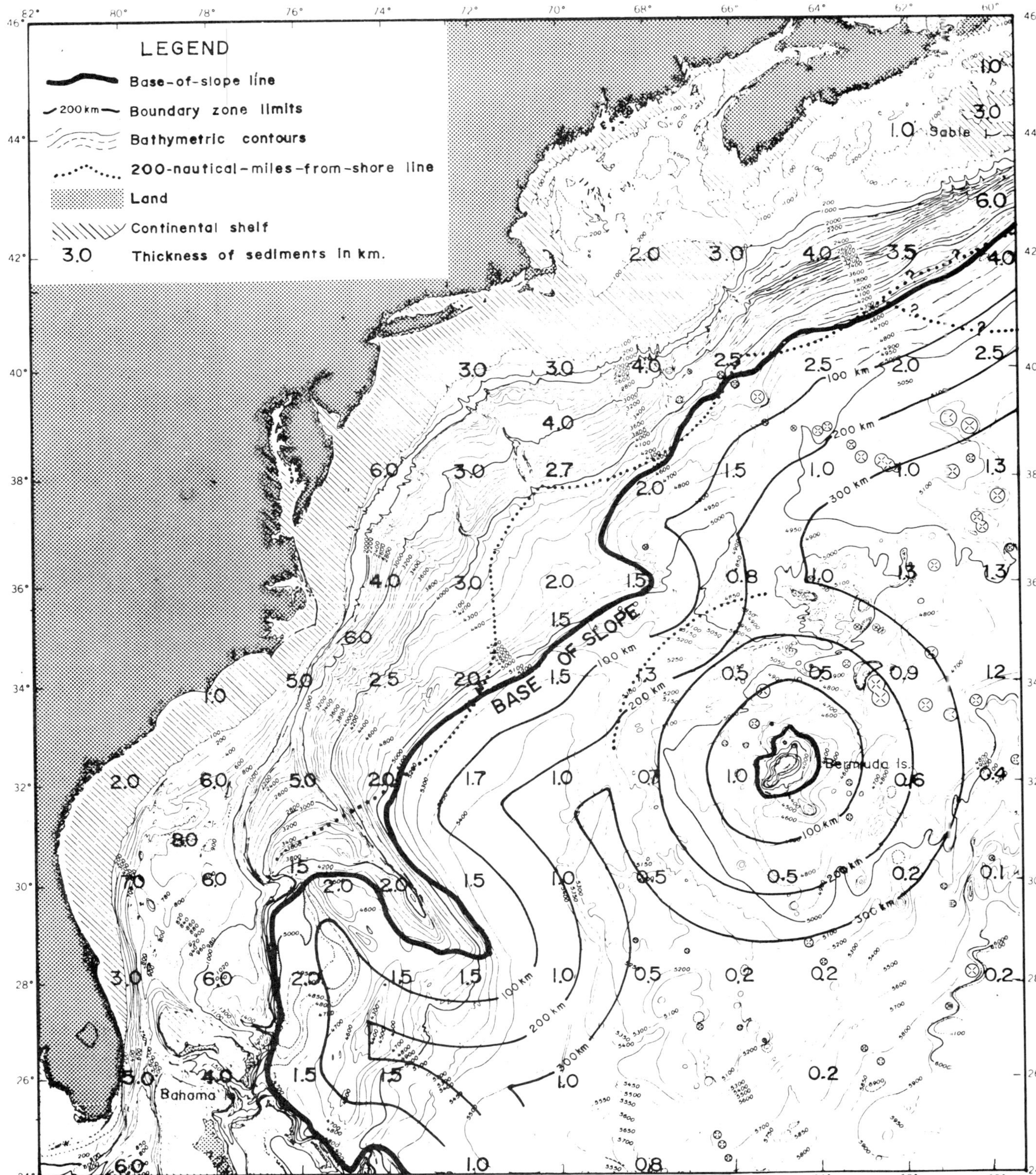

Fig. 5. Bathymetry (meters), sediment thickness (kilometers), and effect of various boundary proposals off the Atlantic Coast of the United States. The approximate base-of-slope boundary guide is shown with three alternative boundary zones on its oceanward side (black lines to right of and parallel to base-of-slope line). These lines indicate possible zone widths of 100 kilometers (54 nautical miles), 200 kilometers (108 nautical miles), and 300 kilometers (162 nautical miles). The thickness (in kilometers) of sediments is measured above acoustic basement (*3*).

only pertinent and practical question at this stage, which is: Do the prospects of the region, in the light of all the knowns and unknowns, reasonably justify exploratory drilling? In the case of the Atlantic offshore, the answer is Yes. The sooner we get on with the job of drilling exploration the better, regardless of quantitative estimates, optimistic or pessimistic (*8*).

To some extent the concern shown about the division of petroleum resources under the deep-ocean waters far from shore may seem a tempest in a teapot, so far removed is their utilization from present reality. With the tremendous costs and marginal economics of many current offshore petroleum exploration and development projects even at very moderate water depths, it may be completely impractical to think of ever producing oil and gas from deep-ocean environments. It may also seem only a barren academic exercise to argue about schemes for national-international boundaries in these remote areas that are so difficult to exploit.

However, the same things were once said about shallow-water offshore drilling; yet in only a few decades this has become commonplace. Offshore oil and gas now occupy an important position in our world economy. With inevitable progress toward the exhaustion of our petroleum reserves on land and in shallow coastal waters, the already steady march of exploratory drilling into deeper and deeper waters farther from land may see no limits as far out as geological possibilities of petroleum accumulations exist. It seems true that if the productive possibilities are good enough, there is scarcely a place on the earth where either technology or economics will be allowed to be a permanent barrier.

In view of this trend, the prudent time to settle matters of jurisdiction is *now*, before actual discoveries in the far offshore make it more and more difficult to peaceably establish boundaries on the basis of uniform and equitable principles without the complicating factor of selfish interest in acquiring specific local areas. The principle on which the boundary is to be established should be decided now, even if, locally, implementation into precise and definitely defined boundaries is delayed somewhat by needs for more accurate bathymetric data.

Even assuming that deep water and remoteness from land may not be permanent barriers to petroleum exploration and development, realistically it is still necessary to recognize that petroleum resources near the base of the slope are resources only for the far future. Many years of improving technology and increasing demand must intervene before such resources can be utilized. Moreover, the chimera of a rapid and enormous payoff from an internationally owned and sponsored petroleum development, to be used for the benefit of developing nations, must be abandoned, even in only moderately deep waters. A much more effective alternative was suggested above, but perhaps it should be considered whether a more practical way for all to benefit from petroleum resources under the deep oceans may not be simply to create stable conditions of national ownership of the ocean floor naturally pertaining to each coastal state so that these

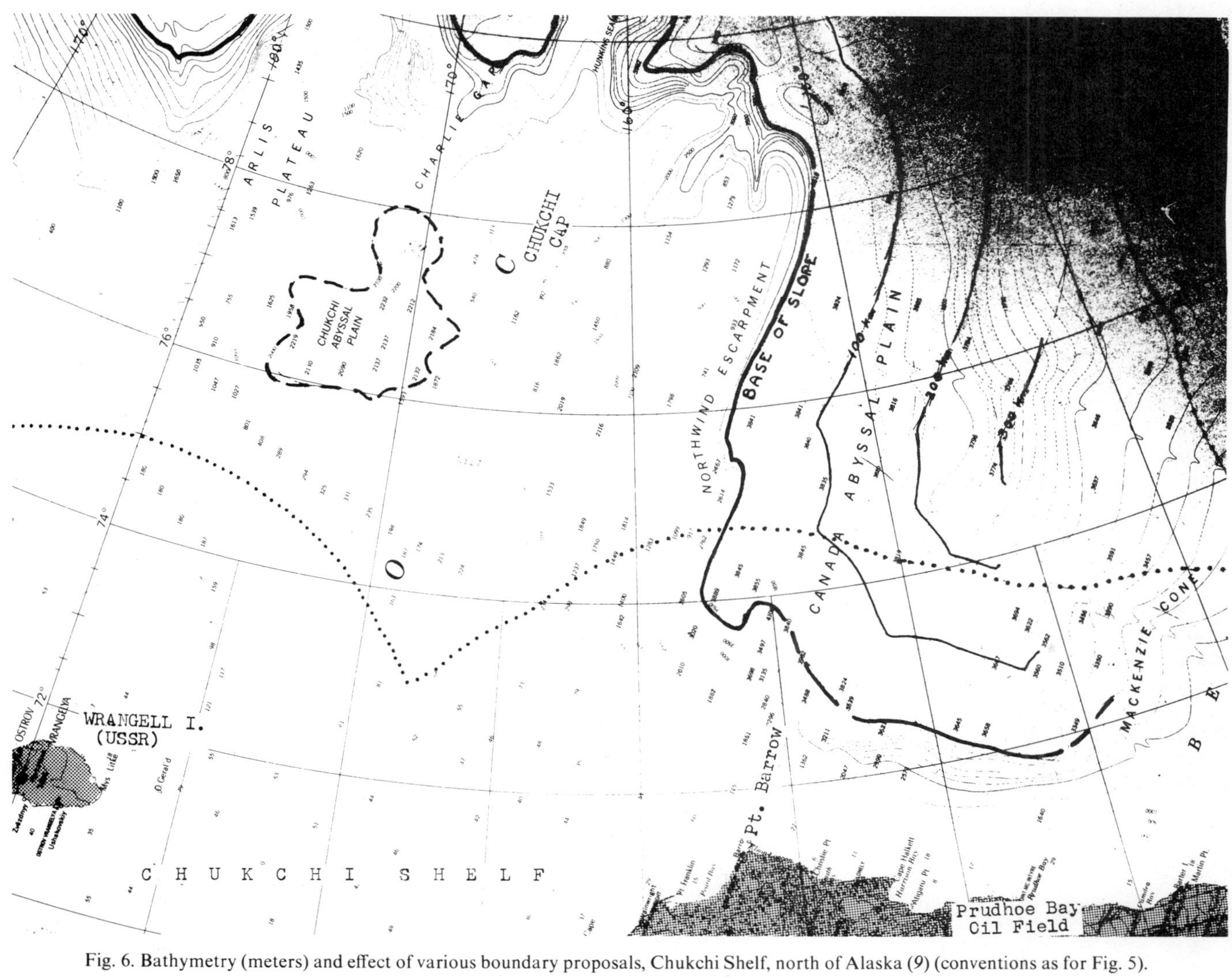

Fig. 6. Bathymetry (meters) and effect of various boundary proposals, Chukchi Shelf, north of Alaska (*9*) (conventions as for Fig. 5).

resources may be developed as rapidly and efficiently as possible, as the need for them arises, by those having the special skills, equipment, and geographic location to do so. A more abundant supply of petroleum more widely distributed among the coastal nations of the world would relieve present monopolies and inevitably work out to the advantage of all countries.

Boundaries off Other U.S. Coasts

The possible effect of proposed national-international boundaries on the U.S. petroleum position off the Atlantic shore has been discussed in detail. It may be of interest to briefly consider three other U.S. coasts similarly affected. It is also worth mentioning that the U.S. Pacific Coast, including Alaska, has a base of slope such that with a 100-kilometer boundary zone all areas of important petroleum promise would probably go to the United States. This would also be true for a 200-mile-from-shore boundary which would, however, result in the needless inclusion also of a broad band of deep-ocean bottom without petroleum prospects and with no other known subbottom values.

1) *Gulf of Mexico*. Geologically the whole Gulf of Mexico may be considered prospective petroleum territory. Extensive offshore production has long been established off the Louisiana, Texas, and Mexican coasts. The base of slope is quite clearly indicated or can be reasonably interpolated around most of the periphery of the Gulf and would serve to mark the limits of *minimum* areas to be assigned to the three bordering countries. With a 100-kilometer boundary zone a thin, elongate international zone would still be left in the middle of the Gulf, but with a boundary zone as wide as 200 kilometers this would be eliminated and the whole Gulf area would be divided between the United States, Mexico, and Cuba (*2*, attachment 3-B).

Use of the 200-mile-from-shore boundary would also leave a small central portion of the Gulf in the international zone, but again, the mode of application of this scheme would be uncertain because of the presence of several small Mexican reef islets at the edge of the shelf about 100 miles off the Yucatan shore. If the 200 nautical miles were measured from these islets rather than from the Yucatan mainland, Mexico's share of the Gulf would be greatly increased at the expense of the international zone and in part at the expense of the United States.

2) *Alaska: Bering Sea*. The Bering Sea not only has a thick sediment cover (locally as much as 10 kilometers) over its very broad shelf area, but also has thick sediments and petroleum potential thoughout its whole deep-water area beyond the base of the slope. A base-of-slope boundary would of course assure to the United States all of the broad eastern shelf of the Bering Sea, and the use of a 300-kilometer boundary zone would probably give essentially all of the deep-water sea floor to coastal state (United States and Soviet Union) jurisdiction.

A 200-mile limit measured from the mainland coast of Alaska and the Aleutian islands would leave much of the potential petroliferous area of the Bering basins, both on the shelf and in deeper water, under international jurisdiction. However, if the 200-mile distance were measured from the small U.S. shelf islands of St. Matthew, St. Paul, and St. George, far distant from the mainland, the United States could claim the entire eastern shelf area and also a substantial part of the deep-water area beyond the base of the slope under this boundary formula. Even so, it would lose much of the deep-sea, potentially petroliferous area that it would have received from the base-of-slope plan with a 300-kilometer boundary zone.

3) *Alaska: Arctic Coast*. Probably nowhere on U.S. coasts does the importance of the base-of-slope boundary stand out more than off the north coast of Alaska. Here the huge shallow-water areas of the Chukchi Shelf extend far out into the Arctic and are underlain by a thick (up to 6 kilometers) section of potentially petroliferous sediments. The situation with respect to the Chukchi Shelf is shown in Fig. 6 (*9*). The proximity of the great North Slope petroleum discoveries adds interest to the Chukchi region.

With a base-of-slope guide and a minimum (100-kilometer) boundary zone, a large part of this extensive and potentially petroliferous shallow to moderate water depth area would come under U.S. jurisdiction, whereas with a 200-mile-from-shore boundary, much of this part would go to the international zone.

Conclusions

In conclusion, the United States has everything to gain and nothing to lose by supporting a boundary formula that gives it its natural subsea territory out to the base of the adjacent continental slope, plus such additional selvage as may be provided by whatever width boundary zone the nations-in-concert decide is desirable. Conversely, the United States would definitely lose by accepting a 200-mile limit. In several places this would deprive it of potential petroleum territory which naturally should belong to it, while in other places it would give it an excess of deep-ocean bottom poor in sediments.

Similarly, for all coastal countries, the base-of-slope formula could give them as a minimum all the potential petroleum territory which they would receive by the 200-mile formula, without the unnecessary addition of inutile deep-ocean bottom better assignable to the international sphere, with the advantage to broad-shelf countries of giving them also all of their rightful petroleum resources beyond the 200-mile limit.

It is of course conceivable that the United States and other broad-margin countries may wish, for political or altruistic reasons, to go along with the 200-mile boundary proposal and donate what may be important potential petroleum resources of the far future to an international regime; but if so, they should know in advance what they are doing, and that they alone of the world's nations are doing it. Another alternative may be for the nations-in-concert to adopt one formula for the narrow-margin countries and another for the broad-margin countries (although differentiation between the two groups might be very difficult); or to adopt a combination of the two proposals, such as 200 nautical miles from shore or the outer edge of the margin, whichever is greater (although implementation of such a plan might be very complicated). Finally, perhaps, the United States and other countries may yet be willing to speak out for the simpler, more effective principle that gives all nations the sub-sea bottom resources that naturally belong to them through the natural prolongation of their land masses beneath the oceans, which treats all coastal states alike, and which still leaves some substantial areas of thick sediment in an international zone.

References and Notes

1. *Ocean Petroleum Resources* (National Petroleum Council, Washington, D.C., 1975).
2. H. D. Hedberg, *Ocean Manage.* **1**, 83 (1973).
3. M. Ewing, G. Carpenter, C. Windisch, J. Ewing, *Geol. Soc. Am. Bull.* **84**, 71 (1973).
4. H. D. Hedberg, in *Methods of Estimating the Volume of Undiscovered Oil and Gas Resources*, J. D. Haun, Ed. (American Association of Petroleum Geologists, Tulsa, Okla., 1975), p. 160.
5. R. M. Pratt, *U.S. Geol. Surv. Prof. Pap. 529-B* (1968). This map was used because it was the only one available to me showing bathymetric contours down to 100-meter intervals all the way across the margin. Offshore boundaries between the United States and Canada and the United States and the Bahamas are not shown because they are still controversial.
6. C. D. Hollister, J. I. Ewing, *et al.*, *Initial Reports of Deep Sea Drilling Project* (Deep Sea Drilling Project, Washington, D.C., 1972), vol. 11, pp. 791–799 and 958–966.
7. B. M. Miller *et al.*, *U.S. Geol Surv. Circ. 725* (1975).
8. An initial exploratory stratigraphic test well for stratigraphic information only, sponsored by a large group of companies and scheduled for a 16,000-foot depth, is reported to have started drilling on 14 December 1975 in 290 feet of water some 75 miles off the New Jersey coast.
9. *Map of Arctic Region, 1 : 5,000,000* (American Geographical Society, New York, 1975).

Liquid Fuels from Coal: From R & D to an Industry

The technology road is clear but the economic driving force is highly uncertain.

L. E. Swabb, Jr.

Today, there is clearly room for synthetic fuels, including liquid fuels from coal, if they could be made available at a competitive price. This is apparent from the fact that in 1977 just over 45 percent of U.S. liquid fuel needs will be supplied by imported oil.

Summary. On the basis of current U.S. oil imports, room now exists for a U.S. coal liquids industry. Unfortunately, technology is not available which can produce coal liquids at a price competitive with imported oil. Direct liquefaction technology is under development, but the prospects are that the technology will not be economic at the time the pioneer commercial plant should be constructed to provide the foundation for a possible coal liquids industry in the 1990's. Government support of coal liquefaction R & D has created the conditions that make possible the development of the technology, and probably government incentives for pioneer plants will be needed. With the proper incentives pioneer plants will lead to lower costs, and this, plus rising prices, will create the conditions necessary to develop a multiplant industry.

Liquid fuels play a critical role in the U.S. economy. They fill needs for which substitutes simply are not available—in transportation, for example, and peak shaving in electric power generation. Also, there are many homes and commercial buildings heated by oil, for which gas is unavailable and coal is impractical. The only alternative in these cases to liquid fuel is electricity, an expensive alternative. Because of the critical role liquid fuels play, they will continue to be in demand for many years.

Petroleum production in the United States peaked in 1970 and began to decline. There is substantial potential for arresting the decline for a time, depending on government leasing policy and successful exploration offshore. However, eventually even world petroleum production must peak and decline because petroleum is, after all, a nonrenewable resource. Many estimates place the inflection point of the world's crude oil supply before the end of this century (*1*). How the world will adjust to declining supplies of petroleum is a major uncertainty of the long-term energy outlook. Clearly, passing the peak of world oil production will restrict sharply the major tool that the United States is now using to manage its own fuel supply-demand gap—imported oil. Thus, although there is room now for a domestic industry producing liquid fuels from coal, the need for that industry could become acute before the end of the century.

One candidate for supplementing current and future petroleum supplies is synthetic liquid fuels made from the vast reserves of coal in the United States. However, there is no commercial industry in place today for making liquids directly from coal, nor, for that matter, is there technology available for producing liquids from coal at an economically competitive price. Developing this technology is the goal of the large coal liquefaction R&D program now under way sponsored by both industry and government.

Over the years, the petroleum industry has developed and commercialized a variety of new refining processes to produce desired products at the lowest possible cost. In general, the development and commercialization of these processes was relatively straightforward. The research and development costs were moderate. There was a clear market for the products, and the economics of commercial plants gave a satisfactory return on investment. However, in the case of liquid fuels from coal, the picture is quite different. The R&D cost is high, and the timing, conditions, and economics of future commercial applications are not clear. This places added challenges on the development and commercialization process.

In this article, I consider how we, as a nation, might proceed from today's R&D efforts to the creation of a commercial coal liquids industry in the private sector, or at least to ensuring that such an option will be available if national policy or needs dictate its use. After a brief description of the state of the art in coal liquids technology, I discuss the major steps needed to develop and commercialize that technology, and then offer some insights into the conditions that appear necessary for these steps to take place.

Coal Liquids and Coal Liquefaction

Basically, there are three ways to liquefy coal: (i) add hydrogen, (ii) take away carbon, or (iii) break it down to individual carbon atoms and rebuild. The third approach is being used today by the South African Coal, Oil and Gas Corp., Ltd. (SASOL) in Sasolburg, South Africa (*2*). Coal is gasified with steam by the Lurgi technology to produce carbon monoxide and hydrogen, which are then reacted over a catalyst in the Fischer-Tropsch process to produce liquid hydrocarbons. The combination of tearing down and rebuilding results in the highest cost route to liquids from coal. Another indirect approach now under investigation would gasify the coal, convert the gases to methanol, and convert the methanol to gasoline. The economics of this route are uncertain.

The other two methods of liquefying coal, adding hydrogen and removing carbon, should be more economical. Adding hydrogen, or direct liquefaction as it is usually called, will probably become the major source of liquid fuels from coal. Pyrolysis of coal (removing carbon) may also become significant, but processes of this type must contend with low liquid yields and economical use of high volumes of char.

The author is vice president for Synthetic Fuels Research, Exxon Research and Engineering Company, Florham Park, New Jersey 07932.

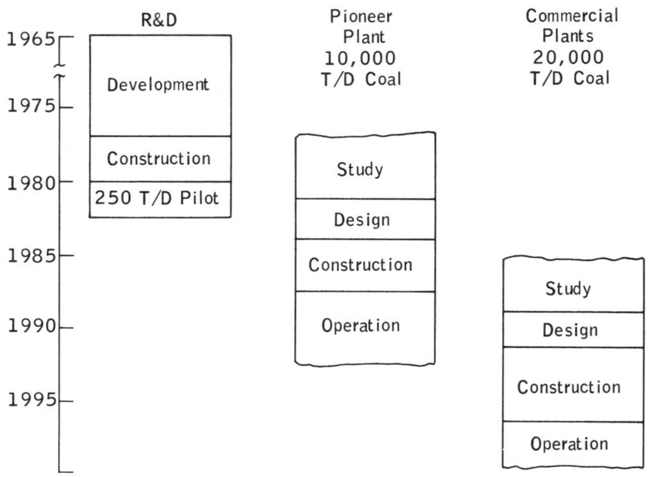

Fig. 1. Timing for the three major steps in the commercialization process. This particular scenario illustrates how the EDS process might be taken to commercialization.

There is no commercially proven technology today for the large-scale, direct liquefaction of coal to produce high yields of distillate liquid products. Several processes for producing yields up to three barrels of liquids per ton of coal are now under large-scale development. Among these are the Exxon Donor Solvent (EDS) process, the HRI H-Coal process, and the Gulf SRC-II process. Small-scale coal liquefaction R&D on other processes is also being done by Pittsburgh Energy Research Center (Department of Energy, DOE), Dow Chemical, Conoco, Lummus, and others.

In general, all direct liquefaction processes now under development involve slurrying the coal in a recycle coal liquids stream and then heating the slurry in the presence of hydrogen or hydrogen donor molecules (or both) to cause the organic molecules in coal to break down and pick up hydrogen. This process of molecular weight reduction and increased saturation with hydrogen converts organic solids in the coal into liquids. The various processes differ in the reactor configuration and conditions, the way catalysts are used, and the way the products are separated from unreacted coal and mineral matter (*3*).

Steps for Commercialization

From industrial experience, we know that commercialization of new process technology involves three major steps. The first step is R&D. This is basically a process of reducing technical uncertainty to the point where a full-sized commercial plant can be designed and built at some reasonable and acceptable level of technical risk. The second major step is the transition from the experimental technical activity to the reduction of the uncertainty concerning the commercial aspects of the technology. The third step is the construction of multiple commercial plants in accord with commercial needs and opportunities.

There are two principal routes for passing through the transition step from R&D to a full-fledged industry; one is through a pioneer commercial plant and the other is through a demonstration plant. On the basis of our experience with developing and commercializing new petroleum and petrochemical process technology, we believe that the pioneer plant approach is preferred. I describe this route to commercialization and then discuss how the demonstration plant approach differs.

Starting with the first step, R&D is progressed to the point where there is sufficient data to design a full-scale commercial plant with acceptable technical risk. How and when this point is reached depends on factors such as the nature and complexity of the new technology; the cost and time required for alternative development routes; the capabilities of the organization that will design, build, and operate the pioneer commercial plant; the level of technical and economic risk that is acceptable to the investor; and whether there is some special value for an earlier commercialization date.

In the case of the EDS technology, the R&D will include the operation of a large pilot plant of the minimum size needed to provide the critical design and the mechanical and process data for scale-up to full commercial size. This minimum size was determined to be 250 tons of coal feed per day, based on a technical analysis of the chemical and physical phenomena occurring in each of the process steps. The EDS project illustrates the timing and costs involved. Exploratory R&D started in 1966. By 1971, a promising process sequence was defined and was under development in the laboratory and in small pilot plants of roughly ¼ to ½ ton of coal feed per day. Further development is being carried out in a 1-ton-per-day pilot plant to confirm the design of the 250-tons-per-day pilot plant, and to define quantitatively the effects of process variables. Construction of the large pilot plant will begin in 1978. Operation will start in late 1979 and is scheduled to be completed by mid-1982. Laboratory and small pilot plant R&D will continue throughout this period. This is the R&D step illustrated in Fig. 1. The estimated cost of the total program to reach the point of designing a commercial scale plant is in the range of $275 million to $300 million.

In addition to having an adequate technical data base for commercial plant design, a new coal liquefaction process must also meet certain other criteria. It must be operable and reliable on a commercial scale and produce products of desired quantity and quality. It must meet current and projected future standards for impact on the environment, and on the health and safety of workers and users of the products. Finally, it must meet the cost criteria which will make commercial viability possible.

Confirmation that these criteria will be met is best achieved in a pioneer commercial plant. A pioneer plant is defined as a commercial-scale, stand-alone facility intended to operate as a commercial venture for a normal plant life of 20 to 30 years. It is a first of its kind. Its location and size are based on commercial and technical considerations. By definition, it is constructed, operated, and managed by the private sector. It might be designed in modules to permit later expansion. However, each module would be to full commercial scale.

The definitive planning and the initial design work for the pioneer plant can be started before the total process development program is completed. The definitive planning for a large commercial project, such as a coal liquefaction plant, is a major undertaking, and it must be started early. Significant areas to be defined are the plant site; the necessary permits; the feed, water, and utility supplies; customers for the products; waste disposal; and economics and financing.

If it is deemed urgent to do so, definitive design work can be completed as soon as the data base for scale-up is available from the large pilot plant. If the overall development program is well planned, this should be available after 1 to 2 years of large pilot plant operation. Thus, the design phase of a pioneer plant for EDS could start sometime in 1981.

Assuming a 3-year design effort, construction of the pioneer plant could start in 1984 and the start-up operations could begin in 1987 (Fig. 1). Current estimates place the cost of a pioneer plant capable of processing 10,000 tons of coal per day at around $75 million (1976 dollars). Such a plant would be one train of a commercial plant and would produce 25,000 to 30,000 barrels of coal liquids per day. It would be designed for possible expansion later, to twice the size by the addition of a second process train, if and when justified.

The specific goals of the pioneer plant, as we see them, are to establish the cost and value of the products, to reduce the business risk, and to reduce the cost of future plants. In meeting these goals, the pioneer plant confirms solutions to whatever community and social effects are encountered. It charts a path through the regulatory and environmental requirements. It sets the basis for the development and growth of a supporting infrastructure and logistics system for feed and products. It provides the opportunity to train a cadre of operating and maintenance people. Especially important, the pioneer plant starts the "learning curve." Experience with a variety of industrial processes has shown that investment and operating costs fall as experience is accumulated in commercial-sized plants. Cost reductions as great as 30 or 40 percent (in constant dollar terms) have been experienced for commercial plants built about 10 years after the start-up of the pioneer plant provided that R&D is continued and plant tests of potential improvements are conducted. Achieving significant cost reductions as a result of the pioneer plant step will be an important factor in establishing a viable coal liquids industry.

The last step in the commercialization process involves the construction and operation of a number of commercial plants. Again, definitive planning for such plants would have to start several years in advance of the design stage. Design of commercial plants might start after the pioneer plant operates for a year. In the case of the illustrated development schedule for the EDS coal liquefaction process, this would mean 1988, and construction could begin in 1991 with start-up in 1996. Present estimates are that a commercial coal liquids plant capable of processing 20,000 tons of coal per day into 50,000 to 60,000 barrels of liquids per day will cost between $1 billion and $1.5 billion (1976 dollars). This estimate incorporates the cost reduction benefits from the pioneer plant.

Pioneer Contrasted with Demonstration Concept

A significant portion of the DOE fossil energy R&D budget provides for demonstration plants. There are some major differences between the pioneer plant route and the concepts underlying the DOE demonstration program.

Demonstration plants are usually intermediate in size between pilot and full-scale commercial plants. Their primary purpose is to reduce technical risk associated with scale-up, plus providing an improved definition of operability, reliability, costs, and environmental acceptability. This contrasts with the pioneer plant concept, which is based on the premise, supported by extensive experience, that scaling directly to commercial capacity from a pilot plant involves acceptable technical risk and does not justify an intermediate step if the pilot plant is properly sized and appropriate supporting data and scale-up correlations are developed in the R&D stage. Further, an intermediate-sized demonstration plant is unlikely to be economic and therefore will not provide adequate answers to questions about commercial viability. Its high cost and short life are likely to lead to shortcuts that result in omitting some of the facilities that would be included in a commercial plant and limiting some of the activites required to develop and implement a commercial venture.

A significant difference between the DOE demonstration plant program and the pioneer plant approach that we favor is related to the degree of responsibility assumed by the private sector and the federal government. In the DOE demonstration program, there is a significant federal role including a strong voice in and the right to control such matters as selection of technology to be employed, project scope, site location, technical involvement by the National Laboratories and the Corps of Engineers, and cost-sharing arrangements. Usually demonstration projects are developed in response to a request for proposal and then awarded as a result of competition on the basis of criteria established by DOE.

Our concept of the pioneer approach contemplates a predominant private sector responsibility with respect to project initiation, selection of technology, scope, location, project planning, development, and implementation. This includes negotiation of all commercial agreements, securing of all permits, arrangements with regulatory and local authorities, and the long-term operation of the pioneer facilities. In this concept, the primary role of the government is to establish the climate and broad framework that will provide appropriate encouragement and adequate incentives for the private sector to undertake pioneer projects. The private sector has the responsibility to advise the government with respect to the adequacy of available or possible incentives for undertaking pioneer projects.

A strong, dominant role of the private sector in the pioneer plant step is necessary to ensure successful commercialization. In the business world, research, development, and commercialization are a combination of closely entwined technical and business activities geared to the market. They involve considerable knowledge and judgments about market acceptance and economics for new technology. Federal agencies, on the other hand, have limited access to experience and judgments honed by competitive markets. Also, in government activities political considerations sometimes overshadow economic factors in the decision-making process. These significantly inhibit the government's ability to transfer its R&D results into commercial ventures.

Conditions Required

The process of creating and commercializing new technology is driven by economic considerations. As the private-sector investor proceeds through the various steps, he is continually reassessing the economics. His willingness to go ahead is based on his expectation of benefits in view of the apparent costs and risks involved.

During the 1960's, the early stages of research on coal liquids were pursued largely on the conviction that eventually they would be needed to supplement petroleum. While no accurate assessment of costs was really possible early in the research, a strong driving force was the hope that small-scale, applied research would result in a liquefaction process with acceptable costs. Also, the business environment appeared reasonably predictable in terms of the overall economy, the mechanism for establishing prices, and the applicable government regulations. While there was uncertainty, it was at a level which industry felt competent to handle.

The events of the early 1970's introduced a great deal more uncertainty. Prices began to be set by political forces.

Inflation caused plant construction costs to soar. Environmental, health, and safety regulations began to change rapidly. Overall, changes in the business environment became less predictable. Also, the coal liquefaction processes under development began showing estimated costs that were much higher than prices for petroleum fuels. In the face of increased uncertainty and poor process economics, private sector investors hesitated to commit all of the large sums of money required for the large pilot plant stage of development of coal liquefaction.

Coal liquids technology is proceeding now into the large pilot plant stage because of government support of the R&D. All the coal liquids processes now in the large pilot plant phase involve at least 50 percent government cost sharing, and, for many, the government share is higher. At the present level of uncertainty, the high program costs, and the relatively poor economics, such government support is an essential condition for proceeding.

Turning to the next stage, that of the pioneer plant, again the private investor's decision to proceed will hinge on economic considerations. Economic studies, based on detailed conceptual designs of a pioneer plant in the mid-1980's, yield estimates of discounted cash flow (DCF) returns ranging from none at all to 10 percent, depending on the assumptions made. The most critical assumption is the rate of increase of the price of coal liquids relative to the rate of inflation of plant construction costs. In our studies, coal liquids are assumed to command a price equivalent to the marginal supply of liquid fuel, that is, imported oil. If coal liquids price and plant construction costs are assumed to escalate at the same rate, estimated DCF returns on a pioneer plant are in the 0 to 5 percent range. If coal liquids price is assumed to escalate faster than construction costs, DCF returns up to 10 percent can be estimated. Even this level, however, is too low to attract the private investor, and thus incentives of some sort appear to be needed to encourage the private sector to proceed with a pioneer plant.

The most preferred incentives would be single plant incentives designed to have their greatest impact early in the project life. Such incentives could include (i) allowing coal liquids to be priced competitively with the marginal supply, (ii) providing a higher investment tax credit, (iii) providing for more rapid depreciation, and (iv) providing a cash grant that would be convertible to a loan at a later date. The precise combination and level of these incentives that will be needed cannot be predicted today. When the time comes to build the pioneer plant, the company developing the project will need to define the necessary incentives in cooperation with the responsible agency of the government.

Loan guarantees have been suggested as an aid for pioneer plants. Loan guarantees may be useful where part of the plant investment is to be borrowed. However, a loan guarantee cannot raise the DCF return on a project; it cannot make an uneconomic project economic. Thus loan guarantees may supplement but not replace the incentives mentioned above.

While the outlook is that incentives will be needed to encourage pioneer plants in the mid-1980's, the longer term prospects for the formation of a multiplant industry are not clear. The private sector will invest in coal liquids plants if there is an expectation of a reasonable return on the investment. The essential conditions for such an expectation appear to be some combination of lower costs and higher prices. Pioneer plant experience plus continuing R&D, if pursued, will undoubtedly result in lower costs to some extent. Thus, pioneer plants appear to be essential to eventual commercialization. However, based on current cost estimates and estimates of learning curve benefits, the cost of the technology now under development is not likely to fall enough to make coal liquids economic at the price coal liquids might command today (that of imported oil). Thus, barring an unexpected breakthrough in liquefaction technology, the second essential condition for commercialization is that the real price coal liquids command will have to rise.

This is not to suggest the subsidization of a commercial coal liquids industry. Commercialization will occur if the technology has been proved workable and acceptable through the pioneer plant phase and if coal liquids command a sufficient price. If coal liquids do not command such a price it means that less expensive alternatives are available and coal liquids are not needed or desired.

Despite the uncertainties of future economic conditions, it is in the national interest to pursue the development of several coal liquefaction processes through R&D and the pioneer plant stage. Multiple technologies are needed because it is probable that no one process will be best in all circumstances. These developments will provide increased options for the nation and insurance against economic disruption caused by severe limitations in supplies of liquid fuels. Since the United States may have a compelling need for a supplemental source of liquid fuel before the end of the century, these developments should proceed now, as rapidly as possible.

References and Notes

1. M. K. Hubbert, "U.S. Energy Resources, A Review as of 1972," Committee on Interior and Insular Affairs, U.S. Senate, Serial No. 93–40 (92–75) (1974); J. D. Parent and H. R. Linden, "A Survey of United States and Total World Production, Proved Reserves and Remaining Recoverable Resources of Fossil Fuels and Uranium as of December 31, 1975," (Institute of Gas Technology, Washington, D.C., 1977); "Oil prices may rocket uncontrollably in the 1980's," *Energy Daily* (23 September 1977), p. 3; C. L. Wilson, *Energy: Global Prospects 1985–2000* (McGraw-Hill, New York, 1977), p. 20.
2. J. G. Krouseder, "Sasol II: South Africa's oil-from-coal plant," *Hydrocarbon Process* (July 1976), p. 56.
3. L. E. Furlong, E. Effron, L. W. Vernon, E. L. Wilson, "Coal liquefaction by the Exxon donor solvent process," national meeting of the American Institute of Chemical Engineers, Los Angeles, 18 November 1975; B. K. Schmid, "Advanced SRC-II processing modes for liquid and solid fuels," Fourth Annual International Conference on Coal Gasification, Liquefaction and Conversion to Electricity, University of Pittsburgh, 2–4 August 1977; C. A. Johnson, E. S. Johanson, M. C. Chervenak, A. G. Comolli, R. Parthasarathy, "Progress Towards Commercialization of the H-Coal Process," Fourth Annual International Conference on Coal Gasification, Liquefaction and Conversion to Electricity, University of Pittsburgh, 2–4 August 1977; J. B. O'Hara, "Coal Liquefaction," *Hydrocarbon Process* (November 1976), p. 221.

Nuclear Waste Disposal: Two Social Criteria

Technical irreversibility and site multiplicity are suggested as criteria for safe nuclear waste disposal.

Gene I. Rochlin

Del carratere degli abitanti d'Andria meritano di essere ricordate du virtú: la sicurezza in se stessi e la prudenza. Convinti che ogni innovazione nella città influisca sul disegno del cielo, prima de'ogni decisione calcolano i rischi e i vantaggi per loro e per l'insieme della città e die mondi.—ITALO CALVINO (*1*).

There is a consensus that radiologically hazardous wastes from the nuclear fuel cycle should be separated from the biosphere to a sufficient degree and for a long enough time so that they present no significant risk to life. But this consensus does not extend to the definitions of "sufficient," "long enough," or "significant risk." Our ability to predict material or geological stability over the containment times required for long-lived components has been questioned (*2–4*). Moreover, the impossibility of predicting socially relevant factors over such relatively short periods as a few hundred years precludes accurate estimation of either the probability of an accidental or deliberate breach of containment or the effects of such a breach on society (*5*).

Recent attempts to organize technical options for management and disposal of nuclear wastes (*6, 7*) have been constrained by a number of factors. The lack of attention received by the "back-end" of the fuel cycle (*8*) resulted in periods when waste management research and development were neglected (*9*). Utilities have had no incentive to absorb waste management costs (*10*). Increased public concern over waste disposal brought pressures on the government to resolve the issue quickly, which, in turn, tended to restrict consideration to those few methods whose techniques and performance had been relatively well studied (*11, 12*). Thus, while criteria, regulations, and techniques for the handling and transportation of wastes after their production have been based on established health and safety standards (*13, 14*), solutions for long-term waste management have largely focused on the selection and promotion of an available method or technology (*15*).

In the absence of coherent goals or comprehensive regulatory standards, suggested methods of waste management are usually divided into three categories: (i) short-term storage, (ii) long-term storage, and (iii) disposal (*16*). Short-term storage, such as in shallowly buried metal tanks (*7*, section 21), requires active maintenance and guarding, and is intended primarily as an interim procedure (*17, 18*). Long-term storage, such as emplacement in underground caverns mined in salt (*7*, section 25; *19*), requires little or no active maintenance, but is susceptible to accidental or deliberate breaching of isolation barriers (*3, 4, 12*). Disposal, such as ejection into outer space (*7*, section 26; *20*), implies that there are no circumstances that could result in the return of the wastes to the terrestrial environment (*21*).

The division of what is properly a continuum of possible waste management methods into three distinct categories is purely arbitrary and appears to have evolved primarily as a way to organize technical thinking in response to political pressures. As each individual waste management method was promoted, heated controversy erupted about its efficacy. A typical response was to aver that the suggested method was a form of storage, pending the development of an acceptable final disposal scheme (*22*). This emphasized the differences between alternative methods rather than the continuity of the problem. The need for perpetual care of a retrievable surface storage facility is difficult to weigh against the probability of reentry of a space shuttle package or the possibility of an undiscovered water source in a salt formation. In the first case the dominant failure mode is social, in the second case technical, and in the third case informational. The tendency to focus on individual methods and their idiosyncratic deficiencies led to incoherent debate (*4*), owing to the absence of a shared basis for comparison. The commonality of the goals and objectives of all waste management methods was obscured (*23*).

Two criteria, technical irreversibility and site multiplicity, are suggested here for use in organizing waste management options in terms of ensuring continued isolation from the biosphere in the face of both social and geological uncertainties. They also reflect the possible consequences of technical or judgmental errors. These criteria translate the goals of waste management—public health and safety, ethical and moral responsibilities, obligations to the future, concern over imperfections in present technical and social knowledge—into standards against which the performance of any suggested scheme can be judged (*24*). The purpose of the classification is not to substitute quantitative social analysis or purely technical measurements for proper consideration of the ethical and social issues associated with a waste management decision. Rather, it is to provide a clear basis for open and conscious policy choice.

Long-term safety is taken to be the overriding concern. Operational and short-term risks and present costs are held to be of secondary importance in determining an acceptable method (*25*), and are to be examined after a desired option has been selected. If this option seems excessively hazardous to present populations, or prohibitively expensive, the next best alternative can then be examined. A minimum ethical requirement is that this choice be made explicitly and self-consciously (*26*), and with an open acknowledgment that the well-being of future generations may depend upon our choice. Unexamined or implicit value judgments, ethical choices, and evaluations of risk should be avoided.

The author is a Research Specialist at the Institute of Governmental Studies, University of California, Berkeley 94720.

Wastes from the Nuclear Fuel Cycle

Radioactive materials of no immediate or foreseeable value are produced at every stage of the nuclear fuel cycle. Collected and contained, they constitute the wastes. Gaseous emissions and liquid effluents, after treatment to remove the more hazardous components (*27*), are discharged into the environment and, although their activity has not decayed away, they are no longer subject to regulation as wastes (*28*). Wastes with low levels of activity comprise large volumes of material containing low concentrations of radionuclides, and arise from activation or contamination of solids and liquids used in routine operation, for example, gloves, wiping cloths, effluent filters, and coolants. The radiological hazard per unit volume is comparatively low, and by regulation only a very small concentration of alpha-emitting transuranics is to be allowed (*29*). Wastes with intermediate levels of activity consist largely of the products of effluent cleanup and of chemical wastes with a higher concentration of fission products and other short-lived radionuclides (*30*).

Transuranic-contaminated wastes primarily come from fuel processing and fabrication, where some portion of the actinides being handled is lost as waste in chemical extraction or machining operations. The very long half-lives and very low permissible concentrations (*13*) of most alpha-emitting actinides requires that these wastes be disposed of by a method that provides long-term guarantees of containment integrity.

The wastes that have most often served as the focus of public debate over long-term containment and disposal are the wastes with a high level of activity (high-level wastes, HLW) that originate in the reprocessing of spent reactor fuel (*31*). It will be necessary to wait a few months for the shortest-lived isotopes to decay away, after which time the spent fuel is to be sent to a reprocessing facility where the uranium and plutonium will be chemically extracted. The recovered uranium may be returned to an enrichment facility to be recycled again as fuel. The extracted plutonium can be used directly (*32*). The HLW generated as liquid wastes from the solvent extraction process (*33*) will contain almost all of the nonvolatile fission products, a residual percentage of the uranium and plutonium (*34*), and the remainder of the actinides (*35*). Under present federal regulations, this liquid must be solidified within 5 years of reprocessing and shipped to a federal waste repository within 10 years (*13*, section 50, appendix F). Shipment as a liquid is prohibited (*36*).

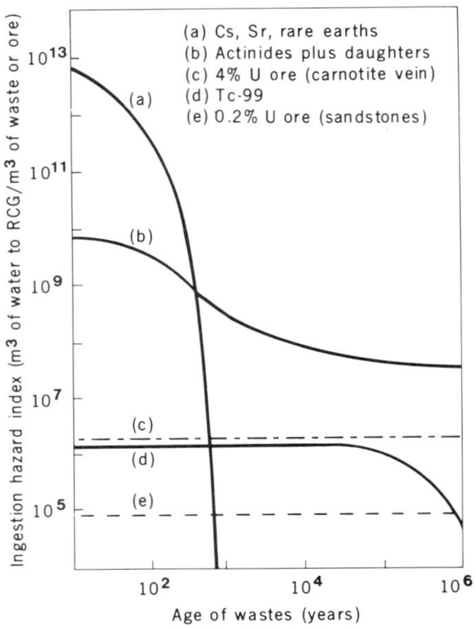

Fig. 1. The ingestion hazard index [volume of water needed to dilute a volume of waste or ore to allowable concentration guidelines (RCG)] for solidified HLW from the reprocessing of enriched-uranium light-water reactor fuel. It is assumed that 99.5 percent of the uranium and plutonium in the spent fuel and all of the krypton and iodine are removed before solidification. For comparison, the hazard indices for a fairly rich Colorado carnotite vein ore and for more typical 0.2 percent concentration sandstone ores are also indicated. [Adapted from ERDA 76-43 (*7*)]

Figure 1 shows some of the constituents of HLW as a function of time from reprocessing. The components usually referred to as short-lived have half-lives ranging up to a few tens of years (*37*). They decay away sufficiently rapidly so that in times less than or of the order of 10^3 years their contribution to waste activity is comparable to or below background radiation. The long-lived components, on the other hand, have half-lives upwards of 10^3 years. As they are generally emitters of alpha rather than beta or gamma radiation, they present a greater carcinogenic potential per curie of activity (*38*). They may have to be kept contained and isolated for times up to 10^6 years. This separation according to half-lives is not entirely arbitrary. As shown in Fig. 1, it reflects the character of the waste stream.

From a social or political point of view, it is difficult to think of 10^3 years as a short time. Figure 2 compares the half-lives of some of the radionuclides present in wastes with times of social or political relevance. It should be recalled that many half-lives must elapse before the net activity due to any specific isotope is appreciably reduced. For beta- and gamma-emitting radionuclides that decay to benign levels in times less than 10^3 years, the decrease in potential risk occurs over times that allow us at least to conjecture about relevant social and political conditions. This is clearly not true for the long-lived transuranics.

It has been suggested that HLW, at least, be further separated into long-lived and short-lived components by additional chemical treatment (*39*). Such partitioning might simplify some aspects of disposal. The mass and volume of the fraction containing the alpha-emitting transuranics would be reduced considerably, which would facilitate more exotic waste disposal methods such as in space and by transmutation (*15*). The shorter-lived fraction could be handled differently, and some of the isotopes might even be in demand for other purposes as they were separated out (*40*).

The definitional problem of determining which components should be treated as wastes and which as potentially recoverable resources is not a trivial one. Present opinions range from the conviction that almost all of the radionuclides have potential value and should be stored retrievably against the day when the need for them is fully developed (*41*), to the opinion that any by-product of the nuclear fuel cycle for which there is no immediate use should be permanently disposed of (*42*). Therefore, no fixed definition of isotopic composition or mass is assumed. Where quantities or specific activities of the wastes are determining factors in the operation of a given waste disposal method, they must be taken into consideration when the feasibility of the method is evaluated, because both the scale of the operation and the level of risk entailed will be affected.

Criteria for Waste Disposal Methods

All methods for management of nuclear wastes must take into account long-term risks, short-term and operational risks, and cost. In principle, these criteria are empirically determined and then used to establish standards (prescriptive norms) for performance that reflect normative judgments as to safety and affordability (*43*). In formulating standards, the weighing of the relative importance of each of these criteria is a social decision involving both social and ethical values. Nevertheless, there is a persuasive case for the subordination of both immediate risks and present costs to potential long-term hazards when selections are being made among alternative options for the disposal of long-lived wastes.

Cost is taken to be least important. It is used here not as a technical determinant for defining an acceptable method, but as an elastic boundary condition to be satisfied. Once a method is selected according to consideration of risks, it is then to be examined to determine the social and economic costs of operation (*44*). As these costs are unlikely to be prohibitive (*45*), the question is what level of safety will society be willing to pay for. Affordability is a flexible social and political decision.

A similar argument is made for subordinating short-term and operational risks to long-term ones. The immediate risks of disposal operations will be borne by the same population that benefits from the nuclear power that generates the waste. That population can weigh both risk and benefit and make its own decisions. Waste disposal, however, poses a difficult problem in that immediate risks and costs may be decreased by exporting risk to other populations, or to the future (*46*). There will be a natural desire to minimize present risk and current costs. The only constraints on doing so at the expense of the future are ethical and moral ones (*47*).

I suggest that there are two principles of ethical behavior to be followed. The first is that the public should be given the fullest information possible about long-term risks and future costs (*48, 49*). That future populations may be unable to act upon this information in no way absolves the present of the responsibility to provide it. If risks are to be exported to the future, a minimum ethical standard is that this should be done openly. The second principle is that present generations should act so as to minimize the amount of irreparable harm that could occur as a result of present decisions (*47*). It is certain that every action has uncertain consequences for the future (*50, 51*). This does not argue against the right to act, but against the refusal to take responsibility for the consequences (*52*). The ethical problem of how to balance the needs of the present against the rights of the future, of justifying the export of risks and costs, is far too large to encompass here. Nor is it clear that there exists a basis on which to resolve it (*49; 51*, p. 232). But to act so as to minimize exported risks, particularly when to do so imposes no great burden upon the present, is the minimum ethical requirement.

Therefore, primacy has been given to minimization of long-term risks in establishing a framework for organizing alternatives for waste disposal. If these risks could be precisely determined, the options could be simply ranked. But there are great technical and social uncertainties as to the integrity of waste containment over the long times that such containment must be maintained. An acceptable method for the disposal of high-level wastes must be proof against technical failures such as corrosion of materials, against scientific failures such as overestimating the efficacy of natural barriers to migration, and against geological changes such as glaciers and earthquakes (*53*). Few disposal methods can, with confidence, be guaranteed to be permanent over the hundreds of thousands of years that isolation is needed. A more reasonable condition is that, should the containment fail, the time to return the wastes to the environment is of the same order of magnitude as the time necessary for the toxicity to be reduced to a level equal to or below background radiation at a comparable site. Since even the short-lived components of the wastes will remain hazardous over times that are long on social or political time scales, no guarantee of future ability to repair, clean up, or even recognize a breach of containment can be assumed.

The amount of radioactivity in the waste as a function of time can be combined with technical and scientific estimates of the probability of failure to generate a set of numbers that express the long-term risk in terms of the probable material release in any given year (*54*). But to translate these into even a rough measure of social impact requires knowledge of available pathways and population distributions and habits. Such numbers are too imperfect and incomplete in the face of social, technical, and geologic uncertainties to provide useful guidelines for the evaluation of alternative disposal methods. What I suggest in the next section is a method for extending the risk evaluations into a pair of criteria that reflect not only technical paths for returning the wastes to the environment, but also the possibility of active intervention by more or less intelligent beings. As the impacts of a given release cannot be adequately determined for times far in the future, the focus is on minimizing the probability and quantity of a breach of containment in the face of a wide range of uncertainties as to the causal factors.

Technical Irreversibility

I define technical irreversibility as the degree to which emplaced wastes are resistant to recovery or release either by accident or by the deliberate application of technology. Its significance as a criterion is that the more irreversible a waste disposal method is, the more confident

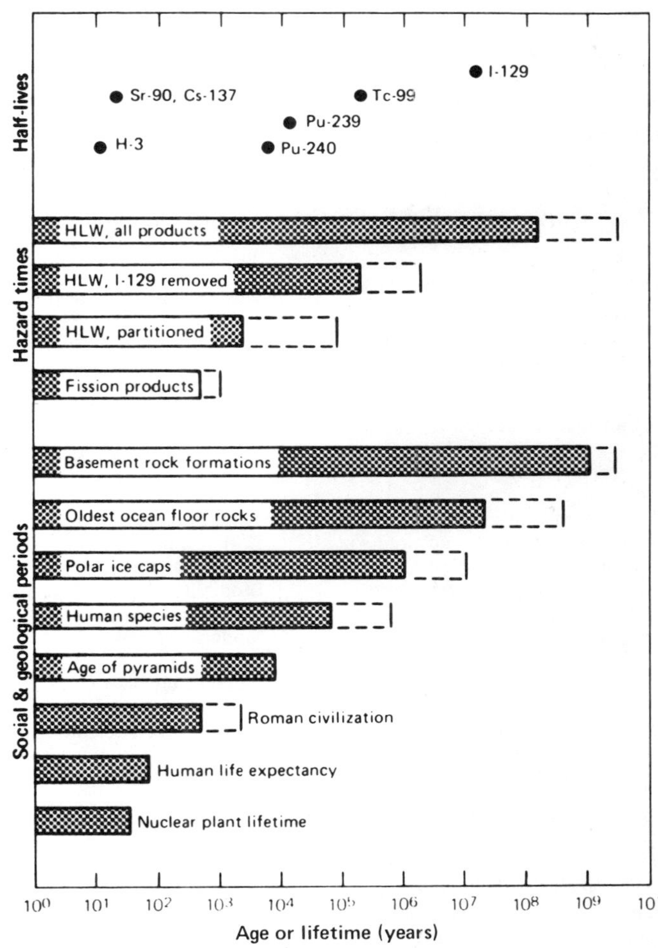

Fig. 2. The duration of hazards from various forms of nuclear wastes compared to times of social or geological relevance. The hazard duration is arbitrarily defined as the time for a given waste to decay to the same ingestion hazard index as that of a 4 percent uranium ore (carnotite) (see Fig. 1). Partitioning is assumed to remove 99 percent of the transplutonic elements from the HLW. The half-lives of some of the more significant constituents are also included for comparison.

we may be that the wastes will remain isolated in the face of social, technical, and geological uncertainties. If technical irreversibility were high, then neither cataclysmic natural events nor the activities of intelligent and technologically adept beings could readily return the wastes to the environment. Retrievable surface storage, for example, is highly reversible: vulnerable to accidents and easily accessible for recovery. Ejection into deep space is almost completely irreversible. Melting the wastes into a solid rock matrix would be highly irreversible: a geologic event that would result in large releases of toxic radionuclides would be very improbable; the application of fairly advanced and sophisticated mining technology would be required to deliberately reextract them from the rock.

Technical irreversibility measures resistance to both social and physical intervention (*55*). It does not correlate precisely with scientifically defined irreversibility. Irreversibility can be expressed mechanically, as with a ball rolling down a hill set in the middle of a flat plain. The application of a little intelligence and energy can easily restore the ball to the top of the hill. The irreversibility embodied in the second law of thermodynamics is based on the difficulty of restoring an initial situation in the face of statistical improbabilities, the unlikelihood that a specified event or set of conditions will spontaneously occur if it is but one of a large number of accessible outcomes. The presence of intelligence, however, allows the creation of improbable circumstances (*56*); reversibility may be expensive, but it is not in principle impossible.

There are parallel examples of social irreversibility. It is easier to create a bureaucracy than to destroy it (*57*). Increases in the perceived quality of our lives are not readily foregone (*58*). An example of almost purely social irreversibility that is more to the point here is the fabulous pirate practice of burying a treasure in a remote or obscure location and then killing those who know of it. Mechanically, the burial is very reversible; retrieving the treasure is simple once its location is known. But it is socially irreversible, since accidental discovery is highly unlikely and a deliberate, but unguided, search has a very low probability of success.

Irreversibility is proposed as a criterion to provide some degree of security against breaching of containment and failure of isolation in the face of unknown social, political, and cultural developments; to provide the greatest possible security against their release or misuse by an agent not equipped to recognize or cope with the dangers (*59*). Stability against geological change is a minimum requirement. But the degree of reversibility also depends on the amount of attention that might be drawn to the site by geological features or identifiable artifacts. Intelligent life is notoriously incautious in indulging its curiosity. Construction of a large concrete mausoleum, for example, would almost guarantee that concerted efforts would be made to breach it by intelligent, but uninformed, life. On social grounds, such a method is held to be quite reversible. Additional irreversibility cannot be provided by warning messages, symbols, or labels. We cannot assume that even a society that has the technology to undo rather irreversible storage will know enough about radioactivity to proceed cautiously, or that it will be able to decipher a message it cannot read (*60*). Indeed, the presence of such an indecipherable message would only arouse additional interest. "Interesting" geological formations such as salt domes are equally likely to draw attention. The society that drills into them may know nothing of radiological hazards, but still be sufficiently advanced technologically and scientifically to be curious about the formation itself and its possible contents (*61*). A condition for site location that aids irreversibility is that it be as uninteresting as possible, and so draw no attention for other reasons (*62*).

Table 1 is a preliminary classification of several waste disposal methods according to the degree of technical irreversibility possessed by each, as derived from consideration of both social and technical factors. The categorization is deliberately broad, since a more precise distinction would not only require more detailed analysis, but is limited by technical and social uncertainty. In addition, many of the suggested waste disposal methods could be made more irreversible by a judicious alteration to provide additional technical or social barriers to prevent breaches of the containment and isolation. For example, emplacement in geological formations would be more irreversible if chemical means could be found to immobilize the wastes against uptake into biological systems, since

Table 1. Technical irreversibility of selected methods for the management of wastes with high levels of activity. [Adapted from (*7*)]

Disposal method	Irreversibility		
	Physical	Social	Technical
Retrievable surface storage	Very low for water cooling to moderately low for above ground convection cooling	Very low	Very low
Sealed mausoleums	Low to moderate	Very low	Low to very low
Mined caverns in salt	Moderately low	Low	Low
Drilled- or solution-mined cavities in salt	Depends on ground water; low to moderate	Low for domes to moderate for bedded salt	Low to moderate
Seabed, emplacement in bottom sediments	Moderate, depends on nature of sediments	Moderate to moderately high	Moderate to moderately high
Mined rock cavity, partial melting	Moderately high	Moderate to moderately high (away from other minerals)	Moderate to moderately high
Mined rock cavity, complete melting	Moderately high	Moderately high (away from other minerals)	Moderately high
Seabed emplacement in basement rock, no melting	Moderately high to high, depends on geologic activity	High	Moderately high to high
Deep rock melt, drilled hole	High; depends on geologic activity	High (located away from other minerals)	High
Deep rock melt, self-descending capsule	High to very high, depends on geologic activity and sinking depth	High to very high	High to very high
Space disposal, outer space mission	Very high to complete	Very high to complete	Very high to complete
Transmutation	Complete	Complete	Complete

such uptake can both increase waste mobility and provide for subsequent reconcentration of the wastes in the food chain (*63*). Disposal on the ocean bottom would be more irreversible if the canisters were randomly placed so that a deliberate and informed search would be necessary to recover them in significant numbers.

Technical irreversibility, then, is defined by a combination of social and physical elements that measures both the size and the sophistication of the technology or natural mechanism that would be necessary to return the wastes to the biosphere in quantities or at rates that would be radiologically significant. It tends to correlate fairly well with the degree of scientific and technical aptitude that would be required for deliberate waste recovery by a society of intelligent beings, and with the size and cost of the necessary effort. The greater the degree of technical irreversibility, the greater the confidence that any failure of isolation and containment will occur only through the intervention of those fully capable of understanding the risks involved. In that sense, it is a useful criterion for establishing standards for waste management that reflect our ethical obligations to the future.

Multiplicity

There will always be some uncertainty as to whether a chosen method for waste disposal is technically sound, or whether we are capable of thinking through all possible circumstances by which containment might be breached. If a single site, single technique method were used, an appropriate question would be: How strong does your basket have to be before you are willing to place all of your eggs into it? The provision of additional baskets has two dimensions—multiplicity of sites and diversity of options. The purpose is, in either case, to provide redundancy as a hedge against error and uncertainty (*64*).

For example, the irreversibility of many types of terrestrial geological disposal methods could be increased by making the number of sites very large, thus reducing the potential risk due to the breach of any single one. This measure could then be augmented by random emplacement in unrecorded locations. The increased probability of accidental discovery must be balanced against the lower radionuclide inventory to see whether this strategy would in fact reduce net risk under a wider range of geological and social factors. An alternative approach would be to collect many years production of waste into a single giant container and then to emplace this so deeply and with such redundant barriers that any breach would seem highly improbable. This approach would significantly increase the probable consequences of a release. Under the specified conditions, multiple emplacement is held to confer more technical irreversibility in the face of uncertainty as to social, technical, and geological futures (*65*).

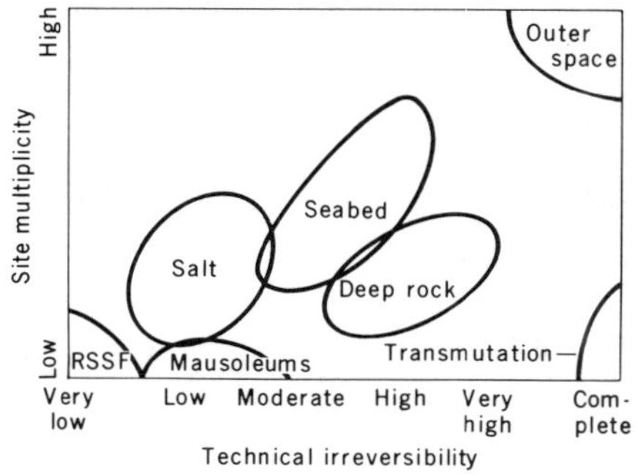

Fig. 3. Qualitative diagram of several of the waste disposal methods listed in Table 1 classified according to site multiplicity and technical irreversibility. The salt, seabed, and deep rock loci include all of the possible options listed in the table.

Multiplicity of sites does not, of course, provide any security against fundamental conceptual or design errors. It does help to minimize the consequences of such errors if the failures are random and widely spaced in location and time. But if confidence in the performance of a single site were high, multiplicity would not necessarily provide an advantage on technical grounds. Its primary advantage would be the reduction of the consequences of the deliberate or inadvertent action of intelligent life.

One aspect is damage limitation. If the opening of a single site caused minimal harm, and if the discovery of one site did not automatically provide the key to uncovering others (*66*), catastrophic releases would be less likely to occur. Furthermore, this could provide time for effects to be connected to the proximate, if not the ultimate, cause. Given the large uncertainties in predicting future social patterns, such provisions for damage limitation should be a leading factor in considering alternatives even if irreversibility is somewhat compromised. For increased site multiplicity is not necessarily identical to increased technical irreversibility. Although the two tend to correlate for many waste disposal methods, there are some (such as retrievable surface storage) for which the two criteria are nearly independent.

Figure 3 locates a number of waste disposal options on a two-dimensional plot that treats technical irreversibility and site multiplicity as independent criteria (*67*). The scale of the axes does not imply any attempt to predetermine their relative importance. It must be emphasized that this is a qualitative map. Not only the absolute but sometimes the relative location of any option is a matter of informed judgment. It is not only difficult but unwise to try and localize any method too narrowly. Even if the axes could be accurately and quantitatively labeled, inherent uncertainties in predicting the future would negate any attempt to pin a given method down precisely.

Applying the Criteria

For these criteria to be useful for organizing alternative approaches to nuclear waste management, they must be translatable into normative standards to guide decisions. A central hypothesis concerning the utility of the two criteria and two suggestions for applying them to guide the formulation of waste management policy are offered for this purpose.

The hypothesis is as follows: Emphasizing the continuity of goals in formulating waste management policy and doing away with arbitrary classifications (such as short-term versus long-term storage) increases the possibility for reasoned and ethically sound policy choices. The two criteria were developed, and the case for them is argued here, specifically to facilitate this procedure.

The first suggestion is this: Both technical irreversibility and site multiplicity are desirable goals for waste management. Given equal uncertainty that two different methods for the management of wastes will suffer from gross conceptual or design faults, the one that maximizes an appropriate weighing of the two criteria is preferable (*68*). Although estimating the relative weights is a part of the decision process not to be preempted here, Fig. 3 suggests how this might be done. The further into the upper right-hand corner a method lies, the greater the reduction of potential future risks in the face of social, technical, and physical uncertainties. Conversely, the greater the confidence in social and physical stability over the time scale during which the wastes must be kept isolated, the closer to the lower left an acceptable method will lie (*69*). Note that this method is applicable to all forms of waste.

The joint application of the two criteria provides a crude measure of reduction of risk in the face of uncertainty. But uncertainty increases with time. With

Fig. 3 being used as an illustrative device, there is an effective containment time scale running from the lower left (reversible, single site) to the upper right (very irreversible, high multiplicity). The shorter lived the wastes, the less the necessary containment time, and therefore the more uncertainty that can be tolerated.

This leads to the second suggestion: For any type of nuclear waste, a set of combinations of the two criteria can be determined that bounds the region of acceptable waste management methods. With reference once again to Fig. 3, this suggestion can be graphically interpreted as saying that there are lines of equal preference that can be (fuzzily) drawn upon the diagram to separate the acceptable from the unacceptable regions of performance. Because minimization of long-term risk is the dominant concern, the desirability of any given option will be measured by the degree to which it lies beyond the region of minimum acceptability. If there were several equally preferable options, the secondary criteria of reducing operational risk and cost could be freely used to select among them (*70*).

It should be kept in mind that technical irreversibility is meant to provide a criterion for choice and not to preempt it. Complete irreversibility that precludes all possibility of recovery may not be the most desirable outcome. It can be argued that our obligation to the future extends to the preservation of options as well as the prevention of harm, that we have an obligation to try to avoid irreversible consequences of our actions (*71*). It may then be considered more desirable to dispose of the wastes by a method roughly as irreversible as the dispersal of uranium in present ores. This would at least partially correct the irreversible depletion of natural supplies of fissionable material. The provision of an artificial ore bed is intended to make these materials accessible only to those who understand what they are mining and why. In that regard, the artificial beds could be somewhat more secure against accidental mining than natural beds have been if care were taken to make sure that the wastes were not located near other desirable minerals.

A locus of minimum acceptability has been plotted on Fig. 4, the same axes being used as in Fig. 3 (*72*). Let us assume that this represents emplacement that presents a hazard roughly the same as that of the uranium ores we are now mining (*73*). A second curve has been drawn at somewhat higher values of irreversibility and multiplicity, representing emplacement that is not beyond potential recovery but would entail considerable cost and effort to recover with present technology. Nevertheless, the wastes could be recovered if the need or desire were great enough. By selecting a waste disposal option that lies between these two limits, we could do our best to ensure that the future would be exposed to no more risk than if we had not used the ore for power at all, while still doing our utmost to avoid irreversibility foreclosing future options.

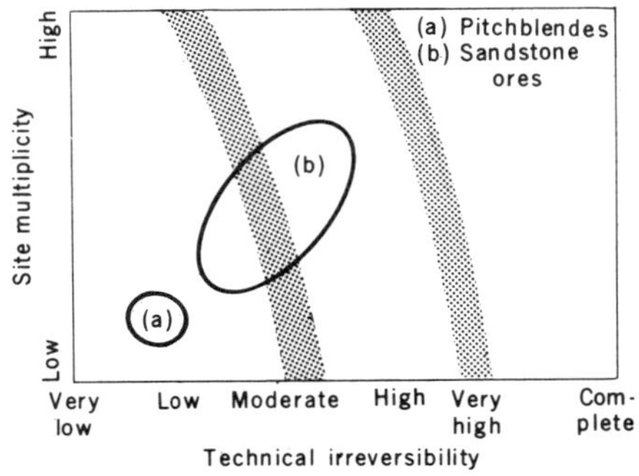

Fig. 4. Hypothetical loci of equal preference for choosing among alternative methods for nuclear waste disposal. The curve toward the left intersects the sandstone ores at roughly the irreversibility and multiplicity of 0.2 percent uranium ores (sandstone). The curve toward the right is intended to represent an emplacement that would be economically difficult, but not impossible, to reverse with present technology.

Diversity of Options

In the past, waste management policy has consistently been drawn to the search for the single most desirable alternative. The method suggested here facilitates the pursuit of several equivalently desirable options. To return to the previous metaphor, it is of little avail to put your eggs in many identical baskets if they all fail at once. Diversity of options can provide protection against such gross failures (*64*).

As with multiplicity of sites, statistical reasoning alone does not necessarily lead to the conclusion that diversity of options reduces risks (*74*). The parallel pursuit of more than one method for waste management is an explicitly normative recommendation, based on the ability of intelligent life to respond to failures. If one of the disposal methods failed at a time when there was a society capable of understanding what had happened and taking remedial action, it would be extremely important to have at least one alternative storage method that was trusted and immediately usable. If a method had been chosen that had fairly high site multiplicity, and if not all sites failed simultaneously, it is even possible that the transfer could be effected before any large fraction of the stored wastes was released (*75*).

If no other proved method were available, it would be far less probable that an acceptable alternative could be developed and proved before the number of site failures increased far past the point at which failure was noted. Furthermore, a society placed in the position of having sites fail with no available alternatives would be more likely to attempt remedial action than to develop new methods with unknown risks (*76*). This would be as likely to multiply the difficulties as to reduce them. The patched or modified sites might have a risk that would have been unacceptable by the original selection criteria. Patches or repairs might also begin to fail, and the modified system might be much more difficult to correct than was the original (*77*).

If the impact of any release were potentially catastrophic, it might be preferable to hold to a single waste disposal method to maximize the probability that no untoward events at all would occur during the required storage period. But a properly chosen combination of several multiple-site options would ensure that an event even of the worst kind would not be catastrophic, particularly for a method with high irreversibility. Furthermore, a most careful monitoring and testing program should be an integral part of waste management procedures to keep track of the condition of the sites and to detect any early signs of failure.

This safeguarding procedure would surely not outlast knowledge of the sites, and would not, therefore, compromise their long-term integrity. It is needed because the highest risk would be from gross failure in the early years of storage, and technologies that do fail are likely to give early warning signs. This should not be only a monitoring procedure, but part of an ongoing program of technical and social research to find and identify procedures and techniques that would increase site integrity and further minimize both the degree and the consequences of uncertainty. Our obligation to the future is not discharged simply by determining the level of risk to which they will be exposed. An ethically sound waste management policy will continuously and determinedly seek methods to reduce that risk.

Conclusion

Two criteria—technical irreversibility and site multiplicity—have been suggested for use in establishing standards for the disposal of nuclear wastes. They have been constructed specifically to address the reduction of future risk in the face of inherent uncertainty concerning the social and political developments that might occur over the required periods of waste isolation, to provide for safe disposal without the requirement of a guaranteed future ability to recognize, detect, or repair errors and failures (*78*).

Decisions as to how to apply or weigh these criteria in conjunction with other waste management goals must be made by societies and their governments. My purpose has not been to preempt this process, but to construct a framework that facilitates consideration of its ethical and normative components. As with many other human activities, the production of nuclear power entails consequences and risks for future generations who can have no voice in present decisions. On that account, their welfare must be carefully considered. It is not within our power to pass on to the future a world unchanged by our residence in it. Nor do we have an obligation to do so. But, as our every act has the potential to profoundly alter future lives, our minimum ethical obligation is to examine most thoroughly the potential consequences of present actions, to acknowledge them openly, and to minimize the potential for irremediable harm.

This obligation would not be satisfied if, in the disposal of nuclear wastes, we imposed upon future societies an obligation to provide for a stability of institutions unprecedented in history (*79*), if we attempted to transfer the responsibility for accidents from our shoulders to theirs. There is no ethical or moral basis for placing social and technical requirements and obligations on future generations for the sole purpose of protecting them from the consequences of present activities and decisions. The obligation to consider the effects of errors in technology or judgment, to provide for our inability to guarantee future technical performance, social stability, and cultural continuity rests with the present. An ethically sound waste management policy must reflect not only our knowledge and skills, but our limitations as well.

References and Notes

1. I. Calvino, *Le Città Invisibili* (Einaudi, Torino, 1972); translated into English by W. Weaver, *Invisible Cities* (Harcourt Brace Jovanovich, New York, 1974). "As for the character of Andria's inhabitants, two virtues are worth mentioning: self-confidence and prudence. Convinced that every innovation in the city influences the sky's pattern, before taking any decision they calculate the risks and advantages for themselves and for the city and for all worlds."
2. P. P. Micklin, *Bull. At. Sci.* **30** (No. 4), 36 (1974); A. B. Lovins and J. H. Price, *Non-Nuclear Futures* (Ballinger, Cambridge, Mass., 1975).
3. T. C. Hollocher, in *The Nuclear Fuel Cycle*, prepared by the Union of Concerned Scientists (MIT Press, Cambridge, Mass., 1975).
4. Transcript of hearings on the California Nuclear Initiative before the California State Assembly Committee on Resources, Land Use and Energy, 4 and 5 November 1975—summarized as *Reassessment of Nuclear Energy in California*, 10 May 1976.
5. By "society" is meant both intelligent life and its organized activities. The social costs of evacuation and contamination of land, as well as the effort needed to relocate and decontaminate are rarely assessed in computing the effects of radiological accidents.
6. K. J. Schneider and A. M. Platt, Eds., *Advanced Waste Management Studies, High-Level Radioactive Waste Disposal Alternatives* (Publ. BNWL-1900, Battelle Pacific Northwest Laboratories, Richland, Wash., 1974); summarized in U.S. Atomic Energy Commission, *High Level Radioactive Waste Management Alternatives* (Publ. WASH-1297, Government Printing Office, Washington, D.C., 1974).
7. Energy Research and Development Administration, *Alternatives for Managing Wastes from Reactors and Post-Fission Operations in the LWR Fuel Cycle* (Publ. ERDA-76-43, National Technical Information Service, Springfield, Va., 1976).
8. ———, *The Nuclear Fuel Cycle* (Publ. ERDA-33, National Technical Information Service, Springfield, Va., 1975). The "back-end" of the nuclear fuel cycle is defined in this document as consisting of spent fuel storage, reprocessing, mixed-oxide fuel fabrication, recycle of plutonium in reactors, and waste management.
9. An analysis of past Atomic Energy Commission (AEC) budgets bears this out. As is also pointed out in (*8*), budgets for commercial waste management have been particularly neglected.
10. Because operating costs are presumed to be a small fraction of total power generation costs, there has been no open conflict over the assumption that they will be passed on to the utilities and through them to the consumer. But, as pointed out in (*8*) and in the testimony of Liverman [J. L. Liverman, in *Oversight Hearings on Nuclear Energy Part I—Overview of the Major Issues* (hearings before the Subcommittee on Energy and the Environment, Committee on Interior and Insular Affairs, U.S. House of Representatives, Government Printing Office, Washington, D.C., 1975), p. 541] the AEC and its successor agencies hold that not only regulation, siting, and the provision of interim storage and final repositories are the responsibility of the federal government, but also support of commercial waste management research and development. This policy is held to derive from the requirements of 10 CFR 50, appendix F (*13*).
11. An examination of past testimony on waste management at AEC authorization hearings before the Joint Committee on Atomic Energy over the period 1960 to 1974 supports this. For a more concise description of the evolution of AEC waste management policy, see Boffey (*12*).
12. P. Boffey, *The Brain Bank of America* (McGraw-Hill, New York, 1975), chap. 5.
13. *Code of Federal Regulations, Title 10, Energy* (Government Printing Office, Washington, D.C., 1976).
14. Compare, for instance, the detailed rules and regulations for various isotopes and shipping containers embodied in 10 CFR 71 and 10 CFR 73 with the general and ambiguous definitions set out in 10 CFR 50 appendix F (*13*).
15. A. S. Kubo and D. J. Rose, *Science* **182**, 1205 (1973).
16. The distinction is that wastes are in principle retrievable from storage, but not from disposal. Recent documents such as ERDA-76-43 (*7*) have discontinued the confusing use of the term "ultimate storage" for disposal.
17. The Energy Research and Development Administration has not only withdrawn the draft environmental impact statement (*18*) on the retrievable surface storage facility (letter from R. C. Seamans, Jr., to J. O. Pastore, 9 April 1975) but has also indicated that it does not now intend to proceed further on it; see *Nucl. News* (*Hinsdale, Ill.*) (April 1976), p. 60.
18. Atomic Energy Commission, *Management of Commercial High-Level and Transuranic-Contaminated Radioactive Waste* (Publ. WASH-1539, Atomic Energy Commission, Washington, D.C. 1974).
19. *Siting of Fuel Reprocessing Plants and Waste Management Facilities* (Publ. ORNL-4451, Oak Ridge National Laboratory, Oak Ridge, Tenn., 1970).
20. A more comprehensive discussion of the prospects and problems is given in BNWL-1900 (see 6).
21. Transmutation and space disposal are classified by ERDA-76-43 (*7*) as "elimination" to distinguish them from terrestrial disposal. This introduces another new category and suggests that other disposal is not "ultimate."
22. See, for example, the statement of F. K. Pittman, in *ERDA Authorizing Legislation Fiscal Year 1976*, hearings before the Joint Committee on Atomic Energy, February 1975 (Government Printing Office, Washington, D.C., 1975), part 2, p. 1258.
23. D. Metlay (private communication) has pointed out that the AEC originally set the problem up well in 1959, but succumbed to the notion that technical fixes could resolve the problem. Initially the AEC used two approaches in dealing with hazardous materials: dilute and disperse; concentrate and contain. See, for instance, the testimony of Lieberman [J. A. Lieberman, in *Industrial Radioactive Waste Disposal*, hearings before the Joint Committee on Atomic Energy, 29 January 1959 (Government Printing Office, Washington, D.C., 1959), p. 989].
24. This approach is in marked contrast to the usual organization of waste management alternatives by one or another technological parameter. For examples of the traditional approach, see (*6*) and (*7*); for a less traditional but still primarily technical approach, see (*16*).
25. In a report [W. S. Maynard, S. M. Nealy, J. A. Herbert, M. K. Lindell, *Public Values Associated with Nuclear Waste Disposal* (Publ. BNWL-1997, Battelle Pacific Northwest Laboratories, Richland, Wash., 1976)] of a survey conducted by the Human Affairs Research Center of Battelle Memorial Institute, the attitudes of 465 persons from five regions of the United States toward waste disposal were analyzed by means of a poll. The respondents were then aggregated into six groups: environmentalists; high school students; nuclear technologists; public utility employees; university students; church and civic group members. The opinions of five of the six groups were in agreement with the rank ordering of the concerns by importance suggested here. Only the nuclear technologists rated short-term above long-term safety. All groups rated costs as being the least important consideration.
26. That is, an action is a candidate for moral choice only if it is voluntary and based on all available information. See, for example Aristotle, *Nichomachean Ethics*, W. D. Ross, Transl. (Clarendon, Oxford, 1925), book III, vol. 9, pp. $1109^{B}30$ to $1115^{A}7$.
27. At present, the more hazardous effluents from the back-end of the fuel cycle are considered to be ^{3}H, ^{14}C, ^{85}Kr, and ^{129}I. Their classification as "potential hazards" is based not only on their activity and half-life, but on the quantities to be released and the available pathways to humans.
28. Although the Nuclear Regulatory Commission has presumptive statutory authority over all gaseous and liquid radioactive effluents, only constituents confined on site by filters or as trapped liquids fall under waste management regulations.
29. Some difficulty has been encountered in attempts to define the level of contamination that distinguishes ordinary from transuranic-contaminated low-level wastes, and in establishing procedures for dealing with the latter. The currently proposed standard of treating all wastes with an alpha-activity exceeding 10 nc/g as transuranic contaminated and shipping them to a federal repository within 5 years as set out in WASH-1539 (*14*) is in dispute between the Nuclear Regulatory Commission and the industry.
30. The distinction between low- and intermediate-level wastes is not specified by quantitative guidelines. One common interpretation is that wastes that are safe to handle without special precautions are low level, and vice versa.
31. Few, if any, industry or government agencies are prepared to consider spent fuel as a waste. It is held to be a valuable (if unnatural) resource, owing to the large inventory of fissile material contained.
32. Atomic Energy Commission, *Final Generic Environmental Statement on the Use of Recycled Plutonium in Mixed Oxide Fuel in Light Water-Cooled Reactors* (Publ. NUREG-0002, Nuclear Regulatory Commission, Washington, D.C., 1974).
33. High-level liquid wastes are defined in 10 CFR 50, appendix F (*13*) as "those aqueous wastes resulting from the operation of the first cycle solvent extraction system, or equivalent, and the concentrated wastes from subsequent extraction cycles, or equivalent, in a facility for reprocessing irradiated reactor fuels."
34. In (*32*) it is stated that about 5 percent of the uranium and plutonium in the spent fuel will be lost to high-level wastes. There will be additional losses to other than high-level wastes at the reprocessing plant and at other facilities.
35. Primarily neptunium, americium, and curium.

36. The container must meet the specifications of 10 CFR 71 and 10 CFR 73 (*13*).
37. There are also very active components that decay away in times of only a few months to a few years. At present, the spent fuel is stored for a long enough time for these decays to occur. For some waste management schemes, such as deep rock melt, it is necessary to ship and reprocess the fuel very quickly to preserve these short-lived isotopes. This may significantly alter the balance between short-term and long-term risks.
38. That is, under specific exposure conditions, such as inhalation of insoluble fine powders.
39. Nuclear Regulatory Commission, *The Management of Radioactive Waste: Waste Partitioning as an Alternative* (National Technical Information Service, NF-CONF-001, Springfield, Va., 1976).
40. The thermal power of wastes that have been aged for the requisite several years is comparatively small, and cannot be recovered without an effective energy subsidy roughly an order of magnitude greater than the usable output. For a summary of possible beneficial uses of radioisotopes, see G. P. Dix, in *Waste Management '75*, R. G. Post, Ed. (Univ. of Arizona Press, Tucson, 1975), p. 153.
41. For an analysis of the present (unfavorable) economics of extracting usable isotopes, see (*22*, p. 1304). Aside from the hazards of storing these materials, carrying charges will add greatly to their price.
42. This could conceivably be extended to include all spent fuel for which there is no currently available reprocessing capacity. But see (*31*).
43. I follow here the general outline of the analysis by W. W. Lowrance [*Of Acceptable Risk* (Kaufman, Los Altos, Calif., 1976)]. Matters of empirical fact are: risk (the measure of adverse effects); efficacy (the measure of beneficial effects); cost (both internalized and external); and the distribution of these matters. These translate via personal and social values into normative measures: safety (the degree to which risks are acceptable); benefit (the degree to which efficacies are desirable); affordability (the degree to which costs are reasonable); equity (perceptions of just distributions). Criteria are taken to be based upon empirical data. They provide only the quantitative basis for choice. Standards are used to screen for acceptability, and therefore are based on the normative factors. Thus the maximum permissible concentration for a given radionuclide in air or water may be expressed in empirical terms such as curies per cubic meter, but are determined both by pathways to life, and by decisions as to safety and equity that are not entirely based on technical data.
44. With the exception of the civil liberties implications of safeguards, potential social costs of either successful operation or of failure have rarely been considered in analyzing the nuclear fuel cycle. Among the factors that need consideration are patterns of employment and land use, the consequences of evacuations and other dislocations, and anxieties and fears raised by both real and potential accidents. See, for example, R. Budnitz and J. Holdren, in *Annual Review of Energy I* (Annual Reviews, Palo Alto, Calif., 1976), p. 553.
45. There appears to be widespread agreement by all parties that operating costs for any waste management method now conceivable will be so small, on a percentage of power costs basis, that safety considerations should predominate in the decision. See also (*23*).
46. It is assumed here that nuclear electric power is a common good—one that distributes its benefits over all persons in a society by virtue of their participation in it—as well as a private one. Similarly, waste management, because of its risks and costs, may be described as a common "bad." In this context, exporting risk means exposing persons who derive neither individual nor common benefits. These may be persons in the future or members of other contemporary societies.
47. M. P. Golding and D. Callahan, *What Is Our Obligation to Future Generations?* (Hastings Center Institute of Society, Working Paper Series No. 2, Ethics, and the Life Sciences, Hastings-on-Hudson, N.Y., 1972).
48. K. Arrow, *Public Policy* **21**, 303 (1973). Arrow argues that the provision of full information about potential costs and risks is the minimum ethical obligation of the seller. H. Jonas (*49*) argues further that modern technology has conferred upon its users the power to have such enormous impacts that traditional ethics will not suffice. Given that uncertain outcomes can destroy the very context in which ethics operates, ignorance can no longer serve as an alibi. The ethical obligation thus extends to making the search for knowledge a prime duty.
49. H. Jonas, *Social Res.* **40**, 31 (1973).
50. "Action" is used here in the sense of Arendt (*51*, chap. 1, p. 7). Arendt states that, in contrast to labor and work: "Action, the only activity that goes on directly between men . . . corresponds to the human condition of plurality, to the fact that men not Man, live on the earth and inhabit the world. While all aspects of the human condition are somehow related to politics, this plurality is specifically *the* condition . . . of all political life." To act is to set something into motion in the context of human plurality, and because of this plurality every action is a beginning whose process is irreversible and whose outcome is unpredictable.
51. H. Arendt, *The Human Condition* (Univ. of Chicago Press, Chicago, 1958).
52. The future is by definition uncertain. Attempts to find a substitute for action and avoid the frustration of unpredictable outcomes must lead to either the suppression of human plurality through tyrannical control or the insistence that one's activity is "worldly" or self-contained, rather than political and interactive. The former reflects the refusal to allow consequences, the latter the refusal to acknowledge them. See Arendt (*51*, chap. 5).
53. Imperfections in human knowledge and the fundamental uncertainties associated with the probabilistic distribution of the frequency and severity of cataclysmic events can be somewhat compensated for by the provision of secondary barriers that ensure slow diffusion and return of the wastes to the environment even if the primary containment is breached.
54. This much, at least, is within the scope of mathematical risk analysis.
55. A good example of such mixed irreversibility is dropping a cracker spread with peanut butter and jelly face down on a sandy beach. In principle, the effects can be reversed. In practice, both social and physical costs are *usually* too high. I thank R. Budnitz for this example.
56. In fact, the continued existence of intelligent life requires it.
57. See, for example, H. Kaufman, *Are Government Organizations Immortal?* (Brookings Institution, Washington, D.C., 1976).
58. Note that only the perceived quality of life is referred to. This makes clear the political nature of the assertion. Translated into economic terms, this is equivalent to the price elasticity of demand being greater for falling prices than for rising prices. At times of great stress, such as during major wars, perceived quality of life may be sacrificed willingly. At other times, changes in perception are simpler.
59. This argument could be extended as follows: Knowing what we do of the dangers of nuclear weapons, we should not leave any fissionable materials for the future, on the assumption that we have been lucky and they could easily do far worse. This presents an ethical problem of even greater complexity than those set out in this article.
60. A. J. Evans excavated the first of the Mycenaean tablets inscribed in Linear B at Knossos in the year 1900. More than 50 years elapsed before they were deciphered. See J. Chadwick, *The Decipherment of Linear B* (Cambridge Univ. Press, Cambridge, 1970). The 3500 years that have elapsed since the inscriptions were made is only about one-seventh of the half-life of plutonium-239. Yet, almost all the information about the culture and language of Minos had been lost.
61. If we depleted existing beds of uranium ores, a future society would, in all probability, develop to a fairly advanced industrial stage before discovering the existence of natural radioactivity.
62. One of the advantages of seabed disposal in the center of the North Pacific plate is that the site is not only geologically stable, but barren of resources and scientifically boring (W. Bishop and C. Hollister, private communication).
63. E. A. Martell, *Actinides in the Environment and Their Uptake by Man* (Publ. NCAR-TN/STR-110, National Center for Atmospheric Research, Boulder, Colo., 1975).
64. M. Landau, *Public Admin. Rev.* **29**, 346 (1969).
65. More detailed method-specific analysis is needed to examine this somewhat intuitive conclusion.
66. That is, by other than informed and sophisticated actors who know just what it is they are seeking and for what purpose it is to be used. Under these conditions, we can be fairly certain that they would be aware of the risks.
67. Technical irreversibility and site multiplicity are taken to be independent variables. It is assumed that for each method of waste disposal there is what amounts to a functional equation that expresses the interrelationship of the variables for that specific method.
68. In the absence of contrary information, a priori equal probability of gross failure is assumed.
69. It should be kept in mind that Fig. 3 is a conceptual and heuristic device, not a quantitative map. The significant information is the relative, not the absolute, position of any given option.
70. This is, of course, only one of the possible ways to select an option. The major point here is the bounding of a region that contains the unacceptable methods, so that they may be discarded.
71. D. E. Boeyink, "Finitude and irreversibility: The duty to avoid irreversible consequences," paper presented at a regional meeting of the American Academy of Religion, April 1975. Irreversible consequences are to be distinguished from irreparable harm. The former involves uncertainty as to the harm or good of our actions in the face of imperfect knowledge and a moral finitude. Note that it is the irreversibility of consequence that is suggested as being avoidable. All action is inherently irreversible. But see Arendt (*51*).
72. Waste disposal methods have been deliberately omitted from Fig. 4 to avoid even the appearance of preempting social choice. The placement of the indicated regions relative to the various options is not purely a technical problem.
73. That is, 0.2 percent uranium concentrations in sandstone.
74. Let us assume that all methods chosen have equal rates of failure. If two options were selected, and the probability of a certain release during the required storage time were ½ for each of the methods, the probability that both would have such a release would be ¼. For three methods, the probability would be ⅛. This would be advantageous only if the procedure resulted in effective damage limitation. But suppose that this release could result in a catastrophe. In that case, the selection of three methods at the stated probabilities would result in a ⅞ assurance that at least one such catastrophe would occur. For two methods, the probability would decrease to ¾, and for only one method to ½. For any combination of methods there would be a distribution of failures in time and a distribution of radionuclide inventory. It would be instructive to have some strategies played out by mathematical analysis to examine over what period the hazards entailed would actually be increased by diverse options in the absence of remedial action. For long times, damage limitation would be expected to dominate.
75. The information needed for monitoring the sites and locating them would not compromise the requirement of technical irreversibility if properly done. The original design consideration should provide for the storage and handling of emplacement data, and the monitoring of disposal areas, in such a way as to provide inaccessibility of both sites and information to a naive actor.
76. This would be particularly true if remedial action could be rapidly effected, since a thorough program to develop a new method from scratch would take a minimum of several years.
77. For example, potential leakage from corroding carbon steel tanks at the Hanford reservation was prevented by solidifying the contained wastes into salt cake. There is no existing method for removing the solidified wastes without risking a potentially serious spill, because the tanks can no longer be checked for integrity. As ERDA itself put it, in *Creating Energy Choices for the Future* [(Publ. ERDA-48, Government Printing Office, Washington, D.C., 1975), vol. 2, p. 119]: "If it is determined that the salt cake must be removed from these tanks before the level of radiation decays substantially (several hundred years), unique fully remote techniques for removing the salt cake from the storage tanks will have to be developed."
78. To the extent that one holds the contrary belief that social stability is more assured than present technological aptitude, it would be better to store the wastes in a small number of accessible sites so that performance could be monitored and errors corrected. Of course, this assumes that future technologies will be an improvement and that operational errors such as that discussed in (*77*) will be avoided.
79. A. Weinberg, *Science* **177**, 27 (1972).
80. The comments of T. Bradshaw, D. Metlay, B. Schiff, K. Smith, and P. Windham are gratefully acknowledged. I thank T. La Porte and A. Middleton for their advice and support.

The CANDU Reactor System: An Appropriate Technology

CANDU combines a proved reactor, a simple industry, and outstanding potential for resource conservation.

J. A. L. Robertson

The Canadian CANDU (Canada Deuterium Uranium) power reactor does not fit easily into categories of U.S. reactors. Pickering nuclear generating station, which is operated by Ontario Hydro and consists of four 500-megawatt (electric) CANDU reactors, has produced more electricity than any other nuclear generating station, at roughly half the cost of electricity from the available alternative, a coal-fired plant. CANDU reactors are therefore "currently commercial," and comparable to the U.S. light-water reactors. In the United States, however, heavy-water reactors, along with molten-salt and high-temperature reactors, are usually considered as "advanced" because of their low uranium consumption. Neither description does the CANDU reactor full justice. Good neutron economy gives CANDU reactors the ability to operate on a thorium fuel cycle at, or near, breeding. This means that the energy costs are virtually independent of the fuel supply cost, making CANDU comparable to the fast breeder reactor. As a commercially proved reactor able to adapt to changing circumstances, CANDU has a useful role to play in the world's future energy supply.

Summary. CANDU power reactors are characterized by the combination of heavy water as moderator and pressure tubes to contain the fuel and coolant. Their excellent neutron economy provides the simplicity and low costs of once-through natural-uranium fueling. Future benefits include the prospect of a near-breeder thorium fuel cycle to provide security of fuel supply without the need to develop a new reactor such as the fast breeder. These and other features make the CANDU system an appropriate technology for countries, like Canada, of intermediate economic and industrial capacity.

To a physicist, the essence of a CANDU reactor is its use of heavy water as the moderator. In any thermal reactor the moderator must slow down neutrons efficiently while capturing as few as possible. Both requirements are factored into a figure of merit called the moderating ratio. Table 1 shows how heavy water stands out from other potential materials. To an engineer, pressure tubes are the major distinguishing feature of a CANDU reactor. In light-water reactors (Fig. 1) there is no distinction between moderator and coolant, and the core is contained within a large, thick-walled pressure vessel. The CANDU moderator, essentially unpressurized and cool, is contained in a reactor vessel (calandria) through which pass some hundreds of identical pressure tubes containing the fuel and coolant. Each tube, made of a zirconium alloy, is about 6 meters long and 10 centimeters in diameter and has a wall 4 millimeters thick.

The pressure-tube design facilitates on-power fueling, achieved by mobile machines which connect to opposite ends of the pressure tube to be fueled [a diagram of the fueling machines, together with further details of CANDU reactors, is provided by Haywood (*1*)]. On-power fueling minimizes the inventory of neutron-absorbing fission products. Heavy water, on-power fueling, and constant attention to neutron economy at all stages of design enable CANDU reactors to enjoy the economies of natural uranium fuel now and the prospect of a near-breeder thorium fuel cycle for the future.

Other attractive features result from these primary characteristics.

1) The use of natural uranium in a very simple fuel design (Fig. 2) provides exceptionally low fueling costs.

2) On-power fueling contributes to a high net capacity factor (the energy actually delivered in a particular period divided by the energy delivered if the unit had operated continuously at its maximum rated output) by eliminating the need for fuel changing during reactor shutdowns.

3) Any fuel assembly that develops a leak during operation can be traced to a particular pressure tube, and even to its position in that tube. On-power fueling allows the leaking assembly to be removed before it can release significant radioactivity to the coolant.

4) The separation of coolant and moderator means that their chemistries can be independently optimized. Specifically, neutron poisons can be added to the moderator to control or shut down the reactor without jeopardizing the corrosion control of the primary coolant circuit.

5) The separate moderator at relatively low temperatures constitutes a very large heat sink capable of absorbing any energy that might be released in postulated accidents. It also facilitates the design of control and shutdown devices.

History

The selection of reactor types for national power systems is not determined by physics alone. The United States, Britain, and Canada cooperated in the wartime development of atomic energy, and all three were aware of the merits of heavy water as a moderator. Indeed, all three have had development programs for heavy-water power reactors, but only Canada has pursued the concept to the commercial stage. History shows why.

Naval propulsion was the first application of nuclear power in the United States. Minimum physical size was the dominant criterion and light-water reactors were a reasonable choice. Enriched fuel was available from diffusion plants built during the war, and construction of the large pressure vessels was within U.S. industrial capability. In Britain the first application was to plutonium-production reactors for a weapons program. Again, long-term economy was not the primary criterion, and gas-graphite reactors were a reasonable choice. In the

The author is assistant to the general manager, Atomic Energy of Canada Limited, Chalk River Nuclear Laboratories, Chalk River, Ontario, Canada K0J 1J0.

early 1950's, when both countries recognized an early need for civilian nuclear power, these two reactor types had the tremendous advantages of demonstrated engineering viability and established industrial infrastructure.

In postwar Canada, abundant hydroelectric resources postponed the need for nuclear power and provided a tough economic challenge. (Even in 1977 domestic customers in some parts of Canada pay as little as 2 cents per kilowatt-hour.) The widespread use of hydroelectricity had accustomed Canadian utilities to capital-intensive plants with very low operating costs. It was recognized that a low fueling cost would be crucial to the success of a reactor and that neutron economy was the key (*2*). Heavy water was therefore the first choice for the moderator when Canada began civilian power reactor studies in the middle 1950's. The choice was reinforced by two practical factors: good experience with the heavy water–moderated NRX research reactor (*3*) and successful operation of heavy-water production plants in the United States. These two factors, incidentally, illustrate the mutual benefits of early U.S.–Canadian collaboration. The design of the Savannah River production reactors drew heavily on NRX experience, while the development of the process to supply heavy water to the production reactors helped the Canadian program.

Table 1. Merit ratings of potential moderators.

Material	Moderating ratio
Zirconium hydride	49
Organic liquids	60-90
Light water	72
Beryllium	159
Graphite	160
Beryllium oxide	190
Heavy water	
Pure	12,000
99.8 atom percent	2,300

The initial design of our first, 20-MWe, nuclear power demonstration (NPD) reactor (*4*) was similar to that of a light-water reactor. Since scaling up would be limited by the industrial capacity for fabricating large pressure vessels, the advantages of a modular pressure-tube design were becoming apparent. The development of Zircaloy-2 by the U.S. Bettis Laboratory in Pittsburgh, made this concept practicable. Even so, the designers would probably not have had the confidence necessary to base a reactor design on Zircaloy-2 had not Bettis workers been testing the material in NRX as part of a collaborative program. This same collaboration also contributed greatly to an early selection by both groups of uranium dioxide and Zircaloy as fuel and cladding materials, respectively.

The NRX reactor and later the NRU (*5*) and WR-1 (*6*) research reactors played a vital role. Their outstanding capabilities for testing fuel and structural materials have attracted U.S. groups, as well as some from other countries, to conduct development programs. Full-size CANDU fuel bundles can be tested under realistic operating conditions in actual pressure tubes in these experimental reactors.

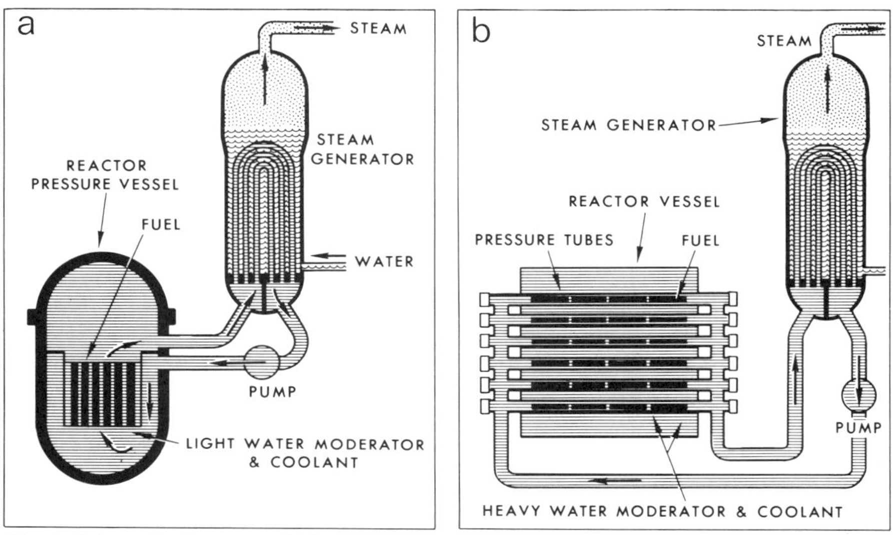

Fig. 1. Schematic diagrams of (a) a light water–cooled, pressure-vessel reactor and (b) a heavy water–cooled, pressure-tube reactor.

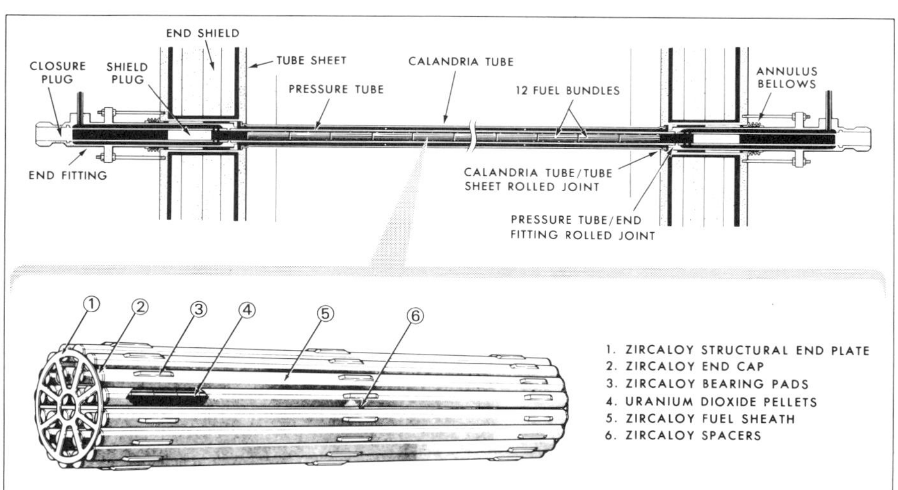

Fig. 2. Individual fuel channel and fuel bundle.

Operating Experience

The NPD reactor, in operation since 1962, demonstrated the engineering viability of the CANDU design, especially the concept of on-power fueling, which had not previously been attempted for a water-cooled power reactor. Its net capacity factor has exceeded 80 percent for the past 3 years, despite its use to train utility staff and as a test-bed for new fuel designs and other reactor equipment. If NPD's good operation failed to alert the designers to possible problems, the same cannot be said of the 200-MWe Douglas Point prototype reactor (*7*), which began operation in 1967. Douglas Point's performance has improved substantially over the years, but its greatest contribution, as befits a prototype, has been to identify design and operational improvements for future reactors.

The Douglas Point experience contributed greatly to the exceptionally smooth commissioning of the four Pickering reactors (*8*) between February 1971 and June 1973. The Pickering net capacity factor to July 1977 has averaged 78 percent; during the last fiscal year (April 1976 to March 1977) all four units exceeded 90 percent and were in the top five of an international table of 67 reactors of comparable size ranked by capacity factor (*9*).

The fuel for all CANDU reactors consists of short (50 cm) bundles of Zircaloy tubes containing uranium dioxide pellets (Fig. 2). The simplicity of the design and the use of natural uranium result in exceptionally low fuel supply costs. Although the design burnup (energy produced per unit mass of fuel discharged) is much lower than for light-water reactors, the resulting fueling costs (fuel cost per unit of energy produced) are less than half those for any other nuclear station in North America. Pickering's fueling costs have varied from 0.9 mill/kWh in 1973 to 1.2 mill/kWh in 1976 (*10, 11*). The near-continuous fueling of CANDU reactors serves as a coarse reactivity control and allows the burnup to be measured to a high accuracy (Fig. 3), confirming design estimates (*12*).

The high capacity factors are partly due to good fuel performance. Of more than 84,000 bundles irradiated at Pickering, less than 0.03 percent have exhibited any leaks (*11*). In 1971, observation of a slightly higher than average failure rate for Douglas Point fuel led to the identification of a new failure mechanism—and the means of controlling it (*13*). When a fuel bundle that has had prolonged exposure at low power is subjected to a sudden increase in power its Zircaloy cladding is liable to fail by stress corrosion cracking. This mechanism was responsible for the failure of about 100 bundles in the first Pickering unit to operate, during its first year. Since the introduction of remedial measures in 1972, only about 20 bundles have failed in all four units.

Another common cause of reactor incapacity, boiler leaks, has had very little effect on CANDU reactors. In nearly 50 reactor-years of operation only five tubes have leaked, one each at NPD, Pickering, and Bruce, and two at Douglas Point (*14*). No failures have been attributed to corrosion, probably because of selection of appropriate materials combined with careful control of the water chemistry on both primary and secondary sides of the boilers. Before the most recent boiler-tube failure at Pickering in 1974, a method for rapidly locating and plugging a leaking tube had been developed; this leak was fixed during a reactor shutdown lasting only 5 days.

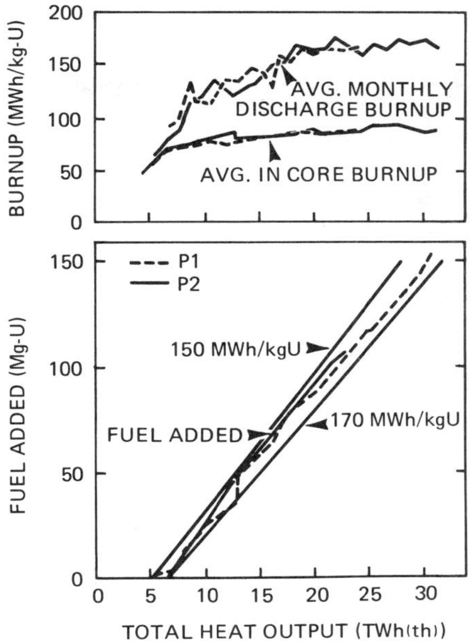

Fig. 3. Fuel consumption and burnup for Pickering units 1 and 2 (*12*).

The single most important cause of incapacity at Pickering was an installation error that led to the development of cracks in some of the pressure tubes of two of the reactors. These cracks did not prejudice the safety of the reactors, and there was never any risk to the public or the operating staff from the resulting coolant leaks, which were contained and collected within the reactor vaults. The cost of replacing 69 pressure tubes, including the cost of supplying energy from an alternative source during repairs, amounted to $75 million. Even after these costs are debited, operation of the Pickering station resulted in a net saving for the utility's customers during the period in question. (During 1976 the saving exceeded $100 million.) The investigation of the cracking (*15*) provided not only an immediate remedy to prevent further cracks but also a much better understanding of the process used to roll pressure tubes into end fittings, thereby providing improved techniques for the future.

The demonstrated ability to replace the NRX and NRU calandrias (*16*) and Pickering pressure tubes provides confidence that even the most radioactive parts of CANDU reactors can be repaired, if necessary.

Other Distinctive Features

The design, construction, and operation of CANDU reactors have other distinctive features.

1) At present, the Pickering station consists of four identical reactor-turbine-generator units, each of which operates independently; a second group of four units is now under construction at the same site (Fig. 4). The multi-unit design has proved most effective in saving costs and in preventing delays in construction and commissioning (*8*). The period from first critical to full power progressively decreased from 95 days for the first unit to 12 days for the fourth.

2) The design of the reactor and associated equipment was undertaken by the federal agency Atomic Energy of Canada Limited, while the design of the remainder of the plant, the project management, and the on-site construction were the responsibility of the provincial, publicly owned utility, Ontario Hydro. This arrangement exploited the complementary expertise and experience of the two organizations. Although completion of units 1 and 2 was late by a few months according to the original schedule set at the start of construction, units 3 and 4 were completed within schedule (*8*).

3) While each reactor is within its own containment building, the multi-unit design features a negative-pressure containment system that minimizes stresses in the containment building in the event of a loss-of-coolant accident. The four containment buildings are connected through a pressure relief duct to a large vacuum building (Fig. 4) normally held at about 10 kilopascals, which contains a pressure-actuated dousing system for pressure suppression. The same vacuum building will serve the four additional reactors now under construction.

4) The vacuum building, with a 1-m-thick concrete wall, was poured by a slip-form method in 7½ days of continuous pouring. The method has subsequently been further developed for reactor buildings, with the formwork of the roof dome being assembled at ground level and raised with the forms (*9*).

5) CANDU reactors pioneered the use of direct computer control. The operative computer is monitored and backed up by an identical standby computer; if both fail the reactor is shut down automatically.

6) As a result of a strong incentive to prevent leaks of valuable heavy water, significant improvements have been made in "conventional" equipment. As an example, packed-stem valves have been modified so that the packing is maintained in compression by spring-loading (*17*). This very simple modification greatly reduces leakage and thereby

minimizes maintenance requirements in areas with high radiation fields.

7) Measures to minimize losses of heavy water also prevent radioactive releases. As a result the present situation is satisfactory; releases from the Pickering station (*18*) have been within the design target of 1 percent of the regulatory requirements based on recommendations of the International Commission on Radiological Protection.

8) Radiation fields that impede maintenance are mostly due to corrosion products in the primary coolant circuit that have been activated in the reactor and subsequently deposited on out-reactor components such as the boilers. Several methods for controlling and reducing these fields without the necessity of removing the fuel or draining the coolant have been investigated (*19*). The most attractive is a mild chemical process in which the cleaning agents are continuously regenerated and the radioactivity is removed in solid form rather than as bulky acid solutions. In 1971 and 1972, fields (behind shielding) at the Douglas Point reactor were reduced sixfold by simple changes in physical and chemical conditions of the coolant, and were subsequently maintained at the reduced level. More recently, the chemical cleaning process was applied during a 3-day shutdown of the reactor to reduce fields in certain areas by a further factor of 6.

9) Exposure to tritium generated in the heavy water of the moderator and coolant is a possible impediment to maintenance work. A new process, combined electrolysis and catalytic exchange (CECE), offers great promise for removing tritium from heavy water and also for upgrading heavy water that contains traces of light water (*20*).

Fig. 4. Two of the four reactor buildings for the Pickering B station under construction next to Pickering A. (The single, large, cylindrical building in the foreground is the vacuum building.)

Table 2. CANDU reactors in operation, under construction, committed, and planned.

Name*	Location	Power (MWe net)	Nuclear designer†	Date of first power
NPD	Ontario	22	AECL and CGE	1962
Douglas Point	Ontario	208	AECL	1967
Pickering A	Ontario	514 × 4	AECL	1971–1973
Gentilly 1	Quebec	250	AECL	1971
KANUPP	Pakistan	125	CGE	1971
RAPP 1	India	203	AECL	1972
RAPP 2	India	203	AECL	
Bruce A	Ontario	745 × 4	AECL	1976–1979
Gentilly 2	Quebec	600	AECL	1979
Point Lepreau	New Brunswick	600	AECL	1980
Cordoba	Argentina	600	AECL	1980
Pickering B	Ontario	514 × 4	AECL	1981–1983
Wolsung 1	Korea	600	AECL	1981
Bruce B	Ontario	750 × 4	AECL	1983–1986
Darlington	Ontario	800 × 4	AECL	1986–1988
		Total 16,703		

*NPD, Nuclear Power Demonstration; KANUPP, Karachi Nuclear Power Project; RAPP, Rajasthan Atomic Power Project. †AECL, Atomic Energy of Canada Limited; CGE, Canadian General Electric Company Limited.

An Appropriate Technology

The CANDU reactor system is appropriate for countries of intermediate economic and industrial capacity, such as Canada. Producing heavy water is basically simpler than enriching uranium, and commercial heavy-water plants have been built in smaller sizes than would have been possible for uranium enrichment plants.

Serious problems were, nevertheless, encountered in developing a Canadian heavy-water industry (*21*). Perhaps because the process used—countercurrent exchange between liquid water and gaseous hydrogen sulfide—had operated well in U.S. plants, the difficulties of transferring the technology to Canada were underestimated. However, Ontario Hydro's Bruce heavy-water plant has been operating well since 1974, and by 1983, when all plants now under construction are expected to be in full operation, Canada will have a capacity sufficient to support a CANDU installation rate of 3000 MWe per annum. In developing its heavy-water industry Canada has introduced the first application of isotope separation under commercial conditions.

The use of natural uranium facilitated the establishment of a domestic fuel industry, while the short bundles and relatively low burnup allowed volume production at an early stage. Similarly, the production of several hundred pressure tubes, with associated end fittings and calandria tubes, is better suited to Canadian industry than would be the fabrication of a single large pressure vessel. The CANDU calandria is as large as a light-water reactor's pressure vessel but, not having to withstand high pressures, its structure is much simpler and it can be manufactured in Canada. Indeed, about 80 percent of each CANDU generating station can be made in Canada (*22*).

Commercialization of the CANDU reactor system has been achieved at a cost between \$1 billion and \$2 billion, including demonstration plants. This sum compares favorably with those for other systems, including what has already been spent on fast breeder reactors.

Future Developments

All CANDU reactors, including those under construction and committed, are listed in Table 2. The first three units of the Bruce station are operating, while the fourth is due to be commissioned in early 1979. The present installed nuclear capacity in Canada of roughly 4000 MWe approximates Ontario Hydro's total electrical capacity in 1954, when development of the CANDU system began.

The separation of moderator and coolant offers scope for future developments, while still preserving the essential characteristics of the design. An organic liquid could replace heavy water as the coolant. This would allow higher operating temperatures, and hence higher efficiency, as well as some other advantages. Also, the temperatures would then be high enough to provide high-pressure steam for extracting oil from tar sands and heavy-oil deposits (*23*).

The technical feasibility of an organic-cooled CANDU reactor has been demonstrated by the WR-1 (*6*) research reactor, which has been operating very successfully at the Whiteshell Nuclear Research Establishment in Manitoba since 1965. Much more work on an engineering scale would be needed to make this modification of the CANDU reactor commercially available.

Another possibility is using ordinary light water as coolant, taking the coolant direct to the turbine without an intermediate boiler. The greatest attraction of this approach is that it would mean an appreciable reduction in capital costs at a time when all energy projects are straining capital resources. The technical feasibility of using a light-water coolant in a direct steam cycle has been demonstrated, this time by a prototype reactor at Gentilly in Quebec (*24*).

Despite the prospect of lower costs from the use of alternative coolants, the excellent performance of the Pickering station is likely to make utilities reluctant to change from the heavy-water coolant. Some total unit energy costs (*25*) estimated by Ontario Hydro (Fig. 5) illustrate several significant points. First, oil is just too expensive to compete, apart from any question of its future availability. Second, the Pickering A nuclear generating station and the Lambton coal-fired plant are both four-unit stations of the same capacity and were built at the same time, so they are closely comparable. Recently, the energy costs from Pickering have been about half those from Lambton. Third, a comparison of pairs of nuclear stations, such as Pickering A with Pickering B and Bruce A with Bruce B, shows an annual capital cost escalation rate of about 13 percent, which may cause concern but which is not out of line with other major construction projects. Fourth, because the capital component dominates nuclear energy costs, future increases in fuel costs are likely to have much less effect on nuclear than fossil-fired plants. Projections such as this lead to the expectation that up to half of Canada's electricity can be supplied by nuclear energy by the end of the century, using CANDU reactors of basically the present design.

Fuel Supply

Canada is well endowed with uranium, and present reserves, amounting to about one-quarter of total world reserves (exluding the Soviet bloc, for which data are not available), would be sufficient to satisfy domestic demand well into the next century. However, an increasing worldwide dependence on nuclear energy as oil and gas supplies dwindle has put increasing pressure on Canada to export uranium; the pressure is likely to increase if the introduction of fast breeder reactors in other countries is delayed. To protect domestic interests while recognizing international obligations, the Canadian government announced a uranium policy in 1974. Canadian utilities are assured of at least a 30-year reserve of uranium for all existing, committed, and planned reactors in any 10-year forward period. In return, the utilities must maintain supply contracts for a 15-year forward period, to provide market stability for uranium producers. These arrangements should provide utilities and producers with the confidence needed to make major investment decisions.

From the start, the fueling of CANDU reactors has been based on a once-through fuel "cycle." The prospect of recycling spent fuel, when required to conserve uranium supplies, has always been envisaged, but there is no incentive to do so until it is shown to be economically attractive under conditions that are demonstrated to be safe. The present simplest possible fuel cycle, which does not depend on fuel reprocessing, is likely to be retained as long as uranium remains plentiful and relatively cheap. Thus the Canadian program is much less immediately affected than many others by recent calls for reassessment of fuel reprocessing.

Internationally, there is considerable uncertainty over how much economically recoverable uranium will be discovered during the next few decades and how rapidly production facilities will be installed. There can be little argument that the availability of fuel cycles more conserving of uranium would be welcome, provided that they can be shown to be acceptable on grounds of safety and security. Their availability would contribute to the world energy supply, to

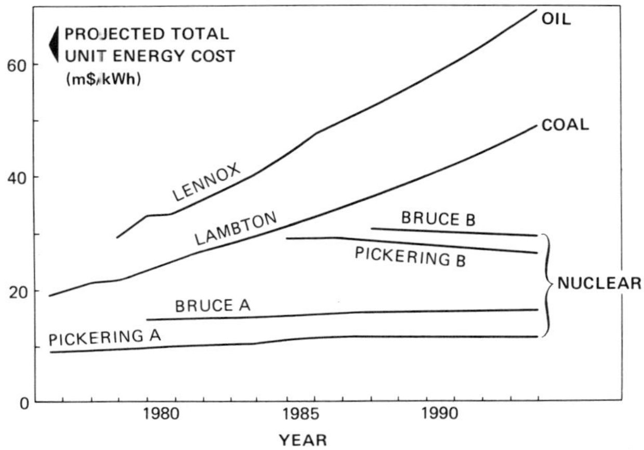

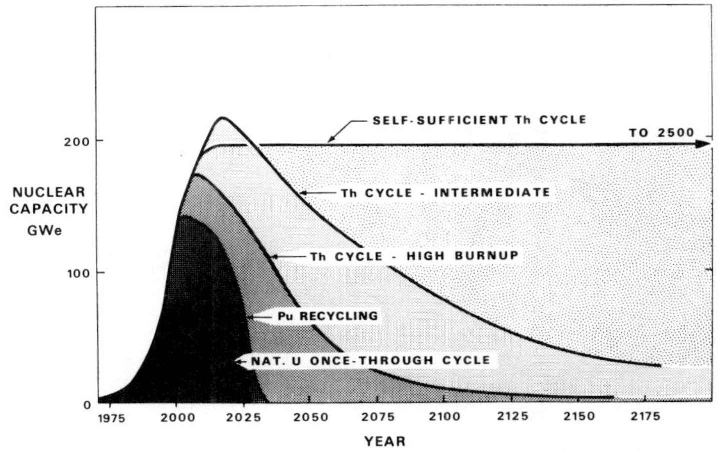

Fig. 5 (left). Projected costs of power from three different fuels (*25*). Fig. 6 (right). Energy that could be obtained from 3×10^5 megagrams of uranium with various CANDU cycles, to illustrate the degree of resource extension possible (*10*).

the security of the national supply, and to the confidence needed for utilities to commit future plants to nuclear fuel and hence save scarce fossil fuels. The good neutron economy of CANDU reactors permits them to exploit a thorium fuel cycle that is exceptionally efficient in its use of nuclear fuel (*26*). This ability to switch when necessary to a more conserving fuel cycle without having to develop a completely new reactor system, such as the fast breeder, is another very attractive feature of the CANDU reactor system.

Current estimates indicate that a "self-sufficient thorium cycle" may be practicable in CANDU with only minimal modifications. At equilibrium, this cycle would require no further uranium and only small amounts of thorium, which is at least three times as abundant globally as uranium. Uranium required to initiate the cycle does not have to be maintained as virgin uranium but can be in the form of spent CANDU fuel. The fast reactor fuel cycle is similar in this respect, and even the amount of uranium needed, 1 to 2 megagrams per electrical megawatt of installed capacity, is roughly the same. Figure 6 illustrates the energy available from various fuel cycles in CANDU reactors, and demonstrates the very large extension in fuel resources possible by switching from the present natural-uranium, once-through cycle to the self-sufficient thorium fuel cycle.

"Electronuclear breeding," the conversion of fertile to fissile material by use of electric power, is of special interest to neutron-economic reactors such as CANDU, where small increases in fissile content can yield disproportionate savings in fuel consumption. The proposed "fusion-fission symbiosis" is probably the best-known example of electronuclear breeding, but it is at least arguable that the spallation process can be applied to produce neutrons and hence fissile material before fusion is available (*27*). Spallation has the obvious advantage that its scientific feasibility has been demonstrated, and demonstration of its engineering feasibility looks promising. In Canada, a relatively small effort has been devoted to spallation for several years, and recently there have been signs of increasing interest in the United States (*28*). Considering the magnitude of the undertaking and the long-term nature of the objective, this subject seems ripe for international cooperation.

Nuclear Waste Management

The management of nuclear wastes is a subject in which international cooperation is already very effective. Canada's program (*29*) is basically similar to that of most other countries, but has some aspects that usefully complement other programs. Since the CANDU system has assumed once-through fueling from the start, there is underwater storage at each station for 5 to 10 years of spent fuel. A centralized, interim facility for storing spent CANDU fuel safely and securely for periods up to a century is now being considered. The options include underwater storage (Fig. 7), as used at present, or the more passive dry storage. Figure 8 illustrates the concrete canister concept for dry storage: four such canisters are already on test, two with electric heaters and two with spent fuel. Whichever design is adopted, the land requirement is insignificant—5 square kilometers or less—for all Canadian spent fuel expected to the end of this century.

For the permanent disposal of nuclear wastes many countries are considering mined depositories deep underground in stable, dry geologic formations. The Canadian program includes examination of salt deposits, the formation currently favored in the United States, but is devoting more effort to mined cavities in certain granitic, mineralization-free structures that are widespread in the Canadian Shield, which constitutes much of Canada's land mass. The expectation is that both types of formation will prove acceptable, and the world as a whole can only benefit from having a choice.

Canada, in common with other countries, expects to immobilize the wastes in a highly insoluble solid before disposing of them permanently. In 1960, representative separated fission products were dissolved in glass blocks, 50 of which were buried in loose sand below the water table in a controlled area of the Chalk River Nuclear Laboratories (*29*). Monitoring of the very small amounts of ra-

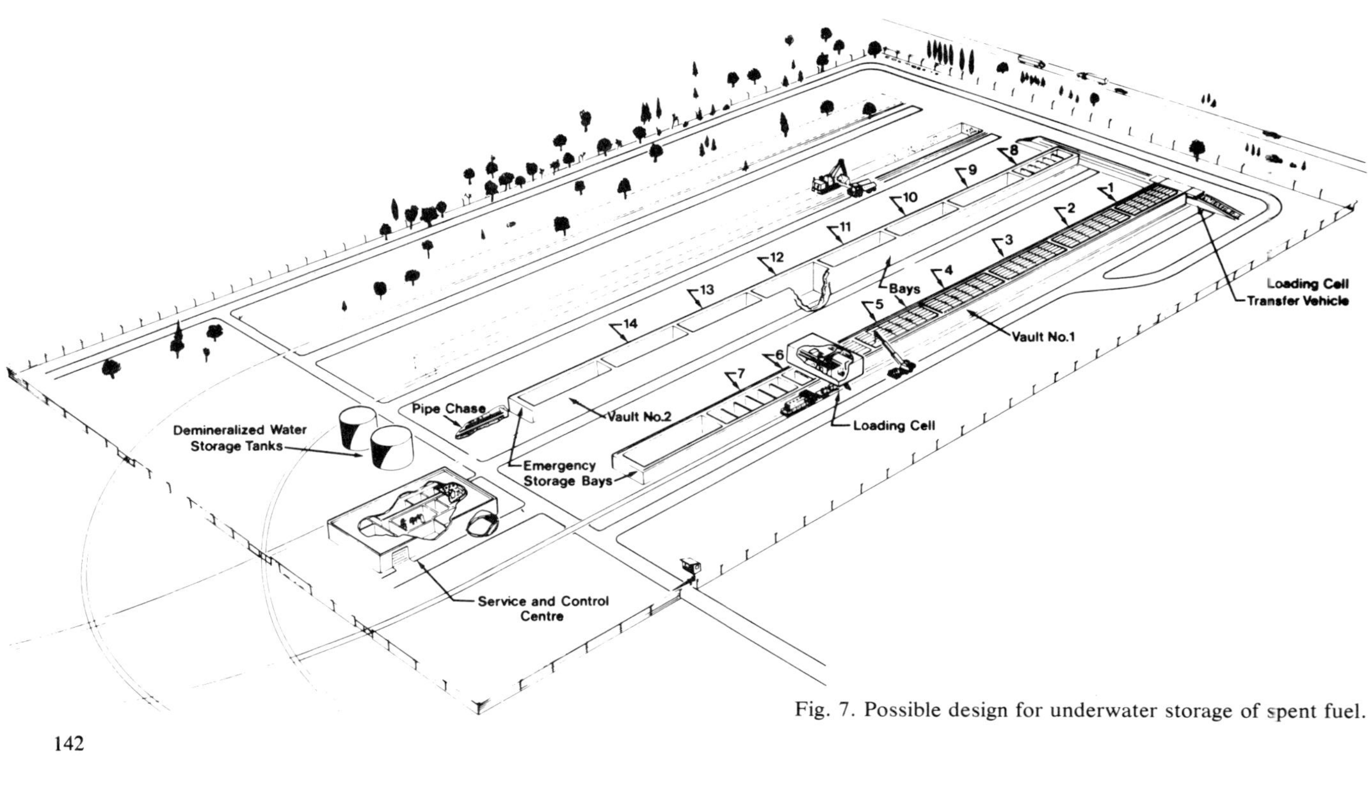

Fig. 7. Possible design for underwater storage of spent fuel.

dioisotopes in downstream groundwater shows that the blocks are dissolving more slowly than the contained radioactivity is decaying. Similar work is required on other compositions and under other conditions, but these results indicate the potential of this approach to immobilizing the wastes.

The overall objective of the waste management program is to provide safe storage of spent fuel until society determines whether the large amount of latent energy it contains should be exploited through fuel recycling. Methods are being developed and will be demonstrated for the permanent disposal of either the separated wastes or the unprocessed fuel. Parallel work will establish the economics and acceptability of fuel recycling, information that will be needed for reaching a sound decision on the disposition of spent fuel.

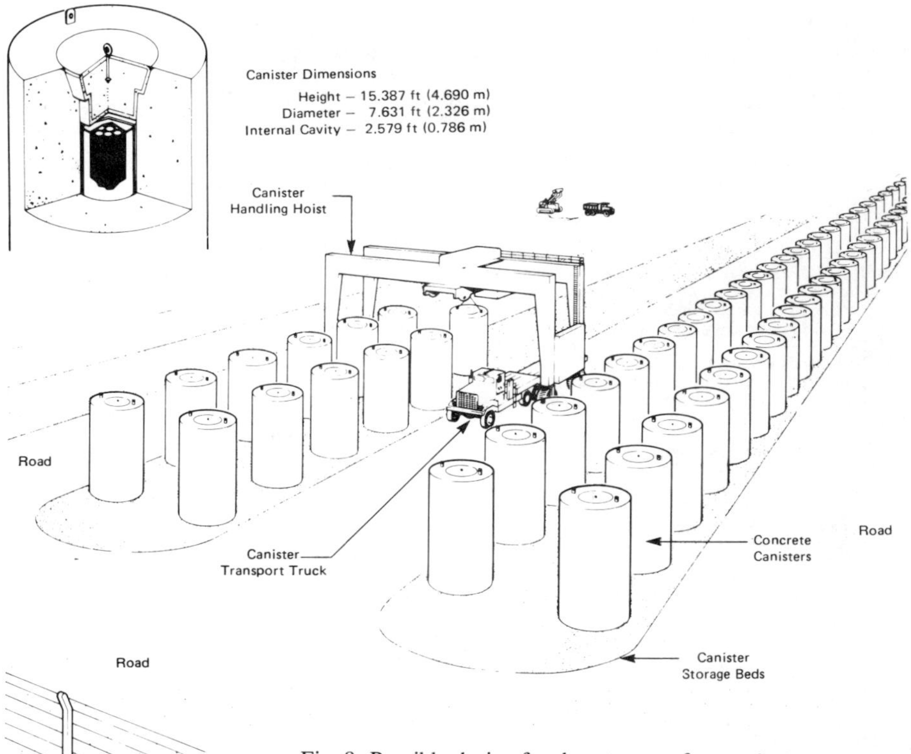

Fig. 8. Possible design for dry storage of spent fuel.

International Aspects

The present CANDU fuel cycle, without uranium enrichment or spent fuel reprocessing, is virtually impregnable to any terrorist group seeking to acquire fissile material. If fuel recycling were introduced some increased protection would be advisable, but there would be no need for shipment or stocks of separated fissile material. Any fuel for CANDU reactors, either before or after recycling, would still be very dilute in fissile content, around the 1 percent level. Thus several factors of concern to critics of fast breeder reactors would remain absent from the CANDU fuel cycle.

The absence of enrichment and reprocessing simplifies the international safeguarding of the CANDU system, but the reactor's capability for continuous on-power fueling offsets that advantage. However, techniques and instrumentation for safeguarding CANDU reactors to International Atomic Energy Agency requirements have been developed and are being demonstrated (*30*). A thorough study (*31*) argues that CANDU reactors are no more likely than light-water reactors to be used for the clandestine production of weapons material, while other studies (*32*) have shown that no power reactors are likely to be so misused.

Table 2 includes CANDU reactors that are in operation or under construction outside Canada. Canadian government policy is that nuclear materials and equipment may be exported only to countries that accept full-scope international safeguards. Canada is also participating in the current International Nuclear Fuel Cycle Evaluation, resulting from the seven-nation summit meeting in London in May 1977, to review over a 2-year period methods by which the proliferation of nuclear weapons capability can be impeded without jeopardizing the role that nuclear power can play as a secure source of energy worldwide.

Establishing confidence in international safeguards and associated political measures is of the utmost importance in making available to the world the benefits of nuclear energy. Assuming a successful outcome, the CANDU system has a useful contribution to offer. The major industrialized countries are unlikely to adopt CANDU reactors with once-through fueling, if only because of their massive investment of capital and experience in the infrastructure for their existing systems. However, a growing number of less industrialized countries are beginning to turn to nuclear energy as supplies of fossil fuels dwindle. These countries could well see in the CANDU system the attractive features that have made it an appropriate technology for Canada. The two systems would complement each other in their use of the world's fuel resources. The light-water reactors, leading to fast breeder reactors, could eventually convert and use the world's uranium, while CANDU reactors are capable of extending these resources to include the even more abundant thorium. During the period of once-through fueling, the existence of CANDU reactors with their greater fuel economy will help to reduce the pressure on uranium supplies.

For the longer-term future it is reassuring to have available two distinct systems, each dependent on a different resource base, either of which is capable of supplying the world's energy needs for many generations. Until the commercial availability of alternatives, such as the renewable energy sources or fusion, has been demonstrated, this degree of redundancy could be regarded as essential.

References and Notes

1. L. R. Haywood, paper presented on behalf of Atomic Energy of Canada Limited at the International Atomic Energy Agency Interregional Training Course on Nuclear Power Plant Project Planning and Implementation, Karlsruhe, Germany, 1975; *At. Energy Can. Ltd. Rep. AECL-5321* (1976).
2. W. B. Lewis, *At. Energy Can. Ltd. Rep. AECL-382* (1956; revised 1961).
3. R. E. Manson and H. E. Smyth, *At. Energy Can. Ltd. Rep. AECL-2692* (1967).
4. "Nuclear power demonstration reactor," *At. Energy Can. Ltd. Rep. AECL-1634* (1962); reprinted from *Nucl. Eng.* **7**, 398 (October 1962).
5. R. E. Manson, *At. Energy Can. Ltd. Rep. AECL-1897* (1974).
6. D. G. Turner, *At. Energy Can. Ltd. Rep. AECL-4763* (1974).
7. C. L. Moon, *At. Energy Can. Ltd. Rep. AECL-1515* (1962); reprinted from *Can. Nucl. Technol.* **1** (No. 4), 15 (1962).
8. W. G. Morison, *J. Br. Nucl. Energy Soc.* **14**, 319 (1975).
9. "Atomic Energy of Canada Limited Annual Report 1976–1977," *At. Energy Can. Ltd. Rep. AECL-5837* (1977).
10. J. S. Foster and S. H. Russell, paper presented at the IAEA (International Atomic Energy

Agency) International Conference on Nuclear Power and Its Fuel Cycle, Salzburg, Austria, May 1977; *At. Energy Can. Ltd. Rep. AECL-5707* (1977).
11. G. R. Fanjoy and A. S. Bain, paper presented at the IAEA International Conference on Nuclear Power and Its Fuel Cycle, Salzburg, Austria, May 1977; *At. Energy Can. Ltd. Rep. AECL-5711* (1977). [Performance statistics have been updated to August 1977 by A. S. Bain (private communication).]
12. C. H. Millar, paper presented at the European Nuclear Conference, Paris, 1975; *At. Energy Can. Ltd. Rep. AECL-5090* (1976).
13. J. A. L. Robertson, in *Proceedings of Joint Topical Meeting on Commercial Nuclear Fuel Technology Today* (CNS ISSN 0068-8517, Canadian Nuclear Association, Toronto, 1975), pp. 2-2 to 2-14.
14. R. I. Hodge, J. E. LeSurf, J. W. Hilborn, paper presented at the IAEA International Symposium on Application of Reliability Technology to Nuclear Power Plants, Vienna, Austria, 10–14 October 1977.
15. E. C. W. Perryman, *J. Br. Nucl. Energy Soc.*, in press.
16. The NRX calandria was changed in 1953 and again in 1970; the NRU one in 1973. See, for instance, G. W. Hatfield, *Mech. Eng.* **4**, 124 (1955); *At. Energy Can. Ltd. Rep. AECL-164* (1955).
17. P. A. Ross-Ross, E. J. Adams, D. F. Dixon, R. Metcalfe, paper presented at the European Nuclear Conference, Paris, 1975; *At. Energy Can. Ltd. Rep. AECL-5092* (1975).
18. L. W. Woodhead and L. J. Ingolfsrud, in *Proceedings of Canadian Nuclear Association 15th International Conference* (Canadian Nuclear Association, Toronto, 1975), vol. 2, pp. 39–70; L. W. Woodhead, in *Proceedings of Canadian Nuclear Association 16th Annual International Conference* (Canadian Nuclear Association, Toronto, 1976), vol. 6, pp. 46–48.
19. J. E. LeSurf, *J. Br. Nucl. Energy Soc.* **16**, 53 (January 1977); *At. Energy Can. Ltd. Rep. AECL-5643* (1977).
20. M. Hammerli, W. H. Stevens, J. P. Butler, in *Separation of Hydrogen Isotopes* (American Chemical Society, Washington, D.C., in press).
21. A. Dahlinger, W. E. Lockerby, H. K. Rae, paper presented at the IAEA International Conference on Nuclear Power and Its Fuel Cycle, Salzburg, Austria, May 1977; *At. Energy Can. Ltd. Rep. AECL-5710* (1977).
22. E. C. W. Perryman, paper presented at the Iran Conference on Transfer of Nuclear Technology, Shiraz, Iran, April 1977.
23. R. O. Sochaski and D. W. Smith, *At. Energy Can. Ltd. Rep. AECL-5682* (1977).
24. G. A. Pon, in *Proceedings of the IAEA Symposium on Heavy Water Power Reactors* (IAEA, Vienna, 1967), pp. 201–215.
25. D. E. Anderson, in *Proceedings of Canadian Nuclear Association 16th Annual International Conference* (Canadian Nuclear Association, Toronto, 1976), vol. 5, pp. 12–30.
26. E. Critoph, paper presented at the IAEA International Conference on Nuclear Power and Its Fuel Cycle, Salzburg, Austria, May 1977; *At. Energy Can. Ltd. Rep. AECL-5705* (1977).
27. S. O. Schriber, J. S. Fraser, P. R. Tunnicliffe, paper presented at the Tenth International Conference on High Energy Accelerators, Serpukhov, Soviet Union, July 1977; *At. Energy Can. Ltd. Rep. AECL-5903* (1977).
28. *Proceedings of the U.S. ERDA Information Meeting on Accelerator Breeding, Brookhaven, January 1977* (CONF-770107, National Technical Information Service, Springfield, Virginia, 1977).
29. M. Tomlinson, S. A. Mayman, H. Y. Tammemagi, W. F. Merritt, J. A. Morrison, H. S. Irvine, G. A. Vivian, J. Gale, B. Sanford, P. J. Dyne, paper presented at the IAEA International Conference on Nuclear Power and Its Fuel Cycle, Salzburg, Austria, May 1977; *At. Energy Can. Ltd. Rep. AECL-5706* (1977).
30. A. Waligura *et al.*, paper presented at the IAEA International Conference on Nuclear Power and Its Fuel Cycle, Salzburg, Austria, May 1977; *At. Energy Can. Ltd. Rep. AECL-5712* (1977).
31. R. W. Morrison, *Sci. Forum No. 60* (December 1977), p. 3.
32. C. Starr and E. Zebroski, paper presented at the American Power Conference, organized by the Illinois Institute of Technology, Chicago, Ill., 18–20 April 1977; *Nuc. News* **20** (No. 8), 54 (1977).

C Solar Energy

Solar Heating and Cooling

Solar energy for buildings is developing rapidly in the United States.

John A. Duffie and William A. Beckman

Thermal energy for buildings, supplied at temperatures near or below 100°C, constitutes an important segment of the U.S. energy economy and accounts for about one-quarter of the nation's energy use. Energy at these temperatures can readily be delivered from flat-plate solar energy collectors, and the solar energy incident on most buildings is more than adequate to meet these energy needs. Flat-plate collectors are manufactured and sold on a small but growing scale in the United States; they have been in use for more than a decade in heating water for buildings in Australia, Israel, and Japan. We expect that solar heating and cooling for buildings, with energy collected by flat-plate collectors, will be the first large-scale application of solar energy.

The basic problem with solar heating and cooling has been that the energy could not, except in special cases, be delivered at costs competitive with costs of energy from other sources. This situation is rapidly changing, and interest in solar energy is increasing almost daily as fuel costs rise. In areas where new natural gas connections are no longer available, where oil is not distributed, and where electrical resistance heating is the only alternative among conventional sources, solar heating is economically attractive.

In addition to technical and economic considerations, several social factors will influence the course and pace of developments. Two examples are worth examining. First, architectural constraints are imposed by the need for collectors to be oriented within rather narrow limits. This will make it difficult to fit solar heating systems to many existing buildings; thus new residential construction will be the easiest starting place for conversion to solar heating. Solar cooling may first be installed in existing low-rise, flat-roof buildings such as schools and shopping centers, where cooling is usually more important than heating. Second, tax policy is important. Today the installation of solar heating or cooling systems brings an increase in property valuation in most states, and a corresponding modest increase in real estate taxes. Government encouragement to invest in solar energy systems in the form of tax write-offs or other inducements (as are provided for investments by other energy producers) could very rapidly change the competitive position of solar energy in relation to conventional energy sources.

In all buildings, intelligent practices for energy conservation are worth following. The basic advantages of reducing energy needs by good thermal design apply whether buildings are supplied with solar or conventional energy. If solar energy costs the same as an alternative energy source, the value of energy conservation techniques, such as extra glazing on windows and doors or added insulation, is the same whether solar energy or the alternative is being used.

The authors are directors of the Solar Energy Laboratory at the University of Wisconsin, Madison 53706. Dr. Duffie is professor of chemical engineering and Dr. Beckman is professor of mechanical engineering at the University of Wisconsin–Madison.

Solar Radiation

The solar constant, that is, the intensity of solar radiation outside of the earth's atmosphere at the mean distance between the earth and the sun, has been determined by measurements from satellites and high altitude aircraft to be 1.353 kilowatts per square meter (*1*). This extraterrestrial radiation, which corresponds closely to that of a blackbody at 5762°K, is 7 percent in the ultraviolet range (wavelength less than 0.38 μm) and 47 percent in the visible range (wavelengths from 0.38 to 0.78 μm), with the balance in the near infrared (largely with wavelengths of less than 3 μm).

Solar radiation is depleted as it passes through the atmosphere by a combination of scattering and absorption; the radiation that reaches the ground—the raw material of this energy source—can vary from almost none under heavy cloud cover to 85 to 90 percent of the solar constant under very clear skies. Energy rates on surfaces normal to the radiation during good weather are not very high, and are typically about 1 kilowatt per square meter (a little more than 1 horsepower per square yard). Solar radiation on the ground consists of a diffuse component that has been scattered by molecules and particulate matter in the atmosphere and, when the atmosphere is sufficiently clear, a beam component that is unchanged in its direction of propagation from the sun. Its spectral distribution is altered in a manner dependent on atmospheric composition, with the major changes due to absorption of ultraviolet radiation by ozone and infrared radiation by water vapor.

There are several sources of solar radiation data. The National Oceanic and Atmospheric Administration weather service measures total (beam diffuse) radiation on a horizontal plane at more than 100 stations. Some stations report daily values, and some report hourly values. These data are available from the National Climatic Data Center (*2*). Monthly averages of daily radiation on horizontal surfaces are available for many locations (*3*). Daily integrated energy quantities at particular locations vary widely during the year. In Madison, Wisconsin, on a clear January day, energy on a horizontal surface is typically 3 kilowatt-hours per square meter, and July clear-day energy is typically 9 kw-h/m^2; the corresponding monthly averages of daily radiation on the horizontal surface are 1.8 kw-h/m^2 and 6.2 kw-h/m^2. Flat-plate collectors sloped toward the south in Madison, with a slope equal to the latitude, will have incident on them an average daily radiation of 3.4 kw-h/m^2 in January and 5.6 kw-h/m^2 in July. These data illustrate the gains to be obtained by orienting a collector in a favorable manner.

Although solar energy intensity is low, integrated energy quantities may be large. For example, in Madison the annual average solar energy incident per day on an acre of ground is the equivalent of about 10 barrels of oil, and on a 200-m^2 house is equivalent to about 25 gallons, which is far more than enough to meet the needs of the building for thermal energy.

Current Status

Two major reasons may be cited for the failure of solar energy to be a serious competitor in the energy market in past years. First, the costs of delivering solar energy have been substantially higher than those of other energy sources. Solar energy has not been able to be a competitor to inexpensive natural gas or petroleum. Second, there was no constituency pressuring for solar energy development in a manner similar, for example, to that of the nuclear industry that existed at the close of World War II and gave a substantial impetus to the development of peacetime uses of nuclear energy. The environmental movement of the last 5 years, the realization that the United States is dependent to an undesirable degree on foreign energy sources, and increasing fuel costs have served to establish a broad base of interest in developing solar energy.

The contrasts between the development of nuclear energy and solar energy are striking. After the destruction wrought by the atomic bombs in World War II, a large, concerned constituency, backed by the nation at large, pushed for development of peaceful uses of atomic energy. The result was a program supported by billions of dollars of federal funds over the course of three decades. Solar energy, in contrast, had no such support and it was only the persistence of a few individuals that kept interest in solar energy alive. Outstanding in this group was Farrington Daniels, who, through his publications (*4*) as well as through his support of the struggling International Solar Energy Society, served as an elder statesman for solar energy.

During the 1960's support for solar energy research for applications in the United States was essentially nonexistent. However, one program resulted in economic studies that have become part of the current interest in solar energy. Tybout and Löf, with support from Resources for the Future, developed a series of cost analyses of solar energy for heating and cooling (*5*). They indicated that solar heating could be competitive with conventional energy sources in high energy cost areas in 1968. They also showed that the combination of solar heating and cooling, which results in higher use factors on the solar energy equipment, is, in most places, more economical than heating or cooling alone. Their two studies were based on optimistic projections of the cost of solar energy equipment ($20 and $40 per square meter of flat-plate collector), but also on 1968 and 1970 energy costs. Later and more detailed studies of cost and thermal performance, based on more realistic collector and energy costs, bear out the same general conclusion that solar heating can now compete with expensive fuels.

By 1972, several dozen solar-heated residences or small laboratory buildings had been constructed and operated. A few of these have been studied, evaluated, and reported (*6*). The few air conditioning experiments were confined largely to experimental operation of 3-ton LiBr-H_2O absorption machines or analytical studies of system performance (*7*). In contrast, the manufacture and sale of solar water heaters to provide hot water for residences and some institutional buildings (hotels, dormitories, and the like) has been a commercial enterprise in Australia, Japan, and Israel for more than a decade. Perhaps a million solar water heaters are in use in these countries.

During the past 3 years, the availability of funds for experimental programs from the NSF Research Applied to National Needs (RANN) program and the Energy Research and Development Administration (ERDA), coupled with private and industrial investment, has led to many new experiments and applications of solar heating and cooling in the United States. Public buildings, schools, and a variety of residential buildings with heating or combined heating and air conditioning capacity are being planned and built. Quantitative information is now beginning to come in from these new experiments (*8*). The Solar Heating and Cooling Demonstration Act of 1974 should lead to many new solar buildings. In addition to research and development activities, there are now a few sales of solar heating systems which are installed as operating heat delivery systems rather than as experiments.

Solar Building Architecture

Several approaches to solar building architecture are evident. Several solar-heated houses and a school are illustrated in Fig. 1. The basic problem faced by architects and engineers is to integrate collectors into or onto the building in such a way that thermal performance is adequate, while obtaining an esthetically satisfactory structure. In this context, the major variable is the area of collector that must be integrated into the building. Collector area is central to the fraction of heating loads to be carried by solar energy and, ultimately, to cost.

The solutions are mixed. Some collectors have been mounted above flat-roof buildings (Fig. 1, a and b). To obtain structural or esthetic advantages, other collectors have been built into vertical walls in higher latitudes (Fig. 1c) or placed flat on horizontal roofs in lower latitudes. In addition, collectors have been integrated into the envelopes of buildings at orientations that are near optimum for the best thermal performance (Fig. 1, d to f).

Solar Energy Systems

Systems for producing service hot water, space heating, and cooling are based on the concept of the flat-plate collector. This unique heat exchanger uses a black absorber plate to absorb solar energy. Ducts or tubes carry air or liquid that remove energy from the plate. Layers of air provide transparent insulation between plates and their covers (usually made of glass) and thus reduce upward heat loss. Conventional insulation is provided on the backs and edges of the plates. The collectors are mounted in a fixed position according to the desired use of the energy. Figure 2 shows cross sections of air and water heaters.

The other major component in the system is the energy storage unit which is designed to accumulate solar energy when it is obtainable and make it available to meet energy needs at other times. Liquid systems usually use insulated water tanks for storage, and air systems usually use pebble beds. A third method of storage takes advantage of the latent heat of a phase transition, and has been the object of considerable study (*9*). Early work on house-heating applications concentrated on $Na_2SO_4 \cdot 10H_2O$, which undergoes a phase transition when heated at 32°C. Because phase separation of this hydrate occurs on cycling, other chemical systems are being sought which can undergo thousands of cycles without loss of storage capacity.

Schematic diagrams of liquid and air solar heating systems are shown in Fig. 3.

Fig. 1. Solar building architecture. (a) Löf home in Denver, Colorado. (b) Retrofitted collectors on school in Boston [photo courtesy of General Electric]. (c) Collector built into south wall of house of Odeillo, France. (d) MIT house IV, Massachusetts. (e) CSU house I, Fort Collins, Colorado. (f) Retrofitted house in Denver, designed by R. L. Crowther.

Both show an auxiliary energy source, which is included in most solar energy systems. In climates where a high degree of reliability is required of a heating system, the auxiliary source must be capable of carrying the full heating load of the building. If auxiliary energy is added in parallel with solar energy, then the maximum amount of energy can be obtained from the solar system and the balance from auxiliary. Other methods are possible.

For the liquid system the heat exchanger between collector and storage tank allows the use of an antifreeze solution in the collector loop, which is one of the methods to avoid freezing and reduce boiling problems. The diagram shows an additional heat exchanger to transfer heat to the building, and another to provide service hot water. The technology of solar liquid heaters is very well established, and most of the systems built recently have used liquids for heat transfer.

Air systems avoid boiling and freezing problems in the collector. In most air systems energy is stored as sensible heat in a pebble bed. A well-designed pebble bed has good heat transfer between air and pebbles, a low loss rate, and a high degree of stratification. Mechanical energy for pumping air can be a significant item of cost, and care is required to design for minimum pressure drops. The design of air systems and the balancing of good heat transfer characteristics against pressure drop are problems that are now receiving adequate attention.

Many of the scores of solar-heated buildings that have been constructed so far have provided reduced fuel bills as well as satisfaction to their owners. The performance of a few of these has been carefully measured, and provides a firm base of data on long-term thermal performance. Experiments up to 1961 were very well reported in papers presented at the U.N. Conference on New Sources of Energy in 1961 and summarized by Löf (*6*).

Massachusetts Institute of Technology (MIT) house IV, built in 1959, was the last in a series of experiments carried out by H. C. Hottel and his colleagues and represented a cooperative effort of architects and engineers to develop a functional, energy-conserving home with a major part of the energy for space heating and water heating to be supplied from the flat-plate collector. Figure 4, from Engebretson (*6*), is a schematic diagram of the heating and hot water system. The collector had an area of 60 m^2 for the 135-m^2 floor area, two glass covers, and a flat, black paint, energy-absorbing surface. To avoid freezing, collectors were designed to drain into an expansion tank. The main storage tank capacity was 5700 kilograms. Means were provided for

Table 1. Performance data for MIT house IV for two heating seasons [summarized from Engebretson (*6*)]. Numbers are expressed in gigajoules.

Item*	1959 to 1960	1960 to 1961
Space heating		
Demand	72.5	70.7
From solar energy	33.6	40.2
Water heating		
Demand	14.7	17.6
From solar energy	8.4	9.7
Total heating		
Demand	87.1	88.3
From solar energy	41.9	49.9
Percent from solar energy	48.1	56.6

adding auxiliary energy, extracting hot water for household needs, and transferring heat to air that was circulated to the rooms. This solar heating system was operated for three seasons, during which its performance was carefully measured. Data for the first 2 years are summarized in Table 1, which shows how energy requirements for space heating and water heating were met by solar or auxiliary energy. During the first two heating seasons solar energy supplied 52 percent of the energy for hot water and heating.

The Denver solar house, built by Löf (*6*) in 1958, uses air as the heat transfer medium and a pebble bed storage unit. The ratio of collector area to house area is about 1 to 5, a proportion much smaller than that of MIT house IV. This house has served as the Löf family residence since its construction, and the equipment has been routinely operated with only nominal maintenance. The system performance was measured in 1959 to 1960, and again in 1974 to 1975. For the period from December to April, 22 percent of the heating and hot water loads were carried by solar energy during the earlier season, and 20 percent during the later season.

Solar air conditioning technology is not as advanced as the heating process, since an additional thermodynamic process is needed for cooling. Several current experiments use absorption cooling cycles that are operated by heat from flat-plate collectors. These coolers are the analogs of the gas-fired refrigerators used in campers, but due to the lower temperature of fluid from the collectors (compared to a gas flame), water cooling is required rather than air cooling. Figure 5 shows a diagram of a solar-operated absorption cooling system. The same collector and storage units that provide winter heating thus can provide summer cooling.

Colorado State University (CSU) house I (which serves as an office building) uses a heating system that differs in some details from that of the MIT house, and also includes an absorption air conditioner. A glycol-water solution is used in the collector to avoid freezing problems and permit collector operation at higher temperatures. A heat exchanger is used to transfer solar heat into the water storage tank, and additional heat exchangers serve for heat transfer from the tank to hot water and the building. Thus, the collector supplies energy for three purposes: space heating, water heating, and air conditioning. A gas furnace provides auxiliary energy for both heating and cooling. The experiments started in August 1974 (*10*), and for the first 6 months of the heating season 86 percent of the space heating loads and 68 percent of the hot water loads were met from solar energy. Integrated performance statistics of a summer's air conditioning operation are not yet available.

A mobile laboratory (*11*) developed by

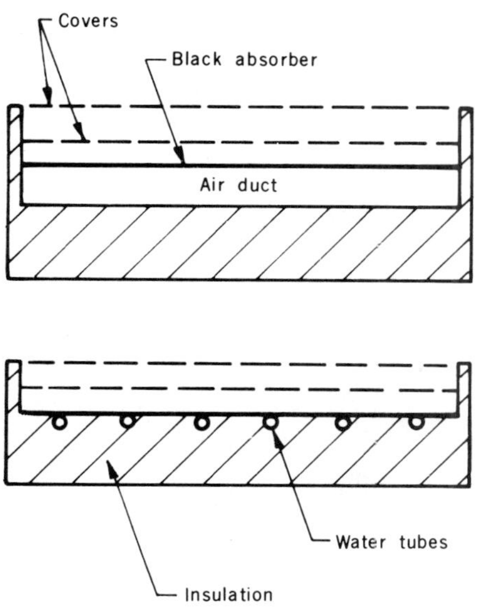

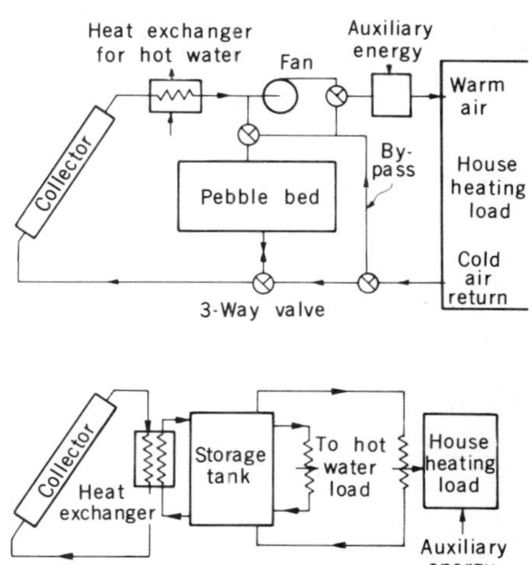

Fig. 2 (left). Cross sections of a solar air heater and a solar water heater. Fig. 3 (right). Schematic diagrams of solar heating systems based on air and liquid heat transfer media.

Honeywell under NSF-ERDA sponsorship includes a heating and absorption air conditioning system similar to that of CSU house I. In addition, solar heat can be used to vaporize a fluorohydrocarbon which then expands through a turbine to drive a mechanical air conditioner. Thus, solar energy is converted to mechanical energy which is then used to provide cooling by conventional means. The mobile laboratory is being operated in several locations to gather data and provide a public demonstration.

In addition to closed cycle absorption cooling, open cycles are of potential interest for solar technology. For example, desiccants can be used to absorb water vapor from room air, which can then be evaporatively cooled; the desiccant is regenerated and recycled. Löf suggested the use of triethylene glycol as the desiccant, with solar-heated air for regeneration (*12*); this system is now being evaluated for use on the Citicorp Building in New York. Dunkle has designed a cycle with rotary beds of silica gel and rotary heat exchangers (*13*). In the Munters (M.E.C.) system LiCl is used as the desiccant; the system is being adapted for solar operation (*14*).

It is also possible to use a collector as an energy dissipater by designing it to lose heat by convection and by radiation to clear night sky. To accomplish this, the collector must have opposite properties to those needed for efficient collection; thus compromises are necessary or movable insulation must be used. Hay designed such a system for a clear, mild California climate. He achieves combined collector-radiator-storage capabilities in the horizontal roof of the building with movable insulation, and thereby provides heating in the winter and cooling in the summer. This system was evaluated for a year (*15*), and kept conditions inside the house within acceptable ranges throughout the period.

Another class of systems combines solar collectors and heat pumps. The heat pump can serve as an independent (auxiliary) source of heating energy, or the collector-storage system can serve as the energy source for the evaporator of the heat pump. The latter system has the apparent advantages of lowering mean collector temperature and raising the mean evaporator temperature of the heat pump (thus improving the performance of each). Systems of this type have been studied experimentally (*16*). A simulation study by Freeman compares these methods in one climate, and indicates little choice between them (*17*), but there remain many unanswered questions on how these combined systems should best be constructed and operated.

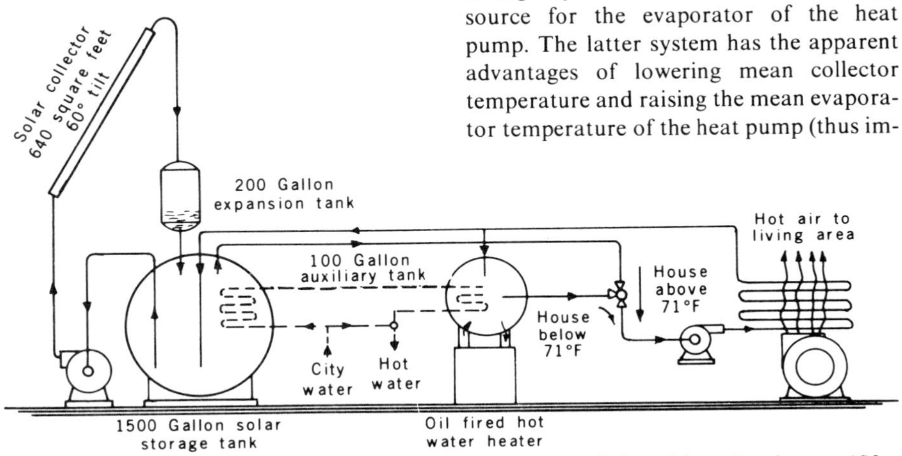

Fig. 4. Schematic diagram of the heating system in MIT house IV [adapted from Engebretson (*6*)].

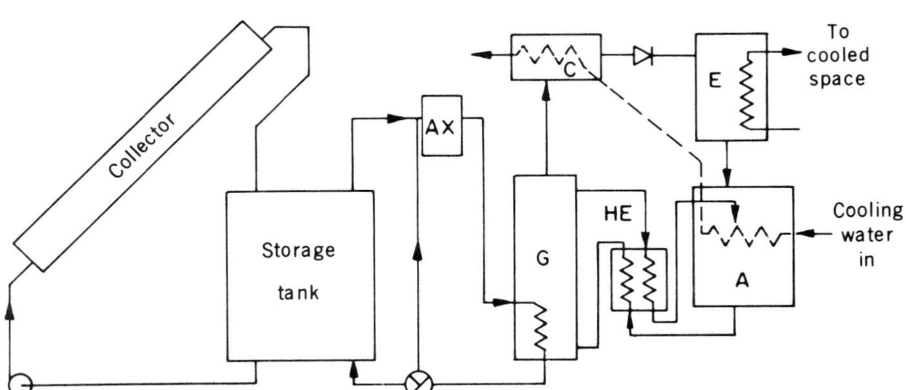

Fig. 5. Schematic diagram of a solar-operated absorption air conditioner. *AX* is the auxiliary energy source. The cooler components are as follows: *G*, generator; *C*, condenser; *E* evaporator; *A*, absorber; *HE*, heat exchanger to recover sensible heat (*18*).

Performance Calculations

The general approach to calculating the thermal performance of solar energy systems is to write the equations that describe the performance of each of the components in a system (including collector, storage, controls, pumps, and the like, as well as the building itself), and simultaneously solve the equations, usually hour by hour. Meteorological data for the location in question, which affects both collector and building heating and cooling loads, are used as forcing functions. The solutions are time-dependent temperatures and energy rates. The energy rates can be integrated to give energy quantities over the period of the simulation. The amount of energy a system is expected to deliver over a year can then be the basis for an economic analysis. These procedures are outlined by Duffie and Beckman (*18*).

The most critical and unique component is the collector. Thanks to pioneering studies of collectors by Hottel and his colleagues, beginning almost 40 years ago and carried on by others since (*19*), methods of predicting collector performance are well established. Based on a detailed analysis, the useful gain of most collectors can be written as

$$Q_u = A_c F_R[S - U_L(T_{f,in} - T_a)]^+ = \dot{m}C_p(T_{f,out} - T_{f,in})$$

where F_R is equal to the ratio of actual energy gain to gain if the whole plate were at the fluid inlet temperature, $T_{f,in}$, and accounts for the material properties and configuration of the plate. This collector heat removal factor takes into account fluid flow rate and temperature gradients along and across the plate and enables the calculation to be made on the basis of $T_{f,in}$ (a very convenient variable). Also, A_c is the collector area, a major design parameter. S equals the absorbed radiation per unit area of collector. It is the product of incident radiation on the plane of the collector, the transmittance of a cover system, and the absorptance of the plate for solar radiation. It is a function of the orientation of the collector, the number of covers, and the properties of the covers and plate for solar radiation. The thermal loss coefficient U_L is a function of the number of covers, cover and plate properties for longwave (thermal) radiation, wind speed, and temperatures. Correlations and charts are available to determine this coefficient (*18–20*). Finally, $T_{f,out}$ is the outlet fluid temperature; T_a is ambient air temperature; and $\dot{m}C_p$ is the product fluid of mass flow rate and heat capacity. The plus sign on the bracket indicates that only positive values are taken. This simulates a controller that turns on the pump or blower whenever useful energy is to be gained from the collector, that is, when the fluid outlet temperature is higher than the inlet temperature.

Included in the equation are a wide range of design parameters and materials properties. For example, the effects of selectivity of the energy-absorbing surface, that is, the absorptance of the surface for solar radiation and emittance for longwave radiation, are implicit in S and U_L.

The equation also illustrates an important determining factor in solar energy system performance. As the collector temperature ($T_{f,in}$) rises, the thermal loss term approaches the absorbed radiation term and collector output diminishes. For most practical designs today, zero output collector temperatures are typically 150° to 175°C above ambient, and normal operating temperature ranges are less than 75°C above ambient. So, collectors are uniquely sensitive to temperature and must be designed to operate at minimum temperatures above the levels required.

New collector developments are aimed at increasing absorbed radiation S and reducing thermal losses. Extensive efforts have gone into development of selective surfaces with low longwave emittances to reduce U_L (*21*). The practical problem has been to maintain desirable combinations of properties over very extended periods (20 years or more) in oxidizing atmospheres. Many of the surfaces studied are metal substrates with semiconductor coatings, for example, chrome oxide on a bright nickel base. Another approach to control of thermal losses is to evacuate the space between the absorbing surface and the cover, thus reducing or eliminating convection and conduction across the gap. This is done by enclosing the absorbing surface in tubes (*22*); elimination of convection and conduction coupled with selective surfaces of low emittances results in very low loss coefficients and allows energy delivery from collectors at substantially higher temperatures than from conventional designs.

Equations based on standard energy and materials balances, rate equations, and equilibrium relationships are available for energy storage, heat exchangers, heating and cooling loads, controls, and other components of solar energy systems. While the models of some components can be based on physical principles, it may be necessary to fall back on empirical models of coolers, heat pumps, and other complex equipment. The combination of the models of each component provides the basis for system performance calculations.

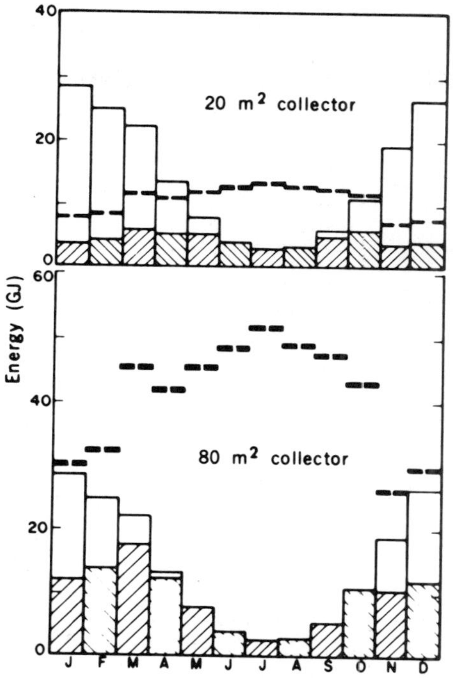

Fig. 6. Month-by-month performance of heating systems of two collector areas on a Wisconsin house with a floor area of 180 m². Incident radiation on the collector is shown by the heavy broken lines. Total heating and hot water load is in- is indicated by the bars; the shaded portion represents the load met by solar, and the unshaded portion, the load met by auxiliary; *GJ*, gigajoule.

Simulations

Physical experiments on solar heating and cooling are indispensable. However, numerical experiments, such as simulations, can yield much of the same kind of information quickly and inexpensively. The effects on long-term system performance of changes in system configuration, materials properties, and component design can readily be assessed in a way that is not practical in experiments. Simulations are also useful in understanding the dynamics of systems (which never operate at steady state) and in selecting and planning experiments. We have developed a modular solar process simulation program, TRNSYS, in our laboratory, and other simulation programs have been described (see *23*).

The two most obvious design variables of a solar heating system for a particular building are collector area and storage size. To see the effects of these parameters on energy delivered to a building, let us consider the following example. A house in Madison, Wisconsin, is to be provided with solar heating and hot water, from a system similar to that of the CSU experiment. The house is a typical, moderate size house with a heat loss rate corresponding to a floor area of 180 m² and with conventional insulation. A liquid heating collector has two glass covers, a high-absorptance, flat black paint for absorption of solar energy, and is sloped toward the south with a slope equal to the latitude (*24*). The storage tank is to be located within the building so that losses from the tank are uncontrolled gains to the building. Hot water demands are typical of a family of four or five.

The results of simulations of this system, with forcing functions of hourly weather data for an average Madison year and a fixed ratio of storage mass to collector area of 75 kg/m², are shown in Fig. 6. Incident radiation, total loads, and the load carried by solar energy for two collector areas are shown by months. Monthly collector efficiencies are the ratio of solar energy delivered to the building to incident solar energy on the collector; these are high when heating loads are high relative to the size of the collector, and low when loads are low. Thus, these systems tend to be overdesigned for part of the year and underdesigned for part of the year.

Annual performances, expressed as the fraction of loads supplied by solar energy, are shown in Fig. 7. Since the total loads are nearly independent of collector size, these data also indicate the total amount of solar energy delivered. These numbers are useful in deciding how much collector area should be used on the house, and indicate that very large collector areas (relative to the heating loads on the house) are needed to approach 100 percent solar heating. In other words, the larger the solar energy system, the larger the fraction of the year that it is overdesigned.

What should the storage capacity be? Figure 8 shows the effects of storage capacity on annual performance of this system. Below about 50 kg/m², system capacity drops off rather sharply as tank size decreases. Above 100 kg/m², there is a slow increase in annual performance as tank size increases. Cost studies by Tybout and Löf (*5*), and others, which take into account the cost of tanks as a function of their size, indicate that a slight cost penalty is incurred on going beyond about 100 kg/m².

There remains a question of seasonal storage from summer to winter. If very

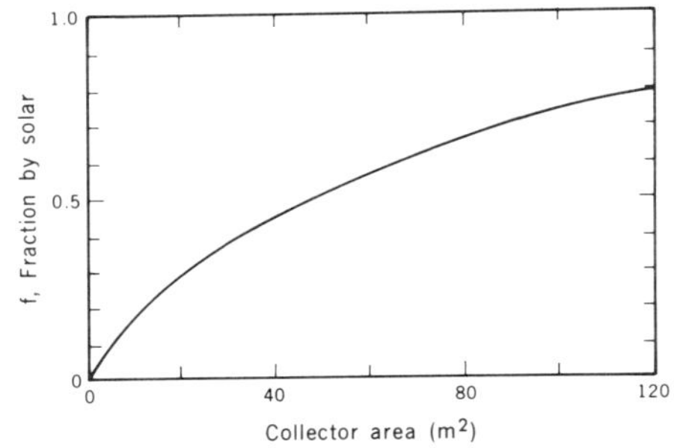

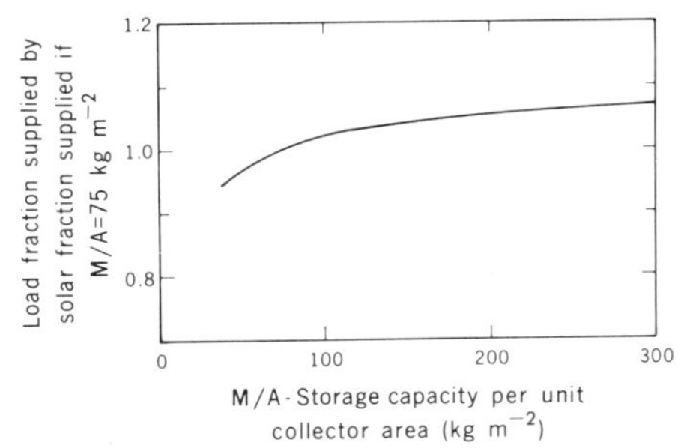

Fig. 7 (top). Variation of the fraction of the annual total load carried by solar energy with collector area for the Wisconsin example. Fig. 8 (bottom). Variation of the fraction of the annual total load carried by solar energy with storage capacity for water storage tanks.

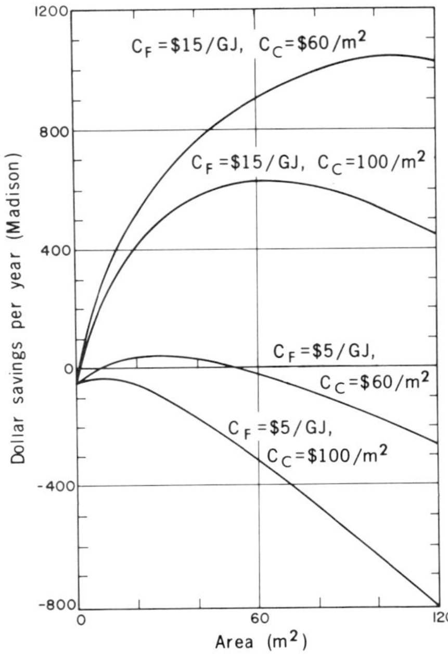

Fig. 9. Annual savings as a function of collector area for the Wisconsin example. Two collector costs and two conventional energy costs are plotted. C_F, cost of fuel; C_C, cost of collector.

large storage systems were used (probably with a volume of roughly the same size as that of the heated space if heat-capacity storage is used) smaller collectors might be possible. Speyer (*25*), in 1959, concluded that this is uneconomical; reexamination of this possibility with simulation methods would be of interest.

Economics

Solar energy processes are generally capital-intensive; large investments are made in equipment to save operating costs (that is, fuel purchases). The essential economic problem is balancing annual cost of the extra investment (interest and principle, based on reasonable estimate of lifetime) against annual fuel savings. Thermal performance predictions, with estimated equipment and fuel costs, show the effects of major design decisions on annual costs.

An example of annual savings as a function of collector area, on the basis of the performance calculations noted in the previous section, is shown in Fig. 9. We assume two collector costs, \$60 and \$100 per square meter, and two conventional energy costs, \$5 and \$15 per gigajoule. Delivered energy costs in the United States today range from less than \$2 per gigajoule for natural gas in the Southwest to more than \$15 per gigajoule for demand electric resistance heating in some parts of the Northeast. The collector cost is the major investment and is proportional to collector area. Storage cost is only slightly dependent on collector area, and there are other equipment costs that are essentially independent of collector area. Here we have used \$500 for the storage and other equipment costs and an annual charge on investment of 12 percent, corresponding to 10 percent interest over 20 years.

The curves show distinctly different behavior, with the maximum "savings" at small collector areas for the expensive collector and cheap fuel, and at a collector area of 100 m^2 for the \$60/m^2 collector and expensive fuel. The savings for the collector cost of \$60/m^2 and the fuel cost of \$15 per gigajoule are positive over a range of collector areas from 10 to 50 m^2. Significant deviations from the optimum values do not greatly affect savings; thus the selection of a precise value for collector area is not very critical.

There are many assumptions inherent in these curves. Costs of taxes, maintenance, and insurance have not been included. Conventional energy costs were assumed to be fixed over the lifetime of the system. The nature of the equipment for supplying auxiliary energy (as indicated on Fig. 3) is assumed to be independent of the amount of auxiliary required during a year, while in fact it may change substantially. Costs associated with the time dependence of auxiliary energy needs are ignored; this implies that the auxiliary energy source is stored on site, since utilities could be subjected to unacceptable peak loads by large numbers of solar buildings that draw on them simultaneously only during periods of bad weather.

Nevertheless, some generalizations can be drawn from these analyses. As fuel costs rise and as the supplies of low-cost natural gas become increasingly more difficult to obtain, solar energy will become more competitive and optimum fractions of annual loads to be carried by solar energy will increase. As collector and other solar energy system costs decrease as a result of mass production, by improved technology, or by users "doing it themselves," similar improvements in the relative economics of solar energy will occur.

Finally, political decisions may be made that will affect the extent to which solar energy can be competitive. Deregulation of natural gas prices or further increases in the cost of imported oil will increase their costs to consumers and make solar energy more competitive. Tax incentives, such as write-off of investments in solar energy–producing equipment, could make an incremental improvement in solar energy economics.

Summary

We have adequate theory and engineering capability to design, install, and use equipment for solar space and water heating. Energy can be delivered at costs that are competitive now with such high-cost energy sources as much fuel-generated, electrical resistance heating. The technology of heating is being improved through collector developments, improved materials, and studies of new ways to carry out the heating processes.

Solar cooling is still in the experimental stage. Relatively few experiments have yielded information on solar operation of absorption coolers, on use of night sky radiation in locations with clear skies, on the combination of a solar-operated Rankine engine and a compression cooler, and on open cycle, humidification-dehumidification systems. Many more possibilities for exploration exist. Solar cooling may benefit from collector developments that permit energy delivery at higher temperatures and thus solar operation of additional kinds of cycles. Improved solar cooling capability can open up new applications of solar energy, particularly for larger buildings, and can result in markets for retrofitting existing buildings.

Solar energy for buildings can, in the next decade, make a significant contribution to the national energy economy and to the pocketbooks of many individual users. Very large aggregate enterprises in manufacture, sale, and installation of solar energy equipment can result, which can involve a spectrum of large and small businesses. In our view, the technology is here or will soon be at hand; thus the basic decisions as to whether the United States uses this resource will be political in nature.

References and Notes

1. See, for example, M. P. Thekaekara, *Sol. Energy* **14**, 109 (1973).
2. Radiation and related weather data are available from the National Climatic Data Center, Asheville, N.C.
3. B. Y. H. Liu and R. C. Jordan, in *Low Temperature Engineering Applications of Solar Energy* (American Society of Heating, Refrigerating, and Air Conditioning Engineers, New York, 1967); G. O. G. Löf, J. A. Duffie, C. O. Smith, *Univ. Wis. Eng. Exp. Stn. Rep. No. 21* (1966).
4. The best example is F. Daniels, *Direct Use of the Sun's Energy* (Yale Univ. Press, New Haven, Conn., 1964).
5. R. A. Tybout and G. O. G. Löf, *Nat. Resour. J.* **10**, 268 (1970); G. O. G. Löf and R. A. Tybout, *Sol. Energy* **14**, 253 (1973); *ibid.* **16**, 9 (1974).
6. C. D. Engebretson, in *Proceedings of the U.N. Conference on New Sources of Energy* (United Nations, New York, 1964), vol. 5, p. 159; G. O. G. Löf, M. M. El-Wakil, J. P. Chiou, *ibid.*, p. 185; R. W. Bliss, *ibid.*, p. 148; G. O. G. Löf, *ibid.*, p. 114.
7. R. Chung, J. A. Duffie, G. O. G. Löf, *Mech. Eng.* **85**, 31 (1963); J. A. Duffie and N. R. Sheridan, *Mech. Chem. Eng. Trans.* **MC-1**, 79 (1965). The NH_3-H_2O cycles have also been studied; six papers at the International Solar Energy Society (ISES) meetings were concerned with NH_3-H_2O systems.
8. The July 1975 ISES meetings also included 12 papers on performance of solar heating and cooling systems.
9. M. Telkes, in *Solar Energy Research*, F. Daniels and J. A. Duffie, Eds. (Univ. of Wisconsin Press, Madison, 1955); final summary report to NSF-RANN on *Conservation and Better Utilization of Electric Power by Means of Thermal Energy Storage and Solar Heating* (National Center for Energy Management and Power, University of Pennsylvania, Philadelphia, 1973).
10. D. S. Ward and G. O. G. Löf, annual report to NSF-RANN on *Design, Construction and Testing of a Residential Solar Heating and Cooling System* (Colorado State University, Fort Collins, 1975).
11. Honeywell report to NSF-RANN, *Design and Test Report for Transportable Solar Laboratory Program* (Minneapolis, Minn., 1974).
12. G. O. G. Löf, in *Solar Energy Research*, F. Daniels and J. A. Duffie, Eds. (Univ. of Wisconsin Press, Madison, 1955). The triethylene glycol system is used in commercial, fuel-fired air conditioning equipment.
13. R. V. Dunkle, *Mech. Chem. Eng. Trans.* **MC-1**, 73 (1965).
14. W. F. Rush, J. Wurm, L. Wright, R. Ashworth, paper presented at the International Solar Energy Society meeting, Los Angeles, Calif., 1 August 1975.
15. P. W. Niles, *ibid.*
16. F. H. Bridgers, D. D. Paxton, R. W. Haines, *Heat. Piping Air Cond.* **29**, 165 (1957).
17. T. L. Freeman, thesis, University of Wisconsin, Madison (1975).
18. J. A. Duffie and W. A. Beckman, *Solar Energy Thermal Processes* (Wiley, New York, 1974).
19. H. C. Hottel and B. B. Woertz, *Trans. Am. Soc. Mech. Eng.* **64**, 91 (1942); H. C. Hottel and A. Whillier, in *Transactions of the Conference on the Use of Solar Energy* (Univ. of Arizona Press, Tucson, 1958), vol. 2, part 1.
20. S. A. Klein, *Sol. Energy* **17**, 79 (1975).
21. For example, see H. Tabor, in *Low Temperature Engineering Applications of Solar Energy* (American Society of Heating, Refrigerating, and Air Conditioning Engineers, New York, 1967).
22. E. Speyer, *J. Eng. Power* **86**, 270 (1965); D. C. Beekley and G. R. Mather, paper presented at International Solar Energy Society meeting, 28 July 1975; U. Ortabassi and W. Buehl, *ibid.*
23. S. A. Klein *et al.*, *Univ. Wis. Eng. Exp. Stn. Rep. No. 38* (1975); see also S. A. Klein, P. I. Cooper, T. L. Freeman, D. M. Beekman, W. A. Beckman, J. A. Duffie, *Sol. Energy* **17**, 29 (1975).
24. As a rule a solar collector should be sloped toward the equator with a slope of latitude +10° for winter use, a slope of latitude −10° for summer use, and with a slope equal to latitude for year-round use. Deviations of 5° or 10° usually make little difference in annual performance.
25. E. Speyer, *Sol. Energy* **3** (No. 4), 24 (1959).

Photovoltaic Power Systems: A Tour Through the Alternatives

A family of new technologies, rich in ideas, may provide the basis for many useful energy systems.

Henry Kelly

We probably know only slightly more about generating energy from photovoltaic devices than James Watt knew about producing mechanical energy from steam. Like Watt, we know that the technology works, we know something about the principles which govern it, and we can dare to speculate about a promising future. Indeed, it is very likely that photovoltaic arrays will be a common sight in less-well-developed nations and in remote parts of the developed world during the coming decade. With only a little more courage, we can envision photovoltaic-generating equipment becoming a major part of new U.S. generating capacity by the turn of the century. These promises, however, are clouded by a host of distressingly relevant questions.

In particular, we have only recently begun to think seriously about designing practical photovoltaic systems and about how these systems can best be integrated into national patterns of energy supply and demand. The number of alternative approaches turns out to be enormous, and the analytical basis for choosing between them surprisingly primitive. We do not know, for example, what the optimum size for such systems will be, we do not know whether we should emphasize the development of low-cost devices (which could perhaps eventually be used as integral parts of buildings) or the development of tracking apparatus or both, and we do not know whether cogeneration or total energy systems should be attempted.

It will be difficult to improve the present photovoltaic development program without answers to these questions. Criteria for components, for example, cannot be established without a clear understanding of the ways in which these components can contribute to integrated systems meeting real loads in real operating environments. Taking advantage of the opportunities presented by the torrents of emerging ideas will require a great deal of imagination and flexibility.

The most immediate barrier for all photovoltaic systems, of course, is the present high cost of the devices. In its latest procurement, the federal government was able to purchase flat plate arrays in quantities of tens of kilowatts for about $11 per peak watt of output (*1*); electricity from systems using such cells costs $1 to $2 per kilowatt-hour.

Developmental work to reduce the cost of photovoltaic energy can be divided into three general categories: (i) reducing the cost of manufacturing the single crystal silicon cells that are now on the market; (ii) developing techniques for mass producing and increasing the performance of cells made from thin films of materials such as CdS/Cu_2S or amorphous silicon, and (iii) developing high-efficiency cells which can be installed at the focus of magnifying optical systems.

There is little doubt that it is technically possible to use any of these approaches to reduce costs to $1 to $2 per peak watt (electricity costing $0.10 to $0.40 per kilowatt-hour) during the next 3 to 5 years. Further cost reductions are almost certainly possible without any fundamental innovations, but costs below $1 to $2 per watt will, at a minimum, require a considerable amount of engineering development. Progress in any of a number of current research programs

The author is technical assistant to the director, Office of Technology Assessment, U.S. Congress, Washington, D.C. 20510.

Table 1. Market forecasts for photovoltaic devices at different prices (in megawatts of annual sales). Some of these projections include estimates of delays from an inability of the industry to grow rapidly enough to meet demands and an inability of the market to react to cost-effective applications as soon as they are theoretically available. The estimates are therefore not fully complete without a careful analysis of the impacts of these assumptions. Abbreviations: FEA, Federal Energy Administration; DOD, Department of Defense; DOE, Department of Energy.

Marketing study	Array prices in dollars per peak watt					Reference
	10	3	1	0.5	0.1–0.3	
BDM/FEA						
DOD market	10	75	100			(*6*)
Worldwide commercial market	1.5	20	70	100		(*3*)
Intertechnology Corporation	0.5	13	126	270		(*4*)
Motorola	1.5	20–30				(*52*)
Texas Instruments	0.4	2.6	30	100	20,000	(*53*)
RCA	0.8	13	200	2,000	100,000	(*53*)
Westinghouse					96,000	(*7*)
DOE planning objectives	1.0	8	75	500	5,000	

would give us greater confidence about meeting the lower cost goals.

A set of goals for reducing the cost of silicon photovoltaic devices was established somewhat arbitrarily during the crash "Project Independence" studies conducted in 1973. Officials in the Department of Energy believe that, with some relatively minor adjustments, these goals are achievable and are using them for planning purposes. The present goals for flat plate arrays (with 20-year+ life expectancies) are \$2 per watt by 1982, \$0.50 per watt by 1986, and \$0.10 to \$0.30 per watt in the 1990's (*2*).

Markets

The level of private interest in manufacturing photovoltaic equipment will clearly be based primarily on perceptions about the commercial markets available at different prices. Photovoltaic sales are difficult to predict, however, because the market differs substantially from the market for other types of electricity-generating equipment. Since this equipment can provide power in remote areas where conventional alternatives are extremely expensive, a demand for photovoltaic devices exists at a large range of prices. The unique features of the equipment may lead to the discovery of markets for energy where no market now exists. The large price elasticity of markets, coupled with the fact that individual installations can be very small, has allowed an evolutionary growth in sales and a gradual reduction in cell prices.

The free world market for photovoltaic cells at 1976 prices was about 380 kilowatts (*3*), of which about 280 kW were sold by U.S. manufacturers. The U.S. government during this period purchased about 108 kW, of which about 50 kW were used in satellites (*4*). Major commercial markets have appeared in communications equipment (68 kW), corrosion protection for bridges, pipelines, and like applications (28 kW), and aids to navigation (20 kW) (*4*). Sales during 1977 were about twice those of 1976.

Several market surveys for photovoltaic equipment have been completed during the past several years, and some of these are summarized in Table 1. The considerable differences in the forecasts reflect differing judgments about the future cost of conventional energy and other forms of solar energy, about the rate at which an industrial infrastructure capable of supporting large-scale production can be established, and about the potential costs of support equipment (storage, controls, and other types of equipment) and installation. Some manufacturers, for example, have been skeptical about the high forecasts for small remote applications since many of these applications require a considerable amount of expensive marketing and engineering.

One point on which the surveys seem to agree, however, is that a significant fraction of sales during the next few years will occur in developing countries. Photovoltaic equipment is ideally suited to places where no utility grid is available and where labor for installing the equipment is relatively inexpensive. Consumers in the capital cities of many developing nations now pay as much as \$0.20 to \$0.25 per kilowatt-hour for electricity, and prices in more remote areas are often higher (if power is available at all) (*5*). The modular nature of photovoltaic equipment has the additional advantage of allowing functioning power sources to be installed quickly and in sizes appropriate for each application. Moreover, an investment in the photovoltaic power source does not commit a nation to finding a reliable source of fuel or to maintaining a highly trained group of operators.

Other possible areas where sales of photovoltaic equipment may increase rapidly during the next few years include applications in the U.S. Department of Defense (where large purchases may be possible) (*6*) and the armed forces of other nations; the agricultural sector, for irrigation and other pumping applications; and the transportation industry, for markers and lighting (*3, 4*).

If prices fall below about \$0.50 per watt, an explosive growth in sales could occur since at this price photovoltaic equipment might be able to provide electricity which is competitive with residential and commercial electricity rates in many parts of the United States (*7, 8*). By the time prices fall to \$0.10 to \$0.30 per watt, photovoltaic electricity may be competitive with electricity sold at bulk rates to large industrial consumers. Estimating sales at these levels is extremely speculative since generating large amounts of power from photovoltaic devices would require a fundamental change in the ways in which this country now supplies and consumes electric energy. Moreover, when prices of photovoltaic arrays reach these low levels, the overall cost and attractiveness of photovoltaic systems are likely to be dominated by factors other than the cost of the cells themselves. Before turning to an analysis of integrated systems, however, it will be useful to examine the costs and capabilities of the assortment of photovoltaic devices which are or may be available in the near future.

Photovoltaic Cells

All photovoltaic installations will consist of arrays of small, individual generating units—the photovoltaic cell. Individual cells will probably range in size from a few millimeters to a meter in linear dimension. A surprising variety of such cells is available or is in advanced development since only during the last 5 years has serious attention been given to designing cells for use in anything other than spacecraft. Devices are available with a wide range of efficiencies, voltages, tolerance to high temperatures and high light intensities, and other characteristics. The variety should not be taken to suggest that this is an area where large amounts of private or public research funding has been directed; indeed, many of the projects that I will describe are being conducted by one- or two-man research teams in laboratories where one is likely to find the inventor hunched over a table soldering leads to his latest cell design.

The photovoltaic equivalent of power engineering is something of an anomaly in the power-generating industry since it involves manipulations in the miniature world of semiconductor physics instead of steam tables, gears, and turbine blades. The following discussion can provide only a very brief excursion through a complex field; more complete discussions of the topics covered can be found in several recent publications (*9–12*). The state of theoretical work in the field is quite uneven. Some areas have been examined extensively; in others, important effects remain to be explained.

The energy in light is transferred to electrons in a semiconductor material when a light photon collides with an atom in the material with enough energy to dislodge an electron from a fixed position in the material (that is, from the "valence band"), giving it enough energy to move freely in the material (that is, into the "conduction band"). A vacant electron position or "hole" is left behind at the site of this collision; such holes can "move" if a neighboring electron leaves its site to fill the former hole site. A current is created if these pairs of electrons and holes (the holes act as positive charges) are separated by an intrinsic voltage in the cell material. Creating and controlling this intrinsic voltage is the trick which has made semiconductor electronics possible. The most common technique for producing such a voltage is to create an abrupt discontinuity in the conductivity of the cell material (typically silicon in contemporary solid-state components) by adding small amounts of impurities or "dopants" to the pure material (this is called a "homojunction"). An intrinsic voltage can also be developed by the joining of two dissimilar semiconductor materials (such as CdS and Cu_2S) creating a "heterojunction" or by the joining of a semiconductor to a metal (for example, amorphous silicon to platinum) creating a "Schottky junction."

A fundamental limit on the performance of all of these devices results from the fact that light photons lacking the energy required to lift electrons from the valence to the conduction bands (the "band gap" energy) cannot contribute to photovoltaic current and from the fact that the energy given to electrons which exceeds the minimum excitation threshold cannot be recovered as useful electrical current. Most of the photon energy not recovered as electricity is converted to thermal energy in the cell.

The bulk of the solar energy reaching the earth's surface falls in the visible spectrum where photon energies vary from 1.8 electron volts (deep red) to 3 eV (violet). In silicon only about 1.1 eV is required to produce a photovoltaic electron, and in GaAs about 1.4 eV. Choosing a material with a higher energy threshold results in the capture of a larger fraction of the energy in higher energy photons but the loss of a larger fraction of lower energy photons. The theoretical efficiency peaks at about 1.5 eV, but it remains within 80 percent of this maximum for materials with band gap energies between 1 and 2.2 eV (*13*).

Electrons actually leave cells with energies below the excitation voltage because of losses attributable to internal resistance and other effects, not all of which are understood (*14*). (An electron leaves a typical silicon cell with a useful energy of about 0.5 eV.)

The same kinds of fundamental limits apply to photochemical reactions in which a light photon with energy about some fixed excitation threshold is able to produce a chemical reaction or a structural change which can be assigned a fixed energy. The theoretical limit to the performance of several types of cell designs are shown in Table 2.

The performance of real cells (also listed in Table 2) falls below the theoretical maximum for a number of reasons. One obvious problem is reflection of light from the cell surface (which can be reduced with special coating and texturing) (*15*) and reflection from the electrical contacts on the front surface of the cell (which can be reduced with careful design of the contacts). Losses also result from the fact that the photogenerated electrons and holes which fail to reach the region in the cell where they can be separated by the intrinsic voltage cannot contribute to useful currents. Photogenerated charges can be lost because of imperfections in the crystal structure of the cell, defects caused by impurities, surface effects, and other types of imperfections. Losses are minimized if a perfect crystal of a very pure semiconductor material is used, but producing such a crystal can be extremely expensive. Manufacturing costs can probably be greatly reduced if cells consisting of a number of small crystal "grains" can be made to operate with acceptable efficiencies. The size of the grains that can be tolerated depends on the light-absorbing properties of the cell material. If absorptivity is high, photovoltaic electron-hole pairs will be created close to the cell junction

Table 2. Photovoltaic cell efficiencies. Techniques for reporting efficiencies differ. Wherever possible, efficiencies were chosen which assume air mass 1 and include losses due to reflection and contact shading.

Device	Probable maximum achievable efficiency	Maximum measured efficiency	Performance of commercial cells	Reference
Silicon devices				
Single crystal homojunction	20–22	19	10–13	(*9, 54*)
Single crystal Schottky with indium-tin oxide	20	12		(*55*)
Polycrystalline homojunction	?	7–14 (?)		(*18, 19*)
Amorphous Schottky with platinum	15	5.6		(*22, 23*)
Thin films				
CdS/Cu_2S (chemical vapor deposit process)(heterojunction)	15	8.6	2–3	(*12, 56*)
CdS/Cu_2S (spray process) (heterojunction)	8–10	5.6		(*40*)
$(Cd/Zn)S/Cu_2S$ (heterojunction)	15	6.3		(*56*)
$CdS/CuInSe_2$ (single crystal) (heterojunction)	24	12		(*57*)
$CdS/CuInSe_2$ (thin film) (heterojunction)	15	6.9		(*58*)
GaAs (homojunction)	25–28	15		(*9, 59*)
Cells for use in concentrated sunlight				
Optimized silicon cell (single crystal homojunction), 200 times concentration	22	18	12.5	(*26, 29, 60*)
Interdigitated back-contact silicon, single crystal homojunction, 100 times concentration	26–27	15 (20?)		(*27*)
Thermophotovoltaic	30–50	13		(*31*)
$Ga_xAl_{1-x}As/GaAs$ (200 times)	25–26	24.5		(*61*)
$Ga_xAl_{1-x}As/GaAs$ (1700 times)		19		(*32*)
Multicolor cell (GaAs/Si/Ge)	40			(*35*)
Vertical multijunction (silicon)	31	9.6		(*62*)

where the voltages exist and relatively small grain sizes can be tolerated since the charges need only drift a short distance before being sorted by the field. It is important that the grains be oriented with the grain boundaries perpendicular to the cell junction so that charge carriers can reach the junction without crossing a grain boundary.

Because silicon is a relatively poor absorber of light, silicon cells must be 100 to 200 micrometers thick to capture an acceptable fraction of the incident light (*16*). This limitation places rather rigorous limits on the sizes of crystal grains which can be tolerated, and all current commercial silicon cells are manufactured from single crystals of silicon. Some investigators believe that, if polycrystalline silicon is to be used, individual crystal grains must be at least 100 μm on a side if efficiencies as high as 10 percent are to be achieved (*17*). A number of research projects are under way to develop inexpensive techniques for growing such polycrystalline materials (*18–20*) and for minimizing the impact of the grain boundaries (*21*). Efficiencies as high as 6 to 7 percent have been reported for vapor-deposited polycrystalline silicon cells with grains about 20 to 30 μm on a side (*18*), and a proprietary process capable of producing grains nearly a millimeter on a side reportedly can be used to produce cells with efficiencies as high as 14 percent (*19*). Work is under way to improve crystal-growing techniques and to enlarge grains with lasers and electron beams.

Perhaps the most intriguing recent development is the discovery that an amorphous silicon–hydrogen "alloy" can be used to construct photovoltaic cells with useful efficiencies. Efficiencies of 5.5 percent have been measured (*22*), and 15 percent efficiency may be possible (*23*). The hydrogen apparently attaches to "dangling" silicon bonds, minimizing the losses that would otherwise result at these sites. Acceptable performance is possible in spite of the large number of remaining defects because the amorphous material is an extremely good absorber of light; test cells are typically 1 μm or less thick (*24*). The properties of this complex material are not well understood.

Cells made from GaAs or CdS/Cu_2S are also much better absorbers of light than crystalline silicon; cells made from these materials can also be thinner and smaller crystal grains can be tolerated than is possible with crystalline silicon. Commercial CdS/Cu_2S cells will probably be 6 to 30 μm thick (*12, 25*), and the crystal structure produced with a relatively simple spray or vapor deposit process is large enough to prevent grain structure from significantly affecting cell performance. The primary drawback of most of the "thin film" cells is that their efficiencies are quite low. Research is proceeding rapidly in a number of areas, however, and a number of thin film cells may be able to achieve efficiencies greater than 10 percent.

Enthusiasm about cells based on materials other than silicon must be tempered to some extent by uncertainties about the health hazards that they may present and about the domestic and world supplies of component materials. Both CdS and GaAs contain toxic materials, and, although it may be possible to reduce the hazards they present to manageable proportions, it clearly will be necessary to examine this issue with some care before their widespread use can be contemplated. Domestic supplies of cadmium will be sufficient to supply annual production rates in excess of several thousand megawatts a year, but production beyond this level could tax domestic supplies (*12*). Gallium supplies will probably not present problems since GaAs devices are likely to be used primarily in concentrating collectors. Silicon, of course, is nontoxic and plentiful.

Cells Designed for Use in Concentrated Sunlight

Achieving high cell efficiencies is important to reducing the cost of photovoltaic systems, since many of the costs of producing and installing the cells are related to cell area instead of power and high efficiencies reduce the land or building areas required for a given level of output. High efficiencies are, however, particularly important for cells used in concentrated sunlight since an increase in cell performance leads directly to a reduction in the area which must be covered with the magnifying optical equipment which is the most costly aspect of the system. Devices designed to optimize performance in concentrated sunlight are not inherently more expensive than ordinary cells, but high-intensity devices are always likely to be somewhat more expensive per unit of cell area because production rates will be lower and because more care will be taken in manufacturing them. Since the cells cover only a fraction of the receiving area, of course, much more can be spent on any individual cell.

The current from a photovoltaic cell increases almost linearly with increasing sunlight intensity, and the voltage increases slightly faster than the logarithm of the intensity. These effects would lead to an increase in overall cell efficiency except for the fact that the increased current densities in the cell lead to increased resistive losses and other effects. The design of standard silicon cells can be optimized for operation in intense sunlight if one carefully designs the wires used to draw current from the cells, optimizes the resistivity of the cell material, changes the thickness of the cell junction, and otherwise takes pains in cell manufacture. Efficiencies as high as 18 percent have been reported for silicon cells operating in sunlight concentrated about 300 times (*26*).

Several ingenious techniques have been suggested for improving the performance of silicon devices used in intense sunlight with novel designs. An "interdigitated back-contact" cell exposes an unobstructed wafer of pure silicon crystal directly to the sunlight. The junctions that produce the cell voltages, and that are attached to electrical leads, are entirely on the back of the cell. An efficiency of 15 percent at a sunlight concentration of 400 times has been reported for a preliminary version of this cell. It is believed that straightforward design improvements will result in cells that have an efficiency of at least 20 percent (*27*).

The resistance losses of cells operating in high concentrations can be reduced if one connects a number of silicon homojunctions in series (thereby reducing the current produced at a given power output) and illuminates the resultant multijunction device from a direction parallel to the plane of the junction (*28*). Designs for horizontal multijunction devices have also been proposed (*29*). Although in theory these devices should have high efficiencies, very little experimental work has been done. The highest efficiency measured to date is 9.6 percent (*29, 30*).

The "thermophotovoltaic" cells may be able to achieve efficiencies as high as 30 to 50 percent by shifting the spectrum of light reaching the cell to a range where most of the photons are close to the minimum excitation threshold for silicon cells. The sun's energy is used to heat a thermal mass to 1800°C (the effective blackbody temperature of the sun is about 5700°C). A large fraction of the surface area of this mass radiates energy to a silicon photovoltaic device. (Reradiation to the environment can occur only from the small aperture through which the sunlight enters.) A surface behind the photovoltaic cells reflects unabsorbed photons back to the radiating mass, thus preserving this energy in the system (*31*).

High efficiencies in intense radiation

can also be achieved with GaAs cells, particularly if these cells are covered with a layer of $Ga_xAl_{1-x}As$, which has the effect of reducing surface and contact losses (*9*, p. 195). Efficiencies as high as 24.5 percent have been measured for such devices operating in sunlight concentrated more than 200 times (*32*).

Loferski has suggested that cell efficiencies could be increased if a number of cell junctions were used, each sensitive to a different color (*33*). An ingenious scheme for producing such a multicolor device has been recently proposed which uses a series of dyes capable of absorbing sunlight and reradiating the energy in a narrow frequency band matched to the band gap of each of a series of cell junctions (*34*).

Manufacturing Problems

With this background in some of the fundamentals of cell design, it will be possible to understand some of the difficulties faced in reducing the cost of manufacturing photovoltaic devices on a commercial scale.

Silicon. The vast majority of the photovoltaic cells now being sold are single-crystal silicon devices; the bulk of federal funding to reduce the cost of cells is being directed to silicon technology. The federal low-cost silicon project, managed by the Jet Propulsion Laboratory (JPL) of the California Institute of Technology, has made a careful analysis of the component costs of each step in the manufacturing process and is systematically examining techniques for reducing costs in four major areas: (i) production of the pure silicon feedstock; (ii) preparation of a thin sheet of single crystal silicon; (iii) fabrication of cells; and (iv) arrangement of the cells in a weatherproof array.

The purified polycrystalline silicon used as the raw material in commercial cell manufacture now costs about $65 per kilogram. Estimates from JPL indicate that, if the goal of $0.50 per watt is to be reached, the cost of the silicon material must be reduced to about $10 per kilogram and the amount of silicon wasted in the manufacturing process considerably reduced (*35*). Perhaps even more important, present techniques for manufacturing silicon are extremely inefficient in their use of energy; approximately 7000 kWh of energy is required to manufacture a cell with a peak output of 1 kW (assuming that cells are 100 μm thick and 82 percent of the silicon entering the manufacturing process is wasted) (*36*). This means that the device must operate in an average climate for about 4 years before it produces as much energy as was consumed in manufacturing the component silicon.

Several promising techniques for improving the purification and reducing silicon wastage have been experimentally verified. It should be possible to develop devices capable of producing all the energy used in their manufacture in 3 to 4 months (*36*). A significant amount of chemical engineering and process development is needed, however, to demonstrate that these laboratory experiments can be scaled up by many orders of magnitude to form the basis of a commercial facility.

Silicon costs probably represent the single greatest technical barrier to meeting JPL's cost goals for nonconcentrating arrays in the early 1980's. This is so because plants capable of manufacturing silicon in quantities large enough to achieve the required cost reductions would need to be established very quickly—probably within the next year—and the investments required will be large, compared to previous spending in photovoltaic manufacturing. New plants are likely to require more capital investment per unit of cell production than any other stage in the cell production process, between $20 million and $40 million for a single plant. There is no incentive to invest in such equipment on this magnitude solely for the purpose of selling silicon to the semiconductor industry because material costs for these devices are already a small part of the device cost. Silicon prices, therefore, are unlikely to fall by 1982 unless the government takes some action. Development of an amorphous silicon cell with adequate performance would dramatically reduce silicon requirements in cells since these cells will probably require less than 1 percent of the silicon used in commercial cells. Silicon requirements can also be greatly reduced if concentrator devices are used.

Growing pure silicon crystals and sawing them into thin wafers now represents about 25 percent of the price of the arrays (*37*). There are a number of active programs for improving the batch processes in which crystal ingots are currently being produced and sawed (*38*); techniques have also been designed for drawing single crystal sheets or ribbons directly from molten silicon, but a considerable amount of engineering work must be done before a commercial process is available (*35*, pp. 71–72). Development of an ingot technique adequate to meet the cost goal of $1 to $2 per watt appears to be assured, and improved ingot techniques may even be adequate to meet the 1986 cost goal. The technological problems remaining appear to be largely ones of improving mechanical designs; this is another area where the program could be accelerated by the government. Although crystal-growing equipment is relatively expensive, unsubsidized commercial interest in this kind of equipment in the next few years is likely to be greater than commercial interest in advanced silicon refinement processes. Research progress which makes it possible to use polycrystalline or amorphous materials would substantially reduce the cost of this step in the production cycle.

The process of converting silicon sheets into an operating photovoltaic cell involves a number of individual steps which now require hours of hand labor (about 35 percent of the price of arrays). Processes include the creation of the photovoltaic junction, the addition of electrical contacts, and the application of antireflective coatings. Studies of mass production techniques conducted for JPL by Motorola, Texas Instruments, and RCA all indicate that this cost could be reduced to between $0.40 and $0.60 per watt (*35*, pp. 71–72) with existing production apparatus if plants capable of producing about 5 to 30 megawatts (electric) annually could be constructed. Projecting further price reductions, however, requires a fair amount of optimism. However, cell prices must fall to $1 to $3 per watt before there will be markets large enough to support several competing fabrication plants of the size envisioned in the JPL studies (Table 1). It is likely that manufacturers will show greater desire to invest in cell fabrication equipment than in the more capital-intensive devices required to manufacture silicon wafers because of the smaller investments required and because it is much less likely that fabricating plants would become obsolete even if breakthroughs dramatically reduce the cost of manufacturing wafers.

Finally, the cost of the process to connect cells together into arrays and encapsulate them to protect them from the weather must be reduced by a factor of 10 to about $0.02 per watt (*35*). A variety of techniques have been proposed and the cost reduction seems feasible, but the exact technique that will be used is not yet clear.

Thin films. The thin film technologies which show the greatest potential for reaching the cost goals early in the 1980's are all based on CdS/Cu_2S heterojunction cells. Photon Power, Inc., of El Paso, Texas, is now completing a pilot plant which should begin making such cells in the near future. If the process works as well in large-scale production as it has in laboratory tests, the facility will be expanded into a small manufacturing plant which, it is hoped, will be able to produce arrays which can be sold for \$2 to \$5 per watt by 1980 (*39*). Low costs are possible because all of the processes in cell manufacturing (cell growth, junction formation, the application of contacts, and encapsulation) involve spraying a series of chemical layers onto hot glass moving through the plant on a continuous conveyor. Laboratory devices produced by a similar process have yielded efficiencies of 5.6 percent, and cells with efficiencies of 8 to 10 percent appear possible (*40*). Even higher efficiencies may be possible if the promising results of experiments in which zinc is mixed with cadmium can be integrated into the process.

Photon Power, Inc., is working with Libby-Owens-Ford (a part owner of the company) to design a process which can be attached to a float glass plant. It may be possible to produce cells for as little as \$0.05 to \$0.15 per watt in such a facility (*41*). The major technical challenge in increasing output will be finding a way to increase the speed of the spray application process from 2 centimeters per minute (which will be achieved in a pilot plant) by about an order of magnitude to match the rate at which glass is produced from a float glass facility (*39*). The SES Corporation (a Shell Oil subsidiary) has also been developing a CdS cell.

A number of other thin film technologies are now being examined. Many of these may eventually compete with CdS/Cu_2S but are now only laboratory processes.

Concentrator systems. The contribution of the cost of photovoltaic cells to the overall cost of an installed photovoltaic system can be greatly reduced if an optical system is used to concentrate sunlight on the cell. If such systems are used, problems of reducing cell fabrication costs are replaced with problems in designing low-cost mechanical equipment.

Concentrator designs can vary from a simple booster system consisting of an inexpensive reflecting surface placed adjacent to a flat plate array to massive devices that look like radar dishes. Concepts are emerging at an astonishing pace.

Most concentrating systems must move to follow or track the sun, but a design called the compound parabolic collector has been developed which can achieve concentration ratios of 9 to 10 for at least 7 hours each day with only ten position adjustments per year. The mirror shape developed for this design can also be used at the focus of other concentrating systems to reduce the requirement for high tracking precision (*42*).

A concentrator system based on use of a plastic doped with fluorescent dyes has recently been suggested which may be able to achieve concentrations on the order of 100 times with no tracking at all (*34, 43*). The dye molecules reradiate light that they absorb at a random angle. The reradiated light striking the receiving surface at an angle greater than the angle of total internal reflection and continues to be reflected until it reaches the edge of the sheet covered with the receiver. (The frequency of the reradiated light can be selected to optimize photovoltaic cell performance.) If the dyes have zero absorptivity at the reradiated frequencies, collection efficiencies as high as 60 to 75 percent are theoretically possible in 1- to 2-meter plexiglass cells (*43*).

In most other types of concentrators thin acrylic Fresnel lenses or inexpensive parabolic mirrors are used to achieve the needed concentration. Continuous mirrored surfaces or a series of segmented surfaces may be used. These surfaces can either be rigidly mounted together or rotated separately. In some tracking systems line-focusing optics are used to bring about rotation about a single axis (this axis can be the polar axis or it can be an east-west or north-south line normal to the zenith at the collector location).

The technology of concentrating systems is too primitive to permit any confident judgments about the cost goals that can ultimately be achieved. In principle, system costs can be quite low since the bulk of the receiving area of such devices need only be covered with a thin mirror or lens surface which may cost as little as \$5 to \$20 per square meter. Tracking system costs can also be quite low, particularly if many units can be driven from a single tracking motor. Wear on moving parts should not present a major problem since a system that would follow the sun for 30 years would only need to rotate 11,000 times. A major part of the cost of current structures results from the need to steady the equipment against strong winds.

One can obtain some feeling for the current state of the art by examining the concentrating devices now on the market for use with solar air-conditioning systems or with other solar thermal installations. Northrop Inc. is marketing a one-axis Fresnel device which costs (F.O.B.) \$130 to \$180 per square meter, and the Acurex Corporation sells an east-west single-axis tracking mirror system for about \$150 per square meter (F.O.B.) in large quantities. The Albuquerque-Western Company sells a one-axis tracking device for about \$50 per square meter (F.O.B.). Sunpower Systems sells a two-axis tracking unit for \$150 per square meter.

There is clearly much room for imagination and development work in collector design, and it is impossible to estimate future costs with any confidence. Projected collector costs (F.O.B.) on the order of \$50 to \$80 per square meter are optimistic but probably not unreasonable.

Assuming that concentrating systems cost 50 percent more than flat arrays to install and three times as much to operate on an annual basis, concentrator systems would be competitive with flat plate arrays costing \$500 per kilowatt-hour and having an efficiency of 15 percent if the concentrating collector costs about \$70 per square meter and concentrating cells had an efficiency of 20 percent. The concentrator could cost about \$180 per square meter if concentrating cells had an efficiency of 40 percent. (In this calculation I have assumed that the optical efficiency of the concentrators is 80 percent and that the concentration ratio is at least ten times the ratio of the cost of the concentrator cell to the cost of flat plate arrays.)

Most types of tracking collectors can only use light from areas of the sky close to the sun's disk; they cannot use diffuse sunlight scattered from clouds or atmospheric turbidity. They compensate for this deficiency to some extent by maintaining a better angle between the collector and the sun through the day. As a result, in many parts of the country there is no great difference between the amount of energy reaching a stationary flat plate system and the amount reaching a focusing device. In Albuquerque, for instance, a fully tracking collector with perfect optics would receive about 13 percent more energy annually than a perfect stationary collector tilted at an optimum orientation; in Boston or Omaha the tracking system would receive 6 to 8 percent less energy than the flat plate system.

Very little operating experience with tracking systems has been accumulated: information about maintenance costs (particularly in adverse climates), cleaning difficulties, and life expectancies is inadequate to permit confident judgments about the practical potential of the systems. The devices may not be appropriate in the remote, unattended installations that will probably constitute the bulk of U.S. installations during the next few years, but they may have applications in developing nations since their repair and cleaning would not require any special skill. The devices have the disadvantage of being rather ugly and ungainly, so that a great deal of imagination would be required to integrate them gracefully into a building or landscape. They clearly have enough potential, however, to be considered serious candidates for photovoltaic applications both in the near future and in the long term. Their importance could increase if it proves difficult to meet cost goals with flat plate arrays.

Although the cost of mounting cells on tracking collectors is clearly a major concern, there is also cause for concern about the cost of mounting flat plate arrays. If array prices fall below $300 to $500 per kilowatt, the cost of supporting and installing the arrays will begin to exceed the cost of the arrays themselves. General Electric has proposed a design for a photovoltaic shingle which it may be possible to substitute for roofing material (*44*). If costs reach $100 to $300 per kilowatt, it may become practical to use photovoltaic sheets as a part of a wall surface. In most locations in the United States, the output of a collector mounted vertically on a wall facing south, east, or west is 40 to 60 percent lower than the output of a collector fixed at an optimum orientation. There have also been proposals for lifting flat plate arrays into synchronous orbit (*45*).

Cogeneration

Thus far I have considered only the electrical output of collector systems, but the attractiveness of photovoltaic devices can be increased significantly if effective use can be made of the thermal energy carried away by water pumped over the back surfaces of collecting cells. Such systems are the photovoltaic analogs of cogeneration systems, and such devices have advantages similar to those of conventional systems. In both cases the net efficiency of the systems can only be understood if one examines the combined demands for electricity and thermal energy in each proposed situation. There are clearly many useful applications for thermal energy with temperatures of 60° to 170°C, which can be extracted from photovoltaic cogeneration devices: the heat which can be provided by such systems can be used for space heating, domestic hot water heating, and operation of absorption air-conditioners. About 2.6 percent of the industrial process heat used by U.S. industry is applied at temperatures below 100°C, and about 40 percent is applied at temperatures below 175°C (*46*).

With photovoltaic systems the critical question is whether the electricity-generating efficiency that is lost because of the high-temperature operation of the cells needed for a cogeneration application can be compensated by the value of the thermal energy produced. Cell performance degrades almost linearly with temperature at temperatures in the range from 50° to 200°C, primarily because of a drop in the operating voltage of the cell (at high temperatures, thermally excited electrons begin to dominate the electrical properties of the semiconductor device). High-efficiency cells such as GaAs cells are less affected by high-temperature operation than silicon devices (*9*, p. 166). A commercial silicon cell operating with an efficiency of 12 percent at 27°C has an efficiency of only about 8 percent if operated at 100°C (*47*), whereas a GaAs cell with an efficiency of 18.5 percent at 28°C can operate with 16 percent efficiency at 100°C and about 12 percent efficiency at 200°C (*48*). In most cases, if a use for low-temperature thermal energy exists, it is preferable to accept these losses of efficiency and to use the thermal output from cells directly rather than to maximize cell performance and attempt to use a heat pump to produce thermal energy.

It is possible to operate flat plate collectors at elevated temperatures, but cogeneration will probably be easier to justify for concentrating systems. Care must be taken to cool concentrator cells even if waste heat is not used, and much less area needs to be covered with cooling apparatus. The long-term prospects for concentrator systems may be significantly improved if photovoltaic cogeneration proves to be attractive.

Backup Power

Selecting an optimum approach to providing backup power for photovoltaic equipment is in some ways the most subtle and difficult problem that must be confronted in analyzing different photovoltaic system designs since it involves institutional as well as technical issues. The problem is particularly complex when photovoltaic systems are integrated with a conventional electric utility.

A utility connection makes it necessary to evaluate the performance of the photovoltaic apparatus as a part of an integrated system which includes a variety of types of generating and storage devices. The problem is not simplified if the photovoltaic equipment is owned by individuals or organizations other than the utility since the costs of less than optimum systems will be communicated to these owners through higher rates for backup electricity purchased and lower rates for any energy sold to the utility.

If the photovoltaic equipment is tied into an electric utility, charging storage equipment from photovoltaic generators seems almost always to be a mistake. A photovoltaic system without a grid backup would charge batteries during the day when utility loads peak and when a utility would be discharging storage, and photovoltaic energy would be withdrawn from storage at night when utility energy is least expensive and when utility storage is being charged. There will be some overlap between the two operating strategies since both types of storage would be discharging near sunset and during cloudy days. It is clear, however, that storage equipment can be used to best effect if it is controlled by the utility. The use of photovoltaic electricity to serve daytime utility loads directly has the additional advantage that smaller amounts of the energy generated would need to be cycled through storage devices which typically have an efficiency of only 75 percent. Design of the storage systems by the utility is also preferable because the utility has access to a greater range of storage opportunities than are available to a small photovoltaic system. Pumped hydroelectric facilities, for example, can be used to provide economical storage when sites are available. If low-cost batteries are developed, utilities may choose to locate battery storage installations at many small load-leveling centers in order to minimize transmission and distribution losses, but the small centers would still be best operated in coordination with the utility as a whole. This logic would apply even if a very large fraction of the utility's energy were derived from solar sources, although in this case the strategy of operating individual storage systems would closely parallel the operation of utility storage.

There are several exceptions to this general line of logic. Photovoltaic cogenerating equipment will almost certainly be designed with on-site storage of thermal energy since energy stored in the form of hot water costs much less than electric storage. There may also be instances where on-site cogenerating devices based on the use of fossil fuels would be less expensive than utility backup power in spite of the poor load factors inherent in backup allocations or if photovoltaic systems capable of providing 100 percent of the energy needs of a building or region can be constructed which provide power less expensively than devices integrated with utility equipment. (Uneven load factors are usually not serious problems for gas utilities or oil delivery systems where storage costs are relatively low.)

The argument in favor of connections between photovoltaic systems and electric utilities clearly does not apply if regulatory procedures or other regulatory distortions prevent the utilities from constructing a rate schedule which accurately reflects the cost of backup power at different times or which accurately reflects the value of power sold back to the utility.

The fact that solar energy is available primarily during periods of utility peak demands means that utilities will not necessarily need to maintain generating equipment sufficient to provide 100 percent of the installed photovoltaic capacity as backup. A study conducted by General Electric has indicated that a utility in Phoenix would only need to maintain a backup capacity equal to between 40 and 65 percent of the peak capacity of the photovoltaic devices installed in a utility in that region, whereas a utility in Washington, D.C., would need to maintain 70 to 80 percent of the photovoltaic capacity as backup (*44*, p. 34). The need to maintain backup generating capacity decreases the value of photovoltaic equipment to a utility, and thus reduces the price which it would be willing to pay for photovoltaic electricity when it is available. One of the conclusions of the General Electric study was that the value of photovoltaic equipment to a utility in Phoenix would be on the order of $350 to $500 per watt, with equipment in Washington, D.C., valued about 1/3 less (*44*, p. 33). The value was reduced if photovoltaic equipment represented more than a small fraction of the utility load. Development of low-cost storage of any kind would clearly make photovoltaic equipment somewhat more attractive to utilities. There are a number of advanced battery systems under development which may be able to reduce the cost of storage to $10 to $40 per kilowatt-hour (*49*).

The problem of developing an optimum integration of photovoltaic systems of differing sizes into existing electric utilities is an extremely complex one and will clearly require considerably more thought. There is unlikely to be a single clearly preferred relationship since utility costs are very sensitive to climate, the types of equipment owned by the utility, the design of the photovoltaic system and the strategy with which the system operates, and the amount of photovoltaic power produced in a utility service area. However, although the widely fluctuating backup loads required by photovoltaic power systems are far from ideal from the perspective of electric utilities, many types of on-site energy-consuming equipment (such as air-conditioners and heat pumps) also impose extremely erratic loads. The real

Table 3. A comparison of the performance of several photovoltaic systems associated with an apartment building in Omaha, Nebraska (*30*). The following assumptions were made: that utilities will purchase excess on-site energy at 50 percent of their sale price and that utility power will be purchased at prevailing rates (including demand charge); that thermal storage costs $0.50 per kilowatt-hour; and that two-axis tracking collectors cost $140 per square meter installed. Details of other assumptions can be found in (*8*). For case 1, the cost of nonsolar energy is assumed to increase only with inflation (5.5 percent); for case 2, nonsolar energy prices increase 40 percent by the year 2000 and solar devices are given a 20 percent investment tax credit. Low-cost arrays are associated with lower cost installation and support equipment.

Equipment used	Cell efficiency (%)	Collector area (m^2)	Percentage of annual building energy requirements met with solar energy	Case 1*: Effective cost of solar energy† (¢/kWh)	Case 1*: Levelized monthly cost to consumer ($/month)	Case 2: Effective cost of solar energy† (¢/kWh)	Case 2: Levelized monthly cost to consumer ($/month)
Conventional all-electric system				3.9§	81– 87‡	4.9§	107–114‡
$100/kW thin film (on vertical south wall)	10	2,000	5.2	3.3– 8.3	79– 84	2.3	103
Thin film on racks							
$300/kW	8	4,000	11.5	9.7–15.2	91–101	6.7	109
$100/kW	10	4,000	14.8	6.2– 9.9	86– 95	4.3	104
Flat plate silicon on racks							
$1000/kW arrays	14	4,000	19.2	16.3–23.6	120–143	11.2	126
$500/kW arrays	14	4,000	19.2	9.2–13.7	97–112	6.3	110
$500/kW arrays with (3000 kWh of batteries at $37/kWh)	14	4,000	18.4	11.6–16.7	103–119	8.3	115
Two-axis tracking							
Silicon cells (cogeneration)	20	2,500	43	3.7–6.0	86–103	2.6	94
High efficiency cells	40	2,500	49	3.9–6.1	90–108	3.0	105
Two-axis tracking with seasonal thermal storage (1.1×10^6 kWh), GaAs cells	22	2,756‖	74.1	5.0– 7.8	102–138	3.3	88
Two-axis tracking, GaAs, seasonal thermal storage (0.6×10^6 kWh) and seasonal electric storage (3.3×10^4 kWh at $11/kWh)	22	11,500‖	100	8.6–12.4	166–231	6.1	123

*Lower cost assumes that the solar device was financed with the apartment building; higher cost assumes utility ownership. †The effective cost of solar energy is computed by dividing the difference in electricity purchased by the difference in capital and operating expenses for buildings operating with and without solar equipment. ‡Range reflects different types of heating equipment. §Levelized cost of conventional energy (cost levelized over 30 years). ‖Collector arrays (including required spacing) will not fit on parking lot and roof.

costs of meeting these loads are also difficult to evaluate.

Analysis of Integrated Systems

Having looked at the problems and benefits of a number of different aspects of photovoltaic power, it is now possible to review the performance of several representative system designs. Table 3 compares the performance of a number of systems designed to meet the energy requirements of a typical 10-story, 196-unit high-rise apartment building located in Omaha, Nebraska. With the exceptions noted in the table, all of the systems listed can be located on the roof or over the parking area associated with the building (assumed to be 4300 m^2). Omaha is a particularly unfavorable location for photovoltaic apparatus because it receives relatively little sunlight and its electricity costs are relatively low. The cost of photovoltaic energy in sunny regions of the United States would be 30 to 40 percent less expensive than the costs shown. In computing costs, actual utility rates in Omaha were used (including demand charges) and it was assumed that utilities purchased excess on-site energy at 50 percent of the sales price. Photovoltaic costs in Table 3 are compared with commercial electric rates and not with the higher marginal costs of electricity from new generating plants.

The great value of cogenerating equipment can clearly be seen in this example. In this particular application, the value of the high-efficiency cells is not exploited since the ratio of electric to thermal output from the device is too high. The high-efficiency cell would be much more attractive in an installation where the ratio of electric to thermal demand was higher. The relatively low cost for the cogeneration devices holds, even though tracking devices are assumed to cost $135 per square meter (installed).

The advantage of using the low-cost photovoltaic equipment as a building material can be clearly seen from Table 3 since this application produces the lowest cost energy of any system shown. The apartment building chosen for this study is not well suited to this application, however, since it has a small roof area in relation to the available floor space and the south wall can only provide about 5 percent of the energy needs of the building. The cost of racks and mounting represents a significant fraction of the costs of the other flat plate systems ($45 to $60 per square meter).

The fraction of a building's energy requirements that can be met from any given collector area depends most critically on the extent to which attention is paid to conserving energy in the building; high-efficiency collectors require about as much land area as lower efficiency flat plate devices since tracking collectors must be spaced so that they do not shadow each other. Storage requirements could be reduced if occupants were willing to defer energy-consuming activities during cloudy periods. (The best strategy for shifting loads will, of course, depend on the pricing relationship with the utility that is providing backup power.)

The analysis of the system shown in Table 3, which is entirely independent of the utility, postulated successful development of the iron redox battery, being developed for the photovoltaic demonstration project at Mississippi County Community College in Arkansas. This battery can store energy inexpensively by means of a tank of iron chloride. The battery system (installed) was assumed to cost $11 per kilowatt-hour (*50*).

Economies of Scale

Apart from the possibility that large purchasers might obtain some discount in purchasing equipment, it is difficult to find clear economies of scale in photovoltaic systems. In fact, smaller systems enjoy a number of advantages. (i) They can be built more rapidly than larger systems, thus reducing the interest paid during construction and reducing the lead time over which the demand for new generating equipment must be forecast. (ii) They can be more effectively tailored to the requirements of the individual load being served (in particular, it is much easier to use the thermal energy produced when collectors are close to the buildings that will use the thermal energy). (iii) Locating systems on the roofs or walls of buildings reduces the requirement for land. (iv) Locating the generating equipment close to the buildings which will ultimately use most of the energy produced reduces the demand for transmission and distribution equipment. (v) The small equipment can be built by local building contractors and local labor much like the conventional heating and cooling systems for new buildings.

Many types of buildings in densely populated urban areas may not be able to utilize the photovoltaic equipment effectively, however, because they are shaded during much of the day. These problems can be minimized in carefully designed new communities, but they represent a major barrier in retrofitting.

It should be understood, however, that the utilization of small photovoltaic systems does not necessarily imply non-utility ownership. There is reason to believe that utilities, which will compare the cost of photovoltaic energy with the marginal and not the average cost of electricity, may find photovoltaic equipment economically attractive before other types of users.

What Happens Next?

In spite of the promising potential of the technology, commercial interest in photovoltaics could grow rather slowly over the next few decades. Prospective manufacturers are likely to be conservative about investments in the area for several reasons. First, there is concern that markets may not prove adequate to support the production volumes required to significantly reduce the prices of photovoltaic cells—market projects are necessarily speculative since photovoltaic devices have no established record in many areas where markets will need to be found. Second, there is concern about investing in manufacturing equipment that may become obsolete before the end of its productive life as a result of progress in manufacturing technology. (It is ironic that the technology is almost paralyzed by the number of alternatives available; prices cannot be reduced rapidly without a large investment in a single approach.) Finally, there is considerable confusion about the commitment of the federal government to photovoltaic technology and about the continuity of federal support. The increased funding for photovoltaic purchases for federal installations that has now passed both houses of Congress in some form will provide some continuity to federal interest in the area, but time will be required to measure industry's response.

The fears leading to hesitation in the private sector are also affecting federal planning since there is concern that a premature decision to support a particular photovoltaic approach may leave the government financing an obsolescent technology. It is becoming increasingly apparent, however, that without a major federal commitment to a specific process for manufacturing silicon devices, the cost goals for the early and mid-1980's are unlikely to be met. But how much should the government be willing to invest in a single approach when it is unlikely that the energy produced from experimental equipment will be able to compete in ordinary commercial markets? One basis for comparison is provided by the proposal to spend about $2 billion in federal funds for the Clinch River breeder reactor design. This project will produce subsidized electricity for Tennessee utility customers with a capital cost of about $5 per peak watt or about $10 per average watt (assuming a 50 percent load factor) (*51*). A substantially smaller federal outlay would almost certainly result in the development of a process for producing photovoltaic arrays for $1 per peak watt (or less) which could be installed in remote areas with all necessary supporting equipment costing about $10 per average watt.

Although the technology of photovoltaics must be viewed as a potentially major energy source in the future, confident predictions cannot be made. Perhaps the greatest uncertainty arises from the fact that there is no reliable way to forecast the future cost of energy from conventional sources or for determining how much society may be willing to pay for the social and environmental benefits of solar energy.

References and Notes

1. In this article photovoltaic prices will refer to the selling price of arrays of cells encapsulated to protect them from the weather, F.O.B. the manufacturing facility in 1975 dollars. The peak output of a cell refers to the output that would be obtained if the cell were exposed to the sun at the zenith on a clear day.
2. Energy Research and Development Administration (ERDA) briefing on the "Photovoltaics Program," 9 September 1977.
3. "Characterization of the present worldwide photovoltaics power systems market" (BDM Corporation, McLean, Virginia, May 1977), p. 1-1.
4. "Photovoltaic power systems market identification and analysis" (Intertechnology Corporation, Warrenton, Va., 23 August 1977).
5. Telegrams received by the Agency for International Development (State Department) from posts in numerous developing nations during July and August 1977 in response to a request for information about local utility rates.
6. *DOD (Department of Defense) Photovoltaic Energy Conversion Systems Market Inventory and Analysis Summary Volume* (BDM Corporation, McLean, Va., 1977), p. 17.
7. "Final report: Conceptual design and systems analysis of photovoltaic power systems, volume I, executive summary" (Westinghouse Corporation, Pittsburgh, April 1977), p. 45.
8. Office of Technology Assessment, "Application of solar technology to today's energy needs," prepublication draft.
9. H. J. Hovel, *Semiconductors and Semimetals*, vol. 2, *Solar Cells* (Academic Press, New York, 1975).
10. C. E. Backus, Ed., *Solar Cells* (IEEE Press, New York, 1976).
11. A. Rothwarf and K. W. Boer, *Prog. Solid State Chem.* **2**, 77 (1975).
12. K. W. Boer and A. Rothwarf, *Annu. Rev. Mater. Sci.* **6**, 303 (1976).
13. M. Wolf, *Energy Convers.* **11**, 63 (1971); J. J. Loferski, *J. Appl. Phys.* **27**, 101 (1956).
14. M. Wolf, paper presented at the 3rd Soviet-American Seminar on General Aspects of Solar Energy Technology, Ashkhabad, Russia, 1977.
15. J. Lindmayer and J. F. Allison, *COMSAT Tech. Rev.* **3**, 1 (1973).
16. J. R. Hauser and P. M. Dunbar, *IEEE Trans. Electron Devices* **ED-24** (No. 4), 305 (1977).
17. A. R. Kirkpatrick, in *Proceedings of the ERDA Semiannual Photovoltaic Advanced Materials Program Review Meeting* (Energy Research and Development Administration, Washington, D.C., 1977), p. 159; A. Baghdadi, in *ibid.*, p. 137.
18. T. L. Chu, S. S. Chu, K. Y. Duh, H. I. Yoo, *IEEE Trans. Electron Devices* **ED-24** (No. 4), 442 (1977).
19. H. Fischer and W. Pschunder, *ibid.*, p. 438.
20. L. D. Crossman, "Research and development of low cost processes for integrated solar arrays" (third quarterly progress report, ERDA EC (11-1)-2721, Energy Research and Development Administration, Washington, D.C., 1975), p. 38.
21. T. H. DiStefano and J. J. Cuomo, paper presented at the National Workshop on Low-Cost Polycrystalline Silicon Solar Cells, Dallas, 18–19 May 1976.
22. D. E. Carlson, *IEEE Trans. Electron Devices* **ED-24** (No. 4), 449 (1977).
23. ——— and C. R. Wronski, *Appl. Phys. Lett.* **28**, 671 (1976).
24. A. R. Moore, in preparation.
25. J. F. Jordan, personal communication.
26. L. S. Napoli, G. A. Swartz, S. G. Liu, M. Klein, D. Fairbanks, D. Tames, *RCA Rev.* **38**, 76 (1977).
27. M. D. Lammert and R. J. Schwarz, *IEEE Trans. Electron Devices* **ED-24** (No. 4), 337 (1977).
28. B. L. Sater and C. Goradia, *NASA Tech. Mem. TM X-71718* (1975); C. Goradia and B. L. Sater, *IEEE Trans. Electron Devices* **ED-24** (No. 4), 342 (1977).
29. H. J. Hovel, paper presented at the Energy Research and Development Administration Photovoltaic Concentrator Systems Workshop, Scottsdale, Ariz., 24–26 May 1977.
30. B. L. Sater, "Current technology status of the edge illuminated vertical multijunction (VMJ) solar cell" (unpublished report).
31. R. M. Swanson and R. N. Bracewell, "Silicon photovoltaic cells in thermophotovoltaic conversion" (EPRI ER-478, Electric Power Research Institute, Palo Alto, Calif., February 1977).
32. J. S. Harris, paper delivered at the Photovoltaic Systems and Applications Workshop, Reston, Va., 13–15 December 1977.
33. J. J. Loferski, in *12th IEEE Photovoltaic Specialists Conference* (IEEE, New York, 1976), p. 957.
34. A. Goetzberger and W. Breubel, *Appl. Phys.* **14**, 123 (1977).
35. H. Macomber, in *Proceedings of the ERDA Semiannual Solar Photovoltaic Program Review Meeting, Silicon Technology Programs Branch, San Diego, California, January 1977* (Energy Research and Development Administration, Washington, D.C., 1977), p. 68.
36. L. P. Hunt, in *12th IEEE Photovoltaic Specialists Conference* (IEEE, New York, 1976), p. 347.
37. A. Kran, paper presented as part of the Proceedings of the Symposium on the Materials Sciences Aspect of Thin Film Systems for Solar Energy Conversion, Tucson, Ariz., 20–22 May 1974.
38. "Low cost silicon solar array project, report-2," (Jet Propulsion Laboratory, California Institute of Technology, Pasadena, 1976), pp. 4–57.
39. G. Rodrick, personal communication.
40. G. A. Samara, paper presented as part of the Proceedings of the ERDA Semiannual Photovoltaic Advanced Materials Program Review Meeting, Washington, D.C., 22–23 March 1977.
41. J. F. Jordan, paper presented at the International Conference on Solar Electricity, Toulouse, France, 1976.
42. "Application of compound parabolic concentrators to solar photovoltaic conversion" (final report, Argonne National Laboratory, Argonne, Ill., February 1977).
43. W. H. Weber and J. Lambe, *Appl Opt.* **16**, 2299 (1976).
44. "Conceptual design and systems analysis of photovoltaic systems: Final report summary volume" (General Electric Corporation, Schenectady, N.Y., 19 March 1977).
45. P. E. Glaser, testimony before the Subcommittee on Space Science and Applications and the Subcommittee on Energy Research, Development and Demonstration of the Committee on Science and Technology, U.S. House of Representatives, 20 February 1976; R. Caputo, project engineer, *An Initial Comparative Assessment of Orbital and Terrestrial Central Power Systems: Final Report* (Jet Propulsion Laboratory, California Institute of Technology, Pasadena, 1977).
46. "Survey of the applications of solar thermal energy systems to industrial process heat" (Battelle Columbus Laboratories, Columbus, Ohio, 1977), vol. 1, p. 10.
47. E. L. Burgess and J. G. Fossum, *IEEE Trans. Electron Devices* **ED-24** (No. 4), 436 (1977).
48. L. James, personal communication.
49. "Workshop, 18–19 November 1975, Palo Alto," Argonne National Laboratories–Energy Research and Development Administration–Electric Power Research Institute–International Lead-Zinc Research Organization, Palo Alto, Calif., December 1975. J. R. Birk, paper presented at the Energy Research and Development Administration–Electric Power Research Institute, Lead-Acid Battery Workshop II, Washington, D.C., 9 December 1976; R. Witaker and J. Birk, *Electr. Power Res. Inst.* **1**, 6 (October 1976).
50. R. Zito, personal communication (the estimated manufacturing cost was increased by a factor of 1.7 to obtain the selling price).
51. Fiscal year 1978 ERDA budget (version current in January 1977).
52. G. Johnson (Motorola Corporation), Testimony delivered to the Subcommittee on Advanced Energy Technology and Energy Conservation, Research, Development and Demonstration of the Committee on Science and Technology, U.S. House of Representatives, 9 September 1977.
53. P. D. Maycock and G. F. Wakefield, paper presented as part of the Proceedings of the 11th IEEE Photovoltaic Specialists Conference, Scottsdale, Ariz., 1975.
54. J. Lindmayer and C. Y. Wrigley, "Development of a 20% efficient solar cell," (NSF-43090, National Science Foundation, Washington, D.C., 1975).
55. J. V. DuBow, paper presented as part of the Proceedings of the Energy Research and Development Administration Semiannual Photovoltaic Advanced Materials Program Review Meeting, Washington, D.C., 22–23 March 1977.
56. A. M. Barnett, J. D. Meakim, M. A. Rothwarf, "Progress in the development of high efficiency thin film cadmium sulfide solar cells" (ERDA E(49-18)-2538, Energy Research and Development Administration, Washington, D.C., 1977).
57. J. L. Shay, S. Wagner, H. M. Kasper, *Appl. Phys. Lett.* **27**, 89 (1975).
58. L. L. Kazmerski, S. R. White, G. K. Morgan, *ibid.* **29**, 268 (1976).
59. R. Stirn and Y. Yeh, *ibid.* **27**, 195 (1975).
60. J. Gibbons, personal communication.
61. J. S. Harris, paper delivered at the Photovoltaic Systems and Applications Workshop, Reston, Va., 13–15 December 1977.
62. C. Goradia, M. G. Goradia, B. L. Sater, paper prepared for the 1977 Photovoltaic Solar Energy Conference, Luxembourg, 27–30 September 1977 (Commission of the European Communities); B. L. Sater, personal communication.

63. I thank D. Claridge, J. Furber, and J. Bell of the Office of Technology Assessment for their assistance in preparing material for this article. I also thank L. W. Yerkes for pointing out the importance of the energy consumed in manufacturing silicon.

Solar Biomass Energy: An Overview of U.S. Potential

Use of wood fuel for small-scale space-heating applications is an important near-term opportunity.

C. C. Burwell

Summary. The U.S. annual biomass production for food, lumber, paper, and fiber, if used exclusively for energy, would provide 25 percent of current energy requirements. The collection of unharvested wood residues and cull trees for direct use as fuel for small nearby space-heating applications—especially for peak winter conditions—is an important near-term solar energy opportunity. Improved management of hundreds of millions of acres of productive forest land is an important opportunity for the long term. Harvest of cropland residues for energy values, new biomass production using intensive short-rotation silviculture, resubstitution of natural products for petroleum-based synthetics, and forest management for large-scale production of electricity and synthetic fuels are judged to be less appropriate directions for the U.S. energy system to take.

Solar energy is an alternative energy source to replace diminishing supplies of oil and natural gas. The U.S. government supports a growing list of research, development, and demonstration activities aimed at an early introduction of promising solar energy processes and technologies. Emphasis is given primarily to two general uses: direct space heating and electricity generation. Suitable technologies are available, although not perfected, to make use of sunlight in these two ways. However, it is generally agreed that, for most locations in the United States, conventional fuels will continue to be less costly than these applications of solar energy for several years to come.

Solar energy is abundant, well distributed, and inexhaustible. To its detriment, it is diffuse; it varies in intensity geographically and seasonally; and it is intermittent. The diffuseness of sunlight imposes energetic and economic burdens on its collection and use—particularly on its immediate use. For example, not only must large areas be covered by energy-collecting devices, but also the energy required to produce the collecting equipment, to transport it from point of manufacture to the solar field, and then to transmit the collected energy back to the point of use can be a significant fraction of the energy gained.

The earth's inconstancy in its attitude toward the sun gives rise both to our need for energy and to the cost of using solar energy. The sun falls unequally on the earth in time and location; thus, about 30 percent of our fossil fuel is used to keep us warm when the sun is inadequate and to cool us when it is too strong. For solar energy, we incur high costs because we must first purchase a great surplus in equipment to capture extra energy when the sun is shining and then we must buy the equipment to store the surplus energy that was collected in order to have it when we need it.

The difficulties of collecting, storing, and distributing solar energy have been generally recognized and provide much of the impetus to an increasing interest in the possibility of biomass production. The collection of solar energy as biomass provides solutions to the problems of high collector cost, of capturing some of the sun's energy in any location whenever it is available, and of storing that energy for use when fuel is needed. In addition, some of the energy costs for energy transmission are reduced by the biomass opportunity because the biomass need be harvested only periodically—at intervals of one or more years.

Two other attributes of biomass energy should be presented although the associated benefits are difficult to assess quantitatively. One is that biomass production, as now practiced in agriculture and silviculture, represents an accepted and desirable pattern of conservative land use. Both agriculture and silviculture are important social endeavors with histories of successful application of research and development results. Thus, biomass energy production can build on and contribute to these important sectors of the economy. The second attribute is that the rural economies of many developing nations, which might otherwise be particularly vulnerable to rising costs for imported fossil fuels, would be adaptable to biomass opportunities for energy production.

In this article, I describe, in energy terms, the present U.S. terrestrial biomass enterprise (that is, agriculture and silviculture) (*1*). Each major category of land use is examined for its present utilization and potential contribution to the production of biomass fuel. The role of specialty crops and the energy relationships of intensive as compared to extensive use of land for biomass production are discussed. To gain perspective on the availability of large areas of land for production of energy, markets for biomass production are reviewed, including a consideration of the importance of exports and livestock production to U.S. agriculture. Finally, some general reflections are provided on environmental issues and the potential for increasing photosynthetic efficiency. From this general overview, opportunities for improving the contribution that biomass can make to U.S. energy requirements are assessed.

Energy Supply and Consumption in the United States

In 1975 the United States used about 71 quads (1 Q = 10^{15} Btu) of energy for all purposes (*2*). Except for the 5 Q contribution from hydroelectric and nuclear, the rest came from the fossil fuels: 13 Q from coal, 20 Q from natural gas, and 33 Q from oil of which about one-third was imported. These data include the oil and gas used as feedstock for the petrochemical industry; thus 1 to 2 Q of the 71 Q total appear in the form of consumer goods (such as plastics, synthetic fibers, and fabrics) rather than as energy.

C. C. Burwell is a research engineer at the Oak Ridge National Laboratory, Oak Ridge, Tennessee 37830.

Traditionally, energy accounting has not included the contribution from biomass. Presumably, the purposes of biomass production are, like petroleum feedstock, primarily for consumer goods (food, paper, construction materials). Even so, about 1 Q of energy is produced annually from burning biomass (*3*). As is shown below, energy values for biomass that appears annually as consumer goods are much larger.

Agricultural and Silvicultural Biomass Production

About 20 percent of the 1900 million acres (1 acre = 0.4 hectare) in the 48 contiguous states is cropland. In 1974, about 330 million acres were harvested (*4*). Major crop acreages, yields, and energy values are summarized in Table 1. All specialty crops (such as fruits, vegetables, nuts, tobacco, flowers, and seeds), although quite important to agricultural economics, represent only a small fraction of our cropland dedication. According to the data and energy values given in Table 1, cropland agriculture produces energy totaling about 12 Q annually, of which about 40 percent is represented by residues left on the land. Also, the gross yield (including residues) averages about 2½ tons per acre and contains 35 million British thermal units (MBtu) per acre for most of the nation's best lands under good management.

Table 1. Major agricultural crops: land use and production in 1974.

Commodity	Acres (million)	Million dry tons: Crop†	Million dry tons: Residue‡	Energy production*: Per acre (MBtu)	Energy production*: Total (Q)
Grains	158	215	303	45(25)§	7.1(3.9)
Wheat	66	48	116	34	2.2
Corn	65	142‖	142	61	3.9
Oats	13	9	24	34	0.5
Sorghum	14	16	21	37	0.5
Green crops	74	153		29	2.2
Hay	61	117		27	1.7
Silage and sugarcane	13	36		39	0.5
Oilseed	68	41	34	18(6)	1.2(0.4)
Soybean	53	34	29	19	1.0
Cotton	13	5	4¶	11	0.2
Peanut	2	2	<1¶	21	<0.1
Fruits and vegetables	9	13		20	0.2
Total	309	422	337	35	10.7

*Based on energy values in MBtu per dry ton given in (*12*, p. 263): grain, 15; oilseed, 18; forage crops, 14; fruits and vegetables, 2 (wet basis); residues, 13. †Crop moisture percentages taken from (*42*): wheat, 10.6; corn, 12.9; oats, 7.7; sorghum, 12.8; hay, 8.0; silage, 75.0; soybean, 8.0; cotton, 8.0; peanut, 6.0; fruits and vegetables, 85.0. ‡Taken from (*3*, p. 6), and adjusted for differences in harvested acreages. §Residue values in parentheses. ‖Includes corncobs. ¶Estimated to be half as much per acre as soybeans.

Table 2. Commercial forest land statistics for 1974 by region. Data for the tree include trunk only to the point of major branching or to a minimum bark diameter of 4 inches (*5*, pp. 26 and 246–247).

Area	Production, Total: Cubic feet	Production, Total: Btu*	Production, Total: Softwood (%)	Production, Per acre: Cubic feet	Production, Per acre: Btu*
		Stemwood inventory			
Northeast	174.4×10^9	57.6×10^{15}	25	980	323×10^6
Southeast	184.5×10^9	55.4×10^{15}	49	959	287×10^6
West	355.6×10^9	99.4×10^{15}	93	2770	777×10^6
Total	714.5×10^9	212.4×10^{15}	67	1430	430×10^6
		Annual growth			
Northeast	5.5×10^9	1.8×10^{15}	25	31.1	10.3×10^6
Southeast	8.6×10^9	2.6×10^{15}	63	44.6	13.4×10^6
West	4.4×10^9	1.2×10^{15}	88	34.2	9.6×10^6
Total	18.6×10^9	5.6×10^{15}	57	37.1	11.1×10^6

*Hardwood basis, 8000 Btu/lb at 44 pounds per cubic foot; values typical for dry oak, hickory, and maple; softwood basis, 8400 Btu/lb at 32 pounds per cubic foot; values typical for dry fir and pine (*43*).

Table 3. Commercial forest land acreage and ownership for 1974, in 10^6 acres, by region (*5*, p. 11).

Area	Acreage	Land ownership: Federal	State and local	Industry	Farm	Private
Northeast	177.9	12.3	19.6	17.6	51.0	77.4
Southeast	192.5	14.3	3.0	35.3	65.1	74.8
West	129.3	80.6	6.4	14.4	15.0	12.8
Total	499.7	107.1	29.0	67.3	131.1	165.1

The Forest Service considers 500 million acres of forest land to be commercial—that is, capable of producing stemwood in excess of 20 cubic feet per acre per year (1 foot = 0.3 meter) (*5*). As shown in Tables 2 and 3, the energy value of stemwood growth equals 11 MBtu/acre-year and totals 5.6 Q. Since stemwood represents 55 to 70 percent of the total tree (*6*), the total biomass growth in commercial forests is somewhat in excess of 9 Q annually, and the average energy yield is about 18 MBtu per acre. However, practical limits may exclude leaves, roots, and small branches from these values, thus reducing them to 6.8 Q and 13.5 MBtu per acre-year, respectively.

In contrast to cropland, most of the forest land is not actively managed for production, except for the 13 percent that is industrially owned. The National Research Council estimates that the forest yields could be doubled if the forests were subjected to improved management (*7*). If so, acre for acre, the forest land energy production potential could become comparable to that for cropland.

The difference, of course, between cropland and forest land production is that the forest yield can be accumulated for many years. For example, the average inventory on forest land is about 715 MBtu per acre (that is, 430/~ 0.6) or about 20 times the annual cropland yield. (The yield of each acre actually harvested would, of course, be expected to be considerably higher than the average inventory for all forest land.) The forest inventory represents more than 38 years of tree growth on the average, and an energy inventory that is three times our annual usage of energy in all forms.

Other Terrestrial Biomass

In addition to cropland and commerical forest, major land classifications include urban area (60 million acres), noncommercial forest (100 million acres), pasture and range (715 million acres), idle cropland (50 million acres),

Table 4. Potentially collectable net yield from U.S. biomass operations under present management practices (1974).

Activity	Gross energy yield (Q)	Collectible net energy yield* (Q)
Agriculture		
Corn	3.9 (1.9)†	3.0 (1.8)
Grains	3.2 (2.1)	2.9 (2.0)
Green crops	2.2	2.1
Oil seeds	1.2 (0.4)	1.1 (0.4)
Fruits and vegetables	0.2	0.2
Other‡	0.7	0.6
Silviculture	9.3 (3.7)§	6.6 (1.2)‖
Pasture and range	7.0	0.7
Total	27.7 (8.1)	17.2 (5.4)

*Energy inputs valued at 1.5 times the value of biomass energy. †Residue values given in parentheses. ‡Taken as 10 percent of the total for all agriculture excluding corn in order to account for minor crop acreages not included in Table 1. §All residues. ‖Excludes tree leaves, small branches, and roots; includes stump, unmerchantable bole, and large branches.

and other (135 million acres of swamp, sand dune, desert, bare rock, and national and state preserves) (*8*). With the exception of idle cropland and cropland pasture, these land areas (without irrigation) would seem potentially capable of making only a limited contribution to biomass production. The largest land category, pasture and range, produces perhaps as much as 7 MBtu per acre (*9*) for a total of 5 Q annually. If the other areas are no less productive, the energy production might total 7 Q for these marginal lands taken all together.

Energy Requirements

Several investigators have estimated ratios of energy output to input for major crops: corn, 2.5 (*10*) to 3.3 (*11*); wheat, 5.4 (*11*); alfalfa, 16.0 (*12*); forest logging, 37 (*13*). These data do not account for the harvest residues. Since the residue yield is substantial and since most of the energy inputs for crop production are for nonharvest operations, a high ratio of output to input energy for incremental harvest of cropland residues would be expected—perhaps 40 : 1 or more—and would be comparable to tree stemwood production where energy requirements are totally associated with the harvest. When this adjustment is made, the ratios of output to input for whole-plant harvest of grains become 5.2 for corn and 13.0 for wheat. When computed in this way, the inverse proportionality between the energy output to input ratio and the yield is apparent (Fig. 1). This relationship is true, of course, because the solar energy input is not included.

For forest land and cropland, energy inputs come from very high quality sources: gasoline, commercial fertilizer, electricity. Even at its best, biomass fuel is not fully dried, is bulky, and has low heat content per pound relative to fossil fuels. To account for this difference, energy inputs to biomass production should be valued considerably higher—say 50 percent (*14*)—than the biomass energy produced unless, of course, the biomass can be directly substituted for oil or gas on a Btu-for-Btu basis.

For pasture and range land, a different relationship exists: the grazing animal provides the harvest energy, but in so doing the harvest yield is reduced by a factor of 10 or more (*15*). A precise value for the factor is not needed to establish the point that the collectable energy yield from rangeland under present management practices is quite low (less than 1 MBtu/acre-year).

At this point, we can make a rough estimate of the current net biomass energy production for the United States that attempts to account for the energy to produce and collect it. These estimates are shown in Table 4. The major differences between gross energy yields and the computed net yield exist because (i) there is a high energy input associated with corn production; (ii) certain forest residues are judged to be relatively uncollectable; and (iii) rangeland production is collected inefficiently in terms of its use for energy.

Biomass Residues Summarized

Opportunities for energy production from biomass exist at one extreme in the use of collected biomass wastes, and at the other extreme in new biomass production. In between lie possibilities for the use of uncollected biomass residues. At first glance, the priority for the use of biomass for energy production would seem to flow downward from the use of collected wastes that would otherwise have to be disposed of, to the collection and use of residues that are available for the taking, and, finally, to new production.

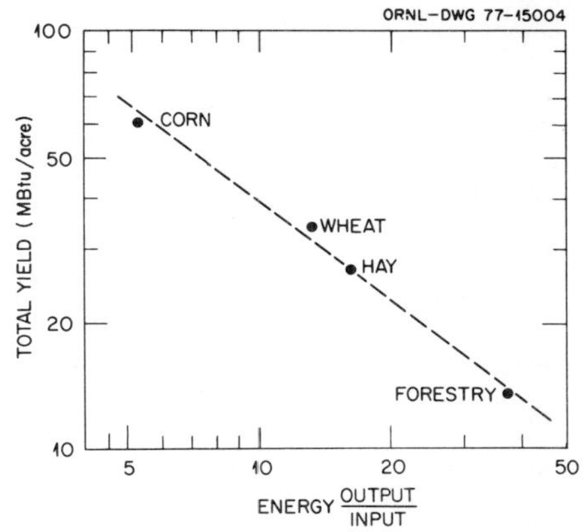

Fig. 1. Inverse proportionality of major crop yields and the ratio of energy output to input.

A summary of biomass residue tonnages and energy contents from major sources is presented in Table 5. Also shown are the estimated residue energy densities per acre for cropland and forest land (*16*). These estimates show that energy density per acre of forest residues is higher, and that wood itself is a much denser form of fuel (for example, 1 cubic foot of wood contains as much energy as 5 to 10 cubic feet of baled field residues). In addition, dead and cull trees from harvested forest land are not included in the figures for logging residues. Dead, rough, and rotten trees, on the average, constitute 9 percent of the forest inven-

Table 5. Major sources of potentially usable biomass residues.

Item	Weight (10^6 dry tons)	Energy* Total (Q)	Energy* Per acre (MBtu)
Collected			
Urban and municipal solid wastes†	160	2.1	
Large poultry and hog operations and cattle feedlots‡	26	0.3	
Large canneries, mills, slaughter houses, and dairies‡	23	0.3	
Wood manufacturing§	15 to 27	0.4	
Total	~230	~3	
Uncollected			
Cereal straw	161‖	2.1	22
Cornstalk	142‖	1.8	28
Logging residues§	50 to 75	1.1	130¶

*Residues evaluated at 13 MBtu per dry ton except for wood residues at 17 MBtu/dry ton. †(*3*), p. 6. ‡(*7*), p. 217 §(*3*), p. 4; (*7*), p. 217. ‖See Table 1. ¶Estimated by the author assuming the large branches, stump, and unmerchantable bole are collected and that the total average aboveground residue is 9.1 dry ton/acre as given in (*6*, p. 75).

Table 6. Major markets for field crop production, 1973 (*4*).

Commodity	Acres harvested (millions)	Market share of net production, (10^6 acres)*		
		Livestock	Export	Other domestic
Corn	62	45 (72)	13 (21)	4 (7)
Wheat	54	7 (12)	32 (60)	15 (28)
Soybean	56	20 (36)	34 (60)	2 (4)
Oats	14	12 (86)	1 (8)	1 (6)
Sorghum	16	12 (75)	4 (25)	
Barley	10	5 (52)	2 (19)	3 (29)
Hay	62	62 (100)		
Silage	10	10 (100)		
Cotton	12†	4 (33)	4 (30)	4 (37)
Total	296‡	177 (60)	90 (30)	39 (10)

*Acreages were calculated based on the fraction of the total weight of the harvest utilized for each market. Values in parentheses are percentages. †About half the weight is lint and linters of which about half is exported. The other half represents oil, cake, and meal; hulls are not included. ‡This represents about 90 percent of the total cropland harvested.

tory (*5*, p. 27). Thus, there is the opportunity to improve the quality of the stand of the remaining growth stock even as dead or deformed standing inventories are collected.

Intensive Tree Farming

The concept of intensive tree farming has recently received much attention (*17*). The concept applies the elements of cropland agriculture (high machinery use, high fertilizer use, short rotation harvesting) to growing trees. It is hard to assign high priority to this technique when there are millions of acres of underutilized forest lands (*18*) that await even rudimentary management. Expected ratios of energy output to input from intensive tree farming are predictably lower than those routinely obtained through traditional forestry operations even if very high yields (*19*) can be achieved as claimed; and, of course, the energy inputs are in the form of oil and gas. While very high yields may be possible, there is a need to distinguish between what is possible and what is practical on a large scale. For example, in agriculture, record yields for small acreages may be four to eight times higher than average yields (*20*).

Finally, evidence to support the view that annual yields obtainable through short rotation tree farming will exceed those obtainable through traditional forestry is not convincing—especially when the comparison is made for comparable levels of management and cultural practice. Indeed, considerable evidence shows that high yields in traditional forestry are obtained after a few years of growth and that these yields can continue for ten or more years (*21*). At present, the choice between coppice forestry, with artificial replacement of nutrients lost in small branches and leaves, and the harvest of stemwood at 10- to 20-year intervals, seems to favor traditional forestry.

Specialty Crops

A massive "Emergency Rubber Project" (ERP) was initiated in February 1942. More than 1,000 scientists and technicians were assigned to guayule production and research. In 3-1/2 years, supported by a work force of 1,000 workers, the ERP planted almost 32,000 acres of guayule at 13 sites in 3 states. It produced one billion guayule seedlings and 3 million pounds of resinous rubber for the war effort.

The preceding excerpt (*22*) is used by its authors as support for the report recommendations calling for government R & D for commercialization of guayule production. The Native Latex Commercialization Act would provide $60 million over the next 5 years to accomplish that purpose (*23*). But 3 million pounds of rubber in 3 ½ years from 32,000 acres of carefully managed irrigated land is less than 0.02 ton per acre per year. By way of comparison, the average oil yield from the peanut crop is currently about 0.5 ton per acre. Thus, whatever other benefits may result from attempts to establish a natural rubber industry, the case for guayule as an energy producer has yet to be made. This is in spite of the views as stated in the introduction to the National Research Council Report (*22*):

> Petroleum, our major source of hydrocarbons, is dwindling and is now widely predicted to run out within a few decades. Thus, today, a plant that produces hydrocarbons—as guayule does—is particularly worthy of investigation.

Agricultural Markets

We gain some appreciation of the possible future availability of large land areas for energy production by considering the current markets for cropland production. In order of acreage, markets for field crops may be divided into three general categories: livestock feed, export, and other domestic uses. These major categories of demand are compared in Table 6 in terms of acreage commitments (*4*). As shown, of major field cropland, 60 percent is dedicated to the production of livestock, 30 percent to production for export, and 10 percent to production for other domestic purposes. The data in Table 6 exclude the contribution that 716 million acres of pasture and rangelands make to livestock support.

Some measure of the overall efficiency of animal production for food is indicated in Tables 7 and 8. These data suggest that, on a dry weight basis, the edible portion resulting from livestock production is about 4 percent of the harvested yield from the dedicated land. Of course, some of the cropland products for animal feed are by-products of other markets (cottonseed and soybean meal), but by and large these are relatively minor activities in the overall livestock enterprise. If we attempt to account for the use of pasture and rangelands, then the efficiency of meat production is further reduced by perhaps a factor of 2 (*24*).

With no intent to judge or influence the actual course of future events, one can note the vulnerability of the U.S. agricultural system to potential changes in its dominant markets. Just as margarine (soybean oil) replaced butter in the 1940's, so might meat substitutes better satisfy future consumer preferences for reasons of both economy and health (*25*). Such a change would take place over decades and, in all probability, would never be complete. Overall implications are illustrated with an extreme example. It can be shown that the basic ingredients to provide 3000 kilocalories per day for 210 million people on a U.S. recommended diet (*26*) could be produced on about 9 percent of the present cropland acreage, that is, 15 million acres each of corn and peanuts, based on current average yields for those crops in contrast to the 240 million acres that are now being used.

The foreign market for grain may not be unlimited. There may be limits in the degree to which importing countries are willing to become dependent on the United States for basic food. There may be financial limits in the ability of developing countries to purchase their food imports or there may be limits in our willingness to subsidize their needs. Finally, there are real questions regarding the efficacy of grain as compared to oil imports for developing countries that have rural economies based on labor-intensive agriculture.

Markets for Industrial Materials from Renewable Sources

Traditional industrial products based on biomass include sawtimber products, primarily for building construction, and pulpwood for the paper industry. Also important, but much smaller markets in terms of volume, include cotton, wool, and cellulose fibers (rayon), and special products and by-products of agriculture and forestry, such as leather, turpentine, and soap.

Beginning in the 1940's, synthetic products, primarily made from petroleum, have rapidly replaced traditional biomass products: first synthetic rubber and detergents, and now a wide variety of synthetic fibers and plastics each formulated to provide the right properties for each specific consumer requirement. More recently, synthetic products are competing favorably with paper and structural products as well (for example, plastic bags and vinyl siding).

Recent demands for industrial biomass materials are given in Table 9. Also shown are the markets for synthetic plastic, rubber, and fiber. The data are given in terms of product tonnage and approximate energy content. These data show that the opportunity for a significant contribution to energy conservation or supply through an attempt to recapture markets lost to the petrochemical industry seems not to exist. This is because the total market opportunity is small, and because little, if any, energy would be saved relative to the use of biomass directly as fuel (*27*). Also, a significant fraction of the energy content of petrochemical feedstocks ends up in the products themselves. Thus, as long as product energy values are ultimately recovered through recycle or as residues for use as fuel, the original form of the feedstock is of secondary importance. Finally, the synthetic products, in general, have a longer useful life than do natural products and, to the extent that our society moves toward recovery of energy values in solid wastes, the biodegradability of natural products is not an important advantage.

Conversely, forestry activities could make a contribution to fuel supply through additional forest production, additional forest residue use, and through recovery of forestry products (wood and paper) at the end of product life. Certainly, direct use of wood and wood products as fuel in place of petroleum and gas seems intuitively much easier to accomplish than does restructuring the petrochemical industry and changing consumer product preferences.

Table 7. Major livestock production in the United States in 1974 (10^6 ton).

Commodity	Weight		
	Live*	Dressed*	Edible†
Beef	19.6	11.0	3.4
Hogs	10.7	6.2	2.4
Sheep	0.4	0.2	0.1
Chicken	6.2	4.3	0.9
Turkey	1.2	1.0	0.3
Milk	57.7		4.4*
Eggs	3.7		0.8*
Total	100.3		12.3

*See (*4*). †Estimated protein, fat, and carbohydrate content of dressed meat weights (dry weights) based on data in (*44*).

Table 8. Major livestock feed crops.*

Commodity	Consumption† (10^6 dry tons)
Corn	102‡
Wheat	4.6
Soybean	13.8
Oats	9.8
Sorghum	17.1
Barley	5.2
Cottonseed	2
Silage	30.6
Hay	124
Total	309.1

*The time periods associated with data in (*4*) for feed disappearance and livestock production are not exactly comparable. †Basic data for year beginning in mid- or late-1973 from (*4*); commodity moisture contents taken from (*42*). ‡Grain only, although corncobs are also used for cattle feed.

Table 9. Demands for industrial biomass materials and for synthetic plastic, rubber, and fiber in 1972 (*7*, pp. 65, 202, and 228).

Commodity	Tonnage (10^6)	Energy (Q)
Round wood requirement		
Structural materials	123	2.1
Pulp, paper, fiber	102*	1.7
Cotton and wool fiber†	3	0.2‡
Biomass extractives		
Turpentine, rosin, tall oil§	1	
Vegetable oils‖	1	
Animal fat‖	1	
Animal hides	2	
Synthetics		0.6¶
Plastics	12.5	
Fibers	3.1	
Rubber	2.8	

*Cellulose fibers account for less than 1 million tons. †Includes exports. ‡Total for cotton and wool fiber, biomass extractives, and animal hides. §Major uses include pine oils, rubber additives, paper size, adhesives, insecticides. ‖Major uses include fatty acids, soap, and drying oils. ¶Total for all synthetics.

The production of nitrogenous fertilizers is not included in Table 9. Total consumption in 1973 was 10 million tons of contained nitrogen (*4*, p. 469). On the basis of 19,000 cubic feet of natural gas used per ton of ammonia produced (*28*), the nitrogenous fertilizer industry consumed 0.2 to 0.3 Q of natural gas in 1973, which represents about 1 to 2 percent of natural gas use for all purposes. Again, the total market is small.

Ammonia could be produced from the direct processing of biomass or, alternatively, the demand for ammonia could be reduced through the development of agronomic crops with an enhanced capability to fix their own nitrogen. Favorable economics for ammonia production require a large-scale operation. For example, an ammonia plant yielding 300 tons per day would require the annual yield from 250,000 acres of forest or else five times that area if woody forest residues were to be the sole raw material source. Thus, biomass production or residues are unlikely to compete with concentrated fossil fuels for the ammonia industry.

The prospects for direct plant fixation of nitrogen would seem promising, but the issue is not simple even if the uncertainties concerning the outcome of the needed research program are disregarded. For example, research on vegetative peas gave the result that the plants consume four weight units of carbohydrate to fix one weight unit of nitrogen (*29*). Synthetic fixation of nitrogen requires only about half as much energy per unit of fixed nitrogen production, suggesting that, from an energy systems standpoint alone, we may be better off using natural gas for ammonia production and biomass for fuel rather than the reverse. At the very best, the results do indicate that changing our present methods may produce only minor gains.

The Market for Biomass Fuel

It is fair to say that a significant demand for biomass fuel is lacking because homes, schools, churches, stores, apartment buildings, and small business establishments are not at present equipped to use such fuels directly. Some biomass industries (sugar, paper, wood products) do use feedstock wastes for some of their fuel requirements.

There are markets for natural gas and petroleum that are satisfied by fuel imports. Because of this, indirect production of gas and oil from biomass, rather than development of direct fuel markets for biomass, has received the major emphasis.

It is not the purpose of this article to review alternative processes for converting biomass residues to electricity or convenient forms of fossil fuels. Favorable economics of electricity generation and chemical processing depend to a degree on large-scale operations. They also depend on favorable conditions for handling raw materials (such as high energy density, pipelining). These attributes are not characteristic of forest and field residues. Biomass residues must be collected in bulky form from large areas at a high cost in labor and transportation; thus, they are unlikely to compete generally with coal or shale oil as feedstock for centralized industrial production of synthetic fuel gas and liquids.

For small-scale operations and under favorable conditions for growth of biomass, production of liquid fuels from biomass may be more competitive. The government of Brazil is embarking on an aggressive program of alcohol production from cassava and sugarcane, as a substitute for imported oil, and at the same time to encourage settlement and development of the Amazon Basin (*30*). Here, the combination of low-cost labor and low-cost land, small rural populations and related energy demands, difficult logistics of supply from the coastal cities, and a long growing season and tropical climate seem ideally suited to the government policy.

The conditions for biomass alcohol production in the United States are almost completely opposite those for Brazil (*31*). In addition, in the United States, the outcome of a net energy analysis is less certain and more complex because of the high energy requirements for corn production. (Corn is the only crop presently grown in the United States with a scale of production large enough to make a significant impact from biomass alcohol.) For example, at the margin, in U.S. corn culture 1 pound of additional nitrogen might be expected to produce an additional 4 pounds of grain yield (*32*). Since it would take the energy from about 3 pounds of grain to produce 1 pound of nitrogen fertilizer, the potential net gain in energy seems marginal even without considering other energy uses for the complete system. Thus, in the absence of the availability of large quantities of surplus agricultural land that would allow corn production for biomass alcohol with lower inputs of fossil energy, it is not obvious whether the U.S. energy balance would benefit or lose from biomass alcohol (*33*).

> A point about solar energy that government planners seem to have trouble grasping is that it is fundamentally different from other energy sources. Solar energy is democratic. It falls on everyone and can be put to use by individuals and small groups of people. The public enthusiasm for solar is perhaps as much a reflection of this unusual accessibility as it is a vote for the environmental kindliness and inherent renewability of energy from the sun.

This view from Hammond and Metz (*34*) appears to be only slightly less valid for solar biomass energy. If so, perhaps a relatively neglected area for consideration is the creation of a demand that substitutes wood residues directly for oil and gas use. This demand could be created among the users of small amounts of energy (for example, schools, stores, residences) that presently depend on oil and gas for space heating. In this sense, solar biomass is a natural complement for direct solar heating. Wood fuel could also supplement the use of electricity for space heating during peak demand periods. Electric resistance heaters are frequently used in electrically heated homes for peak heating requirements on the coldest days, even for homes equipped with more efficient heat pumps. These supplies of electricity are, in turn, typically generated in power stations fueled with gas and oil.

Machinery for whole-tree chip harvesting and equipment for efficient combustion of wood for steam generation and space heating are currently available. Wood fuel is available in many locations at costs competitive with other fuels or nearly so (*35*). Major barriers to increased utilization of wood as a fuel include: acceptance of wood as a practical fuel on a small scale; installation of efficient wood-burning equipment; adequate incentives for harvesting cull trees, deadfall, and sawtimber residues; and incentives to replace trees removed with improved growing stock.

Environmental Considerations

The full range of environmental issues related to greatly expanded production and use of energy from biomass have yet to be addressed. We already have measures of potential problems in the form of the current effects of agriculture—these include stream pollution arising from soil and fertilizer runoff and loss of productive lands through overgrazing, deforestation, and irrigation salting.

Environmental considerations compel us to place high priority on energy production from collected biomass residues. Otherwise, we must accept the high cost and further environmental degradation resulting from disposal.

With regard to uncollected biomass residues from forestry and farming, we note that the annual taking of field crop residues represents the taking of both nutrients and humus from the soil which, at some point, must be replaced if the soil quality is to be maintained. These effects will also occur with the taking of forest residues, but to a lesser extent. For example, partial forest harvests at multiyear intervals ensure that annual residues (leaves, deadfall) are available for maintaining forest floor conditions, and, even at harvest, about one-fourth of the plant—small branches and roots—would be left behind. Woody forest residues contain perhaps only 30 percent as much nitrogen as do wheat and corn residues (*36*), which make up three-fourths of the total field residues.

The carbon dioxide content of the atmosphere is increasing, primarily due to the growing use of fossil fuels since the industrial revolution (*37*). The concentration has increased 5 percent in the last two decades alone. The long-term effects of this trend are uncertain both in time and magnitude but are potentially serious. A new dimension to the problem was introduced by Bolin (*38*) who estimates that the accumulated input of carbon to the atmosphere due to deforestation and expansion of agriculture (that is, the loss of carbon from the soil) is 50 ± 25 percent of the amount transferred to the atmosphere from fossil fuel combustion.

Biomass wastes are soon returned to the atmosphere by natural processes in the form of heat, water, and carbon dioxide. Thus, to the extent that their use as fuel can supplant the combustion of fossil fuel, release of carbon dioxide and sulfur dioxide could be reduced both because the total amount of fuel burned would be lower and because the sulfur content of plant residues is lower than it is for fossil fuels.

The implications for biomass energy seem clear: priority should be given to ways through which we may derive energy from biomass while maintaining or augmenting carbon inventory in land biota and soil. If history is a guide, reforestation and improved forest management seem potentially able to meet these criteria.

Photosynthesis

There is no evidence that the graph in Fig. 1 (showing yield proportionality with energy input) represents immutable biological law. Still, the implications seem clear: the biggest gains in biomass production for the least expenditure of high value fossil fuels are to be made not at the margin represented by corn but, rather, at the other extreme represented by forestry. Indeed, it is discouraging to plot extremely high yields (such as 30 dry tons per acre) on the graph because we would then predict a net loss in energy yield or a mining of the soil resource that could not long be sustained in nature (*39*).

Bassham has estimated the theoretical maximum daily photosynthetic energy conversion of incident sunlight at 6.6 percent and shown that maximum rates during the growing season may reach half this calculated value for C_4 plants (*40*). For the United States, this 50 percent of theoretical value corresponds to 50 tons/acre-year if the conversion rate could be maintained throughout the year. In fact, such yields have been achieved on limited acreages of specialty crops, such as sugarcane, that have a long growing season and nearly optimum growing conditions of sunlight, temperature, rainfall, and rich soil. For much of the United States, however, such conditions do not exist. For example, the maximum measured daily photosynthetic energy conversion rate for corn exceeds that for sugarcane by almost 40 percent but on an annual basis the photosynthetic productivity of corn is only about 10 percent that of sugarcane. The point is not that further improvements in photosynthetic productivity are not possible, but that reasonably high photosynthetic conversion efficiencies are already being achieved (*41*), and that other factors, such as growing season, may be of primary importance in achieving high annual yields for energy crops. Here again, tree crops, with maximum growing season and minimum loss through respiration during the dormant season, are suggested.

Conclusions

Based on this analysis, the total of each year's biomass production in the United States for all purposes—not only fiber and wood but also food—might, if used for energy, suffice for 25 percent of the nation's annual energy needs. We conclude, therefore, that biomass energy must be regarded as a relatively minor contributor to the U.S. energy system at least for the next few decades.

Nearly 90 percent of cropland is used for production of livestock feed and grains for export. Thus, a shift in food preference away from animal protein or a reduction in the demand for grain exports could conceivably release high quality agricultural land for energy crops. Such changes, should they occur, would take place in relatively small increments over many years.

At present, several hundred million acres of public and private forest land is not managed for tree growth—even when multiple use goals are considered. Much of this land is in the eastern half of the United States, in reasonable proximity to rural, town, and suburban areas of energy demand. Forestry has unique characteristics that make it particularly suitable for biomass energy production: the energy harvest per acre can be many times that for annual crops; wood is a dense storable form of biomass fuel; trees grow productively for many years and thus provide live storage from year to year without loss of yield; nutrient losses in forestry are relatively small; wood fuel burns relatively cleanly. These factors suggest that the best present opportunity for new biomass production could be realized through programs that provide incentives for better forest management. In this way forest quality and growth rate could be increased even as low quality forest stands and forest residues are harvested for fuel.

Experience in agriculture shows that a prerequisite for high crop yield is high energy input to production. In agriculture, the increase in yield is in the form of a high value product such as grain; thus, these high energy inputs of fossil fuels seem appropriate. However, it is less apparent that large inputs of high value fuels (gas, gasoline, electricity) can be justified in the production of lower value biomass energy either in the form of agricultural crops or by intensive short-rotation tree-farming concepts. This seems especially true in the light of opportunities to use biomass residues and to increase forest growth rates marginally through less energy-intensive means.

The contribution to energy supply that could be achieved through replacing synthetic rubber, plastics, and fiber with products derived from biomass resources is relatively small in total. In addition, a comparable or even greater energy contribution could be made by using the same biomass directly as a fuel substitute for natural gas and fuel oil. This could be accomplished without the need to reorient consumer preferences away from the many specialized petrochemical products or the need to replace petrochemical industries themselves.

The lack of installed, small-scale equipment for using wood fuel instead of oil and gas is judged to be the primary barrier to expanding the demand for wood. Barriers to development of this market in an acceptable form are: incentives to manage private forest acreages for increased wood production; incentives to harvest forest residues and cull timber; a wood fuel supply industry; and availability and acceptability of products for efficient wood-burning suitable for small-scale users that are now dependent on gas and oil fuels.

References and Notes

1. Aquatic production of biomass is not considered in this paper although its range of technical and biological possibility seems at least as comprehensive as that for terrestrial biomass. Concepts under consideration include (i) use of aquatic plants to feed on nutrient-rich waste waters, thus converting the wastes to more useful products; (ii) harvest of plants growing on unused areas, such as cattails from marshes; and (iii) new production of kelp on ocean farms (suggested by H. A. Wilcox of the Naval Ocean Systems Center in San Diego). In general, aquatic biomass production would be subject to the same overall constraints as land-based activities: photosynthesis, area available, climate, and energy costs for plant culture, nutrient source, transportation, and processing.
2. J. A. Lane, *Consensus Forecast of the U.S. Energy Supply and Demand to the Year 2000* (Report ORNL/TM-5369, Oak Ridge National Laboratory, Oak Ridge, Tenn., May 1977), p. 14.
3. E. L. Ellwood *et al.*, *The Potential of Lignocellulosic Materials for the Production of Chemicals, Fuels, and Energy* (National Research Council, Washington, D.C., 1976), p. 11; available from the National Technical Information Service, Springfield, Va.
4. U.S. Department Agriculture, *Agricultural Statistics 1975* (U.S. Government Printing Office, Washington, D.C., 1975).
5. *U.S. For. Serv. For. Resour. Rep.* (1974), p. 9.
6. K. Howlett and A. Gamache, *Forest and Mill Residues as Potential Sources of Biomass* (Report 7347, Mitre Corporation/Metrek Division, McLean, Va., May 1977), vol. 6, p. 49.
7. National Research Council, *Renewable Resources for Industrial Materials* (National Academy of Sciences, Washington, D.C., 1976), p. 7.
8. U.S. Department of Agriculture Economic Research Service, *Air, Land and Water Resources: Current and Prospective Supplies and Uses* (Bulletin 1290, U.S. Government Printing Office, Washington, D.C., 1974).
9. A. Poole, "Food and fiber project notes" (unpublished paper) (Institute for Energy Analysis, Oak Ridge Associated Universities, Oak Ridge, Tenn., July 1977), exhibit 9.
10. D. Pimentel, in *Energy, Agriculture and Waste Management*, W. J. Jewell, Ed. (Ann Arbor Science, Ann Arbor, Mich., 1975), p. 12.
11. L. D. Hill and S. Erickson, in *ibid.*, p. 118.
12. C. G. E. Downing, in *ibid.*, p. 265.
13. K. C. Hoffman, F. Drysdale, P. R. Eberts, W. McKean, G. Schreuder, G. C. Taylor, N. Bhagat, *Reference Materials System: A Source for Renewable Materials Assessment* (National Research Council–National Academy of Sciences, Washington, D.C., 1976), p. 13; available from the National Technical Information Service, Springfield, Va.
14. By analogy, about one-third of the energy content of coal is used in converting it to higher value synthetic liquid and gas fuels.
15. The assumption is that wastes from grazing animals cannot be recovered.
16. There is, of course, great variation in the per acre quantities of logging residues. For example, J. B. Grantham [in *Status of Timber Utilization on the Pacific Coast* (U.S. Department of Agriculture Forest Service General Technical Report PNW-29, 1974), p. 7] estimates that just the weight of logging residues over 4 inches (1 inch = 2.54 centimeters) in diameter and 4 feet long exceed 57 tons per acre (~ 1000 MBtu) for national forest harvests of Douglas fir.
17. J. A. Alich and R. E. Inman in *Energy, Agriculture and Waste Management*, W. J. Jewell, Ed. (Ann Arbor Science, Ann Arbor, Mich., 1975).

18. In this context the term "underutilized" refers to the practice of minimum management of privately owned forest for many millions of acres. (This is not to imply that minimum management is not rational given current economic conditions.) Some significant fraction of this area is never harvested, and mature trees, both merchantable and cull, topple and decay. For a larger fraction of the acreage, merchantable stemwood is periodically harvested but not replanted, the woody residues and dead stemwood are not removed, and growing cull trees are left to expand their area of coverage.
19. The basis for analysis used in the reference is 30 dry tons per acre-year.
20. S. H. Wittwer, *Science* **188**, 575 (1975).
21. Data from several studies are reported in (*6*).
22. National Research Council, *Guayule: An Alternative Source of Natural Rubber* (National Academy of Sciences, Washington, D.C., March 1977), p. 19; available from National Technical Information Service, Springfield, Va.
23. *Government R&D Report* (U.S. Government Printing Office, Washington, D.C., May 1977), vol. 7, No. 9, pp. 11–12.
24. However, these lands, especially without irrigation, are, in general, only marginally suited for other uses; thus, their harvest by livestock may represent a best use opportunity.
25. C. T. Nisbet, *Technol. Rev.* **79** (No. 7), 5 (1977).
26. The analysis—53 percent carbohydrate, 12 percent protein, 35 percent fat—including consideration of the adequacy of essential amino acids is given by W. C. Yee [*Middle East Study: Nutrition Economics in Desalination Agriculture* (Report ORNL-4489, Oak Ridge National Laboratory, Oak Ridge, Tenn., 1970)].
27. The results of several detailed analyses of the energy requirements to produce competing consumer products using either petroleum or biomass feedstocks are contained in (*7*), pp. 113, 153, and 202. No clear pattern of energy savings emerges. In many cases the energy differences are unimportant relative to product properties that are influential in determining consumer choice.
28. The Conference Board, *Energy Consumption in Manufacturing* (Ballinger, Cambridge, Mass., 1974), p. 266.
29. R. W. F. Hardy and U. D. Harelka, in *Biological Solar Energy Conversion*, A. Mitsui, S. Miyachi, A. San Pietro, S. Tamura, Eds. (Academic Press, New York, 1977), p. 305. Biomass energy is stored in the plant primarily as carbohydrate: cellulose, starch, and simple sugars. A weight ratio of carbohydrate used to nitrogen fixed of four to one corresponds to an energy ratio of about six biomass energy units used as carbohydrate for each biomass energy unit fixed as nitrogen.
30. A. L. Hammond, *Science* **195**, 564 (1977).
31. R. Dubos observes that there is a tendency for developing countries to emulate U.S. high technology regardless of its suitability to their circumstances, and that, if we want to help and encourage these developing countries to adopt and implement other energy policies that are more suitable to their particular economy, we should show them, by example, that these opportunities and related technologies are respectable candidates in the United States as well (René Dubos Environmental Forum, Seven Springs Center, Mt. Kisko, N.Y., 9 to 11 September 1977).
32. L. H. Hill and S. Erickson, in *Energy, Agriculture and Waste Management*, W. J. Jewell, Ed. (Ann Arbor Science, Ann Arbor, Mich., 1975), p. 109. My interpretation of the graph in figure 2.
33. Another approach to U.S. policy governing surplus corn production would consider reduced use of commercial fertilizers in place of restrictions on harvested acreage in exchange for crop price supports.
34. A. L. Hammond and W. D. Metz, *Science* **197**, 241 (1977).
35. P. A. Gnadt and C. D. Murphy, *The Use of Wood as an Alternate Fuel for the ORNL Steam Plant*, unpublished paper, Oak Ridge National Laboratory Continuing Education Program, Course S-600, spring 1977, Oak Ridge, Tenn.
36. W. L. Roller, H. M. Keener, R. D. Kline, H. J. Mederski, R. B. Curry, *Grown Organic Matter as a Fuel Raw Material Resource* (National Aeronautics and Space Administration Contractor Report, NASA CR-2608, October 1975), p. 8; available from the National Technical Information Service, Springfield, Va.
37. C. F. Baes, Jr., H. E. Goeller, J. S. Olson, R. M. Rotty, *The Global Carbon Dioxide Problem* (Report ORNL-5194, Oak Ridge National Laboratory, Oak Ridge, Tenn., 1976).
38. B. Bolin, *Science* **196**, 613 (1977).
39. Discussion at the René Dubos Environmental Forum [see (*31*)] suggests that high sugarcane yields in Florida are due to a one-time opportunity to mine rich muck soils.
40. J. A. Bassham, in *Biological Solar Energy Conversion*, A. Mitsui, S. Miyachi, A. San Pietro, S. Tamura, Eds. (Academic Press, New York, 1977), p. 154. The term C_4 applies to certain plants such as corn and sugarcane that have a special added metabolic pathway. Compared to other plants, the energy efficiency of C_4 plants is higher because they avoid wasteful photorespiration under conditions of high light intensity.
41. Roller *et al.* (*36*, p. 52) conclude, "There are no plant species that will produce biomass more abundantly or more efficiently in the U.S. than our common agricultural crops now grown and/or our forest species."
42. J. H. Martin and W. H. Leonard, *Principles of Field Crop Production* (Macmillan, London, 1967), pp. 1111–1119.
43. O. D. Lorenzi, Ed., *Combustion Engineering* (Riverside Press, Cambridge, Mass., ed. 1, 1952), pp. 25–27; C. D. Hodgman, Ed., *Handbook of Physics and Chemistry* (Chemical Rubber Company, Cleveland, Ohio, ed. 27, 1943), pp. 1238–1240.
44. N. A. Lange, Ed., *Handbook of Chemistry* (Handbook Publishers, Sandusky, Ohio, ed. 8, 1952), pp. 758–768.

Fuels from Biomass: Integration with Food and Materials Systems

Adaptive biomass systems make multiple outputs available in response to demand changes.

E. S. Lipinsky

Biomass is a renewable resource capable of providing nonpolluting, safe fuels. Usually, biomass could alternatively be the source of food, materials, or chemicals. Even when the biomass is useful only for fuel, the same factors of production (such as land and water) could be used to produce food, animal feed, textiles, paper, construction products, or chemicals. The question is whether there are enough of the factors of production to go around.

The purpose of this article is to present some means of integrating the production of fuels from biomass with the production of food and materials, to form adaptive systems. Adaptive systems modify themselves to meet evolving needs and constraints (*1*). These systems contrast with two alternatives: the energy farm and the use of agricultural residues. Either the energy farm concept or the residue approach can evolve into an adaptive system.

The energy farm concept has been analyzed, especially for short rotation forestry (*2*). The output of the energy farm can be used directly as a fuel or indirectly in the production of energy-intensive chemicals. It is processed at a central conversion facility. The farm depends on high yield, frequently obtained by coppicing (regrowing plants from stumps), fertilizing, and irrigating.

At 25 tons per hectare (dry weight basis), almost 3 million hectares are required for a woody biomass crop to produce 1 quad of energy, when the moisture content of the fuel is taken into account (*3*). The Department of Energy's biomass cost target is \$0.95 to \$1.40 per gigajoule (GJ) (*4*). At this cost, the revenue of the energy farm would have to be only \$400 to \$560 per hectare, which is little for intensive silviculture. There are good prospects for attaining the high yields needed when cost is not a factor (*5*). The cost target of \$400 per hectare probably can be attained, but only at the sacrifice of high yield. Therefore, at least part of the energy farm output must be sold at prices higher than the \$0.95 to \$1.40 per GJ. The ability to meet the cost targets will remain in doubt for almost a decade, because many years are required to determine the economics of a new forestry system. This is especially true because low cost is a function of the successful coppicing (*6*) of multiple crops, and selection of appropriate tree species.

The agricultural residue approach avoids the high capital and operating costs (obstacles for the energy farm) on-

Mr. Lipinsky is Research Leader at the Battelle, Columbus Laboratories, 505 King Avenue, Columbus, Ohio 43201.

Summary. The development of fuels from biomass can lead naturally to dispersed facilities that incorporate food or materials production (or both) with fuel production, forming adaptive systems that can be modified to meet evolving needs and constraints. The technology that is appropriate to each system needs to be worked out, taking into account associated food and materials opportunities in order to decrease the ultimate cost of energy delivered to the consumer. I analyze possible systems based on sugarcane, corn, and guayule.

ly at the sacrifice of quantity and complete dedication of the biomass to fuel production. It is rare that residue concentration exceeds 7 dry metric tons per hectare (*7*). The corn stover residues from approximately half of the present crop (or about 13 million hectares) would be required for one quad of energy content. It would be deleterious to the soil to remove more than about half of the residues. Furthermore, once 50 million or more tons of corn stover are mobilized by the adoption of new harvesting, collection, and transportation systems, active competitors for purchase of the residue for food, feed, or materials use may appear.

The concepts of the energy farm, the integrated system, and the residue system blend into each other. An energy farm manager may find that pulpwood prices are high enough to make it profitable to divert part of his chips to the pulp and paper industry. By doing this, he could reduce the unit cost of the remaining biomass. A biomass conversion facility that manufactures steam and electric power may improve its profitability position by establishing facilities that use the steam that otherwise would be suboptimally used or wasted. These facilities could be for the manufacture of materials, such as particleboard or furfural. Thus, normal cost and profit pressures would tend to cause energy farms to become parts of integrated systems.

Agricultural residue systems might evolve into integrated systems for similar cost and profit reasons. Residue biomass conversion facilities would have to offer sufficient payment to farmers when adverse weather or market conditions might reduce residue shipments. The mobilization of residues probably would lead to active competition with the biomass-to-fuels conversion facility.

Systems Flow

Conventional agriculture and silviculture have placed biomass systems in constraints that must be removed if significant fuel production is to result. Before discussing specific opportunities to integrate fuel production with food and materials systems, it is worthwhile to consider the flow of materials in a hypothetical biomass system (Fig. 1) in which steps are taken to integrate fuel production into food or materials systems. The "farm" may be either terrestrial or aquatic. Agronomic practices, such as close spacing (*8*), coppicing, ratooning, and multiple cropping are used to increase yields and thus reduce unit biomass costs. Periodically, separation processes take place on the farm. These include harvesting, the return of selected plant parts to the soil, and the preparation of other plant parts for temporary storage. The separation processes yield primary farm outputs that are transported to central processing facilities. Other, incidental outputs include roots that are left in the ground, leaves and "trash" that are left as field residues, and other residues that remain on the farm. These incidental outputs may be collected and used for livestock feeding, soil improvement, or energy production.

At the central processing facility, the primary factory inputs are separated from those that arrive only incidentally. For example, logs or woodchips are debarked prior to the manufacture of lumber or paper pulp. Bark becomes an incidental factory output. The primary factory input (logs or wood chips) then is subjected to conversion processes to manufacture salable products, which are stored and transported to appropriate markets.

The incidental factory outputs, resulting from the separation processes, and incidental conversion outputs may be of such little value that they are subjected to disposal. Additional conversion processes may upgrade these outputs either to salable products or to products suitable for use at the factory. For example, lignocellulosic residues might be used at a factory to generate process steam and electricity.

The most obvious strategy for seeking fuels-from-biomass opportunities is to examine each incidental output to determine its suitability for manufacture of fuels. Significant quantities of inexpensive fuels can be produced by an astute selection of biomass crop, conversion processes, and end-use markets. The food or materials producer might benefit from reduced disposal costs, increased reliability in energy supply, and possible revenue from outside sales. Another, more powerful, strategy is a flexible coproduct manufacturing system in which key biomass intermediates are processed either into fuels or materials, depending on relative price levels. Application of this strategy to sugarcane is illustrated in Fig. 2. When the price of crystalline sugar is satisfactory, only the molasses by-product of sugar manufacture would be converted to ethanol. When sugar prices are low substantial quantities of sugarcane juice could be diverted to the production of ethanol (*9*). The reduced sugar inventories could quickly lead to sugar price recovery. The feasibility and attractiveness of this strategy depends on having flexible fermentation facilities, and a willingness to suffer the economic consequences of leaving the fermentation facilities idle when sugar prices are high. For sugar crops, molasses is a stable, concentrated, storable material. Therefore, during bad times, sugarcane juice could be fermented during the sugarcane harvesting season and stored molasses could be fermented during the off-season. Thus, the facility would operate for a variable number of days per year, depending on market conditions.

Objectives

Biomass-to-energy systems are of three major types (Fig. 3): (i) direct combustion of biomass materials, generally to manufacture process heat, steam, or electricity; (ii) the manufacture of molecules that are rich in carbon or hydrogen (or both) and poor in those elements (such as oxygen and nitrogen) that do not contribute to high energy content; and (iii) feeding of biomass to work animals. The first two approaches are significant. The first approach requires engineering ingenuity in order to obtain low moisture biomass boiler feed and to conduct direct combustion efficiently. The second approach is of greatest scientific interest. The low efficiency and adverse environmental impact of animals as a source of power removes the third alternative from serious consideration at this time in the United States.

The widespread use of petroleum, natural gas, and coal as sources of fuel arises from their high energy content. The high energy content in turn arises from their chemical composition which generally is low in oxygen, nitrogen, and inorganic matter. Their major chemical

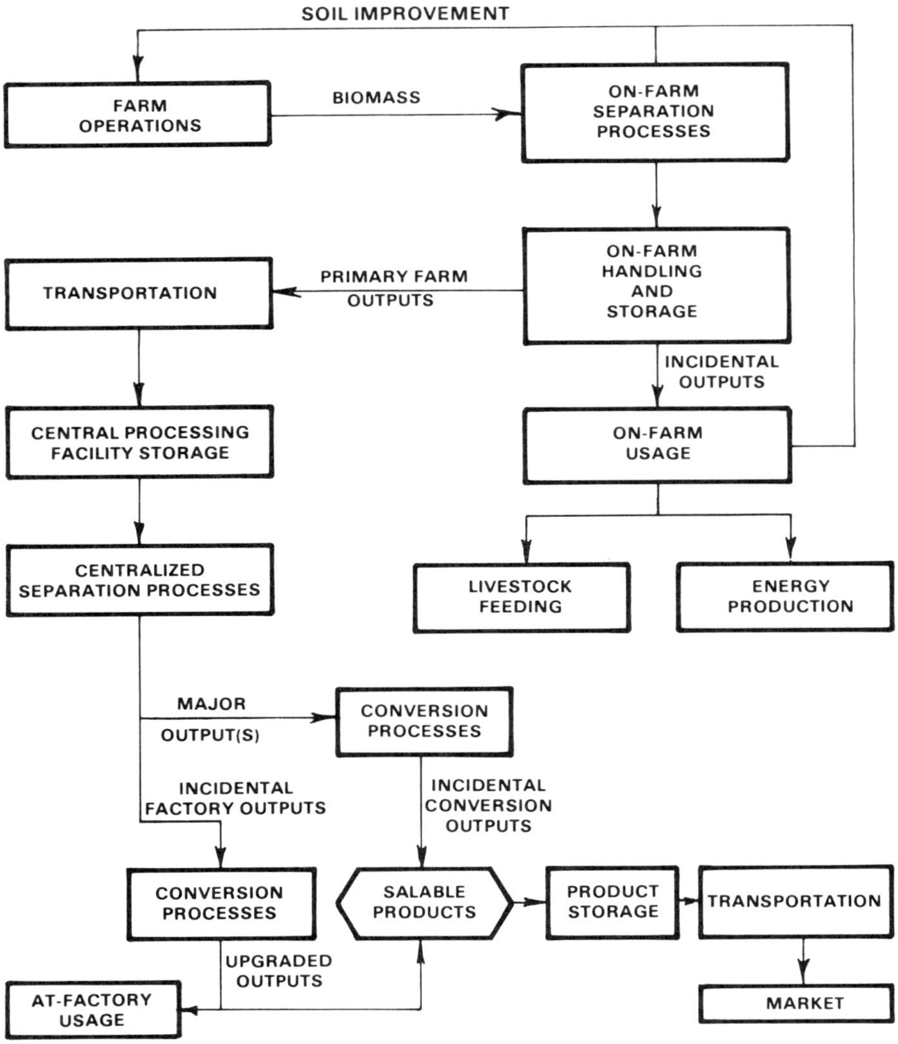

Fig. 1. Systems flow in biomass production and conversion.

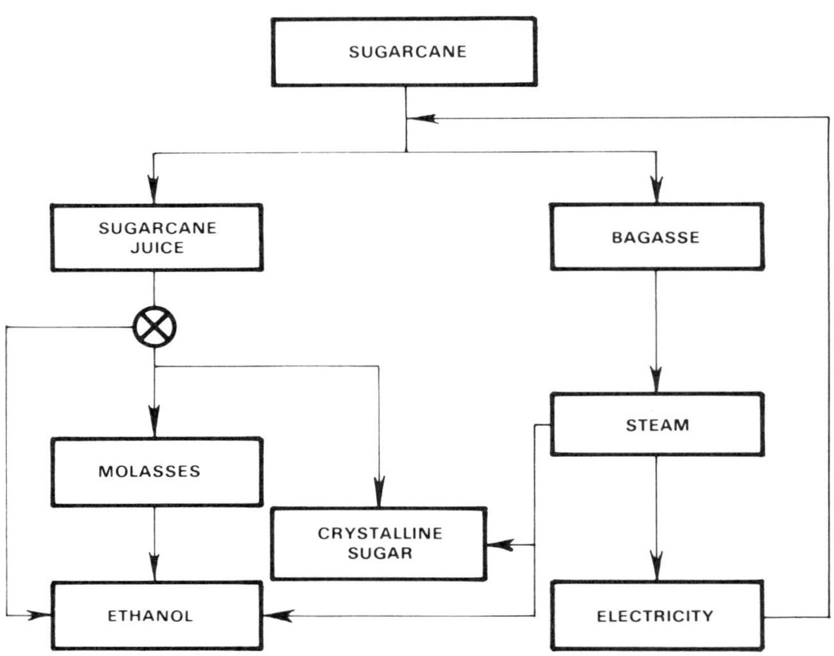

Fig. 2. Use of ethanol production to control crystalline sugar supply.

constituents are carbon and hydrogen, which burn readily, liberating heat that can be used to perform work or to adjust the temperature in an environment.

Some species of plants manufacture energy-rich chemicals by various biosynthetic pathways. This biosynthesis frequently goes through intermediates such as glucose that have as many oxygen atoms as carbon atoms (Table 1). Carbohydrates are inherently low in energy (14 GJ per ton). Storage carbohydrates such as starch have the formula $[C_6H_{10}O_5]_n$ which has a slightly higher energy value (15.4 GJ per ton) due to the loss of a molecule of water from each glucose unit. Cellulose is energetically about the same as starch for all practical purposes. Lignin has a considerably higher energy content than the carbohydrates because (i) many oxygen atoms have been lost from the glucose molecules from which lignin is formed and (ii) energy-rich aromatic structures have been formed (*10*). Lipids, terpenoids, and steroids are manufactured biosynthetically from acetate by means of various secondary metabolic processes (*10*). These products are used primarily for food and pharmaceuticals now, but they are possible fuel sources.

When biomass is subjected to microbiological or thermal conversion, end products can be formed that have more energy content than does the biomass starting material. The additional energy per unit weight arises from loss of oxygen, carbon dioxide, formation of energy-rich carbon-carbon double bonds, or formation of energy-rich ring structures. The energy content of the major candidate fuel products derived from biomass have higher energy contents than do sugars, starch, and cellulose but are no higher and in many instances are considerably lower in energy content than are lipids and terpenoids. In some instances, the advantages in energy content may be small but the liquid or gaseous fuel may be more convenient to transport and store than the biomass starting material would be.

Corn-Based Fuels

Whenever corn or other grains are suggested as sources of fuels, the argument is advanced that conversion of corn grain into fuel is undesirable because it could cause a food shortage (*11*). However, it is at least theoretically possible to reorganize the present U.S. corn biomass system to permit production of 10 billion to 18 billion liters of ethanol (or its equivalent in other fermentation products) while obtaining the same quantity of end-use food products, in the form of beef, poultry, and pork, of a quality equal or superior to present standards. Ten billion liters of ethanol could be used as an antiknock fuel ingredient for approximately 20 to 25 percent of U.S. gasoline, assuming a 10 percent blend with gasoline. This increase in outputs could be achieved without increasing the land area devoted to corn or increasing the yield of corn above that obtained now. Such a change might appear miraculous but is actually based on three simple facts: (i) cattle and other ruminant animals are capable of digesting polysaccharides that are indigestible to nonruminant animals; (ii) corn stover (stalks, cobs, husks, and leaves of the corn plant), which is rich in these polysaccharides that are digestible by cattle, is underutilized now; and (iii) U.S. production of corn stover could provide ample replacement for the corn grain now fed to cattle. Briefly, the concept is to manufacture ethanol and a protein-rich stillage from some of the grain now used to feed ruminant animals in the United States. Figure 4 shows both the present and suggested corn biomass systems along with the valves to adjust product flows. An amount of corn stover equivalent to the grain used for ethanol would be harvested from the cornfields at a rate that does not endanger the soil. The quantity that could be diverted depends on the latitude, terrain, and soil type of each specific cornfield. The stover and the stillage would be fed to cattle so that the same quantity of beef would be obtained. The digestibility of corn stover presently is less than grain, hence the equivalence may be 2 tons of stover per ton of grain diverted (*12*). The U.S. production of stover is ample to supply the needed tonnage. The use of corn grain by nonruminant animals (poultry and hogs) would continue at the present rate, as would industrial conversion of corn into starch and other products.

Approximately 1.7 billion bushels of corn grain are dedicated to domestic cattle feed (*13*). Corn grain is used now because it is highly digestible, relatively cheap, and yields a marbelized effect needed to attain the high meat inspection grades (*14*). Also, corn grain supplies some of the nitrogen compounds from which the ruminant animal manufactures its protein.

The density, size, shape, and waxy outer layer of corn grain make it relatively simple and cheap to transport and to store. When dried properly and stored in a suitable environment, corn grain is stable for years. In contrast, corn stover is low in density, is obtained at a much higher moisture level than is corn grain, and is relatively unstable. Transporting corn stover over long distances to central locations where large-scale conversion could take place over extended processing seasons appears tenuous.

If cattle were fed corn stover locally during the seasons of the year in which it is abundant, some major objections to the use of corn stover could be overcome. The protein not obtained from corn stover would be supplied indirectly from grain or from the stillage.

Researchers at Purdue University have announced a major breakthrough in the treatment of corn stover to obtain simple sugars for fermentation to ethanol

Table 1. Comparison of heats of combustion of biomass constituents, fuels from biomass, and fossil fuels. Values for heat of combustion are approximate.

Heat of combustion (gigajoules/ metric ton)	Biomass constituents	Fuels from biomass	Fossil fuels
50		Methane	Natural gas
45			
40	Turpentine	Butanol	Gasoline
35	Triglyceride oils ~ steroids ~ lignin		Fuel oil > bituminous coals
30		Acetone > ethanol	
25			
20		Furfural > methanol	
15	Wood > starch ~ cellulose > sucrose > glucose		
10		Acetic acid	

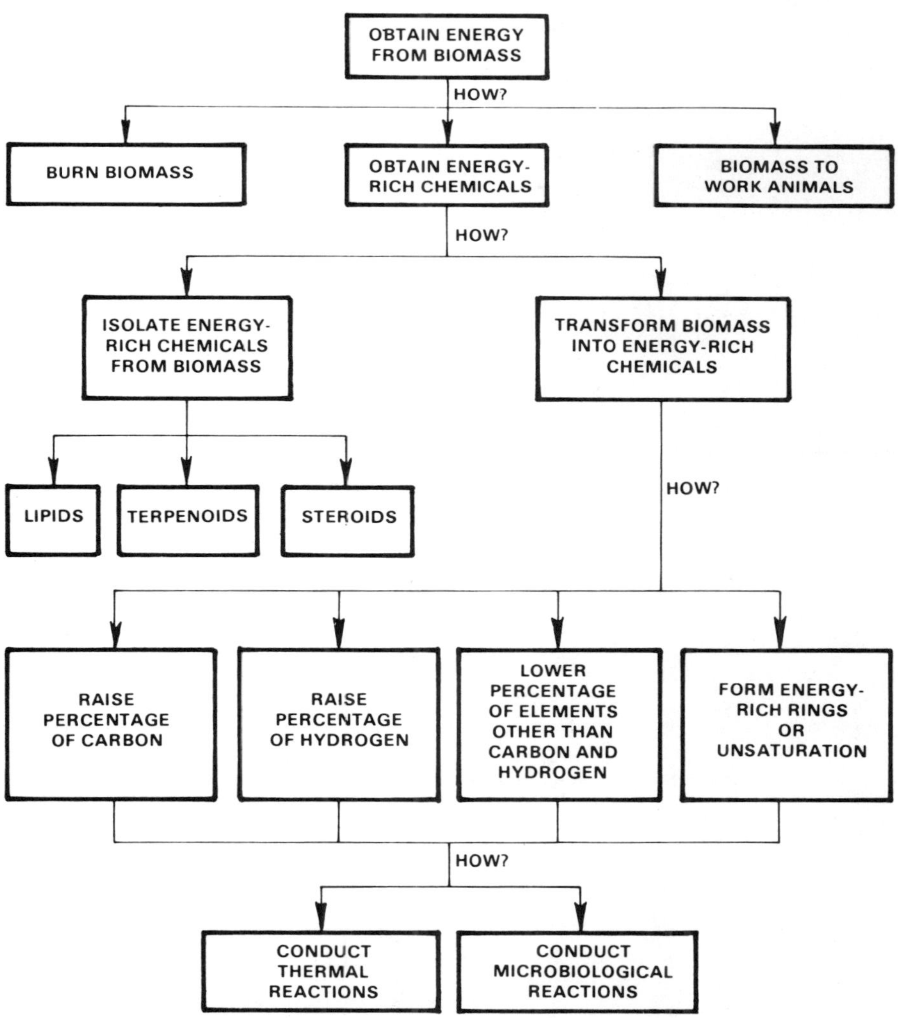

Fig. 3. Alternatives to obtain energy from biomass.

(*15*). An organic acid treatment facilitates the attack of fungal cellulases on lignocellulose. The probability that this process could increase the digestibility of corn stover by cattle is high because the low digestibility arises from crystalline cellulose or cellulose protected by lignin (*16*). Although details of the Purdue process are lacking at this time, it is likely that the same treatment that renders cellulose susceptible to cellulases derived from the fungus *Trichoderma viride* could make the cellulose susceptible to the enzymes present in the cattle rumen. If so, the digestibility of corn stover could rise from about 45 to 70 percent or more.

The stover, with or without pretreatment, would not replace all of the corn grain. Time is money in a modern feedlot and cattle reach market weight faster if they are finished on corn grain.

This suggestion illustrates the interaction of the food and fuels systems to the benefit of all concerned. The corn farmer who previously sold perhaps 250 bushels of corn per hectare at $2 per bushel ($80 per metric ton) and who received no cash for his stover now could expect perhaps $30 to $40 per ton ($80 to $120 per hectare) for his stover, in addition to his corn grain revenue. The cattle producer could hope to use less than 2 tons of stover to replace 1 ton of grain, if the pretreatment were successful. Thus, both the corn farmer and the cattle feeder improve their positions. The ethanol producer would have access to grain that is easy to convert to ethanol or other fermentation-derived fuel. He would not have to deal with the technical uncertainties of the cellulose-to-glucose route. There is a ready market in cattle production for the stillage by-product of the fermentation. Fuel producers could purchase ethanol for antiknock and fuel use, thus reducing petroleum imports.

The skeptic could claim that in times of dire need for food and feed grains the price of corn grain would rise above what the ethanol producer could afford to pay, while the corn is exported in huge quantities. But the suggested system increases the overall biomass resources of the United States. In this situation, the balance of payments and farm income could be raised. The stover cannot be shipped overseas and it would become the U.S. beef consumers' guarantee that beef supplies could remain relatively plentiful.

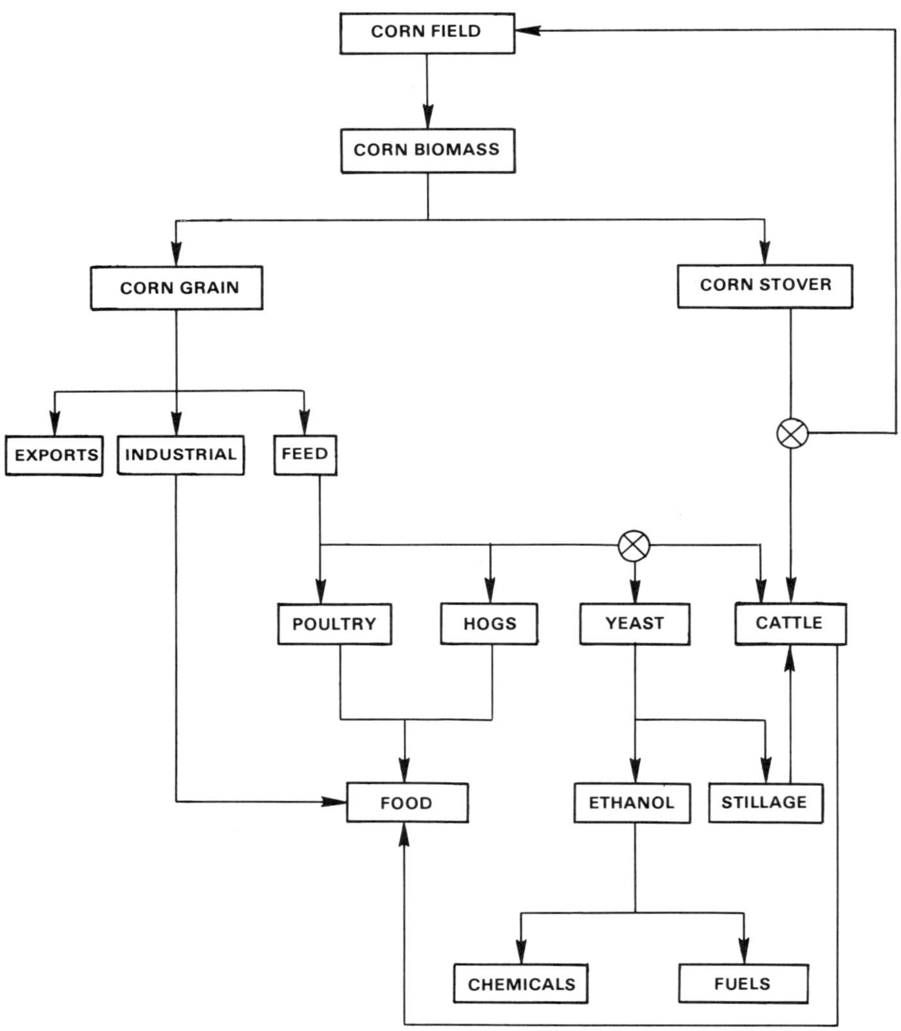

Fig. 4. Hypothetical corn-to-ethanol system, integrated with conventional utilization of corn.

Impact of a New Process

The traditional means of processing conventional crops frequently inhibits production of fuels and chemicals from biomass at attractively low costs. For example, the present milling system for sugarcane results in fermentable sugars (sucrose, fructose, and glucose) at a cost of about $0.132 per kilogram on a dry weight basis (*17*). At this cost level for sugarcane-derived fermentable sugars, ethanol is likely to cost approximately $0.30 per liter (*18*). At $0.30 per liter, industrial applications for ethanol can be tapped, but for substantial penetration of motor fuel markets a lower ethanol selling price is required.

Reduction in the cost of ethanol derived from sugarcane can arise by four approaches. Only one of these four methods is within the major scope of this article but the others are mentioned for the sake of completeness. The cost of the sugarcane delivered to the processing facility can be reduced by increasing the yield of sugarcane without increasing the cost proportionately. The most active and promising yield-increasing research now in progress is the use of narrow row spacings (approximately 0.6 meter instead of 1.83 meters) to increase substantially the plant population and to achieve canopy closure at an early date (*19*). A second method of reducing ethanol costs is to reduce the cost of fermentation process. However, fermentation is a relatively minor factor in the cost of ethanol. Raw materials account for approximately $0.20 of the $0.30 per liter and much of the remainder is provision for profits, taxes, and other capital charges. The third method to reduce ethanol cost is to reduce the processing cost in making sugarcane juice from sugarcane. These costs have been shown to be virtually negligible, compared with the agricultural costs of sugarcane delivered to the mill (*20*).

The fourth approach is modification of the sugarcane process so that coproduct credits reduce the net cost of fermentable sugars. The cost of sugarcane delivered to a processing facility is approximately $66 per metric ton on a dry weight basis. Roughly half of the sugarcane is fermentable sugars; the rest is a lignocellulosic material that presently has a fuel value of $26 per ton or less (*20*). The present sugarcane milling procedure leads to a relatively dilute solution of sugar that requires evaporation and crystallization to obtain a salable product. The sugarcane stalk consists of a soft central core or pith which has over 90 percent of the fermentable sugars, a tough rind, and a waxy epidermis. In the process of removing the sugars from the millable stalk, the pith and rind (which are largely separate in the intact sugarcane stalk) become crushed and ground together, along with the waxy epidermis and adhering dirt. This "bagasse" is of low quality and has only the fuel value that can be assigned to a product with at least 50 percent moisture.

A new sugarcane process is under development in Canada (*21*) that promises to provide coproducts with higher values (Fig. 5). Briefly, sugarcane is cut into billets 1 foot long, which are cut longitudinally by a knife into two halves. A rotating coring device removes the pith from each half without disturbing the rind fibers. The waxy epidermis and any adhering dirt are scraped off the outside of the cane so that both the rind fibers and the pith are clean. The pith contains approximately 92 percent of the sugars in sugarcane. The sugars can be obtained by alternate squeezing and imbibition with water. When the objective is to make both table sugar and fermentation products at the same location, it may be possible to remove a relatively concentrated sugar solution for table sugar production from the pith with the first squeezing and then use the pith with its adhering sugars and molasses as the fermentable sugars in the ethanol plant.

Although this new process probably reduces the processing cost of fermentable sugars, the major advantage will be in the increased value of the rind fiber. This uninjured fiber is initially in the form of sugarcane segments which can be separated into foot-long strands that can be combined with thermosetting polymers to make simulated lumber or plywood products. The development of these products is at too early a stage at this time to assign a value but it surely will be well above that of fuel.

Conventional sugarcane bagasse is a source of fiber for papermaking, especially in countries that have considerable sugarcane and few pulpwood-type trees. To make paper of relatively high quality, it is necessary to depith the bagasse. Depithing is a relatively expensive process that has inevitable fiber losses associated with it. The new Canadian process generates rind fiber that has already been depithed. Furthermore, the rind fiber has not been injured by the crushing and milling operations. Dr. Joseph Atchison has estimated that the rind fiber may attain a value about double that of conventional bagasse for papermaking (*22*). Therefore, the Canadian process appears to meet one of the major fuels-from-biomass goals, namely the goal of a nonfuel coproduct to generate significant revenues that permit assignment of a lower cost to (in this case) fermentable sugars.

Rind fibers also appear promising as a storable fuel source, which would extend the processing season during which ethanol can be distilled with fuel generated at the site. After removal of sugars, the lignocellulosic pith remains. If ethanol and stillage are manufactured at the processing facility, the leftover pith can be used to absorb the concentrated stillage to make cattle feed. If the rind fiber is selling well for high-value uses and fossil fuel is scarce, it may be necessary to use the pith as a source of steam and electricity by direct combustion. Ultimately, this finely divided lignocellulosic product would be an excellent substrate for production of simple sugars by enzyme or acid hydrolysis (or both). The achievement of the high surface area that pith has naturally is a costly process for many other substrates. Enzyme hydrolysis at the present state of the art yields relatively dilute solutions, and concentrating these solutions is an important cost factor. In an integrated facility, the enzyme hydrolyzate can be blended with molasses to achieve a satisfactory concentration of fermentable sugars.

The introduction of the Canadian Separator Equipment Process could lead to greatly reduced net fermentable sugars cost. The fact that the millable sugarcane stalk might yield so much salable coproduct that there would be no fiber left to make steam and electricity to run the process is not a serious disadvantage. In the first place, very little power is required to operate this process compared with one that grinds and mills. Second, the harvesting of the entire aerial portion of the sugarcane plant, with return to the soil only of the amount that is needed to ensure soil conservation, probably would lead to the delivery of an addition-

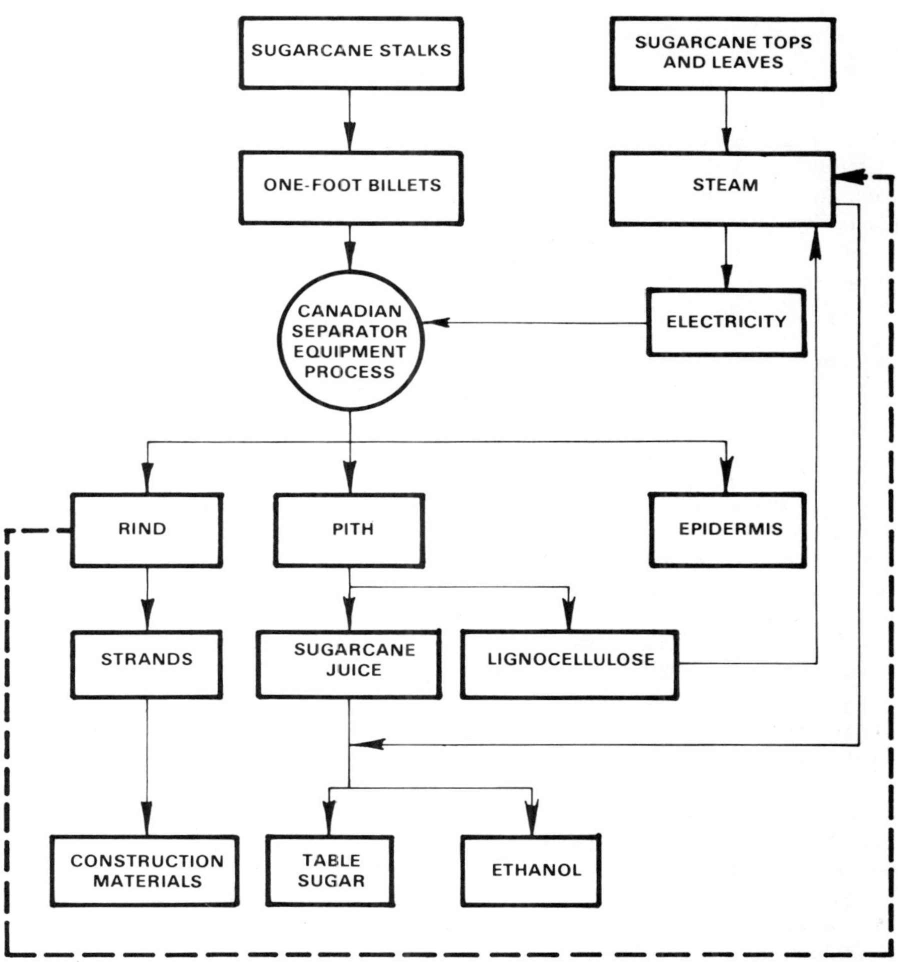

Fig. 5. The Canadian separator equipment process.

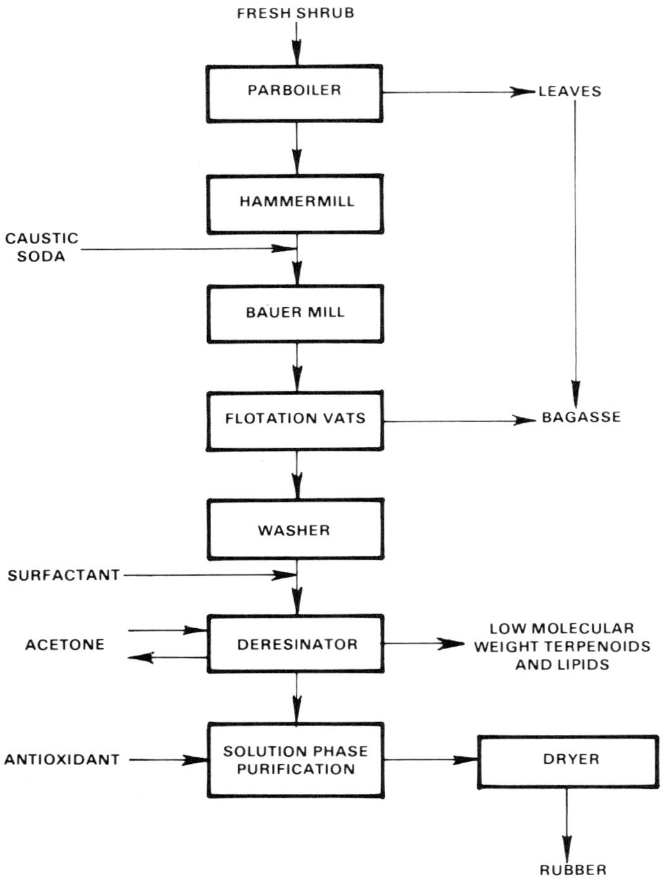

Fig. 6. Mexican process for extraction of rubber from guayule shrubs. Adapted from National Academy of Sciences study.

al 20 percent lignocellulosic material to the central processing facility (*20*). This material (tops and leaves) could be dedicated to the generation of steam and electricity for operation of the sugarcane separation and for ethanol distillation processes.

Guayule

The United States is vulnerable to a natural rubber embargo because production is concentrated in Asia, and the United States produces no natural rubber. Despite the development of a 2-million ton synthetic rubber market, natural rubber is still needed for radial tires for passenger automobiles, for aircraft tires, and for truck tires. Furthermore, synthetic rubber uses petroleum feedstocks that are needed for gasoline production. The development of guayule (*23*) as a domestic source of natural rubber could provide opportunities to integrate critical materials production with fuels production (Fig. 6). The natural rubber would be the "bread and butter" product that would be sold to make the agricultural and conversion ventures profitable. The coproduct mixture of terpenoids and lipids is rich in energy and could be processed into liquid fuels. The bagasse could be used to produce steam and electricity to operate the facility and possibly for outside sale. A process under development in Mexico has been described (*23*) but it may not maximize recovery of lipids and terpenoids because emphasis is on the removal of these contaminants from the natural rubber.

Natural rubber sells for approximately $0.40 per pound. Our consumption is approximately 700,000 tons per year and is expected to grow with the increasing demand for radial tires and as it displaces the increasingly expensive synthetic rubber manufactured from petrochemical sources. In assessing the possible impact of a guayule development on fuels-from-biomass, it is desirable to consider, from this source, a domestic natural rubber production of 1 million tons per year.

Published information regarding the detailed composition and yield of resinous coproducts is sketchy, but they are a complex mixture of lipids with mono-, sesqui-, and diterpenes (*24*). The lipids are triglycerides with both saturated and unsaturated fatty acids. To obtain a high quality natural rubber, it is necessary to deresinate the crude product. The resin coproduct content might equal the natural rubber content of guayule under favorable circumstances. Therefore, the expected direct liquid fuel output from guayule would be moderate. The terpenoids and lipids meet two requirements for fuels from biomass feasibility: (i) they are high in energy, having a carbon-to-oxygen ratio of at least 10 to 1; therefore they equal or surpass gasoline in energy content; and (ii) the mixture of chemicals is so complex that a simple fuel-making process is likely to win out over an attempted separation of valuable chemicals.

To make motor fuel from this crude mixture would require conversion of olefinic unsaturation into aromatic-ring compounds or hydrogenated compounds. The lipids would have too high a boiling point for gasoline but might be converted to diesel fuel. Various approaches to the conversion of such mixtures need to be investigated, not only for guayule but also for the *Euphorbia* species now under study by Calvin *et al.* (*25*).

Policy Implications

The popular press has stressed ethanol as the principal liquid fuel obtainable from biomass but scientists, technologists, and policy-makers need to remain alert to opportunities afforded by other chemicals with higher energy content and potentially lower costs. Terpenes and lipids are two classes of natural products that merit greater attention.

Biomass systems from which variable product mixes of food, materials, and energy can be derived, within changing environmental constraints, are adaptive. The theory of adaptive systems is developing rapidly (*26*). Biomass can and should be used to adapt our resources to concurrent crises in energy, food, materials, and the environment (*27*). The hierarchy of applications for a product or a resource is arranged in an order of displacement based on opportunity cost (*28*). Thus for ethanol, beverage > industrial solvent > chemical intermediate > fuel. These hierarchies act as constraints in the adaptive systems.

A growing forest, the wood from which can become lumber, plywood, pulpwood, or fuel, gives producers and consumers the flexibility needed to survive the multiple future crises that we face (*29*). When the trees are planted, we will not know which uses the forest resource will be put to when it is harvested. To plant trees that are useful only as a fuel represents a suboptimization unless yields far surpass those from multipurpose trees. A restriction of annual crop output to boost prices received by farmers is an archaic notion: grain not needed overseas due to good harvests abroad can be converted to chemicals and fuels.

If or when fuel prices double due to the depletion of fossil fuel supplies, fuels from biomass may possibly supply 10 percent or more of a curtailed U.S. energy consumption. However, in the next 10 to 20 years, the achievement of competitive costs for fuels from biomass would depend primarily on the integration of fuel production with food and materials production. Those engaged in research and development, agriculture, business, and government are challenged to develop not only the know-how but also the know-what, know-where, and know-when, and know-why (*26*). Knowing when and where to switch from emphasis on food and materials to emphasis on fuels is just as important as knowing how to produce the fuels.

References and Notes

1. E. Laszlo, *Introduction to Systems Philosphy* (Harper, New York, 1972), chap. 4.
2. R. E. Inman, *Silviculture Biomass Farms* (contract EX-76-C-01-2081, report to the Energy Research and Development Administration, Washington, D.C., 1977), vol. 1–6; available from the National Technical Information Service, Springfield, Va.; M. D. Fraser, J. F. Henry, C. W. Vail, in *Proceeding of the Institute for Gas Technology Conference on Clean Fuels From Biomass, Sewage, Urban Refuse, and Agricultural Wastes*, W. W. Waterman, Ed. (Institute for Gas Technology, Chicago, 1976).
3. The United States consumes approximately 75 quads (75×10^{15} Btu) per year. One quad is approximately 1.05×10^{9} GJ.
4. R. F. Ward, *Proceedings of Fuels from Biomass Conference at the University of Illinois*, J. T. Pfeffer, Ed. (Univ. of Illinois Press, Champaign-Urbana, 1977).
5. C. L. Brown, *J. For.* **74**, 7 (1976); A. R. Ek and D. Dawson, *Can. J. For. Res.* **6**, 132 (1976).
6. Coppicing is regrowth from old stumps or suckers. Ratooning is regrowth from bud and root points at the base of the stool.
7. J. A. Alich and F. A. Schooley, *Crop Forestry and Manure Residue Inventory—Continental United States* [prepared by Stanford Research Institute under contract E (04-3)-45 for the U.S. ERDA, Washington, D.C., 1976], vol. 1–8; available from the National Technical Information Service, Springfield, Va.
8. R. J. Matherne and J. E. Irvine, "The influence of row spacing on sugarcane stalk population and yield," in *Proceedings of the Seventh Annual Joint Meeting of American Society of Sugarcane Technologists*, in press.
9. E. S. Lipinsky, "Sugarcane vs. corn vs. petroleum sources of ethanol for motor fuels and chemicals," *Proceedings of the Seventh Annual Joint Meeting of American Society of Sugarcane Technologists*, in press.
10. A. L. Lehninger, *Biochemistry* (Worth, New York, ed. 2, 1975).
11. For example, *Feedstuffs* **49**, 10 (1977).
12. H. D. Ritchie, "Beef Cow Management—Crop Residues for Beef Cattle" Michigan State University Fact Sheet 1301 (Michigan State Univ., Lansing, 1977); *Beef* **13** (No. 7), A-18–A-19 (March 1977).
13. U.S. Department of Agriculture, *Feed Situation* (Government Printing Office, Washington, D.C., November 1977), p. 30.
14. G. M. Ward, P. L. Knox, B. W. Hobson, *Science* **198** 265 (1977).
15. *Wall Street Journal* Midwest Edition, 27 October 1977, p. 40.
16. G. J. Hajny and E. T. Reese, "Cellulases and their applications," *Adv. Chem. Ser.* **95**, 245 (1969).
17. E. S. Lipinsky, T. A. McClure, R. A. Nathan, T. L. Anderson, W. J. Sheppard, W. T. Lawhon, "Systems Study of Fuels from Sugarcane, Sweet Sorghum, and Sugar Beets" (final report, prepared by Battelle-Columbus Division under contract W-7405-ENG-92 for the Energy Research and Development Administration, Washington, D.C., 1976), vol. 1; available from the National Technical Information Service, Springfield, Va.
18. E. S. Lipinsky, in *Proceedings of the Twelfth Intersociety Energy Conversion Engineering Conference*, (American Nuclear Society, La-Grange Park, Ill., 1977), pp. 94–100.
19. ———, S. Kresovich, T. A. McClure, E. W. Helper, W. T. Lawhon, "Fuels from Sugar Crops" [second quarterly report, prepared in cooperation with Louisiana State University, Texas A & M University, University of Florida, F. C. Schaffer and Associates, and Department of Agriculture for the Department of Energy (1977); available from the National Technical Information Service, Springfield, Va.].
20. E. S. Lipinsky, R. A. Nathan, W. J. Sheppard, T. A. McClure, W. T. Lawhon, J. L. Otis, "Systems Study of Fuels from Sugarcane, Sweet Sorghum, and Sugar Beets" (final report, prepared by Battelle-Columbus Division under contract W-7405-ENG-92 for the Energy Research and Development Administration, Washington, D.C., 1977), vol. 1; available from the National Technical Information Service, Springfield, Va.
21. F. Rajic and J. Atchison, personal communication.
22. J. E. Atchison, *Preliminary Evaluation of New Cane Separation Process Relative to its Contribution Toward Increased Fuel and Energy Production from Cane Crops* (report to the Department of Energy), in press.
23. *Guayule: An Alternative Source of Natural Rubber* (National Academy of Sciences, Washington, D.C., 1977).
24. A. J. Haagen-Smit and R. Siu, *J. Am. Chem. Soc.* **66**, 2068 (1944).
25. M. Calvin, *Int. J. Energy Res.* **1**, 299 (1977); P. E. Nielson, H. Nishimura, J. W. Otvos, M. Calvin, *Science* **198**, 942 (1977).
26. E. Jantsch, *Design for Evolution* (Braziller, New York, 1975).
27. R. Dubos, *Science* **193**, 459 (1976).
28. W. J. Sheppard and E. S. Lipinsky, in *Sucrochemistry*, J. L. Hickson, Ed. (American Chemical Society Symposium Series **41**, Washington, D.C., 1977), pp. 336–350.
29. J. S. Bethel *et al.*, "Renewable Resources for Industrial Materials," report of the National Research Council (National Academy of Sciences, Washington, D.C., 1976).
30. Dedicated to Professor R. B. Woodward in celebration of his 60th birthday. I thank T. A. McClure and W. J. Sheppard for helpful discussions. Supported in part by Department of Energy contract W-7405-ENG-92.

Solar Energy and Electric Utilities: Should They Be Interfaced?

In many applications adapting solar energy to electric utilities limits its economic potential.

Joseph G. Asbury and Ronald O. Mueller

Solar energy technologies and their applications constitute a broad and diverse array of energy supply options. Solar systems are currently under development for applications in the generation of electric energy, production of industrial process heat, and heating and cooling of buildings. The identification of promising solar applications and the elimination of unpromising ones are necessary for the efficient development and use of the solar resource. In applications where solar systems are interfaced with conventional supply systems, it is important that the solar systems be evaluated in context with the cost and performance characteristics of the conventional systems that they are designed to supplement or replace.

A number of investigators have proposed solar energy systems that interface with conventional electric utility supply systems. Whether considering solar thermal conversion systems to produce electric utility power or solar systems for thermal applications in buildings, the systems' designers have generally concluded that solar energy can reduce electric utility fuel and capital requirements. In both utility power system and utility customer end-use applications, solar energy saves fuel directly by substituting for utility fossil and nuclear fuels. Utility capital and indirect fuel savings occur as a result of the ability of the storage system, always included in

Joseph G. Asbury is senior economist and Ronald O. Mueller is senior energy systems analyst with the Energy and Environmental Systems Division, Argonne National Laboratory, Argonne, Illinois 60439.

the solar system design, to displace the solar system's auxiliary energy requirements to off-peak hours. The effect is to reduce the utility's peak-period loads from those which would otherwise occur, thereby allowing the substitution of base-load plant and fuels for peak- and intermediate-load plant and fuels.

Recent investigations of certain solar heating and cooling systems have emphasized that solar insolation outage can be covered entirely with inexpensive off-peak electric power (*1*). All that is required is the incorporation of a thermal storage system with capacity adequate to meet the design-day building load. What these studies have failed to point out is that the inclusion of thermal storage effectively reduces the function of the solar collector component of the solar energy system to the displacement of off-peak electric energy. This fact greatly diminishes the economic benefits that can be attributed to solar energy systems.

It is the purpose of this article to re-evaluate the economics of solar energy systems that interface with conventional electric utility supply networks. The re-evaluation proceeds along two paths. First, adopting the implicit assumption of many solar system designers that there are abundant supplies of low-cost off-peak electricity, we undertake systems studies of several of the more important solar energy applications. Second, adopting a standard economic approach to the periodic load problem, we examine the general problem of interfacing solar energy and electric utility supply systems.

A central theme of the systems studies is that solar energy systems are most logically compared with the storage-augmented versions of the conventional systems that they are designed to supplement or replace. Adopting this approach, we estimate solar collector break-even costs for solar/electric-resistance and solar/heat pump systems for space heating and solar thermal conversion systems for electric power generation. For solar/electric-resistance heating, the upper bound on solar collector break-even costs is found to be approximately \$30 per square meter ($\approx$ \$3 per square foot) if auxiliary energy is from coal-fired utility generating plants and \$75 per square meter if auxiliary energy is from oil-fired utility plants. For all the solar/heat pump configurations examined, the collector break-even costs are found to be substantially lower than for the comparable solar/electric-resistance heating systems.

The general problem of interfacing solar energy and electric utilities is treated by analyzing the economics of solar collection in terms of two alternative scenarios. In the first scenario, off-peak electricity, even if priced at variable cost (utility fuel cost), remains available for the indefinite future. (This corresponds to the economists' firm-peak case.) In the second scenario, through the introduction of either simple off-peak price discounts or time-of-use pricing, the utility's load curve becomes flat (shifting-peak case). For the first scenario, it is shown that solar collection systems generally will be economical only if they can deliver solar energy at a cost lower than the variable (fuel) cost component of off-peak electricity. In regions of the country where the utility off-peak fuel is coal or nuclear fuel, this implies very low solar collector break-even costs. In regions of the country where the cost of off-peak electricity is higher than that of other auxiliary energy forms—the utility fuel is oil—the most economical and energy-conserving solar systems will not interface with the electric utility. For the second scenario, the economic benefits of solar collection in most applications are found to be approximately equal to the value of the displaced utility fuel, indicating break-even points for solar collectors that are roughly equal to those under the first scenario.

The general conclusion of this article is that solar energy systems and conventional electric utility systems represent a poor technological match. The basic problem is that both technologies are very capital intensive. The electric utili-

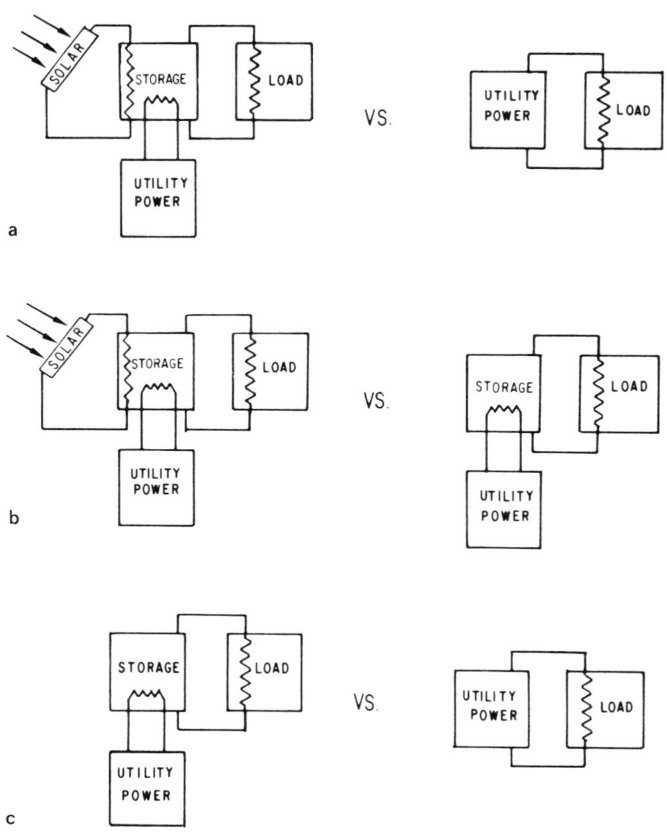

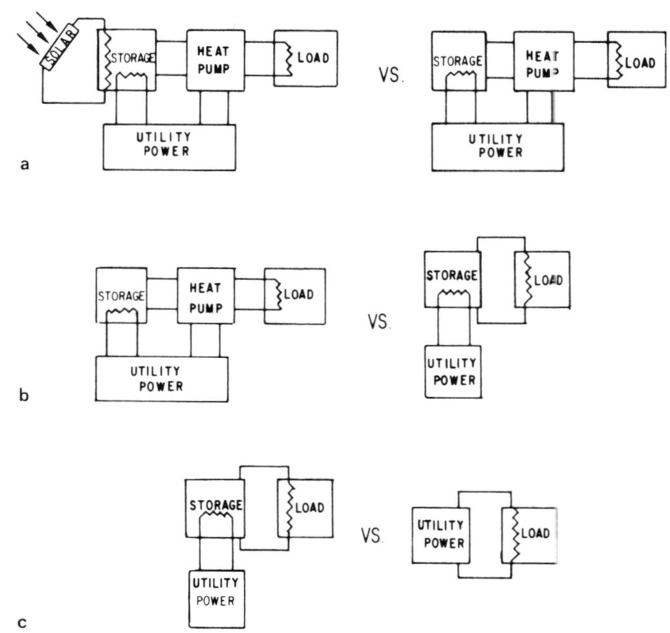

Fig. 1 (left). Comparisons among solar, storage, and conventional space heating systems. Fig. 2 (right). Comparisons among solar-assisted heat pump, storage, and conventional space heating systems.

ty, because of the high fixed costs of generation, transmission, and distribution capacity, represents a poor backup for solar energy systems. On the other hand, the solar collection system, because it represents pure, high-cost capital and because of the periodic nature of its output, should not be considered as a part-load source of auxiliary energy for the electric utility. Viewed in this context, the low break-even costs established for solar collector systems that interface with electric utilities are merely symptomatic of the problem of matching two technologies that in important respects may be incompatible.

Because of our extensive use in this article of the concept of solar collection break-even cost, it is important from the beginning to point out that (i) these costs are always calculated by directly comparing the capital and fuel requirements of conventional systems with those of solar systems providing services of comparable quality, and therefore (ii) break-even cost values, where presented here, are dependent on the costs assumed for the conventional technologies, but (iii) the qualitative conclusions reached are essentially independent of the exact break-even cost values. We do not explicitly examine passive solar building concepts that modify and reduce normal thermal load requirements, nor do we analyze intermittent solar concepts (for example, air-conditioning systems without adequate backup to cover solar outages) for which there are no conventional counterparts. As shown below, the upper-bound break-even costs of solar collection are effectively "pinned" to the value of off-peak utility fuel. By stating break-even costs in terms of current prices of off-peak utility fuel, we effectively neglect: any price-distorting effects of government development or subsidy programs, the many environmental spillover benefits that are commonly associated with solar energy, fuel conservation benefits beyond those reflected in current prices of utility fuel, and the possibility of fuel price escalation beyond the general inflation rate over the lifetime of the solar system (*2*). However, none of these factors are likely to significantly affect the economics of solar systems that interface with electric utilities relative to the economics of solar systems that do not.

Solar/Electric-Resistance Heating

The reconceptualization necessary to properly assess solar heating systems that interface with conventional electric supply systems is shown in Fig. 1. Instead of comparing the cost of the solar heating system including storage with the cost of the conventional heating system (as in Fig. 1a), the solar system designer should compare the cost of the solar-supplemented storage heating system against the cost of the simple storage heating system (Fig. 1b). In practice, the latter comparison will proceed only after the benefits of storage heating relative to direct heating have been determined (Fig. 1c). It is our contention that most of the electricity supply savings claimed for solar energy systems stem from the storage, rather than the solar, component of the systems.

That storage heating systems are cost-effective in service areas supplied by winter-peaking utilities is well established (*3*). A recent study indicates that the utility savings, mostly capital savings, exceed the thermal energy storage costs by a factor ranging from 2 to 4 (*4*). The important consideration for the design of solar systems is that, given a storage capacity adequate to meet the design-day heating load, it makes economic sense to add a solar collector system only to the extent that it is cost-effective to substitute solar energy for off-peak electric energy. This consideration establishes an upper bound on acceptable solar collection costs:

$$C < S \cdot E$$

Here C is the annualized cost (dollars per square meter per year) of the installed collector system (including the costs of piping, pumps, controls, heat exchanger, and the collector itself), S is an upper limit on the amount of solar energy collected (kilowatt-hours thermal per square meter per year), and E is the cost of supplying off-peak electricity (dollars per kilowatt-hour electric). For many regions of the country, a representative value for the cost of supplying off-peak energy is 10 mills per kilowatt-hour (= \$0.01/kwh), which covers base-load fuel costs ($\approx$ \$1 per 10^6 Btu) and base-load operating and maintenance costs. Under efficient cost allocation rules, all utility capital expansion costs should be charged against energy use during other time periods (*5*). An optimistic level of solar collection over the space heating season is 300 kwh m^{-2} year^{-1} (= 100,000 Btu per square foot per year). Because, for a fixed storage capacity, annual collection efficiency decreases with collector area, this figure represents an upper limit on solar collection. Inserting these values for electricity cost and solar energy collection into the relation above, we obtain an upper bound on the break-even cost of the solar collection system of \$3 m^{-2} year^{-1}. If the real capital recovery rate is 10 percent per year, this corresponds to an upper bound on collector system costs of \$30/m^2 ($\approx$ \$3 per square foot). This is a very low break-even value compared with estimates based on either the average or the peak-period cost of electricity supply. The upper bound on the break-even cost is \$75/m^2 if the auxiliary energy cost is 25 mills/kwh.

In service areas supplied by summer peaking utilities, storage space heating generally is not cost-effective. In such service areas, the displacement of daily winter peak loads into nighttime valleys does not reduce the utility's annual peak capacity requirements. That the addition of a solar collector system could be cost-effective in this situation appears highly improbable. Having to support a sizable portion of the investment in the storage system, the return on the investment in the collector system will have to be higher than for the winter-peaking service area. The break-even cost of the solar collector will be correspondingly lower and may even be negative.

Solar-assisted electric hot water heating may justify a higher collector cost than solar space heating. Solar hot water systems enjoy a better duty cycle, displacing off-peak electricity on a year-round basis. On the other hand, solar hot water systems suffer a disadvantage relative to simple storage systems. This stems from the small additional cost of storage hot water heaters over conventional systems. Usually all that is required is a somewhat larger tank with improved insulation. The addition of a solar collection system, however, usually involves the addition of a separate preheat storage tank. A similar cost burden occurs in solar space heating applications when, in order to improve solar collection efficiency, separate storage systems are incorporated in both the solar supply loop and the backup electric supply system.

Solar Energy/Heat Pump Systems

Solar heating systems more complex than the ones described above have been proposed. In particular, the solar-assisted heat pump has received considerable attention (*6*). In one version of this system, the output of the solar collector is first input to a storage reservoir on the cold side of the heat pump (*7*). The solar energy is removed from the reservoir by the compressor action of the heat pump and is delivered to the building load.

The principal advantages of the solar-assisted heat pump are the lower temperature required for the output of the solar collector and the improved heat pump performance because of the solar warming of the input reservoir. According to the system's advocates, this allows the use of a much lower cost solar collector, thus improving the break-even economics of solar heating. However, as shown below, another effect of incorporating a heat pump in the solar system design is to drive down the break-even cost for the solar collection system.

The same techniques used in evaluating solar systems that interface with resistance heating can be used to analyze the solar-assisted heat pump design. Figure 2 presents a set of system comparisons building from direct resistance heating to the solar-assisted heat pump system. One of the steps in the system evolution in Fig. 2 represents an inferior progression and has been included only to facilitate the analysis. The system incorporating a heat pump in Fig. 2b is inferior to the system without a heat pump.

The heat pump operating solely off the electrically augmented storage on the input side is deficient on a number of counts. Because the heat pump can extract no more energy from the storage than is input, supplemental off-peak electricity is more efficiently stored and recovered in the system on the right side of Fig. 2b, where no work of compression is involved. The heat pump system is also more capital intensive, requiring a higher initial customer investment and greater utility capacity to meet the heat pump's contribution to the utility's daily peak load. As will become clear below, the inclusion of the heat pump with an electrically augmented input reservoir in Fig. 2 is simply part of a gedanken experiment to determine the overall cost-effectiveness of the solar-assisted heat pump system.

The system comparison in Fig. 2a lends itself to a particularly simple interpretation of the role of the solar collector subsystem. The system on the right side can be conceived as operating in almost the same way as its solar counterpart, the only difference being that the electrical energy is input to the right-hand storage reservoir during off-peak nighttime hours, while the solar energy is input to the left-hand reservoir during daytime hours. Although, as described above, this use of electricity is extremely inefficient, it amounts to no more than a simulation of the use of the solar energy that is input to the solar system.

Viewed in this context, the gross benefit from the addition of the solar collector to the heat pump system is no greater than the benefit realized when a solar collector is added to a solar-resistance heating system, namely the displacement of off-peak electric energy. However, now the gross benefit must also cover (relative to the solar-resistance system): the added capital cost of the heat pump over the resistance system, the utility capital costs associated with the supply of electricity to the heat pump during on-peak hours, and the cost of utility energy to run the heat pump. For the fraction of time when solar energy is not available, the heat pump can run off ambient air. This does not alter the relative economics of the solar-assisted heat pump and the storage heat pump in the comparison in Fig. 2a, because both systems would benefit equally from this option. However, it would reduce the heat pump cost burden slightly by the amount of the additional off-peak electricity displacement credit.

The additional capital cost of the heat pump system relative to the resistance system, after an air conditioner capital cost credit, is approximately \$1000 (*8, 9*). For a nominal collector area of 50 m^2, this corresponds to a reduction relative to the break-even point for the collector subsystem in the solar-resistance system of \$20/m^2. In service areas supplied by winter-peaking utilities, a more significant cost penalty is the utility capital cost of meeting the design-day compressor load (*10*). This cost, covering utility demand-related capital expansion costs at the generation, transmission, and distribution levels, amounts to approximately \$500 per peak kilowatt, corresponding to \$2000 for the heat pump and \$40/m^2 for the 50-m^2 collector. The break-even point is then $-$\$30/m^2 if auxiliary energy costs are 10 mills/kwh and \$15/m^2 if auxiliary energy costs are 25 mills/kwh.

In a service area supplied by a summer-peaking utility, utility winter peak-day capacity costs are considerably lower, depending on such considerations as reserve margin under scheduled maintenance outage. However, even if utility power costs are assumed to be negligible, the break-even point for the solar collector component of the solar-assisted heat pump system will be lower than that of the solar/electric-resistance heating system by the amount of the added capital cost of the heat pump.

Another solar energy/heat pump design concept is illustrated in Fig. 3. Here, the storage reservoir is on the "hot" side of the heat pump and both the solar collector and the heat pump deliver energy directly to the reservoir. We shall not analyze this system in any detail here (*11*); however, it can readily be shown that this system is always inferior to the solar-assisted heat pump analyzed above. The basic problem with the system design in Fig. 3 is that it quite literally forces two capital-intensive subsystems, the solar collector and the heat pump, to compete rather than supplement one another in supplying the heating load.

The conclusions reached regarding these two generic types of solar energy/heat pump systems appear to be applicable to all solar/heat pump configurations. Our analyses of other heat pump concepts indicate that they can always be evaluated by decomposition into some combination of the solar/resistance-heating and the two solar/heat pump concepts described above. All the solar/heat pump concepts that we have identified have break-even solar collection costs considerably lower than the break-even solar collection costs of the solar/resistance-heating system in Fig. 1.

Solar Electric Power Generation

Many of the same kinds of trade-offs affecting the economics of solar energy applications in buildings affect the economics of solar electric power generation. Figure 4 presents the solar electric analogs of the heating system comparisons presented in Fig. 1.

The usual procedure for determining the cost-effectiveness of a solar electric generating system is the comparison indicated in Fig. 4a. The cost of the solar electric system, including any necessary backup to cover solar outage, is compared with the cost of the conventional generating system. In Fig. 4a the comparison is presumed to be between the solar system including its backup boiler and a conventional intermediate generating plant. (To facilitate later analysis, a base-load generating plant has been included on both sides of Fig. 4a. This facility cancels out in the comparison.) Although the comparison in Fig. 4a is the standard method of evaluating solar electric systems, a more meaningful comparison is that shown in Fig. 4b.

That a solar electric system can survive comparison with a storage-augmented base-load generating plant is very doubtful. Given low-cost storage, a storage-augmented base-load plant can replace a combination of base and intermediate (or peaking) plants (see Fig. 4c). In a recent study, the Public Service Electric and Gas Company (*12*) calculated break-even costs for thermal storage in this application at \$570/kw (10-hour storage device) and \$320/kw (5-

hour storage device) (*13*). On the other hand, in a Fig. 4a–type evaluation of the economics of solar electric power systems, the Aerospace Corporation (*14*) assumed solar storage device costs several times lower than the PSE&G storage break-even costs. For the most attractive solar electric application examined in the Aerospace study (a central station receiver plus 6-hour storage compared to an intermediate generating plant), the heliostat effective break-even cost was about $30/m². Already a difficult cost objective to achieve, the break-even cost would have been considerably lower if the solar electric system had been compared against a storage-augmented base-load generating plant.

Generalization

The preceding analysis was limited to the types of solar systems that interface with electricity supply systems. In considering such systems, we stressed the low break-even costs of solar collection systems whose only effect is to substitute solar energy for low-cost off-peak electricity. The question naturally arises as to how the break-even economics would be affected by the absence of low-cost off-peak electricity.

In an important sense, off-peak electric power is merely a by-product of on-peak electricity. However, its future availability might very well diminish as utilities and regulators begin to price it at cost and customers respond by purchasing it in increasing quantity. The European experience indicates that, given adequate customer price incentives, the time required for complete "valley filling" for a winter-peaking utility system can be as short as 10 to 15 years (*2*). Therefore, to complete our analysis of solar energy systems that interface with electric utility systems, we will generalize our results by considering two alternative scenarios. In the first scenario, off-peak electricity, even if priced at variable cost (utility fuel cost), remains available for the indefinite future; in the second scenario, as a result of the introduction of a new technology or the implementation of some form of peak-load pricing, the utility's load curve becomes flat. The first scenario corresponds to the economists' firm-peak case; the second scenario to the shifting-peak case.

In the first scenario, in most parts of the country off-peak electricity will remain the lowest-cost auxiliary energy available for solar energy systems. Systems that use this form of auxiliary energy will represent the economically (and socially) most efficient solar energy systems. Thus, to be economical, solar collection systems will have to be low enough in cost to supply solar energy that is cost-competitive with off-peak electricity.

Even in the first scenario, there may remain many utility service areas where off-peak electricity is not the lowest-cost source of auxiliary energy. Today, many utilities use oil in base-load generating plants, with the cost of the oil approximately $87 per metric ton ($13 a barrel). After correction for transmission and distribution losses, this gives a cost of approximately 25 mills/kwh ($8.40 per 10^6 Btu). On the other hand, the price of home heating oil is currently about 11¢ per liter (42¢ per gallon), which after correction for furnace conversion efficiency is equivalent to approximately 15 mills/kwh ($5 per 10^6 Btu). Thus, in service areas supplied by utilities using oil-fired base-load plants, home heating oil represents a lower-cost auxiliary energy form. Although natural gas can also be used to supplement solar systems, its price (marginal cost) under national deregulation is likely to be somewhat higher than the price of oil.

To summarize, the findings for the scenario involving indefinitely available low-cost off-peak electricity are as follows. In service areas supplied with electricity from coal or nuclear generating plants, the most economical solar energy systems will be those that use off-peak electricity as the auxiliary energy source. Accordingly, solar collection systems generally will be economical only if they can deliver solar energy at a cost lower than the cost of off-peak electricity. As we have already seen, the upper limit on break-even costs is about $30

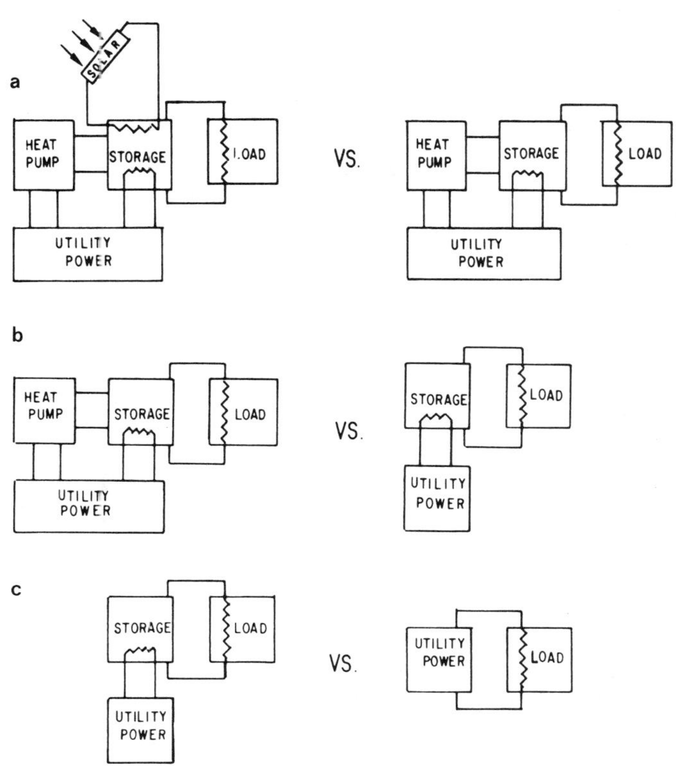

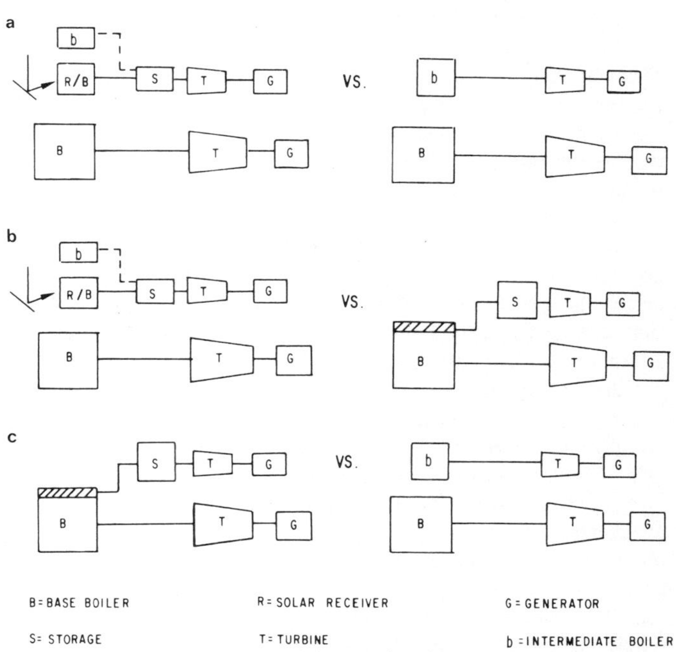

Fig. 3 (left). Comparisons among solar/heat pump, storage, and conventional space heating systems. Fig. 4 (right). Comparisons among utility solar thermal conversion, thermal storage, and conventional power generation.

per square meter of collector area. In service areas supplied with off-peak electricity produced from expensive fuels, such as oil or natural gas, the most economical solar energy systems will be those that utilize auxiliary fuels other than off-peak electricity. In these service areas the most economical solar systems will not interface with the utility supply system.

Probably a more meaningful test of the interfacing of solar systems and electric utility systems occurs in the second scenario, when the utility load curve is flat. This can occur if the utility, through the application of a judicious pricing strategy (higher prices during high-demand hours than during low-demand hours), manages to equalize the rates of consumption during all the subperiods of its demand cycle. In this case, the demanders during each subperiod press against the utility's capacity and are responsible, in varying measure, for the utility's long-run expenditures for capacity expansion (*15*).

In this situation, the effect of a solar heating system on the utility's load curve and on its unit cost of supply can be conceived as occurring in the following way. Let us assume that the utility takes into account the effect of solar outages by setting its high-demand and low-demand subperiod prices such that consumption during the low-demand period, including consumption required to cover worst-case solar outage, is equal to consumption during the high-demand period. It follows, in this case, that utility capacity is underutilized during the demand cycles when solar insolation is available for home heating. Solar energy still serves to displace utility fuel; however, the underutilization of utility capital relative to its rate of utilization in supplying a simple electric storage heating system raises the unit cost of electricity supply. This added cost, which represents a real increase in the capital component of electricity cost, must be incorporated, ex ante, into the higher unit price charged solar heating customers than charged storage heating customers. From the solar heating customer's perspective, the increase in unit price is exactly offset by the zero variable-cost component of energy from the customer's solar collection system. (The utility's capital investment and the total energy consumed by the customer are unchanged.) Thus, just as for the firm-peak case, the net benefit of solar collection for the customer, and for society as a whole, is equal to the value of the displaced electric utility fuel. [This result can be shown to follow from a general analysis of the periodic load problem (*11*).]

Although in the preceding discussion of the shifting-peak case we have implicitly assumed the existence of only one type of utility plant, this is not a serious limitation for most solar applications. In home heating applications, for example, the fraction of the load for which solar substitutes for electric supply will not be sufficient, of itself, to allow the utility to substitute intermediate-load for base-load plants. In central station electric power applications, solar must compete directly against base-load plants, the only plant type in use under the uniform load, shifting-peak case.

Summary

Analyses of the economics of solar collection in the firm- and shifting-peak cases (that is, with off-peak electricity indefinitely available or with a flat load curve) indicate that, for many important applications, solar energy systems that interface with electric utilities can be justified only in terms of the value of the off-peak utility fuels that they displace. In regions where off-peak electricity costs are low, the most economically efficient solar energy systems will be those that use electricity as the auxiliary energy source. This implies extremely low break-even costs for a number of important solar energy applications. In regions where the cost of off-peak electricity is higher than that of competing energy forms, the most economical solar energy systems will utilize auxiliary fuels other than electricity.

The general conclusion is that conventional electric utility systems and most solar energy systems represent a poor technological match. The basic problem is that both technologies are very capital intensive. The electric utility, because of the high fixed costs of generation, transmission, and distribution capacity, represents a poor backup for solar energy systems. On the other hand, the solar collection system, because it represents pure, high-cost capital and because of its outage problems, cannot be considered as a part-load source of auxiliary energy for the electric utility system.

References and Notes

1. See, for example, P. J. Hughes, W. A. Beckman, J. A. Duffie, paper presented at the Sharing the Sun Conference, Winnipeg, Canada, August 1976. See also *Solar Engineering Magazine* (August 1976), p. 24.
2. Basing the break-even cost for the solar collection system on the current price of base-load fuel is equivalent to assuming that the real rate of increase of the price of base-load fuel less the real time rate of discount of money is zero. This appears to be a reasonable assumption. See, for example; Staff Report, Executive Office of the President, Council on Wage and Price Stability, *A Study of Coal Prices* (Washington, D.C., March 1976); J. G. Asbury and K. Costello, *ASME Publ. No. 76-IPC-PWR-7* (May 1976).
3. J. G. Asbury and A. Kouvalis, *Argonne Natl. Lab. Publ. No. ANL/ES-50* (April 1976). Although the storage components on either side in Fig. 1b are drawn the same, the storage devices used in electric storage heating systems are generally of higher quality or lower cost than the storage devices used in solar systems. Under worst practice, the same storage device used in the solar system can be incorporated in the storage heating system. The principal advantage of storage in electric storage heating is the high temperature, up to 1200°F, that can be achieved in magnesite bricks or other materials.
4. J. Asbury, R. Giese, S. Nelson, L. Akridge, P. Graf, K. Heitner, "Assessment of Energy Storage Technologies and Systems (Phase I): Electric Storage Heating, Storage Air Conditioning and Storage Hot Water Heaters," *Argonne Natl. Lab. Publ. No. ANL/ES-54* (October 1976).
5. See, for example, C. Cicchetti, W. Gillen, P. Smolensky, "The Marginal Cost and Pricing of Electricity: An Applied Approach" *Natl. Sci. Found. Rep. STA 74-24370* (June 1976); L. T. Mahoney, Jr., National Economic Research Associates, Inc., testimony at a hearing before the Public Service Commission of New York, Case No. 26806, January 1976; testimony at a hearing before Public Service Commission of Wisconsin, Docket No. 6630-ER-2, December 1975; J. G. Miller, Wisconsin Power and Light Company, testimony and exhibits at a hearing before the Public Service Commission of Wisconsin, Docket No. 2-U-8085, April 1976.
6. S. F. Gilman, Ed., *Solar Energy Heat Pump Systems for Heating and Cooling Buildings–Workshop Proceedings* (Pennsylvania State Univ. Press, University Park, and ERDA Technical Information Center, Oak Ridge, Tenn., 1975).
7. See, for example, V. D. Karman, T. L. Freeman, J. W. Mitchell, "Simulation study of solar heat systems," paper presented at the Sharing the Sun Conference, Winnipeg, Canada, August 1976.
8. Gordian Associates Incorporated, "Evaluation of the air-to-air heat pump for residential space conditioning," report prepared for the Federal Energy Agency under contract FEA-CO-04-50171-00, April 1976.
9. As shown in (*8*), the cost of a heat pump sized to meet the cooling load is approximately $1000 greater than the cost of an air conditioner. This cost difference holds approximately for most regions of the country. Although the heat pump is less efficient in the air-conditioning mode than the standard air conditioner, this charge against the heat pump is ignored here. None of the solar-assisted heat pump concepts considered here use solar energy when the heat pump is operating as an air conditioner.
10. We ignore the resistance component of the heat pump's design-day power demand. For the system design concept examined here, energy can in principle be input to the storage reservoirs during nighttime hours to eliminate the following day's on-peak resistance power requirement.
11. For a detailed analysis see J. G. Asbury and R. O. Mueller, *Argonne Natl. Lab. Publ. No. ANL/ES-52* (August 1976).
12. Public Service Electric & Gas Co., "An assessment of energy storage systems suitable for use by electric utilities," report prepared for the Energy Research and Development Administration under contract E(11-1)-2501 and the Electric Power Research Institute under research project 225, July 1976.
13. These break-even storage costs correspond to PSE&G's "typical values" for generating plant capital and fuel costs. The PSE&G also calculated break-even cost ranges corresponding to variations in input costs. The PSE&G typical value conventional-plant generating costs are similar to those used in the Aerospace study cited below.
14. Aerospace Corp., "Solar thermal conversion mission analysis—Southwestern United States," report prepared for the National Science Foundation under contract NSF-C797, January 1975.
15. For the efficient-price solution, see O. Williamson, *Am. Econ. Rev.* **56**, 810 (1966); M. Crew and P. Kleindorfer, *J. Polit. Econ.* **79**, 1369 (1971).
16. We thank E. Croke for a very helpful suggestion and R. Giese, S. Nelson, and V. Rabl for a number of useful comments. Supported by the Chemical and Thermal Energy Storage Branch, Division of Energy Storage Systems, Office of the Assistant Administrator for Conservation, Energy Research and Development Administration.

The Tokamak: Model T Fusion Reactor

Don Steiner and John F. Clarke

Fusion power holds promise of using cheap, abundant fuel and of being relatively low in radiological hazard potential. To date, experimental results indicate that scientific progress has been greater in the tokamak approach than in other magnetic confinement fusion concepts. For this reason, the tokamak concept enjoys generous financial support and is the main thrust of the American and world fusion programs. But the tokamak scheme for power generation is not without obstacles, which have led to concerns raised by fusion researchers (*1*) and representatives of the electric utility industry (*2*) about the commercial feasibility of tokamak power. These concerns, which were well articulated by Metz (*3, 4*), boil down to two questions.

Summary. During the past several years there have been significant scientific and technological advances related to the tokamak magnetic confinement scheme. These are summarized in the context of a recent tokamak reactor design study which emphasizes reduced size, higher power density, and enhanced plant reliability and maintainability relative to earlier tokamak reactor design studies. The direct plant cost of the proposed reactor is estimated to be in the range \$1000 to \$1500 per electrical kilowatt. A three-phase strategy for demonstrating tokamak fusion power generation at a committed site is outlined. It is estimated that implementation of the three-phase program would require about 20 years and a total escalated expenditure of \$10 billion to \$15 billion. The tokamak power plant described here is not viewed as definitive but rather as a point of departure in the development of a plan to demonstrate tokamak power generation.

1) Can the tokamak concept lead to an economically competitive power system?

2) Is the cost of the program required for developing the tokamak concept acceptable?

Issues raised regarding the economic potential of the tokamak concept include reactor size, power output, plant costs, and plant reliability and maintainability. Associated with these issues was the concern that a demonstration program would require a succession of increasingly large and costly devices. Metz's description of the scope of these issues was based on an examination of tokamak reactor design studies carried out between 1971 and 1976, and in particular on the tokamak reactor design study called UWMAK I evolved by the University of Wisconsin Fusion Technology Group in the early 1970's. These early studies represented initial attempts to identify problems that might be encountered in tokamak reactors. They were not intended to design economically competitive systems. Subsequent studies have benefited from the early studies and have focused more on the requirements for economic competitiveness. Moreover, they reflect scientific and technological advances which have been made over the past several years, primarily as a result of an associated, significant increase in the U.S. fusion research budget (see Fig. 1). The more recent studies indicate that many of the problems identified earlier are tractable and therefore should not be viewed as insurmountable barriers to the economic potential and development of tokamak reactors.

At Oak Ridge National Laboratory (ORNL), for example, we have evolved a conceptual reactor design which suggests that tokamak power systems can be economical without being monstrously large (both in electrical output and physical size), that their plant costs need not greatly exceed those of other advanced energy systems such as liquid-metal fast breeder reactors (LMFBR's) and solar electric plants, and that their reliability and maintainability should not be unacceptably low. Moreover, we have been able to identify the elements of a strategy for demonstrating significant power generation with acceptable program costs. This is not to say that our conception of the tokamak power system should be regarded as definitive. Viewed from the perspective of the ultimate tokamak fusion power system, our present concepts will undoubtedly appear as primitive as a Model T Ford compared to a modern automobile. However, just as that Model T bears a recognizable relation to its distant descendant, we believe that our present design concept of the Model T tokamak reactor can provide a realistic base for assessing the nature of tokamak fusion power systems of the future.

In this article, we will first describe the essential elements of the commercial tokamak power plant that we envision. Then we will examine the scientific and technological basis (that is, the scientific and technological progress during the past few years) for this view of a commercial tokamak power plant. Next we will consider the economic potential of tokamak power plants. And finally, we will suggest the elements of a strategy to demonstrate tokamak power generation.

Commercial Plant

As we envision it (*5*), the commercial power plant would consist of multiple (for example, two to five) tokamak reactor units sharing a number of common

Don Steiner is Advanced Systems Program Manager, Fusion Energy Division, Oak Ridge National Laboratory, Oak Ridge, Tennessee 37380, and John Clarke is Deputy Director, Division of Magnetic Fusion Energy, Department of Energy, Washington, D.C. 20545.

elements, including such key elements as the pulsed power supplies for driving the plasma current and the pulsed power supplies that provide the plasma supplementary heating. Sharing these elements among multiple units would save about 15 percent in the plant costs of fusion power relative to the case where each unit has its own committed power supplies.

Each reactor unit would generate between 500 and 1000 megawatts electric (MWe) of output power. The precise value of the electrical output of each unit would be determined by both plasma physics and cost optimization considerations. Note that the large output powers characteristic of some earlier tokamak reactor studies reflected the assumption that single units of large electrical output would be desired by the utilities. Our studies show that single units of large electrical output should not be viewed as an inherent characteristic of tokamak reactors. For this discussion we will assume that an electrical utility would like to develop a particular site for a total capacity of 1500 MWe. This would be accomplished by a plant employing two tokamak reactor units, each generating 750 MWe.

Reactor unit. The major components of each tokamak reactor unit are illustrated in Fig. 2 and the essential system parameters are summarized in Table 1. The toroidal plasma core has a minor radius of about 1.5 meters and a major radius of about 6 meters. By comparison, note that the UWMAK I design had a toroidal plasma core with a minor radius of about 5 m and a major radius of about 13 m. As indicated in Fig. 2, the plasma cross section is somewhat elongated; in this case the ratio of the plasma height to width is about 1.6. The elongation is expected to result in improved plasma performance compared to the case of a circular plasma cross section. The plasma would operate with a density of $\sim 2 \times 10^{14}$ particles per cubic centimeter, a confinement time of about 1 second, and a temperature of $\sim 10^8$ K. It would be brought to the operating temperature in about 10 seconds, using ~50 to 100 MW of neutral beam injection power. Fueling would be accomplished by injecting solid fuel pellets into the plasma, and spent fuel removal would be accomplished by guiding charged particles out of the plasma chamber along diverted magnetic field lines generated by "divertor" coils.

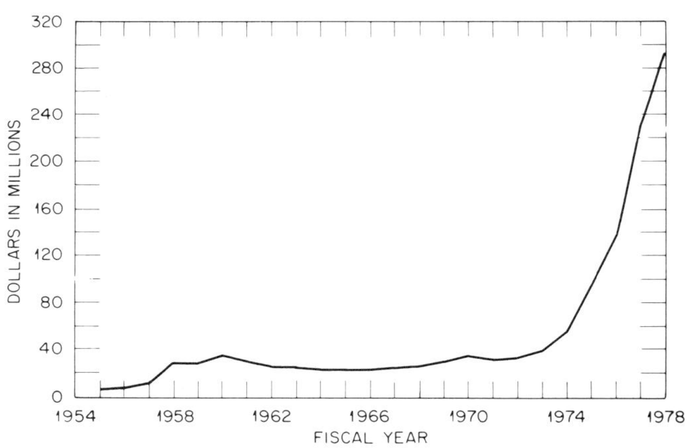

Fig. 1. The U.S. magnetic fusion energy research budget during the past 24 years in absolute dollars. If the effects of escalation were included, the drop in the budget during the period ~1960 to 1970 would be more dramatic. The significant increase in the budget which began in the middle 1970's is yielding substantial scientific and technological progress (see text).

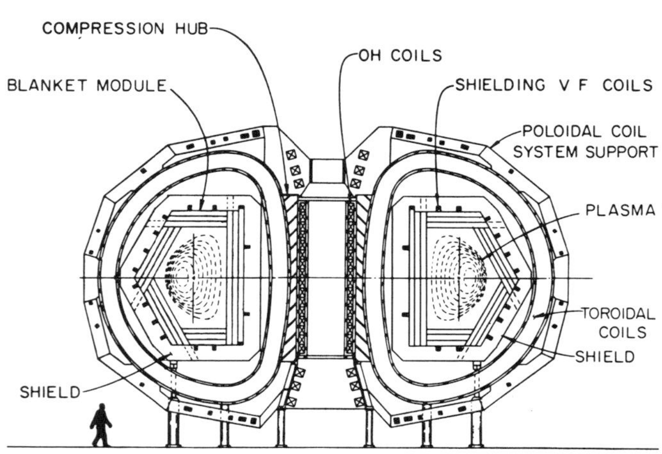

Fig. 2. Cross section through a tokamak reactor as envisioned in the ORNL study (*5*). The toroidal plasma core has a minor radius of ~1.5 m and a major radius of ~6 m. The superconducting toroidal field coils have inside dimensions of ~7 by 10 m. Such a reactor would produce ~750 MWe. Abbreviations: *OH*, ohmic heating; *VF*, vertical field.

As noted by Metz (*4*), the size and cost of the tokamak power system is a strong function of the fusion power density. The fusion power density in the plasma chamber of each unit is about 5.3 MW/m^3 as compared with about 0.8 MW/m^3 in the UWMAK I design. The plasma volume of each tokamak reactor unit is about 15 times smaller than that in the UWMAK I design, and the thermal power of each unit is about 45 percent of that in the UWMAK I design. We emphasize that these comparisons with the UWMAK I design are not made to criticize that design, which, as Rose and Feitag (*6*) said, was a problem-finder rather than a problem-solver. Instead, we wish to point to the significant reductions in projected size and power output which have occurred in the area of tokamak reactor design as a result of scientific and technological progress.

Surrounding the plasma would be a blanket whose main purpose is to recover the fusion energy as heat and to breed new tritium for fueling the reactor. The blanket first wall and structure would be made of an austenitic stainless steel. The amount of steel in the blanket is about an order of magnitude less than that in the blanket of the UWMAK I design. The optimum structural alloy for the fusion reactor environment is yet to be identified, and a materials program to develop such an alloy is now under way. Nevertheless, preliminary results on the performance of austenitic stainless steels under simulated fusion neutron irradiations give us considerable optimism about the prospects for such alloys. These results will be considered further in our discussion of economic potential.

At present, there are several promising coolants for the blanket. One is a salt consisting of sodium and potassium nitrates and nitrites. These salts have been used extensively as heat transfer media in the petroleum and chemical processing industries for more than 30 years. They have relatively low melting points (~150°C) and are relatively inexpensive (~$1 per kilogram), and their thermal stability and compatibility with iron-base alloys seem acceptable to temperatures around 500°C. A development program will be required to determine the ultimate acceptability of such salts in a fusion reactor environment. Other coolants under consideration are helium gas and liquid lithium. The breeding medium must consist of lithium in some form since only lithium offers any promise for tritium regeneration through neutron interactions. At present, liquid lithium seems to offer the greatest potential as the breeding material.

The blanket region is ~0.8 m thick and is surrounded by a shielding region ~0.7 m thick whose main purpose is to protect the superconducting toroidal magnet coils. A number of materials could be used for the shielding region, including iron, boron carbide, and lead. The toroidal field coils employ niobium titanium as the superconducting material. The coils are somewhat D-shaped with inside dimensions of ~7 by 10 m, approximately half the inner dimensions of the coils for the UWMAK I design (~15 by 21 m).

Balance of the plant. Each tokamak reactor unit would be housed in a circular containment building ~50 m tall and ~60 m in diameter. The volume of this containment building would be almost one order of magnitude less than the volume of the containment building suggested for the UWMAK I reactor and about the same as that of the containment building of a fission reactor with comparable output power.

Each reactor unit would have its own power conversion system. This would consist of a primary coolant loop and an intermediate salt loop coupled to a conventional steam system and turbine. The main purpose of the intermediate heat transport loop is to prevent pressurization of the blanket by high-pressure steam should there be a tube leak in the steam generator. Assuming a primary loop exit temperature of around 450°C, a steam-cycle thermodynamic efficiency of ~35 percent would be achieved with this power conversion system. Each reactor unit would operate with a burn time of ~20 minutes and a down time of ~1 minute, thus achieving an overall duty factor of ~95 percent. Salt storage tanks coupled to the intermediate heat transport loop would provide the necessary thermal energy storage to ensure continuous power to the turbine plant during the entire operating cycle.

The commonly shared power supplies for initiating the plasma current and providing the plasma heating would operate in a phased fashion for the two tokamak units. Note that these power supplies operate in a pulsed mode and need to be operating for only 10 seconds at the start of the initial plasma breakdown and heating phase. For the assumed conditions, two tokamak units and a 21-minute cycle time, these pulsed power supplies would have to be pulsed once approximately every 10 minutes. It appears that the primary energy storage requirements for these power supplies can be satisfied with motor-generator flywheel sets that are similar in rating and capacity to those being proposed for the Tokamak Fusion Test Reactor (TFTR) to be built at the Princeton Plasma Physics Laboratory.

Table 1. Essential system parameters of the ORNL reactor design and the UWMAK I design.

Parameter	ORNL	UWMAK I
Plasma radius (m)	1.5	5
Major radius of torus (m)	6	13
Plasma shape	Elongated 1.6/1	Circular
Neutral beam injection energy (keV)	150–300	500
Plasma beta (percent)	8	3
Fusion power density (MW/m^3)	5.3	0.8
Neutron wall loading (MW/M^2)	3.0	1.25
Cycle time (seconds)	1260	5800
Duty factor (percent)	95	93
Average thermal power (MW)	2150	5000
Average electrical output (MW)	750	1500

Scientific and Technical Basis

The tokamak reactor described above is attractive on the basis of electrical output and physical size; the question of economic potential will be considered later. At this point we will address the fundamental question of whether the reactor operating characteristics are consistent with the basic physics of tokamak plasmas. A secondary question, assuming a positive answer to the first, is whether these operating characteristics can be achieved using practical technologies. At this stage of the fusion research and development program, it is not possible to give a definitive answer to either of these questions. However, it can be said that operation of a tokamak reactor in the proposed regime is consistent with our *present* theoretical understanding of tokamak plasma behavior, and that current tokamak experiments are producing results pointing toward the proposed regime of operation. Furthermore, the key technologies used to advance the plasma performance in these experiments seem to be capable of extrapolation to the requirements of reactors.

Progress toward the reactor regime. A tokamak reactor plasma will operate in the "ignition" regime. In this regime the fusion energy deposited within the plasma exceeds the energy lost from the plasma by heat conduction and radiation; thus, the fusion plasma can be self-sustaining. To operate in the reactor (that is, ignition) regime with a deuterium-tritium–burning plasma, the plasma ion temperature must exceed about 40×10^6 K and the plasma energy confinement, measured by the product of plasma density and energy confinement time, must exceed about 6×10^{13} cm^{-3} sec. Note that the plasma in the proposed reactor operates with a temperature of about 100×10^6 K and with an energy confinement of about 2×10^{14} cm^{-3} sec.

Figure 3 illustrates progress toward the reactor regime within the U.S. toka-

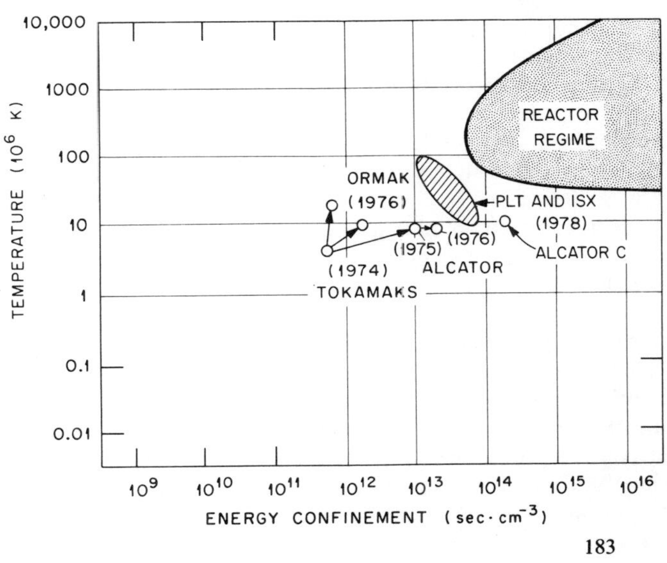

Fig. 3. Recent and projected tokamak experimental performance in terms of plasma ion temperature and plasma energy confinement; the latter is measured by the product of plasma density and energy confinement time. To enter the reactor regime, this product must exceed about 6×10^{13} cm^{-3} sec and the ion temperature must exceed about 40×10^6 K.

mak experimental program. By 1974 tokamaks heated only by the toroidal current flowing in the plasma (that is, ohmic heating) had advanced closer to the reactor regime than any other fusion confinement scheme (*7*). By 1975 the application of extremely high magnetic fields in the ALCATOR experiment at Massachusetts Institute of Technology (MIT) produced a factor of 10 increase in the energy confinement (*8*). By 1976 supplementary heating experiments at Oak Ridge (*9*) on ORMAK with newly developed neutral beam injectors had demonstrated that the plasma ions could be heated to 20×10^6 K. At the same time, experimental data from the heating experiments at Oak Ridge indicated that the improvement in energy confinement seen at MIT could also be produced by the neutral heating technique (*9*). This finding is shown by the second ORMAK point in Fig. 3, which represents operation designed to optimize energy confinement rather than heating.

The performance projected for the next generation of experiments is also shown in Fig. 3. The hatched area indicates the operating regime expected in the neutral beam heating experiments on the PLT (Princeton Large Torus) and ISX (Impurity Studies Experiment, at Oak Ridge). The ALCATOR C point indicates the maximum achievement in energy confinement anticipated in the new high-field experiment at MIT. These experiments collectively should exceed the minimum requirements for temperature and energy confinement in a tokamak reactor. Following closely on these experiments are the DOUBLET III device (at General Atomic) and the TFTR device (at Princeton). The DOUBLET III device has the goal of achieving simultaneously minimum reactor level temperatures and energy confinement (that is, the boundary of the "reactor regime" in Fig. 3) in hydrogen. The TFTR has the goal of achieving simultaneously minimum reactor level temperatures and energy confinement first in hydrogen and then in deuterium-tritium plasmas. It is emphasized that while the TFTR and DOUBLET III will approach the boundary of the reactor regime they will not enter it. It is also noted that similar tokamak experimental programs are being pursued in Europe, Japan, and the Soviet Union.

Although optimism within the program is high, there are still a number of uncertainties to be confronted by these experiments. Although ALCATOR has demonstrated extremely clean plasma operation (*8*), the mechanism for this impurity-free behavior is not well understood. All other devices contain contaminants of either light elements, such as carbon and oxygen, or heavy metals, such as iron and tungsten. These impurities radiate large amounts of energy and, without techniques for their control, could prove to be a serious impediment to reaching the reactor regime. In recent experiments on the DITE tokamak (Divertor and Injection Tokamak Experiment, at Culham, England) impurities have been actively controlled by diverting surface plasma into a burial chamber, thus shielding the inner plasma from wall-originated impurity atoms (*10*). The impurity studies on PDX (at Princeton) and ISX will further examine the evolution, transport, and control of impurities (with and without divertors) in tokamaks.

Furthermore, although the energy confinement of the ions in current experiments is very close to that expected on the basis of classical kinetic theory, the electron energy confinement is one to two orders of magnitude less than classical (*7*). The precise mechanism for this anomaly is not established although several theoretical explanations are available. Without a firm theoretical understanding, we cannot predict with confidence the behavior of the electron energy confinement as our experiments approach the reactor regime. Fortunately, the data from all existing experiments operating over a wide range of parameters indicate that it will improve at least with the square of the plasma radius (*11*). This gives us confidence that the ignition regime can be entered in devices larger than TFTR and DOUBLET III. However, the question remains, how much larger? The answer requires some additional discussion of the scaling of energy confinement.

Scaling to the reactor regime. We can measure energy confinement in present experiments as a function of plasma parameters such as size, temperature, and density. These measurements can then be used to derive a scaling model to predict the tokamak plasma size required to achieve the reactor levels of temperature and energy confinement. This "empirical scaling model" (*9*) predicts ignition conditions in plasmas roughly 1 m in radius if the product of plasma pressure and the magnetic field strength can be made sufficiently large. However, the current empirical scaling model is based predominantly on experiments without supplementary heating. In these experiments there is an implicit dependence of the plasma temperature on the magnetic field, which disguises the actual temperature and magnetic field dependence of energy confinement. Within a few years there will be sufficient data from tokamak experiments with supplementary heating to test this scaling model at reactor temperatures.

An alternative approach is to attempt to predict the energy confinement in the reactor regime from basic principles of plasma physics. The resulting scaling based on classical collision processes indicates that ignition can be attained in very small devices (*12*). Although the ions in our experiments seem to behave in accord with this theory (*13*), the electrons do not. Their energy confinement appears to be dominated by plasma turbulence caused by a variety of fine-grained plasma instabilities. The worst of these instabilities predicted for the reactor regime are known as trapped particle modes (*14*). They have not been clearly identified in experiments to date, and as a result their effect on a reactor tokamak plasma can only be estimated from theoretical calculations. Because of the complexity of the calculations, there is a great deal of uncertainty in the result. However, the scaling model resulting from this theory is remarkably similar to the empirical scaling model in its dependence on the product of magnetic field and plasma pressure. This model also predicts ignition conditions with roughly a 1-m radius if the plasma pressure and magnetic field are sufficiently large.

The net result is that current scaling models based on either empirical or theoretical grounds predict that the reactor regime could be entered with devices only slightly larger than DOUBLET III and TFTR. (Note, for example, that the plasma radius in the TFTR is about 85 cm.) The uncertainties in these scaling laws will be reduced by the current and planned tokamak experimental programs, and within several years the exact size of a reactor plasma should be predictable with a high level of confidence. However, even at present it appears that the reactor plasma radius can be expected to be in the range 1 to 2 m and that the plasma radius of the proposed reactor (~1.5 m) is a reasonable choice for engineering design studies. This statement is based on the strong dependence of energy confinement on plasma pressure and magnetic field as evidenced in the current scaling models. This dependence should suffice to accommodate errors in our present estimates of the coefficients in the scaling models.

Limits on plasma pressure. Plasma pressure is normally measured in a tokamak in units of the toroidal magnetic field energy density. The ratio of plasma pressure to toroidal magnetic field energy density, known as beta, is an important parameter describing the performance of tokamak fusion reactors. Beta is a measure of the efficiency with which the system utilizes the magnetic field provided by the expensive toroidal field coils. Both the empirical and trapped particle mode scaling models referred to above indicate that the plasma radius of a reactor is reduced as beta (*15*) and the applied toroidal field (*16*) are increased. To produce the most economical tokamak fusion reactor, we would like to work at the lowest possible value of magnetic field and consequently the highest practical value of the plasma beta.

Until recently, it was assumed that a simple tokamak plasma could not be maintained at an equilibrium position within the surrounding reactor structure for betas greater than a few percent. At such low values of plasma beta, one would require either a very large plasma or very high magnetic fields to enter the reactor regime according to either of the scaling models given above. The early fusion reactor studies were performed in accordance with this assumed restriction on beta (see Table 1), and therefore it is not surprising that they resulted in large, expensive reactor systems.

Recent work analyzing the behavior of tokamak plasmas undergoing the intense heating required to raise them to reactor temperatures (*17*) has revealed that this equilibrium limit on plasma beta is nonexistent and that a tokamak plasma equilibrium can be maintained for beta approaching 100 percent. Experimental results from fast pinch experiments (*18*) show the existence of these equilibria at betas of ~70 percent for short periods of time.

The beta of the proposed reactor (~8 percent) lies in the range 5 to 10 percent. Although plasma equilibria exist for this range of betas, there has been some concern about the stability of these equilibria (*19*). It is known theoretically that as the plasma pressure is raised in a tokamak, a point is reached at which an aneurism called a ballooning instability develops in the magnetic field confining the plasma, and the equilibrium is destroyed. Calculations obtained from recently developed computer codes indicate that the beta limit set by this instability is in the range 5 to 10 percent for elongated plasmas (*19*). The ORMAK experiment at Oak Ridge has produced an average beta of just over 1 percent (*20*), a factor of about 3 short of the minimum beta required for ignition in typical reactor configurations and a factor of about 5 short of the minimum beta required for economic interest. Plasma

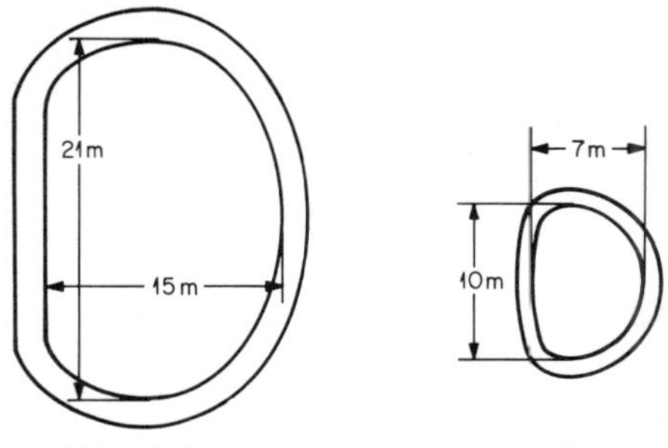

Fig. 4. If the expected value of beta is attained, it is possible to envision a compact tokamak reactor whose toroidal coil size is only a fraction of that assumed in earlier designs. This diagram shows the difference in toroidal coil size when the coil in the UWMAK 1 reactor design is compared with that in the ORNL reactor design.

shapes predicted to be compatible with the beta range of reactor interest will be explored during 1978 and 1979 in the ISX, PDX, and DOUBLET III experiments.

Technological implications. If betas in the range 5 to 10 percent can be achieved, relatively compact tokamak reactors can be constructed with maximum field requirements (that is, the field at the coil) in the range of 8 to 12 tesla (*5*). Figure 4 illustrates the reduction in toroidal coil size that has resulted from the improved prospects with regard to achievable plasma beta. In Fig. 4 the toroidal coil in the UWMAK I reactor design is compared with the coil in the ORNL reactor design. Both coils are assumed to operate with a maximum field of about 8 T at the coil winding; however, a plasma beta of ~8 percent is assumed in the ORNL design while a beta of ~3 percent is assumed in the UWMAK I design.

As the size of the superconducting coils required for the reactor is reduced, the technological extrapolation required to produce these coils becomes much more reasonable and likely to be achieved in the near term (*21*). The Large Coil Program at Oak Ridge has just let contracts for the design and construction of superconducting coils about half the size required for our compact reactor. Coils made of niobium titanium (with the potential of operating in the range 8 to 10 T) and coils made of niobium tin (potential operating range, 10 to 12 T) will be constructed. These coils will be assembled and tested in a toroidal geometry in the early 1980's. The goal of the Large Coil Program is to provide the technological base for constructing coils of the size required in reactors.

The development of tokamak reactors will also require a technique for heating the plasma to reactor temperatures. Injection of powerful beams of neutral hydrogen atoms into a confined tokamak plasma provides such a technique. The proposed reactor would require 50 to 100 MW of neutral beam power at an energy of roughly 150 to 300 kiloelectron volts to achieve sufficient heating (*22*). Because of the reduced plasma radius, the beam energy required to penetrate to the center of the plasma is a factor of 2 to 3 lower in this reactor design than in the earlier reactor studies (see Table 1). To date, low-power 120-keV beams have been operated in initial experiments with the TFTR beam test facilities at Oak Ridge and Lawrence Berkeley laboratories. In addition, neutral beam injectors at 40 keV with 700 kW of power and capable of almost continuous operation have been produced at Oak Ridge (*23*). Data from these experiments lead us to conclude that we will be able to supply the 20 MW of 120-keV neutral beams required for the TFTR experiment and that the higher-energy beams required for the power reactor can be developed from this technology base.

In addition to the injection of energetic hydrogen particles to heat the plasma, one must also deal with the problem of injecting cold hydrogen to refuel the plasma during the extended burn. The most promising technique proposed for fueling is the injection of pellets of frozen hydrogen at high velocity, greater than 10^3 m/sec (*24*). Recent experiments in which hydrogen-ice pellets were injected into ORMAK demonstrated the basic feasibility of this technique (*25*). In these experiments 30-μm pellets were accelerated in a hypersonic gas jet to roughly 10^2 m/sec. One potential technique for producing the higher velocities and larger pellets needed for reactor-size plasmas is mechanical acceleration in an advanced centrifuge. The first test of this advanced fueling technique is scheduled for 1978 (*26*).

Superconducting magnets, neutral beam heaters, and fueling mechanisms have been singled out for discussion because they represent the most novel technologies that will be required for reactor operation. In addition to these technological developments, there are a host of engineering problems which must be solved for successful operation of the reactor. Over the past few years, studies of tokamak devices following the TFTR have been undertaken in order to uncover these engineering problems (*27, 28*). Major components of a device which would demonstrate all of the characteristics of the fusion core of a power reactor were evaluated with regard to engineering feasibility and cost, and comprehensive systems models were developed so that changes in the operating characteristics of the device could be evaluated in terms of their effect on individual components. These systems models provide us for the first time with tools for evaluating the sensitivity of an overall system design to the precise operating details of the tokamak plasma.

As part of these studies, the development problems which will have to be addressed in order to provide the required device components were evaluated in the context of the range of possible plasma operating parameters. A careful study of the technological requirements to build this fusion core demonstration device in the light of these plasma uncertainties was completed (*29*). The general conclusion of this analysis is that, with a few exceptions, the major technological issues are the subject of active development programs which are scheduled for completion in the early 1980's. The exceptions identified in the analysis are such that they could also be addressed in this period if the relevant programs are initiated soon. Consequently, it appears that with adequate support we will be able, by the early 1980's, to develop the technologies necessary to demonstrate the characteristics of the fusion core of a tokamak power reactor.

Economic Potential

Power generation cost is generally divided into three components: capital cost, fuel cost, and operation and maintenance cost. For example, the light-water fission reactor power generation cost is presently distributed with about 70 percent capital cost, 20 percent fuel cost, and 10 percent operation and maintenance cost. Fission power generation is said to be highly "capital-cost intensive"; that is, the power cost is to a large extent determined by capital-cost investment and to a lesser extent by fuel cost. Fusion reactors will be even more capital-cost intensive: about 90 percent of the power generation cost for fusion reactors will be capital expenditures and about 10 percent will be operation and maintenance; the fuel cost of fusion power should be less than 1 percent of the total. The capital cost (in mills per kilowatt-hour) is derived from the installed plant cost (dollars per kilowatt electric) and the plant capacity factor (the ratio of the average generated load for a period to the total rated capacity of the plant). The capacity factor is a measure of the plant performance which reflects system

Table 2. Representative plant cost estimates for a tokamak and an LMFBR. The estimates are in 1976 dollars and do not include interest during construction or escalation.

	Cost ($/kWe)	
	Tokamak	LMFBR
Reactor systems	450	100
Balance of plant	800	700
Total	1250	800

reliability and maintainability. In this section we will consider the plant costs and the reliability and maintainability of tokamak power systems.

Plant costs. At the outset it is emphasized that estimates of plant costs for power systems based on advanced energy sources must be regarded as approximate and somewhat speculative at this time. Nevertheless, such estimates are instructive in comparative assessments of alternate energy sources.

Based on a model developed at Oak Ridge (*5*), the direct plant cost of the conceptual tokamak power plant described earlier has been estimated to be in the range $1000 to $1500 per kilowatt electric, in 1976 dollars. On the same basis (1976 dollars) the direct plant cost of the LMFBR is estimated to be in the range $600 to $1000 per kilowatt electric (*30*) and that of solar electric plants is estimated to be in the range $1500 to $3000 per kilowatt electric (*31*). It is noted that the *direct* plant cost does not include interest during construction or the effects of escalation. Such factors are usually expressed as percentage increments relative to the direct plant cost and should be independent of the energy source. Thus, current estimates suggest that tokamak plant costs would be somewhat higher than LMFBR plant costs and somewhat lower than solar electric plant costs.

Since a tokamak reactor represents a more extensive set of advanced technologies than does the LMFBR, it might at first appear that tokamak plant costs should be much greater than LMFBR plant costs. In this context it is most illuminating to examine the figures in Table 2, which show representative, estimated direct plant costs for a tokamak and an LMFBR broken down into reactor systems costs and balance of plant costs. The following points are emphasized.

1) The reactor systems cost of the tokamak is greater than that of the LMFBR by about a factor of 4 to 5, reflecting differences in the respective technologies.

2) The balance of plant costs is comparable for both systems.

3) The balance of plant costs is the dominant cost of both systems.

Thus, although the reactor systems cost of a tokamak may indeed be much greater than that of the LMFBR, the total plant costs of the two systems might differ by only ~50 percent.

System reliability and maintainability. Although it is too early in the development of fusion power to project a meaningful value for the plant capacity factor, it is very important to consider the areas of system reliability and maintainability in conceptual design studies. In the initial fusion reactor design studies, reliability and maintainability were examined by extrapolating experience gained from the operation of fission reactors and small-scale plasma confinement experiments. Although such an approach was useful in identifying potential problems, it did not allow for creative approaches to system reliability and maintainability based on the unique features of large fusion devices. As discussed below, recent experimental results in the area of materials performance and current design studies that take advantage of the unique features of fusion systems indicate that the prospects for reliability and maintainability of fusion devices are encouraging.

A major issue associated with the question of fusion power plant reliability is the limitation on the service life of the blanket structure arising from neutron irradiation effects. It is well known from fission reactor experience that deleterious changes in physical and mechanical properties occur in structural materials as a result of neutron-induced atomic displacements and transmutation gas products—particularly helium, which has a very low solubility in the metal lattice. Calculations indicate that, in general, atomic displacement rates will be comparable in the first-wall region of a fusion reactor blanket and in the high-flux region of advanced fission reactors; however, the helium production rates will be much higher in the fusion reactor environment because of the difference in neutron energy spectrum between fusion and fission reactors.

Early projections (*32*) of the performance of stainless steel when used as the structural material of a fusion reactor blanket were based on extrapolations of low-fluence fast fission reactor data. Because of the nature of the neutron energy spectrum of a fast fission reactor, the helium generation rates associated with these data were orders of magnitude too low for simulating fusion reactor conditions. These extrapolations suggested that even without a high helium content, loss of ductility might limit the service life of a stainless steel first wall to about 2 years when it is operated at a temperature of ~500°C. Thus, the early projections of blanket structure performance had ominous implications for fusion power reliability.

In 1975 it was noted (*33*) that mixed-spectrum fission reactors (reactors in which both the fast flux and the thermal flux are high) could simulate the ratio of the fusion displacement rate to the helium production rate very closely in alloys such as stainless steel which contain significant amounts of nickel. Helium is produced by the reaction sequence with thermal neutrons

$$^{58}Ni + n_{th} \rightarrow ^{59}Ni + \gamma$$

$$^{59}Ni + n_{th} \rightarrow ^{56}Fe + ^{4}He$$

Simultaneously, displacement damage is produced by the fast neutrons. Extrapolations of recent data obtained in a mixed-spectrum fission reactor, at high fluence and with high helium production, suggest that the service life of a stainless steel first wall might approach 10 years if it is operated at ~400°C (*34*). The reduction in first-wall temperature does not seriously affect the plant thermal efficiency, and an efficiency of ~35 percent should be possible. Thus, the prospects for extended wall life appear encouraging at this point, and although a major alloy development program will be required to identify and qualify the most suitable blanket structural material, structure life should not adversely affect fusion power reliability.

In previous tokamak reactor design studies, it was generally assumed that the first wall of the blanket would also serve as the major vacuum boundary between the plasma and atmospheric pressure. This usually required that the first wall contain hundreds to thousands of lineal meters of welds. Should an operating failure such as a pinhole leak develop in a radioactive first wall, it is doubtful that it could be repaired without unreasonable difficulty. Therefore, in the reference power plant it has been proposed that the tokamak reactor system be enclosed in a vacuum building. This completely changes the character of the first-wall surface from a welded structure with absolute vacuum integrity to one that can have mechanical joints and only requires high pumping impedance (it can be slightly leaky because the pressures on both sides are about the same). It is believed that such an approach has significant assembly, disassembly, and repair advantages over the vacuum first-wall approach.

Vacuum enclosures of the physical size needed to house a fusion reactor are structurally reasonable, as demonstrated by the Plum Brook facility of the National Aeronautics and Space Administration (NASA). A cross section of that facility is reproduced in Fig. 5, which also shows the approximate dimensions of the reference tokamak reactor. The facility volume and the radiation shielding provided by the concrete structure appear to be consistent with the reactor requirements. It is also noted that such a building is comparable in size to the containment building of fission reactors.

To further minimize downtime and facilitate maintenance, a modular approach has been taken for the blanket design, which eases the problems of remote maintenance. Small, easily replaced blanket modules have been emphasized (see Fig. 6) to eliminate the need for remote maneuvering and welding of massive components in tight quarters.

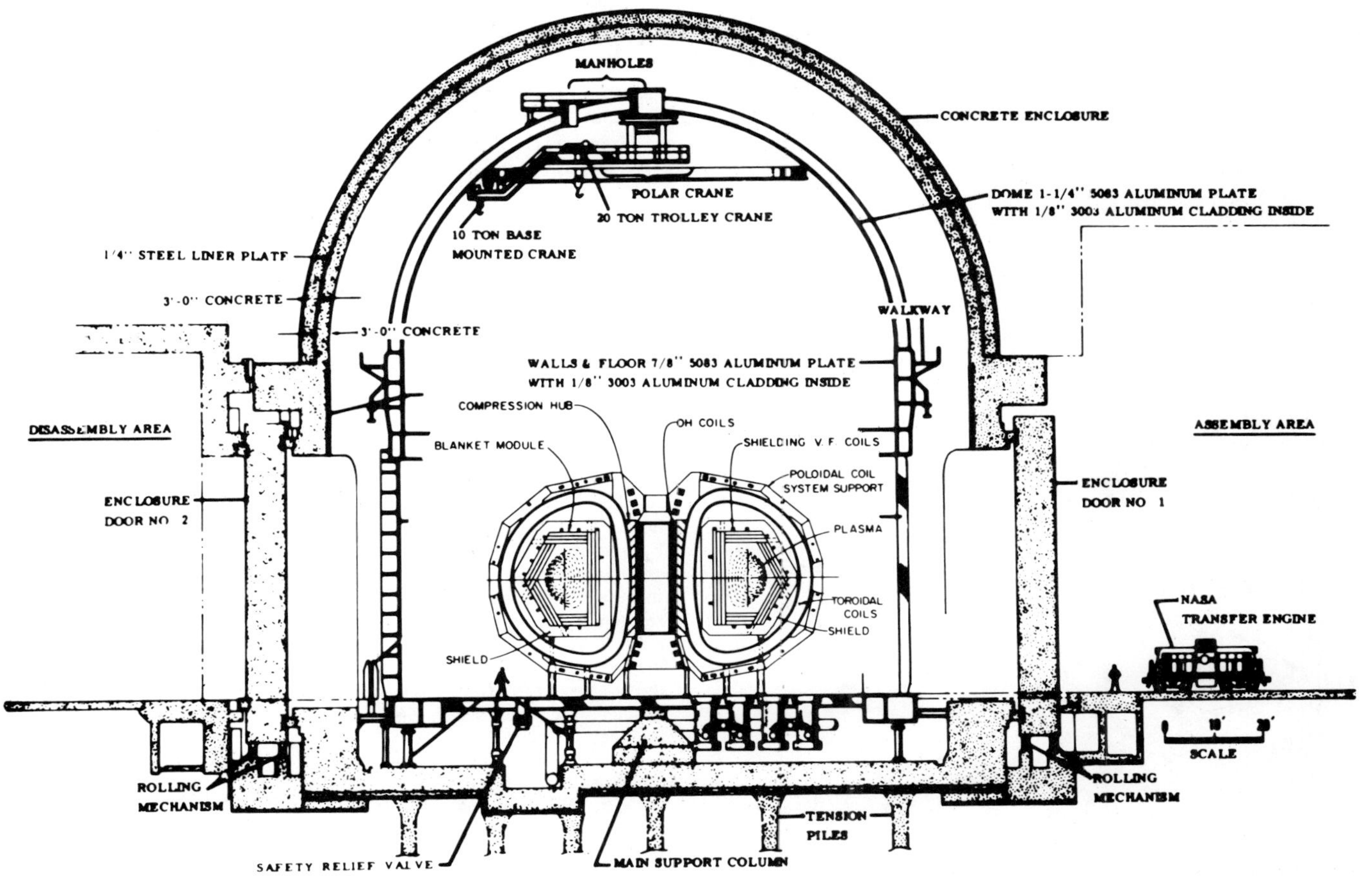

Fig. 5. The ORNL reactor design superimposed on a cross section of the NASA Plum Brook vacuum facility at Sandusky, Ohio (*36*). A vacuum building of the approximate size needed to house a fusion reactor is structurally reasonable. Such a building is comparable in size to the containment building of fission reactors. As discussed in the text, it is believed that a vacuum building approach would enhance the reliability and maintainability of fusion reactors.

Strategy to Demonstrate Power Generation

Initial estimates of the facility and cost requirements for demonstrating tokamak power generation reflected the large size of the early reactor designs and the concerns about first-wall service life. When these early designs (plasma radius, ~5 m) were compared with the TFTR (plasma radius, ~1 m), it appeared that a succession of increasingly larger and more expensive devices would be required to reach reactor-size systems. At the same time, the pessimistic projections about the performance of stainless steels as first-wall materials suggested that a very extensive and expensive materials program would be needed to identify a first-wall material. As part of this program, several expensive high-energy neutron-irradiation facilities would have to be developed and built.

As a result of the continued progress in understanding tokamak physics, the projected plasma radius of power reactors is now about 1.5 m, or within a factor of 2 of that of TFTR. Moreover, recent data on the performance of stainless steels as first-wall materials are encouraging, and it appears that much of the irradiation testing could be accomplished in existing mixed-spectrum fission reactors, although some high-energy neutron facility (*35*) will be required for correlation purposes.

In view of the current situation, we believe that the facility and cost requirements for demonstrating tokamak fusion power can be significantly reduced compared to those reflected in the initial planning exercises. In particular, we suggest that a committed-site, multiple-unit strategy be adopted to demonstrate power generation after the operation of DOUBLET III and the TFTR. This strategy would consist of three phases: (i) a fusion reactor core demonstration, (ii) a power technology demonstration, and (iii) a commercial prototype demonstration. These phases are described below.

During the reactor core demonstration phase, a central pulsed electrical plant would be built to provide pulsed power for all the units. Concurrently, a single tokamak would be built and connected to the central pulsed electrical plant. The device would consist primarily of a torus, a nonbreeding blanket, a shield, toroidal field coils, poloidal field coils, neutral beam injectors, and a divertor system. The scope of this phase is to establish a controlled fusion energy source.

During the power technology demonstration phase, lithium would be introduced into the blanket and a heat transport system, a turbine system, and a tritium recovery system would be added to the basic facility. Electrical power produced during this phase could be fed into a commercial grid. The scope of this phase is to establish technical feasibility.

During the commercial prototype demonstration phase, improved units (for example, ~750 MWe per unit) would be added and tied into the central pulsed electrical plant. The purpose of this phase is to demonstrate system reliability under practical utility conditions. The scope of this phase is to establish economic feasibility.

The strategy outlined above is motivated by the following considerations.

1) The plasma size requirements are essentially the same for the reactor core demonstration device and the commercial prototype. A succession of increasingly larger devices is not required to demonstrate power generation.

2) Capital equipment is conserved during all phases of the demonstration program.

3) Increasing amounts of power are supplied to the electrical grid as each additional unit is added at the site.

The facility cost for the three phases is estimated to be between $2 billion and $3 billion (in 1976 dollars). This does not include engineering or contingency costs, nor does it include indirect capital costs. In addition, it is noted that development costs would accompany the demonstration program. The total program cost including engineering, contingency, and development, but excluding escalation, is estimated to be ~$5 billion to $8 billion (in 1976 dollars), or ~$10 billion to $15 billion if escalation is included. The escalated figures are based on preliminary projections which suggest that prototype operation could commence approximately 20 years after a commitment had been made to the three-phase demonstration program. It appears that a $10 billion to $15 billion expenditure represents an acceptable cost for developing a new energy source.

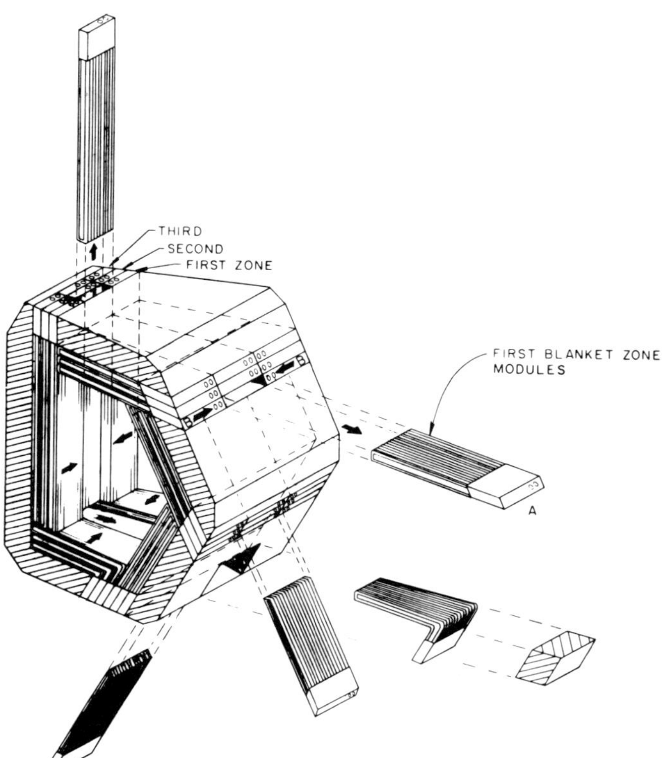

Fig. 6. Remote maintenance of the blanket is receiving major emphasis in current reactor designs. This diagram shows remote removal of individual blanket modules that have been damaged by radiation. In this arrangement there is a five-step sequence to remove the middle module (*A*). The outer modules (*B*) are then moved to the free space and the sequence is repeated. Such a procedure could be carried out without having to move the toroidal field coils.

Concluding Remarks

The tokamak power plant described in this article is based on a specific set of assumed technological directions, design approaches, and plasma characteristics. It is emphasized that there is no unique set of technologies, engineering approaches, and plasma characteristics that will lead to commercial fusion power. Several acceptable sets no doubt exist, and the set described here should be viewed in this context.

The technology base required to realize the reference plant described here is, for the most part, being addressed in ongoing programs. At this point it appears that uncertainties about tokamak fusion power are primarily related to the expected plasma physics performance and not to foreseeable limitations in the areas of technology and engineering. Within the next 3 to 5 years these uncertainties should be resolved.

A three-phase strategy for demonstrating fusion power generation at a committed site has been proposed in this article. It is emphasized that this is a strategy and not a detailed plan. Nevertheless, the strategy outlined here suggests that tokamak fusion power could be demonstrated with reasonable expenditures of money.

References and Notes

1. *Workshop on Fusion Reactor Design Problems* [International Atomic Energy Agency (IAEA), Vienna, 1974]; *Nucl. Fusion*, special supplement (1974).
2. C. P. Ashworth, paper presented at the Atomic Industrial Forum Conference on Energy Development: Probing the Possible, 11 February 1976.
3. W. D. Metz, *Science* **192**, 1320 (1976).
4. ——, *ibid.* **193**, 38 (1976).
5. D. Steiner *et al.*, *ORNL Fusion Power Demonstration Study: Interim Report* [ORNL/TM-5813, Oak Ridge National Laboratory (ORNL), Oak Ridge, Tenn., 1977].
6. D. Rose and M. Feitag, *Technol. Rev.* **79** (No. 2), 21 (1976).
7. H. P. Furth, *Nucl. Fusion* **15**, 487 (1975).
8. R. A. Parker, *Bull. Am. Phys. Soc.* **20**, 1392 (1975).
9. L. A. Berry *et al.*, in *Proceedings of the 6th Conference on Plasma Physics and Controlled Nuclear Fusion Research* (IAEA, Vienna, 1977), vol. 1, p. 49.
10. J. W. M. Paul *et al.*, in *ibid.*, vol. 2, p. 269.
11. M. Murakami, *Bull. Am. Phys. Soc.* **21**, 509 (1976); D. L. Jassby, D. R. Cohn, R. R. Parker, *Nucl. Fusion* **16**, 1045 (1976).
12. F. L. Hinton and R. D. Hazeltine, *Rev. Mod. Phys.* **48**, 239 (1976).
13. L. A. Berry, J. F. Clarke, J. T. Hogan, *Phys. Rev. Lett.* **32**, 362 (1974).
14. B. B. Kadomtsev and O. P. Pogutse, *Nucl. Fusion* **11**, 67 (1971).
15. J. F. Clarke, *EPR and TNS* (ORNL/TM-5418, ORNL, Oak Ridge, Tenn., 1976).
16. D. R. Cohn *et al.*, *Trans. Am. Nucl. Soc.* **26**, 61 (1977); *ibid.* **27**, 25 (1977).
17. J. F. Clarke, D. J. Sigmar, R. A. Dory, Y.-K. M. Peng, *Phys. Rev. Lett.* **38**, 70 (1977).
18. O. Gruber and R. Wilhelm, *Nucl. Fusion* **16**, 243 (1976).
19. G. Bateman and Y-K. M. Peng, *Phys. Rev. Lett.* **38**, 829 (1977); A. M. M. Todd, M. S. Chance, J. M. Greene, R. C. Grimm, J. L. Johnson, J. Manickam, *ibid.*, p. 826.
20. M. Murakami *et al.*, *ibid.* **39**, 615 (1977).
21. M. S. Lubell, in *Proceedings of the Fifth International Cryogenic Engineering Conference* (IPC Science and Technology Press, London, 1974), p. 164.
22. J. A. Rome, Y.-K. M. Peng, J. A. Holmes, *Injection Heating Scenarios for TNS* (ORNL/TM-5931, ORNL, Oak Ridge, Tenn., 1977).
23. H. Haselton *et al.*, in *Proceedings of the International Conference on Plasma Science* (Institute of Electrical and Electronics Engineers, New York, 1977), p. 134.
24. L. Spitzer, Jr., D. J. Grove, W. E. Johnson, L. Tonks, W. R. Westendrop, *USAEC Rep. N.Y.O.-6047* (1954).
25. C. A. Foster, R. J. Colchin, S. L. Milora, K. Kim, R. J. Turnbull, *Solid Hydrogen Pellet Injection into the ORMAK Tokamak* (ORNL/TM-5897, ORNL, Oak Ridge, Tenn., 1977).
26. S. L. Milora, C. A. Foster, G. D. Kerbel, *Bull. Am. Phys. Soc.* **29**, 1052 (1976).
27. M. Roberts, *Oak Ridge TNS Program—1976* [ORNL/TM-5275 (excerpts), ORNL, Oak Ridge, Tenn., 1977].
28. *TNS Scoping Studies* (GA-A14, General Atomic Co., San Diego, Calif., and ANL/FPP/77-2, Argonne National Laboratory, Argonne, Ill., 1977).
29. M. Roberts, *Draft Program Plan for TNS—The Next Step after the Tokamak Fusion Test Reactor, Part I—Summary* (ORNL/TM-5982, ORNL, Oak Ridge, Tenn., 1977); Oak Ridge TNS Program Staff, *Draft Program Plan for TNS—The Next Step after the Tokamak Fusion Test Reactor, Part II—R & D Needs Assessment* (ORNL/TM-5983, ORNL, Oak Ridge, Tenn., 1977).
30. M Levenson, P. M. Murphy, C. P. L. Zaleski, *Nucl. News* **19**, 54 (April 1976).
31. J. C. Powell, *Dynamic Conversion of Solar Generated Heat to Electricity* [NASA-CR-134724 (vol. 1) and NASA-CR-134723 (vol. 2), National Aeronautics and Space Administration, Washington, D.C., 1974].
32. G. L. Kulcinski, R. G. Brown, R. G. Lott, P. A. Sanger, *Nucl. Technol.* **22**, 20 (1974).
33. F. W. Wiffen and E. E. Bloom, *ibid.* **25**, 113 (1975).
34. E. E. Bloom *et al.*, *ibid.* **31**, 115 (1976).
35. A. N. Goland, D. H. Gurinsky, J. Hendrie, J. Kukkonen, T. Sheehan, C. L. Snead, Jr., in *Proceedings of the International Conference on Radiation Tests Facilities for the CTR Surface and Materials Program* (ANL/CTR-75-4, Argonne National Laboratory, Argonne, Ill., 1975), p. 354.
36. *Space Power Facility: Description and Capabilities* (NASA-Lewis Research Center, Plum Brook Station, Sandusky, Ohio, 1974).

37. We are pleased to acknowledge the assistance of Carolyn Krause of the ORNL Information Division for her help in the preparation of this manuscript. Sponsored by the Department of Energy, Division of Magnetic Fusion Energy, under contract with Union Carbide Corporation.

Engineering Limitations of Fusion Power Plants

Problems related to radiation damage and plant costs may prevent the practical application of fusion.

W. E. Parkins

During the past few years, increasing attention has been directed to preliminary engineering studies of possible fusion reactor power plants. These studies have employed the best available theoretical extrapolations of results of confinement experiments in order to predict operational parameters for the power-producing plasma. While no net power-producing fusion plasma has yet been demonstrated, the engineering studies are important in providing guidance to the development program and in pointing out significant practical problems to be faced, once it is learned how to achieve useful thermonuclear reacting conditions. This article does not deal with the difficulty or probability of success of plasma confinement, but instead focuses on engineering aspects of proposed full-scale plants believed to be of critical importance to the future of fusion power.

One problem area that has been stressed is the difficulty of processing and containing the tritium in a fusion plant employing the deuterium-tritium (D-T) reaction. This subject was evaluated in a section of the first report of the Atomic Industrial Forum Committee on Fusion (*1*). The conclusion was that the technology is available to meet the operational requirements and that the principal concern is the impact that plant design features for tritium handling might have on total capital costs. Another problem area treated was related to the acceptability of plant operation from the environmental and safety standpoints. Again, it was concluded that fusion plants will be able to meet all environmental and safety requirements. The greatest difficulty appears to be that of adequately limiting the release of tritium during normal plant operation and as a result of postulated accidents. This then reflects back on the plant design features that will ensure adequate tritium containment, and the effect on capital costs again becomes a principal point of concern.

There are other features inherent in a fusion reactor plant that will force increases in the cost of the initial installation, and the magnitude of the total investment is recognized to be a problem of critical importance to the eventual successful application of fusion power (*1*). In this article some of the engineering factors bearing on capital costs will be evaluated. A second very serious engineering problem area discussed is that of the limited operating life of the reactor vessel, caused by the deleterious effects

The author is Director of Research and Technology at Atomics International, Canoga Park, California 91304.

of the radiation to which it is subjected. Periodic replacement of this vessel will unfavorably affect plant availability and operating costs (*1*).

Most of the conceptual studies of full-scale fusion power plants have been based on the tokamak-type reactor. It is also used for the purpose of illustration here, but it should be noted that most of the engineering considerations treated apply to other fusion plasma confinement methods, whether magnetic or inertial. Reference is made here to the preliminary designs prepared by the University of Wisconsin Fusion Feasibility Study Group. Their series of tokamak reactor plants, designated UWMAK I, II, and III, represent the most thorough effort yet completed to evolve a workable fusion power plant design.

Summary. If conditions for a net power-producing thermonuclear reaction are ever demonstrated, difficult engineering problems must still be overcome. Two such obstacles to any practical application of fusion power are the magnitude of the plant capital cost and the limited lifetime of the reactor vessel. Among the factors contributing to the high initial cost is the constraint heat removal places on reactor size. This results from a fundamental engineering disadvantage of the fusion concept. The problem of limited reactor vessel operating life is inherent in the use of thermonuclear reactions, such as deuterium-tritium, which release damaging energetic neutrons.

Impact of Plant Capital Cost

It must first be recognized that the overriding figure of merit that will determine the acceptability of fusion power plants by utilities will be the competitiveness of the net cost of electrical energy produced. The history of the industry bears out that no other consideration can compensate for an appreciably higher cost of power. Then since the total cost of power produced by a fusion plant would be almost entirely contributed by charges against the capital investment, the future successful large-scale application of fusion in central station electrical generation will be critically dependent on these initial plant costs.

Concern that the magnitude of capital costs might rule out any practical application of fusion, even if and when its feasibility has been demonstrated and operating experience has permitted engineering optimization of plant designs, has been expressed by representatives of the electric utility industry. This and other problems facing the fusion development program have been discussed in a series of articles by Metz (*2*). The need to move in the direction of less expensive plant designs had already been recognized by many workers in the field, and a number of ideas exploring this possibility have been brought forth (*3*).

The portion of the fusion power plant having the greatest capital cost impact, and the portion representing the greatest design unknown, is the so-called nuclear island. It includes the reactor vessel, all internals, vacuum systems, neutron absorbing blanket, shields, all magnet and coil systems, neutral beam injectors or other auxiliary means of plasma heating, devices for fueling, means of impurity control and "ash removal," reactor instrumentation and control systems, and associated remote handling equipment. Whatever the design features of all of these components entail, it is apparent that the most effective approach to cost reduction would be through overall size reduction.

To illustrate, the UWMAK I, with an output of 1500 megawatts electric (MWe) (*4*), required the following amounts of type 316 stainless steel for structure (in metric tons): initial vessel, blanket, and shield, 10,000; vessel and blanket replacements; 21,000; and magnets, 19,000 —giving a total of 50,000 metric tons. At today's prices this amount of type 316 stainless steel, fabricated and field-installed, would by itself exceed the total cost of a present-day fossil-fueled or fission power plant of equivalent output.

By switching from lithium to helium as the primary coolant in the UWMAK II design (*5*), the power was increased from 1500 to 1700 MWe without a change in geometric size. A greater step toward higher power per unit size was taken with the next design, UWMAK III (*6*). Although its output was 2000 MWe, the major radius of the vessel was decreased from the 13 meters used in the earlier designs to about 8 meters, and the minor radius from an average of approximately (noncircular cross section) 5½ to 4½ meters. This smaller size permitted, for example, a reduction in the weight of structural metal in the vessel, blanket, shield, and magnets to about one-third that required in the earlier designs. The reduction in present-day cost of the structural metal would not be proportional, however, since a shift was made from type 316 stainless steel to the molybdenum alloy TZM for the reactor vessel and all hot-coolant ducting.

While the UWMAK III represents the most recently completed and cost-estimated preliminary design of a full-scale tokamak-type fusion power plant, other groups of investigators have been giving consideration to the possibilities of higher power density and more compact configurations. These invariably call for more difficult design of the magnets and the achievement of more difficult conditions in the reacting plasma. It is hoped that the feasibility of these directions can be tested in one or more of the various pilot experimental tokamak power reactors now being proposed.

Effect of Heat Removal Constraint

An aspect of fusion power complicating heat removal is that no heat transfer surfaces can be used within the reacting plasma. All of the energy released by the thermonuclear reaction must pass as heat or radiation through an envelope that surrounds the plasma region. As will be discussed later, since there are practical limits on the amount of power per unit area that can be transferred in this manner, there are then limits on the minimum physical size of the reactor that can be used for a particular plant power output.

Since reducing the size of a fusion reactor for a particular power level is an effective means of lowering the capital cost per unit output, it is important to examine the constraints on such size reductions. As power density is increased the physics of confining a plasma becomes more difficult and, for any confinement concept, a limiting practical condition will eventually be reached. For the purpose of this discussion I will assume that any required plasma conditions can be achieved. A second difficulty that is made more severe by size reduction for a particular power output is radiation damage to the vessel and adjacent components. A third difficulty is related to the problem of usefully removing the heat produced by the reactor, and this is examined in more detail in the paragraphs that follow.

For any fusion reactor employing the D-T reaction, most of the energy released will be carried into the surrounding blanket by the fast neutrons produced. This energy, plus more resulting from nuclear interactions in the blanket and shield regions, will contribute typically about 80 percent of the total heat produced. Since this heat is deposited throughout a large volume, it is not very difficult to remove by use of a circulating coolant. Unfortunately, while this heat removal feature represents an outstanding advantage of the use of the D-T reaction, the neutrons carrying the energy to the blanket are also the principal cause of the radiation damage effects that limit the operating lifetimes of the vessel and adjacent components. The balance of all of the energy released in the D-T fusion reactor, approximately 20 percent, must be absorbed by the vessel walls or be deposited in a divertor. Some designs employ a magnetic field configuration that provides paths along magnetic field lines for ions diffusing out of the plasma, and for unwanted ions that might otherwise enter the plasma, to be swept out of the reactor vessel. These ions are then guided along the field lines to a divertor collector surface where they give up their energy.

For example, the three UWMAK plants were designed *(4-6)* for a nominal power of 5000 MW thermal during "plasma burn." In UWMAK I the reactor vessel received 700 MWt (about 560 at its inner surface and 140 from internal heating), resulting in a requirement for removing an average heat flux of 0.25 MWt/m^2. This could be handled by the flowing lithium coolant, although magnetohydrodynamic effects resulting from interaction with the high magnetic field that was present made the problem difficult. The divertor collector, in the form of a flowing liquid lithium film within the vacuum, was estimated to receive an average heat flux of 1 MWt/m^2 (equivalent to 317,000 Btu's per hour per square foot). As an average heat flux into flowing lithium, this appeared manageable.

At the time of the UWMAK II design, new information indicated that a higher rate of diffusion of particles from the plasma should be expected. As a result, radiation to the inner surface of the vessel was decreased and the heating load into the divertor was increased. Including internal heat generation, the vessel heat flux averaged about 0.08 MWt/m^2, and the liquid surface of the divertor collector was estimated to receive an average of 3 MWt/m^2. The pressurized helium employed as the UWMAK II vessel coolant appeared adequate, but the same was not true of the divertor coolant. Since the charged particles swept from the plasma along magnetic field lines to the liquid metal surface could not be expected to distribute themselves uniformly over the surface provided, the University of Wisconsin group estimated that in this case a peak-to-average ratio of as much as 10 might be expected. It was pointed out that dealing in this situation with heat fluxes of up to 30 MWt/m^2 represented an unsolved problem.

In the UWMAK III preliminary design a similar basis for diffusion from the plasma was followed, but the plasma power density was increased to reduce the size of the reactor. A cross section of the UWMAK III torus is shown in Fig. 1. A 25-centimeter-thick graphite barrier was used near the inside surface (nearest the axis) of the torus to reduce the flux of high-energy neutrons arriving there, and thereby extend the operating lifetime of that part of the vessel. The energy deposited by the neutrons in the barrier appeared as heat, which was radiated with a nonuniform distribution to the walls of the vessel. That heat, plus some due to radiation from the plasma, plus internal heat generation in the vessel gave a total average heat flux of about 0.4 MWt/m^2. This heat was removed at a very high temperature (approximately 1000°C) by pressurized helium flowing in a dense pattern of TZM tubes, thereby increasing the heat removal surface area and reducing the heat flux to an effective average there of about 0.25 MWt/m^2 (~80,000 Btu/hour-ft^2). This is slightly higher than the average heat flux that would be used in a helium-cooled graphite-moderated fission reactor power plant.

For the UWMAK III divertor collector, the lithium vapor pressure in the reactor vessel would be excessive if such a high heat flux of charged particles were deposited directly in a flowing film of the liquid metal. It was therefore decided that the divertor ion stream had to be handled differently than was done in UWMAK II. The charged particles were carried out of the high magnetic field region and made to strike a thin replaceable TZM surface cooled on the back side by flowing sodium. The calculated average heat flux was 6 MWt/m^2, and whether such a high value could be successfully managed is questionable. Certainly, designing for the peak conditions, which would exceed the average, would appear impossible.

It is apparent that satisfying the requirements for peak heat flux removal will be an important problem in the engineering of fusion power plants. The peak conditions will exceed the average for reasons that are inherent in the design and because of abnormal operating conditions that may occur. Both must be anticipated, and the cooling system features must be made adequate to ensure that no melting will result from transient conditions and no deleterious effects will result from excessive local temperatures over long operating periods.

Another related and important aspect of the engineering of the heat removal system is the treatment of "hot channel factors." For whatever heat removal requirement exists locally, the design must be made sufficiently conservative to meet the worst conditions that could realistically occur. Variations in manufacturing tolerances, fouling of heat transfer surfaces, inaccuracies in calculations of

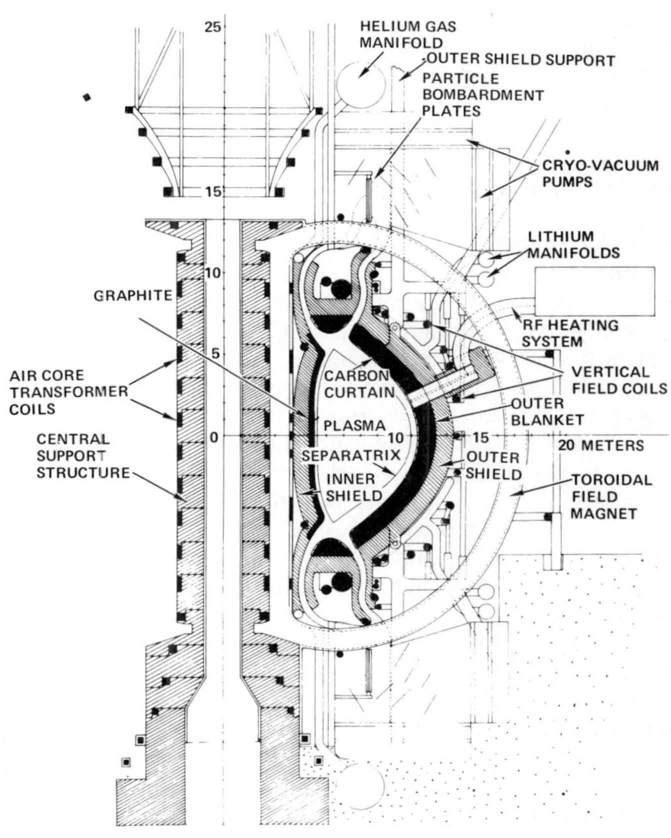

Fig. 1. Schematic cross section of the UWMAK III reactor torus.

magnetohydrodynamic effects on coolant flow, and other departures from ideal conditions will reduce the design value of the maximum coolant temperature in all coolant passages. Engineering consideration of these hot channel factors and of variations in the peak-to-average heat fluxes results in reducing the mixed-mean coolant outlet temperature. This, in turn, has a direct effect in reducing the plant power conversion efficiency and net output and in increasing the plant capital cost per unit output. None of the fusion plant preliminary designs have yet attempted to account for the effect of these considerations on plant size per unit output.

This discussion of heat removal considerations and the requirements of the UWMAK reactors illustrates the nature of the engineering problems involved. Any fusion plant employing the D-T reaction must be designed to remove a minimum of about 20 percent of the total reactor thermal power at the vessel wall and the divertor combined. (Actually, in UWMAK III, because of the effect of the partial internal barrier, the figure was approximately 27 percent.) As a practical matter, heat removal at a divertor is unlikely to permit any reduction in scale because of the difficulty of providing extended surface area for its ion collector. Regardless of the design selected, the coolant system for the vessel and any divertor must be capable of handling the peak heat flux conditions and accommodating all hot channel factors. These engineering considerations inevitably result in lowering the overall plant design performance.

If the thermal power is fixed and the reactor size is considered to be made smaller, a condition is reached where the heat removal requirement is limiting. In order to determine what minimum fusion reactor size for a particular power output is permitted by this heat removal constraint would require optimizing a design that provides for maximum and adequate peak heat removal at the vessel wall and at the divertor. While the UWMAK III heat removal system did not meet this goal, the UWMAK III power per unit size may well represent a maximum practical value. Its vessel surface area was 1600 m^2, and 20 percent of the 5000-MW thermal power would represent an average heat flux at the vessel of 0.62 MWt/m^2 (200,000 Btu/hour-ft^2), assuming no neutron barrier to add to the heat load and no power extracted by a divertor. This figure is typical of the average heat flux in the core of a pressurized-water fission power reactor.

Costs Dependent on Reactor Vessel Area

The importance of the parameter of minimum reactor vessel size for a particular power output results from the fact that the vessel area directly determines the required area of the blanket and shielding, and even determines the size of the various coils that produce the magnetic fields. These components must be external to the reactor vessel. To absorb the fast neutrons from the plasma and reduce external radiation intensities to sufficiently low levels, the blanket and shielding must be 1½ to 2 m thick. They are composed of expensive materials and depend on costly fabrication for their structure and encapsulation. As a rule, the less thick the total layers of blanket and shielding, the more expensive per unit volume are the materials of which they are made.

Again for illustration, the UWMAK I design (*4*) of blanket and shielding outside the vessel called for a 152-cm-thick structure consisting of 14 different zone layers composed of varying mixtures of stainless steel, lithium, lead, and boron carbide (Fig. 2). The UWMAK III (*6*) outer blanket and shielding regions were 194 cm thick, having nine zone layers with different fractions of stainless steel, molybdenum, lithium, graphite, alumina-silica, and boron carbide (Fig. 2). Openings through these layered structures would have to be provided (without permitting streaming of neutrons or other radiation) for the vacuum and divertor systems, equipment for fueling, instrumentation, coolant manifold, and neutral beam injectors or other equipment for plasma heating. It is also anticipated that the design would have to incorporate provision for remote access to the vessel interior for leak detection and repair. The final design will require not only large but complex structures. They will have to be field-assembled and are unlikely to lend themselves to the use of automatic welding equipment. To ensure that air will be kept out and tritium will be kept within the blanket, vacuum leak-tight welding subject to special standards and inspection techniques will be required.

It follows that any tokamak-type fusion power reactor will suffer a significant capital cost penalty because of size, complexity, and choice of materials for the vessel, blanket, shielding, and surrounding magnet coils. There are engineering restrictions on the minimum size per unit power output that will apply to these components. At some value of this parameter, heat transfer considerations will determine a limitation. Increasing the scale to permit higher plant power output will not decrease this penalty because the volumes of the components will increase in proportion to the power produced. The implication of the relationship just stated is that fusion power plants will become less competitive with other types of power plants as total output is made larger. The large cost contribution of the vessel, blanket, shielding, and magnet coils will vary approximately linearly with power output, not to some exponent less than one.

All fusion reactor concepts have the fundamental engineering disadvantage that the energy released cannot be extracted from within the plasma but must be collected by some means at the periphery of the reacting volume. This disadvantage might not be so serious were it not for the fact that for technical reasons the minimum periphery represents a sizable area per unit power output and requires a large capital investment per unit area. This is quite in contrast to power plants based on the combustion of fossil fuels or the nuclear fission process. In both fossil-fueled furnaces and fission reactors, heat can be extracted from within the reacting volume and the cost of the boiler tubing or fuel element cladding per unit area is small. In comparison, the heat transfer surface per unit area represented by the vacuum vessel in a fusion reactor is extremely expensive. As described above, every square meter of such a surface in a reactor based on the D-T reaction requires 1½ to 2 m^3 of expensive materials.

Capital Cost Estimates

The preceding paragraphs gave qualitative reasons why the capital cost of a fusion plant, particularly the nuclear island, will be high. A more quantitative argument was presented in an earlier paper (*7*) in which the cost of a tokamak-type fusion power plant was estimated. As an example the UWMAK I vessel, blanket, shield, and magnet design was used, but the volume enclosed by each of these components was reduced to correspond to the reactor vessel size with the thermal power output of the UWMAK III. This is a favorable case in that the choice of materials and component thicknesses in the UWMAK I comprises a low-cost design, and the thermal power-to-size ratio of the UWMAK III represents what was just described as a practical maximum based on heat transfer considerations.

The estimate was based on unit costs for large-scale procurement and field installation to required standards typical of current nuclear industry equivalent experience. Other direct and indirect costs were based on the UWMAK III estimate prepared by the University of Wisconsin group and the Bechtel Corporation (*6*). The result was a total estimated cost (first-quarter 1977 dollars) of $8590 million for a plant with an average design output of 1750 MWe (*7*). Contingency and interest during construction, but no escalation or special owner costs, were included. Also, the cost of replacements for the vessel and adjacent components over the life of the plant was not included. The plant was assumed to be a first-generation one but not the first of its kind constructed.

A revised estimate for the same fusion power plant has been subsequently prepared with the benefit of more detailed costs for the liquid-metal heat transfer system, steam generator, turbine generator, and entire power conversion installation compiled for the prototype liquid-metal fast breeder reactor (LMFBR) plant (*8*). These costs were scaled according to the 0.7 power of the ratio of electrical outputs (1750 MWe divided by 1000 MWe for the LMFBR). Similarly, the indirect costs were revised by scaling those from the UWMAK III estimate according to the 0.7 power of the ratio of total direct costs (the new direct cost figure divided by the UWMAK III direct cost estimate). These indirect costs include construction facilities and services as well as engineering services. The result, after adding in contingency and interest during construction on the same percentage basis as in the earlier estimate, was a total construction cost of $7780 million.

This revised total cost estimate is equivalent to $4450/kWe. Neglecting operating and maintenance costs, the electrical energy produced would cost 10.8¢/kWh, assuming a plant factor of 0.8 and charges of 17 percent per year against the capital investment. For comparison, capital costs for coal-fueled plants with stack gas desulfurization and for light-water fission reactor plants are $780 and $975/kWe, respectively (*7*). While variations in location, plant size, and other special factors result in a range of capital costs, these figures are typical for plants ordered, assuming 1977 dollars without escalation or special owner costs.

A comparison can also be made with the capital cost of an LMFBR plant, using the recent design study for a 1000-MWe prototype station (*8*). Including a 30 percent contingency and an average interest of 8 percent per year for 3 years during construction, which were employed in the fusion plant estimates, gives a total construction cost of $1820/kWe for the LMFBR. In view of the relative status of fusion and fission breeder technologies, use of the same contingency factor should weight the comparison in favor of the fusion plant.

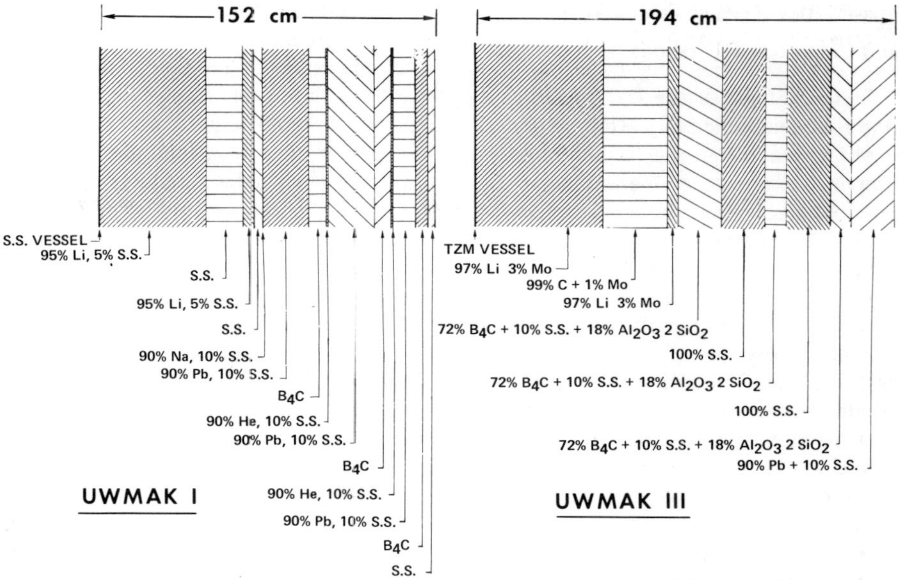

Fig. 2. Materials and their thicknesses comprising the blanket and shield regions surrounding the reactor vessels in the UWMAK I and UWMAK III designs.

Limitations Imposed by Radiation Damage

The second problem area regarded as critical to any successful application of fusion power is related to the integrity of the reactor vessel. There are a number of difficult engineering requirements which this component must satisfy. The vessel must maintain vacuum tightness, operate at elevated temperature, withstand repeated thermal cycles and stresses from external pressure and nonuniform temperature distributions, be corrosion-resistant to the primary coolant and its impurities, retain adequate mechanical properties and dimensional stability while subjected to intense radiation, be capable of field fabrication, and be available in large quantity at an economic price. While each of these performance demands limits the design choices, the most serious requirement is that the vessel retain adequate mechanical properties in spite of the effects of a high fluence (time-integrated flux) of energetic neutrons.

The principal difficulties caused by the high-energy neutron interactions are loss of ductility and swelling in the vessel material. These effects are sufficiently serious that all engineering studies of fusion power plant designs based on the D-T reaction have concluded that periodic replacement of the reactor vessel would be required. Some recent experimental results (*9*) have indicated that the frequency of replacement could be reduced by using cold-worked type 316 stainless steel and operating at a maximum temperature of 300°C. It is possible that a total fast neutron fluence equivalent to 20 MW-year/m^2 could be achieved with such a design. Unfortunately, the need to keep the reactor size as small as possible for a particular power level, to minimize capital costs, results in an intense neutron flux. For example, the UWMAK III thermal power level and size would give rise to an average fast neutron flux of 2 MW/m^2. Designing for a lower maximum vessel wall temperature also has the disadvantage of reducing the plant electrical output.

It is a basic difficulty of the fusion concept that the component that must meet the most stringent design conditions, the reactor vessel, is necessarily the component that is subjected to the most intense damaging radiation. Its replacement is made difficult because the induced radioactivity requires the use of remote means. This difficulty is made more extreme by the fact that the vessel is necessarily designed into the systems for heat removal, vacuum pumping, fueling, plasma heating, and other functions. Furthermore, the vessel is almost inaccessible because it must be completely enclosed by a sufficient thickness of neutron-absorbing blanket and radiation shielding. And, in a tokamak configuration, it is further enclosed within a set of surrounding and interlocking magnet coils.

Possibilities of Other Fusion Reactions and Confinement Methods

Since any use of the D-T fusion reaction is accompanied by certain major difficulties, such as the requirement to breed tritium and the inherent high radiation damage rates, the question of the possibility of employing other fusion reactions invariably arises. Investigators studying this prospect are generally agreed that the tokamak-type reactor is probably incapable of achieving the necessary plasma conditions for any other fusion process. The next least difficult fusion reaction would be D-D. While it would entail no requirement for breeding, there would still be a significant production of high-energy neutrons from the plasma. The limitation that heat removal by the combination of vessel and divertor places on minimum reactor size would be more severe. With the D-D reaction, approximately 60 percent rather than 20 percent of the total thermal power would have to be removed there. Other advanced fusion fuel cycles are not considered feasible with the tokamak-type reactor.

Although this discussion of engineering aspects of possible future fusion power plants has been directed to the use of the tokamak concept, many other approaches to the controlled thermonuclear reactor are being investigated. They include methods based on other magnetic plasma confinement schemes and on concepts dependent on inertial confinement. If the fusion reaction employed is to be D-T, essentially all of the foregoing discussion on engineering problems still applies. The effect of radiation damage in limiting the operating lifetime of the reactor vessel and its adjacent components would be present. A minimum vessel size per unit power output would be established by heat transfer limitations, and a significant plant capital cost would be contributed by the volume of vessel, blanket, and shielding. Inertial confinement methods can avoid the expense of surrounding magnet coils, but they would have their own special equipment, such as a high-power laser system, to provide for pellet heating.

Next, we must ask whether one of these alternative plasma confinement methods might permit the use of an advanced fuel cycle that would avoid the engineering problems inherent in the D-T and even the D-D fusion reactions. It is true that some of the proposed fusion fuel cycles, particularly any based on reactions producing only charged particles, could avoid the fast neutron radiation damage problem. The requirement for a breeding blanket might be removed, and the thickness of the surrounding shielding could be reduced. Such fusion reactions, however, aggravate the problem of minimum reactor vessel size per unit power output. A reaction involving only charged particles requires that 100 percent of the thermal power be extracted at the vessel wall and the divertor, if one is employed, by coolants capable of removing whatever peak heat fluxes occur at those locations.

To carry this sequence of possibilities one step further, schemes have been suggested for removing the power of the fusion reaction by directly converting the kinetic energy of the plasma charged particles into electrical energy, rather than by extracting heat radiated in some form from the plasma. Such an approach would obviously avoid the limitation on minimum reactor vessel size per unit power output imposed by heat transfer considerations. However, the requirement that the energy released by the fusion reaction must serve to maintain the plasma at a high temperature, where an appreciable fraction of power will then be radiated, makes this concept unlikely to be practical. Also of unlikely practicality is another suggestion that a plasma could be maintained at such an extremely high temperature that the radiation would appear as x-rays capable of penetrating the reactor vessel wall and depositing within some high-density flowing coolant, which then would not be subject to surface heat transfer limitations.

Conclusions

The importance of reducing the construction cost of fusion plants through reductions in reactor size for a particular power output has been discussed. It was pointed out that a minimum size limitation is set by the need to remove the heat produced, and that this would result in a minimum volume of expensive construction materials. This is a consequence of a fundamental engineering disadvantage of the fusion power concept, in which all energy must be gathered outside the reacting plasma region and the required area of surrounding surface must have associated with it a thick structure of complex design.

The real significance of the capital investment required for the nuclear island will become more apparent as further engineering studies of fusion plant designs are carried out and cost estimates for the necessary fabrication and field erection are based on more recent procurement experience in the nuclear industry. In the example presented here, although the assumptions related to the fusion plant were favorable, the comparison with fossil-fueled and fission reactor plants indicated that the electrical energy produced would be noncompetitive.

Regarding the other critical problem area, it was pointed out that there is still no structural material and operating condition that will ensure a lifetime for the reactor vessel that will match that of the plant. Although all the proposed designs have suggested periodic replacement of the vessel, the associated costs and shutdown time will probably make this an unacceptable operating requirement for utilities.

Most of the discussion was directed to the tokamak magnetic confinement reactor operating on D-T fuel, but the arguments related to costs were for the most part applicable to other confinement methods and other fuels. The radiation damage limitation on the reactor vessel would apply only to the use of thermonuclear reactions producing high fluxes of energetic neutrons.

It must be concluded that any future commercial application of fusion power is faced with two serious engineering limitations. Present information indicates that even for future optimum plant designs, the magnitude of the required capital investment is likely to lead to a noncompetitive cost for the power produced. In addition, the operational limitation resulting from the deleterious effects of radiation on the reactor vessel remains an obstacle to the successful application of the D-T fusion reaction.

References and Notes

1. *Atomic Industrial Forum Committee on Fusion Report No. 1* (Atomic Industrial Forum Inc., Washington, D.C., 1976), part 2, pp. 18–37.
2. W. D. Metz, *Science* **192**, 1320 (1976); *ibid*. **193**, 38 (1976); *ibid*. p. 307.
3. R. W. Conn, *What is Past is Prologue: Future Directions in Tokamak Power Reactor Design Research* (Report UWFDM-175, Nuclear Engineering Department, University of Wisconsin, Madison, 1976).
4. University of Wisconsin Fusion Feasibility Study Group, *UWMAK-I, A Wisconsin Tokamak Reactor Design* (Report UWFDM-68, Nuclear Engineering Department, University of Wisconsin, Madison, 1973), vols. 1 and 2.
5. ——, *UWMAK-II, A Conceptual Tokamak Power Reactor Design* (Report UWFDM-112, Nuclear Engineering Department, University of Wisconsin, Madison, 1975).
6. ——, *UWMAK-III, A Noncircular Tokamak Power Reactor Design* (Report UWFDM-150, Nuclear Engineering Department, University of Wisconsin, Madison, 1976).
7. "Engineering limitations of proposed fusion power plants," W. E. Parkins, paper presented at the Atomic Industrial Forum Conference on U.S. Options for Long-Term Energy Supply, Denver, Colo., 21 June 1977.
8. *Liquid Metal Fast Breeder Reactor Conceptual Plant Design 1000 MWe* (Report FBR-77-3, prepared by Atomics International and Burns and Row, Inc., May 1977), vol. 5. Copies of this report should be obtainable from the Department of Energy, Washington, D.C., or the Electric Power Research Institute, Palo Alto, Calif.
9. E. E. Bloom, F. W. Wiffen, P. J. Maziasz, J. O. Stiegler, *Nucl. Technol.* **31**, 115 (1976).

Index